THE BOOK OF MEDICINES

THE BOOK OF MEDICINES

ANCIENT SYRIAN ANATOMY, PATHOLOGY AND THERAPUTICS

E. A. WALLIS BUDGE

Routledge
Taylor & Francis Group
LONDON AND NEW YORK

First published 2002 by Kegan Paul Limited

2 Park Square, Milton Park, Abingdon, Oxon OX14 4RN
711 Third Avenue, New York, NY 10017, USA

Routledge is an imprint of the Taylor & Francis Group, an informa business

First issued in paperback 2016

British Library Cataloguing in Publication Data
A catalogue record for this book is available from the British Library

ISBN 978-1-138-96490-7 (pbk)
ISBN 978-0-7103-0707-1 (hbk)

Publisher's Note
The publisher has gone to great lengths to ensure the quality of this reprint
but points out that some imperfections in the original copies may be
apparent. The publisher has made every effort to contact original copyright
holders and would welcome correspondence from those they have been
unable to trace.

PREFACE.

THE present work contains the text of the great Syriac "Book of Medicines", edited from a manuscript in my possession, in an English translation of the same, with Introduction, Index, &c.; this is here published for the first time.

The first section of the Book of Medicines consists of Lectures upon Human Anatomy, Pathology, and Therapeutics, to each of which is added a series of prescriptions of the most detailed character, which the author recommends to be administered in the treatment of the various diseases described in the Lecture preceding. These Lectures were translated from Greek into Syriac by a Syrian physician, who was probably a Nestorian, and who was well acquainted with Greek and Syriac; and he may well have been attached to one of the great Medical Schools, which existed at Edessa (Urfa) and Âmid (Diarbekir), and Nisibis, in the early centuries of the Christian era. The style of the Syriac is fluent and good, and exhibits everywhere the touch of a master hand. Unfortunately the name of neither author nor translator is given, and more than two of the Lectures are wanting in the manuscript.

The system of medicine expounded in the Lectures is, fundamentally, that of Hippokrates, the "Father of Medicine", whose actual words are quoted in many places. The frequent appeals made by the lecturer to the evidence derived from dissections of the human body proves that he did not view the bodies of the dead with the same respect as the Egyptian and early Greek physicians. And they suggest that he held the same views about the importance of dissections as Herophilus of Chalcedon, a pupil of Praxagoras of Cos, physician in ordinary to Ptolemy Soter (B.C. 323—284), who is said to have dissected the bodies of the dead, and vivisected those of convicts. The Lectures afford ample proof that their author must have devoted much time and study to the results that he obtained from dissections and to the application of them to the cases that he was called upon to treat.

The Lecturer, like Hippokrates, regarded diseases as the result of perfectly natural causes, and thought that they were produced by interference with the laws of nature. According to him, the human body in a state of health contains various kinds of bile (red, yellow, reddish-yellow and black), and phlegms, rheums, and chymes, in quantities the proportions of which are fixed by the natural laws that govern the body. When these quantities are present in their proper proportions the body is healthy; when any one of these humours

is present in any member in excess, disease is the in-
evitable result. The effect of the excess of a humour
is to produce an abscess, or to bring about a stoppage
in the veins or arteries, or to change the composition
of a member; when any one of these three things
happens, the natural functions of the brain or heart,
or of some muscle, nerve, tendon or sinew, are interfered with, and disease manifests itself.

In spite of the lecturer's belief in the value of dissections, his knowledge of anatomy is not great, and
he makes little distinction between veins, arteries, nerves,
tendons, sinews, and ligaments. About the primary importance of the brain, heart, and liver he has no doubt.
The brain is the "head" or "governor" of the senses
from the beginning, for by means of the nerves it
sends the power to feel into all the members of the
body. Cut a nerve, and the member served by that
nerve at once loses the power to feel. The heart is
the dominant member of the body, for from it, as from
a fountain, by means of the veins, life is transmitted
to every part of the body. The "throb" of the arteries,
i. e., the pulse, which proceeds from the heart, makes
manifest to us all the diseases that attack the body.
The heart is a kind of governor of the body, the brain
gives sensation and motion, and the liver is the "head"
of the function of nourishment, for it changes food into
blood. Any alteration in the "composition" of the
heart produces death, for when this takes place, the
functions of the brain and liver, and of every member

of the body, are immediately suspended. The lecturer
devotes much attention to diseases of the lungs, liver,
spleen, and bladder, and if his descriptions of these
organs be compared with the cuts and illustrations
given in Sir RICHARD QUAIN's "Elements of Anatomy"
(ed. Schäfer, Symington, and Bryce, London, 1909—12)
they appear, at least, to a layman like myself, to be
generally correct. There is no evidence that the lecturer had ever thought of counting the beats of the
pulse, but that he attached great value to the evidence
of the "throb of the veins" is clear from his remarks
about it in connection with fevers.

Throughout his work the lecturer inculcates per-
sistently the great importance of what physicians in
these days call prognosis, diagnosis, semeiology, aetiology and dietetics. The eye, the ear, the nose, and
the hand must all help the skilled physician in treating
diseases, and the appearance of every member of the
patient will tell him something about the disease that
he is called in to heal. He must be able to weigh
the relative value of every symptom. He who cannot
make a correct diagnosis cannot heal the disease, is
the dictum of the lecturer. And the wise physician
must also take into account when treating a patient
his age, his physique, the kind of life he has led, his
personal habits, the nature of the country in which he
lives, the season of the year in which his sickness
attacked him, and the state of the atmosphere. Equally
important is it to consider what the patient should

eat and drink, and to take care that food suitable to his condition is supplied to him. And the lecturer prescribes a course of diet for those who are not sick, of which the following extract will give a good idea. In the month which includes a part of September and a part of October "It is right to avoid honey and "cream. Eat the flesh of birds of all kinds, and meat "cooked with garlic, and drink wine. Eat no vegetable "except rock parsley, gourds, leeks, carrots, and *babhûlâ* "with honey, after food. And drink the following "medicine every day: one drachm of aniseed, and one "drachm of peppercorns, crushed together and drunk "in wine and honey water. This medicine is good for "stabbing pains and pains in the knees, and it increases "the seed in men, and it hath been well tried, and is "a sure remedy. And also get married in this month". Again, in May and June "Eat no roast meat, and "avoid olive oil and eggs, and fatty substances, and "all kinds of wine, and cream, and fat, and do not "eat gourds" (pp. 644—645). Over-eating and over-drinking are regarded by the lecturer as the causes of many sicknesses, and he quotes with approval the saying of one of the philosophers, "I wonder how a "man who does not over-eat himself can ever die". In treating most diseases the lecturer advocates bleeding and purging. For bleeding the place usually chosen is one of the arms, but the veins in other places, *e.g.*, behind the ears, may be "cut" with advantages. Emetics, purges, and enemas are most important, and

everywhere the physician is advised to empty the
body as a preliminary to general treatment. Fomentations, poultices, and baths are also regarded as potent
means of healing. Next in importance come suitable
food, a cool, airy dwelling place, proper clothing, water
taken from a swiftly running stream, absence of all
mental and physical fatigue, and sufficient sleep. The
medicine prescribed must, of course, be taken regularly.

A very important feature of the Book of Medicines
is the series of prescriptions found in it; these number
nearly one thousand. Many of these are attributed to
the older physicians, *e. g.*, Galen, Dioskorides, Solon,
Philo, Philagrius, Theodoretus, &c., and many are of
Egyptian, Persian, and Indian origin. They resemble
in form those found in the Ebers Papyrus, and in some
of them, *e. g.*, those for diseases of the eye and stomach,
the Egyptians, Greeks, and Syrian physicians prescribed
similar medicines. The lecturer assumes that every
physician is able to write prescriptions, if it be necessary to do so.

From first to last the lecturer proclaims boldly the
independence of the art of healing, and its freedom
from the trammels of magic and priestcraft. He claims
for himself no special knowledge, and no peculiar
power. He does not regard his art as complete or
perfect, for he asks in one place (p. 10), "Do not all
"physicians learn by experiment?" He speaks with
respect of the "ancients", but the men to whom he

refers under this appellation were not the priests who used spells and incantations when administering their "secret" medicines, but the great founders of Greek Medicine, the "Asclepiadae" and Hippokrates and his followers, who were the first to give medicine the status of a science.

The second section of the Book of Medicines is astrological in character, and was included in the manuscript by some student or scribe who could not free himself from the trammels of the beliefs of some of his contemporaries. Not satisfied with the medical system of Hippokrates he had recourse to omens, portents, spells, divinations, and planetary forecasts in order to find out whether a woman had conceived, and if so whether the unborn child would be a boy or a girl. As a complete summary of the contents of this section, and a translation, are given further on nothing more need be said about it here.

The third section contains four hundred prescriptions, many of them of a most extraordinary character; these must have been written by "physicians" who were both ignorant and superstitious. These prescriptions have, however, some value, for they illustrate the folk-lore of a part of Mesopotamia, and preserve a number of popular beliefs and legends about birds, animals, magical roots, &c. The curious enquirer will find many parallels to them in statements made in mediaeval "Bestiaries".

In editing the Syriac text I have followed my

manuscript authority closely. The type was set from
the manuscript itself, and I have only attempted to
correct obvious slips of the pen of the copyist; in a
few words I have supplied vowel points. The passages
and words that are unintelligible to me are indicated
in the translation by dots.

The translation has been made as literal as possible,
and I have made no attempt to introduce into it
modern medical terminology. The author of the
lectures was well acquainted with the branches of
the healing art which are to day called prognosis,
semeiology, symptomatology, aetiology, dietetics, &c.,
but as he had no definite names for them, or at any
rate does not use any, I have not introduced these
modern words into the translation. In the lectures
many diseases are mentioned for the names of which
I can find no exact equivalents; this is due partly to
the very vague descriptions of the diseases which are
found in the native lexicons, and partly to the want
of adequate medical knowledge on my part. Among
the diseases of the eyes that are mentioned by the
author I am sure that ophthalmia, cataract, and glaucoma
are included, because I have heard natives of Mesopotamia, who were suffering from these diseases,
describe them almost in the same words as those
used in the Lectures; but the lecturer only uses general
descriptions, and has no names for these three diseases.
Again, in describing fevers it is tolerably certain that
among those for which he describes the proper

treatment, he includes chicken pox, small pox, measles, German measles, scarlet fever, typhoid fever, typhus, &c., but he has no name for any one of these diseases. He speaks of blotches, "spots", pig-sores, red sores, hard sores, running sores, soft sores, pustules, carbuncles, whitlows, abscesses, ulcers, and, in connection with them, of fever, burning fever, blazing fever, &c., but only a physician can name the various diseases that he describes. A similar difficulty meets the translator in the case of the names of certain drugs, and of plants and herbs used in medicine. And although I have consulted all the native descriptions enshrined in the *Thesaurus* of PAYNE SMITH, and in the Lexicon of Bar Baḥlûl edited by Duval, and the Neo-Syriac Lexicon printed by the Dominican Fathers at Môṣul in 1900 (ܣܘܡܐ ܕܠܫܢܐ ܐܬܘܪܝܐ), there are many names for which I can find no satisfactory English equivalent. I am indebted to Dr. LOEW's invaluable work *Die Aramaeischen Pflanzennamen*, Leipzig, 1881, and to Sir LAUDER BRUNTON's *Pharmacology, Therapeutics, and Materia Medica*, 2 vols, London, 1893, for many suggestions, which I have adopted.

It is now my pleasing duty to thank the Council of the Royal Society of Literature of the United Kingdom for undertaking the publication of this work. Their munificence has placed in the hands of Syriac scholars the complete text of the most important and most perfect Syrian treatise on Anatomy, Pathology, and Therapeutics known to us, and has given to the great world of students of the History of Medicine in Antiquity a mass of material that cannot fail to be of the highest interest to them. To Sir Edward William Brabrook, C.B., Director of the Society of Antiquaries of London, and to Percy W. Ames, LL.D., F.S.A., I am indebted for the kindness and consideration that they have shown me during the performance of a long and difficult piece of work. This edition has been printed by Messrs. W. Drugulin, and read by Dr. Chamizer, to whom my thanks are due for the care that they have shown in the typographical portion of the work.

E. A. WALLIS BUDGE.

British Museum,
April 14, 1913.

CONTENTS.

CHAPTER III. *

    * This is the first chapter

THE BOOK OF NATIVE MEDICINES—PRESCRIPTIONS.

b*

INTRODUCTION.

DESCRIPTION OF THE MANUSCRIPT OF THE "BOOK OF MEDICINES".

THE Syriac text of the "Book of Medicines" used for this volume is edited from a manuscript that was copied at my own expense at Alḳôsh, a town about thirty miles from Môṣul on the Tigris, by Îsâ bar-Esha'yâ, a deacon, the son of Cyriacus, the deacon, whose native village was Eḳrôr, in the district of the Sendâyê. It consists of two hundred and eighty-nine paper leaves measuring thirteen and a half inches by nine and a half inches. Each page is occupied by one column of writing, and a full column always contains twenty-eight lines. The volume is written in a fine, bold Nestorian hand, with numerous vowel-points, &c. The paper is tough and good, and every leaf has been made smooth and shiny by rubbing with a glass bottle. The headings of the Chapters and sections are written in red ink. There is no ornamentation of any kind.

The volume comprises:

I. The greater part of a work on Human Anatomy, Pathology, and Therapeutics, as formulated by Greek physicians, and accepted by certain Syrian physicians, at least. The first two Chapters, and probably two at the end, are wanting; and there are a few gaps in the text near the end, due to the illegibility of the passages in the manuscript from which my copy was made. The following are the Chapters:

Chapter III. On all the sicknesses (or, diseases) that take place in the head, and first of all on injuries [thereto], and on mental functions (or, the operations of the mind and the reasoning powers). Fol. 1 *b*.

Chapter IV. Of the injuries that happen to the organs of breathing, and of all the ailments that affect the nostrils and the symptoms thereof, and of their cure, and of the medicines that are good for them. Fol. 30 *a*.

Chapter V. Of the composition and placing of the eyes, and of the injuries that happen to them, and the symptoms of the same, and of the means to be employed in healing them. Fol. 33 *a*.

Chapter VI. Of the composition and construction of the tongue, and on the injuries that attack it, and on their symptoms. Fol. 46 *a*.

Chapter VII. Of the diseases of the ears and of the hearing. Fol. 48 *b*.

Chapter VIII. Of the faculty of feeling and of the injuries that occur in the nerves. Fol. 53 *a*.

Chapter IX. Of the disease that is called the strangles, and of all the ailments of the mouth. Fol. 74 *b*.

Chapter X. Of the injuries that happen to the organs of speech. Fol. 85 *a*.

Chapter XI. Of difficulty of breathing, and shortness of breath, and asthma. Fol. 90 *a*.

Chapter XII. Of the bursting-forth of blood from the internal organs. Fol. 96 *a*.

Chapter XIII. Of the symptoms of the injuries that take place in the lungs, and in all the organs of the breast. Fol. 104 *a*.

Chapter XIV. Of the injuries that happen to the heart, and of their symptoms. Fol. 121 *b*.

Chapter XV. Of the diseases that attack the stomach, and of their distinguishing features, and of the means to be employed in healing them. Fol. 128 *a*.

Chapter XVI. Of the diseases that attack the liver. Fol. 158 *a*.

Chapter XVII. Of all the diseases of jaundice. Fol. 182 *b*.

Chapter XVIII. Of the symptoms of the diseases of the spleen. Fol. 189 *a*.

Chapter XIX. Of the symptoms that take place in the bowels, and of all the various kinds of looseness of the belly. Fol. 194 *a*.

Chapter XX. Of the symptoms and distinguishing characteristics of the diseases that subsist in the intestines of the colon. Fol. 201 *b*.

Chapter XXI. Of the symptoms of disease that appear in the kidneys. Fol. 209 *b*. [Incomplete.]

The remainder of Section I, and the beginning of Section II are wanting.

II. Forecasts derived from stars. Fol. 211 *b*.

Lunar calculations and tables. Fol. 214 *b*.

A long series of short sections dealing with divinations, planets and planetary influences, the Signs of the Zodiac, and their influences on the lives of men, tables of lucky and unlucky days, calculations as to the probability of sick men recovering, auguries for journeys, the course of the circle of heaven, eclipses, divisions of time, portents from the Treatise of Hermes, weather forecasts, dieting, &c. Fol. 215 *a*—261 *a*.

III. A series of four hundred "native" prescriptions, which were used by physicians who were unacquainted with the theory and practice of Greek medicine. Fol. 261 *b*—285 *a*.

IV. List of the names of medicines arranged in alphabetical order. Fol. 285 *a*.

V. Colophon. Fol. 289 *b*.

The manuscript, from which my copy of the "Book of Medicines" described above was made, was one of a small collection which belonged to a native of Môsul, the famous town on the Tigris opposite the ruins of Nineveh (Kuyûnjik), and which he guarded jealously. After much talk he agreed to allow a copy of it to be made by one of his friends who was a scribe, and he allowed a collation of the copy to be made with his manuscript by another copyist. His manuscript was, it seemed to me, written in the twelfth century. Its size was quarto, and it was bound in the ordinary thick, brown leather covers of the period. It was, unfortunately, incomplete. Three or four quires at the beginning, and three or four in the middle, and several leaves at the end of the manuscript, were wanting. From a few of the leaves portions had been torn away. During the short examination of the manuscript which I was permitted to make I found nothing that shewed where, when, or by whom it had been written. Its size and general appearance suggested that it had been copied in some monastery on or near the Tigris, in which it had formed part of the library, for, had it been intended for the private use of some monk, the manuscript would have been of smaller size. The quires that had been torn away from the beginning must have contained the Introduction and the first two Chapters of the Book of Medicines, and it is very probable that there were in them theories or statements that were not acceptable to the monkish readers into whose hands the manuscript must at one time or another have fallen. The later Chapters that were contained in the quires lacking at .the end of the manuscript probably dealt with the organs of reproduction and their diseases; information of this kind may have been considered to be both unsuitable and unnecessary for ascetics and recluses, and the Chapters may have been torn out in consequence. The contents of the extant Chapters of the Book of Medicines may be now briefly summarized.

In the **Third Chapter** the head and its diseases are treated of. The functions of the body are of two kinds, those that are due to the spirit, or soul, and those that are due to the natural body. The functions of the spirit are of three kinds, which belong to the mind, and senses, and motion respectively. The brain is the source of all these. The operations of the mind are of three classes, each of which corresponds to one of the three cavities of the brain. One cavity produces phantasy or imagination, another the thinking power, and the third the memory. The brain is not an organ of perception only, but the source of all sensation and of all nerve power. If a nerve be severed, the member served by that nerve loses all power of sensation. In sleep the senses are wholly or partially inoperative; in the former case the sleep is "deep", in the latter "light". During sleep the soul rests, but the body works, and digestion goes on satisfactorily. Sleep and food restore the waste of the power of the body caused by working. During sleep the heart beats gently, but the brain never rests. Men fall asleep after exertion, after partaking of food, after drinking much wine, and after bathing in hot water which they pour over their heads. These things refill the brain that has become emptied or "dried up" through exertion. Cold and moisture in the brain produce deathlike senselessness and stupor. "Wet" medicines produce sleep, but "cold" medicines produce stupor only, together with a confused perception and inertness. Excessive heat or dryness causes wakefulness, just as does any acid or bitter medicine. The principal diseases of the brain are apoplexy, epilepsy, or the "falling sickness" and sleeplessness, or insomnia, delirium, and unconsciousness; these affect the whole body. There are also diseases of the brain that affect one, two, or three faculties of the mind only. The faculties of the mind are rendered active by heat, and under its influence rapidity of thought, talkativeness, and versatility are increased; cold and phlegm produce apathy of mind and torpidity in the members. According to Hippokrates, "Excessive "sleep and excessive wakefulness are equally bad". Bile and phlegm produce stupor and torpor, and when mixed together produce a pain in the brain, which is partly hot and partly cold. Pain in the brain is caused by some change in its own composition, or in that of the arteries that are in it, and by an increase of moisture in the brain.

In making the diagnosis of a patient suffering from some disease of the brain, we must first find out if his sleep is "heavy" or "light", for the character of the sleep indicates what kind of change is going on in the composition of the brain. We must also examine the matter that is ejected through the nostrils and mouth. This will inform the physician what kind of change is going on in the brain; if the change is caused by cold, then a hot treatment must be employed, and if by heat, a cold, wet treatment must be prescribed. The brain must be made neither too wet nor too dry. When we know the cause of an ailment of the brain, and therefore of the mind, we can cure it, but when the ailment is associated with a diseased member of the body we cannot.

Epilepsy, or the **falling sickness,** is accompanied by a rigidity of the body which is temporary, and is caused by the blocking up of the exits of the breath, or "soul-spirit", by thick chyme; the roots of the nerves also quake and twitch. The primary seat of the disease is located either in the medial or in the posterior cavity of the brain, probably in the latter. It really does not matter where the seat of the disease is; what we have to do is to effect a "good cure", and to remember that the disease is caused by the thick and viscous chyme, which collects in the cavities of the brain where the "soul-spirit" that works the brain is situated. We must, however, find out what kind of chyme it is that blocks the cavities; it may be one of the phlegmatic chymes, or one of the chymes that are formed from "black bile", or the "living chyme" that is found in urine, or the "glasslike" chyme, or that which is found in saliva. The intensity of the disease depends upon the thickness of the chyme; if it be thick, attacks are frequent and serious, if it be thin, they are less frequent and not serious. Madness is produced when the chyme exists in excess in the brain; this chyme is due to black bile. Moderate insanity is due to the chyme of yellow bile, but the chyme of reddish-yellow bile produces the madness of which violent outbreaks and cruelty are the characteristics. Savage and ferocious madness is due to the burning of the reddish-yellow bile in the brain. People who are "idiots" must not be classed with madmen of this kind, for they experience no pain. The delirium caused by fever is caused by "hot vapours", which go up into the head from the belly, for the belly and the head are partners, and they act in sympathy, because of the great size of the nerves that go down to the "mouth of the belly"[1] from the head. For this reason with all injuries to the head that affect the "milk of the brain", vomitings of bile are associated, and pains in the head are accompanied by nausea and pain in the stomach.

Melancholia is caused by thick chymes when they are in the matter of the brain in excess; these chymes affect the brain like an organic disease, or like homogeneous substances. When the vents in the brain are blocked up the disease is organic, but when the brain itself is diseased melancholia is caused by a change of its substance. This observation was made by Hippokrates in the Sixth Chapter of his Treatise on the Comings of Sicknesses, but the views of this physician, so far as they can be derived from his subsequent remarks about the change of substance causing the disease of melancholia, have been abandoned by all physicians. If melancholia is caused by the circulation of black bile throughout the body, we must, to effect a cure, begin by blood-letting; if the black bile is in the brain only it is useless to let blood from the body. The bodies of persons that are soft, and white, and fat, never contain the chyme

---

[1] Compare the Egyptian *ra àb* ⌣ᅌ "mouth of the belly".

of black bile; but it is produced in great abundance in those who are spare of body, of dark complexion and hairy, and in those who have dilated veins it is produced in abundance. Certain men with fair or ruddy complexions do produce black bile if they be oppressed with the cares of business, or if they do without sleep, and if they overwork, and are unduly anxious about the affairs of life. In treating cases of melancholia it is all-important to obtain information from the patient as to his habits of life, the food he eats, and his natural constitution, and at the same time the physician must take into account the season of the year, his natural surroundings, and the state of the weather before or during the attacks. When the physician is certain that the blood in all the body is bilious, he must let blood from the middle vein in the arm, which comes from the shoulder and passes through the upper arm down to the hand.

In connection with diseases of the head we must note the **sympathetic weakness of the epigastrium**, which is described by the physician Diocles in his work entitled "Of diseases, and the causes why physicians exist". The effect of this weakness is "inflation" and flatulence, eructations, abundant saliva, burning heat in the epigastrium, distention of the bowels, and pains in the belly and shoulders; these pains come on as the result of both eating and fasting. Diocles attributed this weakness to excessive heat in the "veins" that receive the food from the belly, and to thickness of the blood. The food is not assimilated, and inflammation arises, and vomiting takes place. Or it may be that there is a boil or ulcer on or by the "mouth of the belly", and the food cannot

pass down into the body. The observations which
Diocles made on the symptoms of the ailment are far
too short, and he ought to have described fully the
symptoms of disease in the epigastric region.

Again, Hippokrates tells us that the intellect of a
man will become weak if he be oppressed by fear
and sorrow too long, but he says nothing at all about
the cause of the mind becoming weak. This is a
point worth enquiring into. Supposing there is heat
in the veins, and supposing that an ulcer does exist
on the mouth of the belly, there is no reason why
the mind of a man should become weak because of
these things. We know well that the belly becomes
flatulent, and that flatulence is relieved by eructations
and vomitings, but I do not see my way to connect
weakness of the intellect with disease of the belly.
Those who suffer from epileptic fits are attacked by
pains that arise from disease in the brain, as well as
in the belly.

Fear is always present in those who are weak in
intellect. One thinks that he is an earthenware pot,
and is afraid that he will be broken. Another is
afraid that Atlas will become tired, and slipping out
from under the sky will let it fall down and crush
him; another is in constant fear of death, and another
is terrified by darkness. The cause of every kind of
fear in a man is the black bile, for it clouds the
intellect, and changes the nature and constitution of
the soul, or mind. Since mind and body are intimately
connected, it follows that when weakness of mind
is associated with pains in the belly, the mind will
return to its normal state when digestion is complete,
or after vomiting and stool. Physicians may if they

ointments, plasters, &c. which may be employed in the
treatment of headaches and other ailments of the head.]

In the **Fourth Chapter** the injuries that attack the
organs of breathing and the **nostrils** are discussed,
and the treatment of them is described, and a number
of prescriptions are given. The function of breathing
is performed by the front cavities of the brain and
by the nostrils, which serve both for smelling and for
breathing. The function of breathing is stopped by a
boil in the nose, or by a fleshy growth therein, or by
the sore called "polypodium", or by a flow of blood,
or by some injury to the head. If the breathing be
stopped by some injury to the brain, make use of
bathings, and spongings, and plasters, and draughts,
and even of blood-letting. If the breathing be obstructed by haemorrhoids or fleshy growths in the nose,
cut those out with an iron knife, or cauterize them,
and cleanse the wounds with astringent substance which
are to be pounded and blown up into the nose daily.
In cases of obstinate bleeding from the nose, dip plugs
[of wool] into a mixture of aloes, frankincense, and
vinegar and insert them in the nostrils. Opium and
burnt vitriol may also be used. If the bleeding does
not stop, lay the patient on his back, pour cold water
over his face, put plugs dipped in vinegar in his nostrils,
bind his hands, feet, and testicles tightly, and cup him
above either the liver or spleen, and bleed him, and
administer injections or purges.

Cold in the head is caused by a change in the
composition of the brain, which becomes overcharged
with rheum; this it ejects by the nostrils. "Rheum"
is the name given to the moisture that flows from the

d *

brain down into the mouth, and "catarrh" is the name
of the moisture that flows down from the brain into
the nostrils. If rheum enters the throat it causes deafness, and if it reaches the gums and palate a boil is
produced. If it descends to the two sides of the mouth
it causes mumps. Catarrh is cured by inhaling the
fumes of burnt vinegar, and by bathing and washing
the feet in hot water; the patient must also anoint his
head, and anus, and genitals with aromatic oil. He
must abstain from meat and wine, drink hot water,
and eat light and easily digested food.

The **Fifth Chapter** treats of the **eyes**, and of
diseases of the eyes, their symptoms and cure. There
are two optic nerves that proceed from the first opening
of the brain; these "mingle" (*i. e.*, run parallel, or are
twisted together), and then separate, the one turning
to the right and the other to the left. Along these
nerves the power of sight is transmitted. At the end
of each nerve is the fluid which is called "glass", because it is like unto glass. Over this fluid is the grape-
like tunic, and in its centre is a "cavity" (*i. e.*, the pupil
of the eye), wherein the sight is situated. Above this
tunic is the second fluid (*i. e.*, the cornea) which some
say is like crystal, and others like the white of an egg.
The first fluid is called "crystalline", and the second
"horny", because it is like the thin transparent slabs
of talc (?) that are inserted in lanterns, when men wish
the flame in the lanterns to give light and not to be
blown about by the wind, or extinguished by the rain.
Above the tunic of the eye is a net-like membrane,
which is of a fleshy nature and contains multitudes of
small veins, which in diseases of the eye become over-

charged with blood. Outside this membrane are the
eyelids and eyelashes, and the angles formed by them.
The eyes are worked by six muscles which are set in
motion by the nerves of the brain. Injuries to the eyes
are of three kinds and these are caused by: 1. disease
of the eye itself; 2. defect in the power of sight;
3. disease in one of the auxiliary members of the eye.
The crystalline fluid is sensitive, and it owes this sensitiveness to the nerves that come from the brain. An
injury to the tunics, or muscles, or any part of the eye,
causes the sight to be impaired, and even blindness.
Eight changes of composition take place in the eye,
as in the other members. Sometimes the luminous
spririt that comes from the eye is wanting partially
or wholly. Disease in the optic nerve destroys sight.
The upper eyelid is worked by muscles, but the lower
eyelid does not move at all. Phantasm appear in the
eye, and are due either to disease in the eye itself,
or to the stomach, or to fever. "Mistiness" comes
before the eyes as a result of gazing too much on the
sun, or living too long in a dark place, and cataract
is caused by the presence of some opaque humour
between the cornea and the crystalline fluid of the
eye, which congeals in front of the pupil of the eye.
Other diseases of the eye are caused by constriction
of the tunics, which dries up the eye; by smoke, fat,
marrow, and milk touching the eye; by sores that eat
away the eyes, in whole or in part; and by "diarrhoea
and inflammation" (ophthalmia?), which produce such
agony that people who suffer from this disease are
driven to commit suicide. As for the rheum that comes
from the brain, if it enters the eye it produces granu-
lation, swellings, watery humour, pus, and boils, or styes,

in the corners of the eyes. The proper treatment for this disease is to cut the veins of the temples and those behind the ears. There is also a disease of the eye that arises inside the "bowls" of the head, and this is very difficult to cure. In the Ṭepherâ disease (*i. e.*, πτερύτιον) a membrane like a net grows over the eyes, and this must be either removed by medicines or by cutting. Sometimes the eyelashes grow inwards and pierce the eyes. Among sores that attack the eyes are the "shûrnâḳâ", and the "sôrîghâ" or the "nâsôrâ"; the last named grows in the bones of the nostrils, close to the corner of the eye, and if it discharges into the eye it destroy the sight. To cure this, the sore must be opened and then cut right out down to the bone. Eyes that are small and sunken are the most highly esteemed; large, protruding eyes are generally weak.

In treating diseases of the eyes we must make the patient eat and drink sparingly, and abstain from copulation and he must rest; if there be blood in excess, it must be let from the upper vein of the arm, and, if any of the pains in the eye are due to the stomach, we must excite vomiting by means of a dose of hiera pikra.[1] Baths and fomentations are also very useful, but if the pains continue we must cut the veins and arteries in the head as in the case of headaches. Bathing the head with infusions of chamomile, aniseed, trigonella, origanum marjoram, and mugwort are very useful, and to allay irritation in the eyes use oil of roses, or oil

---

[1] Or Pulvis Aloes cum Canella (Hepatic Aloes 16, White Canella Bark 8). See Martindale & Westcott, *Extra Pharmacopoeia*, 13th edit. p. 120; Butler, *Textbook of Materia Medica* p. 114; and on Canella Cortex see Lauder Brunton, *Pharmacology*, vol. ii, p. 867.

of violets, and the tincture of the plant called "live-for-ever".

[Here follows a long series of prescriptions for oint-ments, eye-powders, and lotions of many kinds. In several of these the principal ingredients are white lead (*i. e.*, acetate of lead), and Cadmin, copper and opium, and when I was in Môṣul and its neighbourhood in 1889—1891 these were in common use among those who suffered from ailments of the eyes. The most acceptable gift to many natives was half a pound of acetate of lead or sulphate of copper; opium could be obtained in the bazaar without difficulty.]

The **Sixth Chapter** treats of the composition and construction of the **tongue**, and of its diseases.

The nerves that come to the tongue and produce the sense of taste and feeling issue from the third opening in the brain, whilst those that give it motion issue from the seventh opening, which is close to the beginning of the spinal cord. The tongue is not only an instrument for tasting, but it assists in speech, and its auxiliaries in this function are the lips, the teeth, and the nose. Injury to speech is caused by some defect in the brain near the spot where the seventh nerve-opening is. The principal organic diseases of the tongue are caused by boils, and by hard, obstinate swellings containing pus. The cause of these is often over-fullness of the body, and the remedy is blood-letting, and we must clear away the chyme which is in excess by means of cathartics. If necessary, we must cut the veins under the tongue, after the body has been emptied. Gargles, lotions injected into the nose, wet compresses on the head and neck, and hot bandages on the head and nape of the neck, are very useful. The food of the patient at this time must be light and easily digested, and hiera pikra must be used if necessary. For gargles use infusions of extract of foxgrapes, chicory, lettuce, polygonum, sour milk, lentils, dried roses, rind of pomegranates, worm-root, mulberries, beetroot, rhubarb, myrtle, olives, &c. A gargle made of mustard and honey is very efficacious, and baths are very beneficial.

[This Chapter contains two prescriptions only, and each is to be used for a relaxed tongue. The remedies given must depend upon the symptoms, and must be in accordance with the general rules that have been laid down as regards treatment of diseases of the head and nerves.]

The **Seventh Chapter** treats of diseases of the **ears** and defective hearing. The sense of hearing is effected by the nerves that come from the fourth opening in the brain. As all parts of the ears are plainly visible we can see at once when they are diseased, but sometimes there are in the ears diseases of which the symptoms are not visible, and in such cases we must make a very thorough diagnosis. If the hearing alone is affected, we attribute the defect to disease in the aural nerve, but if defective hearing be coupled with defects in other members, we must then attribute it to disease of the brain itself. Diseases of the ears resemble diseases of the eye, and the operations of the internal limit of the tube of the ear are to the sense of hearing what the crystalline fluid is to the eye. The folds of the ear also may be compared to the tunics of the eye. Diseases of the ear are caused by boils, or ulcers, or abscesses, or by some abnormal fleshy growth, or even by the wax that blocks up the tube. The sounds, singings, and whistlings that are often heard in the ears are due to the over-sensitiveness of the brain, or to wind blowing into them, and heat and cold also produce pain. Sometimes the ear is perforated by a sore that eats into it and causes intense pain. External causes of diseases of the ears are bath-water, or some poisonous substance, or insect, or a particle of stone or some other substance. In cases where disease of the ear is caused by overfullness of the brain, or of the whole body, and where there is blood in excess, we must let blood from the upper vein of the arm, and get rid of the chyme which is in excess by means of hiera pikra, and empty the brain by means of gargles, and excite sneezings. And we must apply medicines to the tubes of the ears, and so free them from any obstruction.

[Here follows a series of prescriptions, which are said to be good for the ailments caused by cold, and stoppages, but which are useless for those that are caused by diseases accompanied by inflammations, and for the deafness that is caused by "emptiness". Next comes a series of general prescriptions for boils in the ears, or any kind of sore that produces pain. Among the remedies enumerated is the treatment of the ear by the steam of a medicinal infusion. Thus it is said:—
"Infuse chamomile, dill, mugwort, laurel leaves, king's "crown (melilotus), mint, and thyme in water in a "vessel, the mouth of which is well plugged up. Place "in the mouth of the vessel a reed tube, and let the "place where the reed is inserted be well covered up "so that the steam of the water may not escape, and

"then let the patient place his ear over the tube so
"that the vapour of the infusion may rise into his ear."
These prescriptions are followed by another series, in
which the ingredients of each lotion or liniment are
carefully given. To remove foreign substances from
the ears we are told to dip a needle that is used for
applying stibium to the eyes into gum or glue, and to
insert it carefully in the ear; in most cases the substance that is causing the irritation will stick to the
needle, and so may be removed. Or we may use a
small pair of tweezers, and take it out. If these means
fail we are told to "blow into the ear some greasy
"substance. And the patient must close his mouth and
"nostrils, and incline his ear towards the side which is
"affected, and lay the palm of his hand over his ear,
"and then leap up into the air"!]

The **Eighth Chapter** treats of the sensation of feeling, and of **diseases of the nerves.** Having discussed the first four senses and the organs by which
they are served, it is right to treat of the function
that is common to them all, *i. e.,* the sense of feeling.
The nerves, which extend throughout the whole body
in fine filaments, are of five kinds, and they are called:
1. **Arterial veins,** 2. **Arteries,** 3. **Nerves,** 4. **Ligaments,** and 5. **Tendons.** 1. **Arterial veins** are those
that emerge from the liver; they resemble pipes, and
are filled with the blood that flows from the liver into
all the body. 2. **Arteries** have their source in the
heart. They are of tubular formation, are filled with
thin, refined blood, and are moved with a motion of
dilatation and contraction by the heart. 3. The **Nerves**
emerge from the spinal cord, and they divide and extend throughout the body in all directions. In them

"then let the patient place his ear over the tube so
"that the vapour of the infusion may rise into his ear."
These prescriptions are followed by another series, in
which the ingredients of each lotion or liniment are
carefully given. To remove foreign substances from
the ears we are told to dip a needle that is used for
applying stibium to the eyes into gum or glue, and to
insert it carefully in the ear; in most cases the substance that is causing the irritation will stick to the
needle, and so may be removed. Or we may use a
small pair of tweezers, and take it out. If these means
fail we are told to "blow into the ear some greasy
"substance. And the patient must close his mouth and
"nostrils, and incline his ear towards the side which is
"affected, and lay the palm of his hand over his ear,
"and then leap up into the air"!]

The **Eighth Chapter** treats of the sensation of feeling, and of **diseases of the nerves.** Having discussed the first four senses and the organs by which
they are served, it is right to treat of the function
that is common to them all, *i. e.,* the sense of feeling.
The nerves, which extend throughout the whole body
in fine filaments, are of five kinds, and they are called:
1. **Arterial veins,** 2. **Arteries,** 3. **Nerves,** 4. **Ligaments,** and 5. **Tendons.** 1. **Arterial veins** are those
that emerge from the liver; they resemble pipes, and
are filled with the blood that flows from the liver into
all the body. 2. **Arteries** have their source in the
heart. They are of tubular formation, are filled with
thin, refined blood, and are moved with a motion of
dilatation and contraction by the heart. 3. The **Nerves**
emerge from the spinal cord, and they divide and extend throughout the body in all directions. In them

by boils, or ulcers, or abscesses, or by some abnormal fleshy growth, or even by the wax that blocks up the tube. The sounds, singings, and whistlings that are often heard in the ears are due to the over-sensitiveness of the brain, or to wind blowing into them, and heat and cold also produce pain. Sometimes the ear is perforated by a sore that eats into it and causes intense pain. External causes of diseases of the ears are bath-water, or some poisonous substance, or insect, or a particle of stone or some other substance. In cases where disease of the ear is caused by overfullness of the brain, or of the whole body, and where there is blood in excess, we must let blood from the upper vein of the arm, and get rid of the chyme which is in excess by means of hiera pikra, and empty the brain by means of gargles, and excite sneezings. And we must apply medicines to the tubes of the ears, and so free them from any obstruction.

[Here follows a series of prescriptions, which are said to be good for the ailments caused by cold, and stoppages, but which are useless for those that are caused by diseases accompanied by inflammations, and for the deafness that is caused by "emptiness". Next comes a series of general prescriptions for boils in the ears, or any kind of sore that produces pain. Among the remedies enumerated is the treatment of the ear by the steam of a medicinal infusion. Thus it is said:—
"Infuse chamomile, dill, mugwort, laurel leaves, king's "crown (melilotus), mint, and thyme in water in a "vessel, the mouth of which is well plugged up. Place "in the mouth of the vessel a reed tube, and let the "place where the reed is inserted be well covered up "so that the steam of the water may not escape, and

is located the whole power of sensation. 4. The **Liga-ments** are simple, like the nerves, but they are harder and there is no sense of feeling in them. They are attached to the bones, and are in the joints, which they bind together, and the bones that are below them are attached to them. They are called "Ligaments" because they tie together the joints. 5. The **Tendons** resemble somewhat the Ligaments and the Nerves, but they are attached to the muscles; they spread in all directions in fine filaments, and are slightly sensitive. The nerves possess the power of reasoning, and are the servants of the wishes of the soul, and they direct all the natural function of the body, and make the members perform all the dictates of the soul (or, mind). When injuries happen to them the powers of volition and the natural functions of the body cease. Cold and compression produce numbness in them, and numbness is also produced by contact with the cramp-fish, or numb-fish, a kind of electric ray-fish. Moreover, in-action and disease of the nerves are produced by over-eating and over-drinking, and by a riotous, dissolute life. Power from the brain passes through the nerves as light passes through water or air, but if there is a fissure in any one of them, or an obstruction, it cannot pass, and disease is the result. Malnutrition, or con-striction, or pressure, or cold produces "dissolution" in the nerves, and, if this takes place in all of them at the same time, the body at once becomes senseless and motionless, and death supervenes. Injury to the brain paralyses all the members of the body; injury to the top of the spinal cord paralyses only the members of the head. Sometimes a member can move, but not feel, and sometimes it can feel, but not move. When

the whole body is paralysed, we know that the common source of all the nerves is diseased. Hippokrates
said that the spasm that produces rigidity is due to
fullness and to emptiness, and though only wise men
understand this remark, I am convinced that this statement is correct. The nerves of the body are like the
strings of a harp: if overstretched too often they snap,
and when they snap the spasm of rigidity is the result.
Nerves may be overstretched by toil, vigil, poverty,
anxiety, drunkenness, and excess of any kind. **Apoplexy** is due to disease of the brain, but in **paralysis**,
though due to the same cause, only one half of the
body is affected, the right side or the left. The manner in which the nerves work may be learned by dis-
secting the breast and portions of the spinal cord.
The man who knows accurately how the nerves come
from the spinal cord, and from which parts of it, will
be able to recognize from the symptoms of his patients
which parts of the cord are diseased.

To cure rigidity and paralysis we must let blood
from the middle or upper vein of the arm, and we
must administer the centaury plant (κενταύριον) and
paganum (πήγανον), and anoint the body with hot oil,
and use fomentations as well as plasters and the proper
medicines. Never forget that rigidity is due to tension
of the nerves, and paralysis to their slackness. And
the patient must rest, take suitable food, avoid all
sexual excitement, and take plenty of sleep.

[Here follows a long series of prescriptions for potions
and unguents and plasters. Several supplementary prescriptions are given that are to be used in cases in
which the nerves have been injured by a blow, or a
wound, or the bite of some animal.]

The **Ninth Chapter** treats of the disease called **the strangles,** and of all the ailments of the mouth. Strangles, according to Hippokrates, is a very difficult and evil disease, and it brings a man quickly to his end. Of this disease there are four kinds. In the first kind there is a red abscess in the mouth. In the second there is nothing to be seen inside the throat or mouth, and only the pain shews that the disease is there. In the third there is great pain in the throat, and an abscess appears. In the fourth there is an abscess both inside and outside the throat. Akin to these is the disease in which the vertebrae of the neck, and the muscles, and the stomach are affected. Sometimes the gums and the tongue become diseased as the abscess moves from one place to another in the throat and mouth. In these diseases the vertebrae of the neck incline inwards to such a degree, that in the outside of the neck a depression can be seen; any pressure on this depression causes pain. Breathing becomes difficult, food can only be swallowed with the greatest difficulty, and in drinking the liquid is sometimes driven upwards and flows out of the nose. The lining of the throat and the mucous membrane of the mouth become perforated, and the neck is bent out of the straight, inclining to one side or the other. Those who suffer from the strangles sometimes have high fever about the temples, but their feet and legs are cold. Of the cases of the strangles which I have seen, some patients died through diseases induced sympathetically in other parts of the body, but others died of suffocation caused by the abscesses in the throat and mouth, which choked them. These abscesses were caused by a bad quality of blood. Blood-letting, purgings, and cooling and purifying medicines are the principal agents in **treating the strangles**; and any excess of chyme in the body must be got rid of, for that forms the nourishment of the abscesses. We must employ medicines to check the rheum, and also medicines to give relief to the pain in the throat and mouth; these may be mixed together sometimes. Among the former may be mentioned dried roses, rind of pomegranates, gall nuts, rhubarb, lentils, aloes, cisthus parasite, purple balaustion, plantain, polygonum, oil ot styrax, glaucium, lycium, mulberries, fresh walnut bark, &c. And among the latter, aphronitron, sulphur, Cyrenean gum, myrrh, crocus, cinnamon, or peppercorns. Gargles made of barley water mixed with honey and vinegar, or sour wine, are very beneficial.

[Here follows a series of prescriptions for potions and powders that are to be blown into the mouth. From one of these it seems that in some cases of the strangles the abscess penetrated the ears; the medicine prescribed for the cure of this abscess might be blown into the mouth in the form of a powder, or painted on the back of the throat with a feather, or administered as a gargle.]

When the **uvula** is diseased it may be cut with a knife, or reduced by means of caustic medicines. The time for cutting it is when its root is thin, and its end is thick, and there is a sore upon it, but the body must be first cleared by means of purgatives. Ailments of the uvula are to be treated by gargles containing vinegar, green gall nuts, alum, rhubarb, salt, vine sap, and to get rid of the rheum which comes from the head down to the throat we must use an ointment made of Cyrenean fat and smelters' dross and honey.

We must now treat of the **teeth**, both **molars** and **incisors**, the aching of which is known to every man. Teeth split, and decay, and become corroded with tartar, and are at times so tender that they cannot bear the touch of anything that is hot, cold, or acid; but sometimes the cause of **toothache** is not always apparent. Teeth become diseased either through overfeeding or through insufficiency of nourishment. If there be an abscess in the gums caused by rheum, they may become diseased sympathetically and they suffer pain when the gums recede, and they are stripped of covering. They become loose and fall out, or they become crooked, and some grow too long or too large; and they suffer from external injuries, that is, they may be knocked out by a blow, and then defective speaking, drinking, and mastication ensue. In treating diseases of the teeth we must use blood-letting, and cleanse the stomach by means of an emetic or purgative, and administer a gargle whereby the tongue will be cleansed also. Superfluous and irregular teeth must be removed, and teeth that are too long must be filed. When the ordinary medicines fail to relieve the pain of toothache, we must apply the following treatment. Boil some sheep-oil or cow-oil in a saucepan. Take a piece of wool and wrap it round a piece of wood of suitable size, and dip it in the boiling oil and rub with it that part of the gums where the teeth are affected until it becomes white, or apply the boiling oil to the cavities of the teeth. When the roots are burnt, apply *surantîcôn* on a piece of linen rag If cauterization be necessary, use a finely pointed iron instrument for cauterizing the cavity of the molar, and an instrument, spatula-shaped, for the gums.

[Here follows a long series of prescriptions for the mouth, teeth, and gums. These were to be used in cases of relaxed mouth, tender gums that bleed at the slightest touch, running sores and gangrene of the mouth, common toothache, cancer of the mouth, stinking breath; recipes for various kinds of tooth-powder are included among the prescriptions. A few of the prescriptions (p. 190) were intended to dislodge teeth painlessly. Thus the bark of mulberry root and pyrethrum, macerated in strong vinegar in the sun until the mixture had the consistency of honey, if applied to the tooth once a day, would eventually dislodge it. Or, we may put a layer of wax round the teeth that cause pain, and then apply to them pyrethrum that has been soaked in vinegar for forty days. Let the medicine work for one day, and we shall then be able to remove the teeth with our fingers, or an iron instrument.

The **Tenth Chapter** treats of the diseases of the organs that produce the **voice**. The most important of these is the tongue; the nose, lips, and teeth help the tongue, as also do the throat and all the nerves and muscles that bring power down to it from the brain. When disease attacks any one of these members, either actually or sympathetically, the voice is injured. Hippokrates says that the voice is due to the indrawing of the breath, and this is true, speaking generally; but it is more accurate to say that it is due to the expulsion of the breath. The relaxation of one of the sides of the breast reduces the voice to one half. If a man does not breathe with two of the organs of the chest he becomes suffocated

immediately. And we must distinguish between breathing with sound and breathing without. Since you
know which muscles perform these operations, it is
quite easy for you to calculate which one of them
is diseased; and if you forget the facts that have
been demonstrated to you in dissections, the facts
about the voice ought to remind you. All the muscles
together produce inspiration, but it is the muscles
between the ribs that produce expiration. The muscles
of the throat produce exhalation with sound, and the
muscles of the throat produce the voice. The voice
is regulated by the tongue, which produces speech,
and the tongue-string, and the teeth, and the lips,
and the nostrils, and the roof of the mouth help.
Stuttering, stammering, and **defective speech** are
due either to a natural defect in the organs of speech,
or to something resembling it that hath happened to
them, or to a polypus in the nose, or to sore lips,
or to the loss of teeth. In treating diseases of the
throat and voice especial care must be paid to the
kind of food that is eaten, and the kind of drinks
that is drunk. Rough, strong wine, and salt and acid
substances must be eschewed, and much talking and
shouting must be avoided. Succulent and well-cooked
food, soft wines, butter, honey, &c. are very beneficial.
For **hoarseness** certain medicines must be placed
under the tongue. Painless remedies are new wine,
gum Arabic, gum tragacanth, and extract of licorice
root. When swallowing medicines that are placed
under the tongue, great care must be taken to prevent
coughing; if a fit of coughing comes on a reed must
be inserted in the throat. [Here follows a series of
prescriptions for diseases of the throat and voice.]

e

The **Eleventh Chapter** treats of **difficulty of breathing,** and shortness of breath, and **asthma.** Shortness of breath and asthma are caused by moisture that is confined between the lungs and the chest, or by an abscess in the lungs or chest, or by weakness of the nerves or muscles, or by wind that cannot escape, or by the filling of the lungs or oesophagus with thick and viscous fluid. The theory of the matter has already been described, but, in brief, the above are the causes that produce asthma. In treating asthma we must use plasters, and medicines that clear out the rheum that is discharged from the head into the body, or that comes there from some other member.

[Here follows a series of prescriptions for medicines, many of which are to be taken in the form of pills or tablets, together with a draught of some digestive medicine, or oxymel, or honey water. Among the prescriptions are three that seem to have crept into the series from some native Book of Medicine. In the first millepedes, or the insects that collect under the waterjars, are to be roasted and powdered, and administered with boiled honey in the form of a linctus when the patient is fasting. In the second the patient is to drink the lungs of a fox pounded in wine; and in the third the roasted body of a river tortoise, worked up with honey and a few peppercorns, is to be administered to the patient as a linctus.]

The **Twelfth Chapter** treats of internal **haemor-rhage.** The **primary** causes of internal haemorrhage are falling from a height, injuries that result from athletic contests, overstraining, anger, violent shouting, and injuries caused by a falling piece of timber or stone. **Predisposing causes** are the superabundance of blood, or something similar, and gangrene. We must be careful to note whether bleeding arises from the head, or from the organs of the belly. Internal bleeding sometimes arises from ulceration of the lungs; about the possibility of healing of this disease opinions differ. When the veins in the chest become torn, the wounds cannot be healed, and in my opinion a rent in the lungs can never be quite healed. A rupture in a blood vessel in the lungs is often due to frequent movement, but it may also be caused by chymes when present in excess, or by a cold, inflating wind. Hot drinks, hot foods, a hot climate, a hot period of the year, and over-indulgence in hot baths will also produce it. Blood from the belly is expelled by vomiting, blood from the kidneys and bladder goes forth in the urine, blood from the bowels descends through the anus, blood from the womb through the vagina, blood from the head through the mouth and nostrils, and blood from the organs of respiration by coughing. To treat haemorrhage successfully we must find a means of rejoining the burst vein or artery, and of repairing the rent or the wound caused by gangrene. When bleeding is caused by the presence of a leech in the belly, the blood that comes forth is thin and inert.

[Here follows a series of prescriptions for pills, tablets, and potions, some of which are highly praised by the author of the "Book of Medicines". These prescriptions are followed by a statement as to the foods on which the patient is to be fed. Herbs that are slightly astringent, milky dishes, partaken of when warm, goats' milk, sheep's milk, honey, &c. are highly recommended. It is especially ordered that the animal from which the milk is taken be properly fed. No exertion of an exciting nature must be indulged in. Plasters of aloes, frankincense, acacia, and purple balaustion are to be used, the bowels must be kept open, and the feet and hands must be massaged. Among the prescriptions are some the object of which was to expel the leech, which a man sometimes swallowed inadvertently. First of all the patient had to be placed in a position opposite the sun. The physician then pressed down his tongue with a specillum, and looked carefully to see if a leech was clinging to his oesophagus. If a leech was there he laid hold of it with the instrument that was used for holding back the uvula, and lifted it out carefully to prevent the creature from breaking. If the leech was beyond the reach of the physician's eye, it was necessary to expel it by means of medicines, i. e., by a gargle made of vinegar, nitre, and garlic, or by mustard and nitre, or by hyssop and madder, or by pounded vitriol, blown into the throat. The patient is also ordered to eat a large quantity of garlic, but to drink no water with it. If he then will go and bend over a tank of cold water, and stir up the water, keeping his mouth open at the same time, the leech will drop out of his mouth.]

The **Thirteenth Chapter** describes the symptoms of the injuries to the lungs, and the organs of the chest. We have already treated of the diseases of the lungs, but I must refer to these once again here, in connection with pleurisy and blood-spitting. A hard ulcer or an abscess produces shortness of breath and a certain amount of pain, and rheum from the head or some other part of the body produces shortness of breath. When shortness of breath is accompanied by oppression, and weight, and fever, and inflammation, there is an inflamed abscess in the lungs; but if the inflammation be great, and the sensation of weight and oppression be little, the cause of the disease is a hard, red abscess in the lungs. Many diseases are caused by the effusion of chymes into the lungs, and these are common. Sometimes we see a man beginning to cough suddenly, and bringing up reddish-yellow chyme, and each day he spits more and more. Then subtle fevers attack him and his body begins to waste away. He next spits pus, and after about four months blood comes up with the pus, and fever and inflammation ensue. His spitting becomes copious, fever increases in proportion as his strength declines, and he dies in the same way as those who die of phthisis. I knew two patients who suffered from this disease, but in spite of my care they died, and before they died they coughed up blood and portions of their lungs, which were rotten. In the case of the second patient I tried to dry the lungs by means of aromatic herbs, and I made him smell the drug hydrargyrum the whole day long. I also administered to him draughts of Metdôritôs,[1] and Ambrosia, and "Immortality"[2] and Theriake,[3] but they only kept him alive for a year. **Perforations of the lungs** is produced by abscesses in the moving membrane of the chest, and it is accompanied by fever

---

[1] For the prescription see p. 274.
[2] For the prescription see p. 407.
[3] For the prescription see p. 409.

and by stabbing pains. The breath is little, and is drawn in gasps, the pulse is hard, there is coughing, and pains are felt in the ribs. An abscess in the membrane called "parnôs" (πρόνοος), or "diaphragm" (διάφραγμα) produces disease of the brain. The symptoms of this disease are light sleep, which is disturbed by visions, during which the patient cries out and sits up in bed, forgetfulness of an unusual character, as for example the asking for a vessel in which to micturate, and then forgetting to use it, or making water elsewhere, sudden attacks of violence and ferocity, dryness of the eyes, which are covered by a film and have their veins filled with blood, a parched tongue, fits of hilarity and of depression, &c. The attacks of phrenitis due to disease in the diaphragm last only a short time, and respiration is very irregular. In treating lung diseases the first things to be done are to let blood and to clear out the body. Next, strict attention must be paid to dieting, suitable medicines must be given, and plasters, cataplasms, fomentations and unguents must be applied to the chest. If there is an abscess in the lung, let blood from the left arm; if there is an abscess in the muscles between the ribs together with pleurisy, let blood from a vein in either arm, or from the same vein in the palm of the hand, from a spot above the little finger and close to the third finger. In the case of pleurisy let blood from the arm on the side opposite to that in which is the disease. But the strength, age, and habits of life of the patient must be carefully considered, as well as the season of the year, the climate, and the country in which he lives. Purging may be effected either by medicines or foods. Broths made of barley, mallows, spinach, gourds, orach, sorrel, chick peas

and sesame oil are very beneficial. The medicines given
must be those that relieve coughing and bring up
pus, and with these the drugs that deaden feeling (*i. e.*,
narcotics) are often administered. These are opium,
hyoscyamus, bark of mandragora root, and hemlock,
but, strictly speaking, the physician ought not to use
them, for the excessive coldness that they produce
causes injury to the patient. Information concerning
diseases in the lungs is often afforded by coughing,
and therefore careful attention should be paid to the
character of the cough. In treating diseases of the lungs
every attention must be paid to the chest.

[Here follows a long series of prescriptions for diseases of the chest, including coughs, colds, hoarseness,
sore throat, "stomach coughs", loss of voice, asthma,
ague, &c. In spite of the author's remarks about the
use of narcotics many of the prescriptions contain
Egyptian poppies, opium, hyoscyamus, poppy seed, &c.
In one prescription the total weight of the ingredients
is four and a half ounces, the amount of opium being
one ninth of the whole (half an ounce, see p. 269); in
another on the same page the opium administered in a
dose formed nearly one third of the whole prescription.
Another prescription is interesting. A spoonful of green
arsenic and ten peppercorns are to be pounded, and
mixed with a little white of an egg and some sesame
oil; this mixture is to be spread on a piece of new
cotton wool, which is to be dried in the shade. When
required for use we are to take a vessel like a thurible,
make a hole in the cover and fix a reed in it, put hot
coals in the vessel, and throw a piece of the cotton
wool on them, and fasten on the cover. The patient
must then put the end of the reed in his mouth, and

inhale the fumes of the arsenic and peppercorns. The operation is to be performed for seven days. If the patient lives upon bread and broth only, and inhales this medicine he will become well.]

The **Fourteenth Chapter** treats of **disease of the heart.** The heart is the dominant member of the body, and from it, as from a fountain, by means of the arteries life is transmitted throughout the body. And the throbbing of the arteries, *i. e.,* the **pulse,** indicates to us the diseases which happen in the body. The heart is composed of nerve matter, and in it are two hollows, or cavities, which are filled with fine blood and living spirit. It is suspended in the lungs, in the middle of the breast, and is surrounded by a tunic of nerve matter. It has two "ears", which are opposite to its two hollows, and it is inclined more to the left side than to the right. The fine point of it is opposite to the spleen. The **heart, the brain,** and the liver are the three chief members of the body; the last-named organ governs the nutrition of the body. In cases of serious injury to the heart, all the other members cease their operations; as long as the heart remains uninjured a man can think that he is alive. A man may be unconscious for several days, and still be alive; but deprive the heart of breath and man is destroyed immediately. **Heart disease** is caused by a change in the composition of the organ, whether it be regular or irregular. The presence of an abscess or of a hard, red sore in the heart causes immediate death. Palpitation of the heart is a symptom of disease, and the action of the heart is sometimes destroyed by the large quantity of liquid that is collected in the tunic

of the heart. We once saw an ape, the body of which was wasting away, and when he was dead, and we examined his members, they were all found to be in a healthy state, with the exception of the tunic of the heart. On this was an abscess of some kind, which was filled with bloody matter resembling that which is found in cysts in the eyes. And again, in a cock which we once dissected, we found a thing like an abscess, and it also was in the tunic of the heart; it was hard, and resembled many layers of fat placed one above the other. As men who die suddenly from heart disease are in full possession of all their reasoning powers up to the moment of their death, it is evident that the opinion of the ancients that the reasoning and thinking powers of the soul are not situated in the heart is correct. Palpitation of the heart comes to both young and old, even when they are in a healthy condition, apparently without cause, and every patient is benefited by the letting of blood, and those who live very carefully after the letting of blood may be cured. I knew a man once who was attacked by palpitation of the heart every spring, and the letting of blood relieved him. Early in the fourth year he was bled before the attack came on, and he did this for many years with the best results; he died, nevertheless, before he became an old man, even as all the others died. Some contracted fevers and died suddenly, some of them died through improper medical treatment, and one or two of them died from causes other than weakness. Most of these men were between forty and fifty years of age. Disease in the tunic of the heart need not cause death, but the heart will suffer sympathetically.

The **conditions of the heart** are seven: 1. Hot.

2. Cold. 3. Dry. 4. Wet. 5. Hot and dry. 6. Cold
and wet. 7. Cold and dry. The **hot condition** is indicated by deep breathing, a light and rapid pulse,
bravery, wrath, boldness, hair on the breasts, and a
whistling sound over the heart. The **cold condition**
is indicated by a slower pulse, timidity, languor, supine-
ness, and hairless breast. The **dry condition** is indicated by a hard pulse, and its possessor is slow to
anger; once roused his anger becomes a violent passion,
and it dies down very slowly. The **wet condition**
is indicated by a soft pulse, quickness to wrath, which
subsides quickly, and a moist body. The **hot and
dry condition** is indicated by a hard, full, and rapid
pulse, deep, quick, and frequent breathing, hairy breast
and epigastrium, fiery and savage actions, a disposition
to wrath, boldness, overbearing acts, and promptitude
in action. The wrath of such is not followed by cruel
deeds. The **cold and wet condition** is indicated by
a soft pulse, timidity, languor, and carelessness. The
cold and dry condition is indicated by a hard, small
pulse, hairless breasts and bodies, and slowness to wrath;
but when wrath is once roused it subsides very slowly,
and the possessor of this condition nurses his ire.
Characteristics other than these are abnormal. The
hot condition of the heart is caused by the presence
of red bile in excess, and gentle, easy exercise will
effect a cure. A dry temperament needs abundant
nourishment, and violent and excessive exercise must
be avoided, as well as anger, anxiety, sleeplessness,
and frequent coition. Possessors of the wet temperament should micturate before taking a meal, and purge
themselves and bathe frequently. Wine that promotes
micturation is beneficial. The cold temperament needs

moderate exercise, warming foods, drinks of hot wine, abundant sleep, and regular motions of the bowels.

[Here follows a series of prescriptions for palpitation and other diseases of the heart. Among them are the Caesar Antidote, the Gold Antidote, the Wild rue Antidote, the Musk Antidote, the Pearl Antidote, and the Coralium Antidote.]

The **Fifteenth Chapter** treats of the **stomach and its diseases.** The ancients divided the body into three parts: the stomach, mouth of the belly, or heart, and the belly proper, wherein are the mesenteric nerves. The food enters first the stomach, where it is "cooked" (*i.e.*, digested) until it is in a suitable state for absorption by the liver. When the liver is injured all the other members suffer. The stomach has two functions; it takes down things from the mouth, and it brings up to the mouth things that ascend from the belly. Dissectors of the body say that the stomach has two tunics or coats. Diseases of the stomach are of two classes, one class arising from the things that enter it, and the other from the things that come from it. The latter class is due to some stoppage or to an abscess of some kind. When the mouth of the belly is diseased, there is loss of appetite, and the food is turned to corruption. This has been the cause of misery to large numbers of men, and patients are, at times, subject to fits of insanity. The appetite, the abominable lust for food, and the lust of the dog, *i.e.*, gluttony, are all due to natural causes, and to certain conditions in the body that can be explained. The longing that is common to pregnant women, and is called "Ḳîṭâ" (*i. e.*, whale), is due to noxious chymes in the coats of the belly. The abominable lust for drink in men is due to bilious chyme in the coats of the belly; some men crave for foul drinks, and in others the longing is so unquenchable that they drink themselves to death. Men have been known to eat the vipers called *danpasdês*, because they produce thirst, and others have drunk wine wherein a viper of this class was drowned. **Bulimy** (βούλιμος), or faintness of the stomach, is due to cold, weakness, or emptiness of the stomach. The conditions, or temperaments, of the belly are the dry, the wet, and the hot, and each has its well-defined characteristics, and requires the special treatment that I set down here. **Hiccoughs** are caused by an excess of the cold chymes in the body, and may be got rid of by taking rue seed and wine, or natron and honey, &c., or castoreum administered in vinegar, or a piece of betel nut about the size of a chick-pea given in wine. Purges should, of course, be given, and a fit of sneezing will often put an end to hiccoughs. **Nausea** is caused by the weight of the chyme or by its acid bitings. This chyme is in the coats of the stomach, like water in a sponge, and it is difficult to dislodge it. It is cured by: 1. Fasting, toil, and sleep, whereby digestion is promoted; 2. vomiting caused by honey water; 3. an emetic made of radishes and honey. It is sometimes necessary to administer hiera pikra. **Abscesses in the stomach** must be treated with cooling medicines and foods, ointments, plasters, and gentle aperients. **Heartburn** and **flatulence,** and **gripings** are relieved by the use of aristolochia, the two sorts, and castoreum macerated in vinegar.

[Here follows a very long series of prescriptions for

the stomach, which fills about forty seven pages. The ailments prescribed for are nausea, looseness of the bowels, heartburn, diarrhoea, weakness of the stomach, thirst, cramp, fever, stoppages of the bowels, dysentery, bilious attacks, indigestion, pain in the sides and liver and spleen, shortness of breath, flatulence, chill of the heart and liver, acute sickness, malaise of the body, gluttony, hiccoughs, hardness of the liver and spleen, inflation, wind, shiverings due to stomach troubles, delirium, vertigo, kidney-disease, gout, pain in the womb, elephantiasis, colic, straining at stool, dropsy, &c. A second section of the prescriptions deals with **emetics**, which are made of salt, elaterium, natron, mustard, fennel, honey, black hellebore and radishes, honey and radishes, mustard and borax and saponaria, and the patient is advised to assist the working of these medicines by thrusting a feather down his throat! The author is afraid that he has given too many prescriptions in this Chapter, but he thinks that the skilful physician will find it quite easy to make up a prescription for himself. He can start with a simple medicine, which he knows is suitable for the case in hand, and he can add to it from the prescriptions before him. If he cannot make a diagnosis of his patient's ailment, he cannot prescribe for him and cannot cure him. In the third section we find prescriptions for fomentations, plasters, and cataplasms for the stomach, and on p. 363 is a prescription for consumption, which it is said will prevent the wasting away of the body. A sleep-producing plaster, which is made of opium, hyoscyamus seed, &c., is given on p. 370, and is attributed to Asclepiades.]

The **Sixteenth Chapter** treats of **diseases of the liver.** The liver is a substance charged with blood, it is situated on the right side of the body, and the whole of it is intersected by large veins. It receives the juice from the stomach by means of the mesenteric veins, and transforms it into blood by the power which it has, and then sends it out again by the large vein that is called the "Door". The function of the liver is dominant, for the nutritive power, which waters the whole body with blood by means of the veins that go forth from it, comes from the liver. The food must be digested three times before it becomes suitable for nourishing the body. The **first digestion** takes place in the stomach, and the residuum is the faecal matter, which is ejected through the anus. The **second digestion** takes place in the liver, and five products are the result: 1. **Phlegm,** which is transmitted to the lungs, chest, and brain. 2. **Red bile,** which is drawn away by the **gall bladder** in the liver, and becomes the bile. 3. **Blood,** which is transmitted throughout the body; the impurity of the blood is removed by the **spleen.** This is the black bile. 4. The **residuum** of the second digestion. 5. A **watery product,** which becomes the **urine.** This is drawn away by the **kidneys,** which transmit it to the **bladder,** and it is emptied into the urinary canal. The **third digestion** transforms the nutritive substance, and makes it suitable for absorption by the members. In the stomach the food is transformed into a white fluid. The residuum of the third digestion passes into the urine. Other waste products are excreted by the nose and mouth, and through the skin in the form of sweat, and some escape in the form of fumes that are imperceptible. Eight states or conditions of the liver have been distinguished, which I describe in the next section (p. 381). These conditions are: 1. The Hot. 2. The Cold. 3. The Dry. 4. The Wet. 5. The Hot and Dry. 6. The Hot and Wet. 7. The Wet and Cold. 8. The Cold and Dry. Liver diseases are of two kinds: The first kind is due to the change in the condition simply, and the second kind is due to a change in the condition and to the presence of a hard or hot ulcer, or pustule. The distention of the liver is due to wind, and obstructions are caused by the blocking up of the ends of the veins by thick, viscous chymes. Ulcers in the liver can be distinguished by the touch, especially those that occur in the four pairs of muscles which are in the upper part of the body, and which ye have seen in dissections. The abscesses that grow in the lower sides of the liver are not so sensitive, or so painful, as those that are found in the upper part of it. When the liver is cold, the food is not transformed into blood, and all the body becomes cold.

Dropsy is due to a defect in the power of the liver, which transforms food into blood: The dropsy that attacks the whole body is called **Besrânâ**, *i. e.*, appertaining to the flesh. The dropsy that attacks the region between the bowels and the peritoneum, in which wind is in excess of water, is called **Tablânâ**, *i. e.*, appertaining to a drum, because when you strike the epigastrium a drum-like sound is heard. The dropsy in which water is in excess of wind is called **Zeḳḳânâ**, *i. e.*, appertaining to a wine-skin, because when the epigastrium is touched it shakes as does a wine-skin when it is touched. The causes of dropsy are excess of cold and excess of heat, and each

destroys the power of the liver to transform food
into blood. Dropsy also arises through excessive
blood-letting, and through a perpetual flow of the
menstrual fluid. Any disease in any member which
interferes with the natural function of the liver can
produce dropsy. The kind of disease from which the
liver is suffering can be diagnosed from the stool
of the patient, as the proofs, which I append here,
show (p. 393). I advise the physician to study the
symptoms of the disease, and the members and their
conditions, otherwise he may do as several unskilled
physicians have done, and confound disease of the
liver with disease of some other member. There may
be two diseases present in the patient, and if so, the
physician must diagnose each carefully and separately.
We must never assign causes to a disease unless we
are certain of our facts, and we must never misread
symptoms. For example, do not think, because faint-
ing fits come on through some disease of the stomach,
that the patient's hands are diseased, or that a man
has heart disease because he cannot move as fast as
he did formerly. Or again, if the hands and the feet
are paralysed through the presence of some disease
in the spinal column it is useless to apply the means
of healing to them. If you study the stool of a
patient carefully you will be able easily to find out
whether his mesenteric veins are diseased or not. In
all such matters general and wide-spread experience
is invaluable, for a physician with limited experience
may be easily deceived when he sees something with
which he is not familiar, and may think that his
patient is in great danger when it is not so. In
treating diseases of the liver, if the patient is strong

and the season of the year suitable, we may let blood. If the stomach is swollen we must not excite vomiting, similarly, if the bowels are swollen we must not employ a cathartic. We must watch the growth of the ulcer, whether in the upper or lower part of the liver, and administer carefully the medicines that will check and purify, and subsidiary helps, both external and internal. When the ulcer is ripe we must apply medicines that quiet and relieve it, and so by degrees get rid of it. If the ulcer is in the "vault" of the liver, medicines that produce urine are very beneficial, *e. g.*, spikenard, petroselinum and ammi copticum. As regards a hard, obstinate ulcer in the liver itself, or in the flanges thereof, I personally have never been able to cure such, and I never knew of a case in which such had been healed. My experience is that such an ulcer always breaks up into corruption and ends in dropsy.

[Here follows a long series of prescriptions for liver plasters and medicines, which fills forty-two pages. The ailments prescribed for are hot ulcers, fever, thirst, obstructions, pain in the spleen, hard liver, hard spleen, pain in the kidneys, vomitings, dropsy, coughs, pleurisy, dysentery, blood-spitting, Philo's disease, wind, strangury, bites of insects and reptiles, delirium, insanity, the falling sickness, palpitation, phthisis, jaundice, twisted gut, fistular pains, womb diseases, retention of urine, all the diseases of women, cold in the stomach, gout, &c.]

The **Seventeenth Chapter** treats of the diseases of **jaundice**. The liver contains a certain source which produces all the chymes on which our bodies

f

subsist, and each of these chymes is transmitted to
its proper destination by means of the veins. The
blood nourishes the body, and makes it beautiful
and ruddy; its power is hot and moist, its colour
is red, and its natural taste is salt. The **red bile**
stimulates digestion, and produces the expulsion of
the urine and of the faecal matter; it opens the pores
of the skin, and produces **sweat**. Its power is hot
and dry, its taste is bitter, and there are three varieties
of it, red, yellow, and green. The **phlegm**, which
is moist in nature, nourishes the bones, nerves, membranes, tendons, ligaments, joints, lungs, and brain,
and cools the red bile; its power is cold and moist,
its colour is white, its taste is sweet, but it may
become acid and saltish. **Black bile** knits the organs
together, consolidates the skin and members, creates
the appetite, stimulates the intellectual powers, and
makes the heart brave and bold. It is in nature
cold and dry, black in colour, and its taste is acid
or bitter. All these are produced by the liver, and
if the liver is destroyed they cease to be; when this
is the case the other members are destroyed. When
the gall bladder is weak, or its veins become stopped
up, black bile permeates all the blood and fouls it,
and then the disease called Yarkânâ (jaundice) arises in
the body. Jaundice also arises from diseases other than
those of the liver, and we must be careful to see
whether it is due to the liver or to some other
cause. The blood can become bilious through some
destroying agent, as, for example, the venom of a
serpent. I knew a certain man who was a catcher
of vipers, and one day he was bitten by a viper,
and when the colour of his body became green, or

yellow, he came to me and described his case. I
gave him several draughts of Theriake, and in a very
short time his body returned to its normal colour.
[This medicine contained equal quantities of gentian,
laurel berries, aristolochia (round or long), and myrrh,
which were to be worked up in skimmed honey. The
dose was one drachm in hot water. See p. 409].
As to the various colours which the body assumes in
liver disease, it is impossible to explain them. On
several occasions I have diagnosed the diseases of
patients from the colour of their skins, and I can
always tell whether the disease is due to the liver
or to the spleen. As you know, when Satsînôs was
sick many physicians attended him, and last of all he
sent for me; and as soon as I saw him I said to
him, "There is nothing wrong with your liver". What
he suffered from was an abscess in the deep-seated
muscles. Jaundice arises sometimes from injuries to
the gall bladder. Sometimes this becomes too full
and is too weak to empty itself, and sometimes it
suffers from obstructions. Here again, if we wish to
know the character of the disease, we must examine
the matter that is excreted. The water in which
men with jaundice bathe contains bile, which comes
through the pores of their skin. At times I have
dosed patients with absinthe, and after they bathed
I had the sweat scraped off them with a leathern
scraper, and I found that it contained bile. One patient
suffered from bilious fever, and jaundice supervened,
but I was certain that this was not due to a diseased
liver. I ordered him to bathe in the water that comes
up hot from the earth, and the jaundice disappeared,
and he was cured permanently. The gall bladder,

like the urine bladder, suffers from over-fullness, and like it becomes distended, and is unable to effect a discharge. In short, there are five kinds of jaundice: 1. jaundice caused by a bilious fever; 2. jaundice caused by the bite of a reptile; 3. jaundice due to change of condition of the liver; 4. jaundice that arises sympathetically with disease of the spleen; 5. jaundice due to some injury in the gall bladder. The first kind is cured by baths, rubbing with natron, and suitable dieting. The second by Theriake, and the third and fifth kinds by the medicines enumerated in the preceding Chapter. The jaundice that is due to a diseased spleen is very serious, and is very difficult to cure. First of all let blood, and then administer oxymel and extract of barley. If there be inflammation and thirst administer extract of chicory, stewed orach, or spinach, or sorrel, and use a purge. And apply plasters containing aloes, absinthe, stacte, crocus, and asarum. [Here follows a series of prescriptions for the simple and compound medicines that are to be administered in cases of jaundice.]

The **Eighteenth Chapter** treats of the diseases of the **spleen.** The spleen draws away by means of its veins the impurity of the blood, namely, the **black bile** from the liver; a little of this it discharges into the stomach, where it causes the desire for food. The spleen is situated on the left side of the body, and round about it are the belly, the membrane of the intestines, and the arteries that nourish it. Disease of the spleen is associated with disease of liver, stomach, and kidneys. When the spleen is diseased the colour of the body tends to become dark, or black. The spleen at times discharges a large quantity of black bile into the body just as the liver discharges yellow bile, and if this be not ejected from the body, sickness, and discomfort, and even mental derangement supervene. A discharge from the spleen produces either lust for food, or a loathing for food. When both liver and spleen are diseased jaundice appears, and the colour of the patient is so dark that you might think the red bile had been mixed with soot. As to the connection of the spleen with dropsy, so much has already been said about it that I need add nothing more. We know well what is the function of the spleen, as well as the symptoms of all the diseases that attack it. In curing ailments of the spleen medicines that are bitter and acid must be administered, and oil and plasters may be used, and draughts of oxymel may be given. [Here follows a short series of prescriptions for ailments of the spleen; one of the prescriptions is attributed to Galen, and another to Asclepiades].

The **Nineteenth Chapter** treats of **diseases of the bowels.** Here we must discuss the various kinds of **diarrhoea** that exist. Diarrhoea arises naturally from the stomach, liver, and bowels. The diarrhoea caused by the stomach is due to weakness and indigestion, or to a collection of chymes, or to bad or insufficiently cooked food, or to inflammation in the belly, which destroys the food, or to cold. Diarrhoea arising through the liver is due to obstructions in the liver, or in the mesenteric veins, and when the liver cannot absorb the juice from the intestines, it goes down undigested, and so produces diarrhoea. Diarrhoea arises also from the bowels, and is a result of weakness, or "coughing", of a collection of bile or phlegm, or of griping, or of a rent in the veins or arteries, or of boils and fistulas of the anus. **Dysentery** begins with evacuations of bile in large quantities, which cause griping pains, and afterwards portions of the bowels come forth, and then a little blood. These portions of the bowels must be carefully examined, for they indicate the place where the ulcer, which causes the dissolution of the bowel substance, is situated. When the ulcer is in the upper part it may be treated by medicines that are drunk, but when it is in the lower part injections must be used. Ulceration of the bowels does not come on suddenly. **Haemorrhoids** produce great straining at stool and a frequent desire to evacuate. Certain physicians have said, "In cases where strong haemorrhoids "have existed, certain men have sometimes evacuated "from the anus stones full of holes, which resemble "those that exist in the bladder." But this I myself have never seen, and I never heard any man say that he had seen such. In treating ailments of this kind the physician should not restrain the discharge, but should assist nature. If the patient vomits, let the physician assist him in vomiting by giving him emetics, and if he has diarrhoea, let the physician give him medicines that will help him to clear himself out thoroughly. Unsuitable food must be replaced by suitable food, which must be given at regular intervals and in moderate quantities. [Here follows a long series of prescriptions for ulceration of the bowels, colic, indigestion, straining at stool, diarrhoea, nausea, stomach pains, gripings, intermittent fevers, &c. Some of the medicines are to be administered through the anus, either in the form of injections, or on wads of cotton wool. Among the prescriptions is one that was used by the author: Tincture of arsenic 12 drachms, Sandarach 6 drachms, Unslaked lime 26 drachms, and Burnt paper 30 drachms; these were worked up into tablets with extract of lamb's tongue.]

The **Twentieth Chapter** treats of diseases of the **colon.** We have already shewn that after the lower mouth of the belly, which is called the "Fingers", comes the intestine "Ṣâwmâ". Next comes the "Little Intestine", then the Blind Intestine, then the Colon, then the Straight Intestine, and last of all the intestine that is next the anus. Costiveness is a disease of the colon and is due to ulcers, both hot and cold, chyme of bile, phlegm, dense wind, dry faeces, and clots of blood. Vomiting is a symptom of disease of the colon. Maw-worms produce languor and vomiting, but they do not cause serious diseases like those of the colon. As all the diseases of the colon have been discussed in other parts of this book there is no need to add further remarks, but I append a list of the medicines that are useful as remedies for them. If you think that the diseases that you are called upon to treat are due to worms, then use the medicines of which I also add a list. [Here are two series of prescriptions. Following these is a short paragraph describing the difficulties that attend the treatment of diseases of the anus, and after this come the prescriptions that are to be used for boils and fissures in the anus, itch in the anus, protruding anus, &c.]

The **Twenty-first Chapter** treats of **kidney disease.** The kidneys perform two functions in the

bodies of the children of men: 1. They withdraw the
urine from the liver and transmit it to the bladder.
2. They stir up the organ which contains the semen
so that it may transmit it to the testicles. The kidneys
lie on the spinal column, near the loins, and they are
enveloped in fat, and they are fixed one on the right
and the other on the left; they and the inner parts
of the spinal column are surrounded by the veins that
bring the seed, and by those that bring the urine.
They are nourished by the great vein called "Ḥalîlâ",
i. e., the "Cavern", which brings the blood from the
liver, and their substance is nervous and fleshy, like
that of the heart. Kidneys suffer from three kinds of
disease. Organic disease of the kidneys is due to an
ulcer, or to some kind of obstruction caused by the
chymes, or to stones. Another kind of disease is
caused by heat, or cold, or dryness, or moisture in
them, and yet another is due to any injury to them
caused by the tearing or splitting of any part of them.
The pain caused by fever and the deposit of stone in
the kidneys, or in one of the veins that extend from
them to the bladder, resembles that caused by pain in
the colon. Kidney disease attacks men after they have
reached maturity, and stones are produced when the
kidneys possess naturally a hot, burning constitution.
When I treat kidney disease I administer certain
draughts that remove pain of the kidneys, and I make
a very careful diagnosis of the disease at the same
time. If, after the patient has drunk the draughts, I
find sandy particles in his urine, I know immediately
that the disease is due to the kidneys. If he has fever
and shivering fits at intervals, we may assume that
he has an ulcer on one of the kidneys; and, if one

kidney be ruptured and pus passes from it by way of the urine, the patient will feel relief. Much information concerning a tumour of the kidney may be obtained from the urine. The bursting of a vein in the kidney, or a fall, or a violent blow, causes a large quantity of blood to pass into the urine. When there is a tumour in the kidneys, small, fine particles of flesh appear in the urine, and these come from the kidney-substance itself; and in cases of cancer long strings or filaments of flesh appear in the urine. These were observed by Hippokrates, who describes them in his book. We have seen them a span in length, and even longer, and on one occasion they were so long that we wondered that the kidneys were large enough for anything of the kind to subsist in them. We therefore thought that they must be produced in the veins. They resemble the nerves in nature, but in their colour and thickness they are like unto maw-worms. Personally I am convinced that these filaments consist of thick and viscous chyme, and that they are transmitted to the kidneys, where they subsist. I thought that they could be cured by means of the medicines that are called "combative", and subsequent experience justified my supposition, and the patient suffered in no way after they were ejected from his body. The kidneys, the bladder, and the urinary veins were in no way injured by the discharge, and in this respect they resemble the other intestines, but under certain conditions they might be injured by the acidity of the urine itself.

Where the author of the Book of Medicines begins to treat of debility of the kidneys, our manuscript

comes to an end, and we have no means of knowing
what the concluding portion of the manuscript contained. The next section deals with divination, forecasts, omens, the influence of the planets and the Signs
of the Zodiac on the characters and dispositions of
men, and on human affairs in general. This section
will be summarized later, and we therefore pass to
the consideration of the third section, which contains
about four hundred prescriptions for diseases, which
were evidently drawn up by "native" physicians, *i. e.,*
physicians who did not adopt the theory and practice
of medicine as laid down in the first section, which
we have already summarized.

Throbbing of the head, or violent headache in one or more parts of the head, is relieved by the application of vinegar and honey, vinegar and ashes, vinegar, walnuts and terebinth nuts, dried grapes and aromatic tincture, a hot barley plaster, and by beetroot juice injected into the nostrils. We are also told to work up roasted asses' dung, and the bones of a crab, and ḥenna with oil, and anoint the head with the mixture. We may also inject the unsalted marrow from the right leg of a sheep into the nostrils of the patient.

For acute pain or **wound in the head** apply frankincense and crocus, or madder and cummin worked up with the white of an egg. For **vertigo** a medicine containing hyoscyamus is to be given. **Itch** in the head is cured by sagîrâ seed and oil rubbed on the head. **Lice** in the head are to be got rid of by a plaster made of dried grapes and myrtle oil, or by washing the head in salt water. For **sores in the head** use a plaster made of barley worked up in white of eggs, or an ointment made of scrapings of *gêrôd* stone; for running sores use old sandals rubbed into powder mixed with oil of pitch. For disease of the head use powdered partridge bones made into a plaster with oil, and pour vinegar over the head. Or use an ointment made of soft soap, dill, and sheep oil, which

has been placed at night out in the open on the fourth
day of three consecutive weeks. For **sores** and **scabies**
use a plaster made of frankincense, ṣôrâ, bread crusts,
and olive oil, then apply white of egg, or human blood,
or camphor leaves steeped in vinegar, or pyrethrum root
and honey.

The next twenty-six prescriptions are for **diseases
of the eyes.** For ailments of the eyes in general use
a lotion (or, "drops") made of tinctures of anethum
foeniculum and sweet pomegranates, and honey, which
have been boiled down to one half of their original
volume. Drop the drops into the eyes when the belly
is empty. To prevent **eyelashes from entering the
eye** use eye-paint made of the skin of a serpent, or
of the gall of a stork, or turbot, or eagle. For the
sleepless eye use thorn leaves boiled in wine and vinegar, or cummin and the white of an egg. Fleshy
growths are removed by eggs beaten up, or by crocus
rubbed down in wine. Hairs in the eyelids are removed
by the blood of bugs, or blood of dogs' ticks, or the
gall of an owl mixed with sal ammoniac. Hare's dung
pounded and antelope's dung pounded and used as an
eye-paint will prevent the eye-lashes from falling out.
Snow-blindness may be cured by applying to the eyes
wheat water, or crushed wheat, or garlic juice. Tincture
of sweet pomegranates poured into the eyes will remove
dust from them; afterwards apply extract of hyoscyamus.
For pain in dry eyes pour in menstrual fluid and the
semen of a man; if the pain continues, smear them
with antimony. **Weak eyes** are to be painted with
lizard's dung and olive oil, or with the milk of an ass
or bitch. To eyes with gangrene apply a lotion made
of grapes, wine, and vinegar, or an ointment made of

almond kernels, cummin, and wine, or the liquid in which the testicles of a fox have been boiled. To open children's eyes that are weak apply black dried herbs and honey; redness of the eyes is removed by syrup of sweet pomegranates and chick-peas; and into the eyes that have an effusion [of blood] pour the blood of doves, or the white of an egg warm. **Streaming eyes** are cured by the juice of bitter pomegranates, or the blood of white doves, or the juice of black thorns, or asparagus seed and roasted lentils and wine, or red thorns and vinegar, or an infusion of anethum foeniculum, and the juice of grass. For **inflammation** apply chewed *pêêmâ* leaves, for **throbbing**, olive oil and olive eye-paint. Severe pain is relieved by tâwlê juice, crocus, salt, and woman's milk, or the flesh of a swallow and Persian gum, or sponge plants dipped in hot water. Drink no wine, and let blood. **Blast** and **blight** in the eyes are cured by purslane juice, barley flour, henna and white of egg; or by a lotion made of dried grapes, hellebore, sweet pomegranates, sumach, and black thorns; or by an ointment made of sweet pomegranates and olive oil; or let the patient drink hot cow's milk. For attacks of **dimness of sight** use narcissus water, or the blood of a fox, or the blood of an antelope's liver, or the blood of a black sheep, or the blood of fish, or human excrement, or leek juice and urine, or sour-grape juice, or gourd juice injected into the nostrils, or vine juice and urine, or the gall of a goat and honey. For **darkness** of the eyes apply the ashes of the head of a young dove. For **weak sight** use burnt crab-shell and the juice of bitter roots, or the ashes of the heads of young swallows and honey, or powdered ass's hoof and asses' milk, or fish fat and honey, or partridge fat and mare's milk, or partridge gall, or crocus and the white of an egg, or the juice of sweet pomegranates. For **night blindness** apply the marrow from a mule's leg, or the roasted liver of a horse mixed with oil of musk. For **watery eyes** apply vulture's gall or the blood of a yellow frog. For **white deposit** in the eyes apply dog's gall, raven's eggs, crushed sea-shells, burnt date stones, and egg-shells, or mint seed and the milk of an ass, or saffron and the milk of a she-ass, or rook's gall, or crushed sugar, gourd sugar, bitter herbs, cat's gall, warânâ tongue, all in equal quantities, made into a powder.

For the nose only two prescriptions are given (p. 664). To stop **bleeding** apply to the bridge of the nose figs dipped in honey, or a paste made of frankincense, sulphur, glass and vinegar to the temples, or dust and vinegar to the face. Foul nostrils are cleansed with oil of vetches and pounded almonds.

Eight prescriptions deal with the ears. The first tells us how to remove a splinter from the ears. The second deals with deafness and partial deafness; the third with ear-ache, the fourth with ringing in the ears, the fifth with children's ears, the sixth with worms in the ears; the seventh is to prevent bleeding, and the eighth to cure the discharge of pus. The following animal and bird ingredients are mentioned, pig's fat, turtle-dove's fat, goat's fat, urine of boys, gall of a partridge, gall of goats and urine of goats, fat of a black cock, pig's gall, milk of a woman, juice of ox kidney, dog's excrement, &c. The last of the eight prescriptions is for the "Egyptian" medicine, which is composed of one litrâ of honey, three litrê of cinnabar, and three measures of vinegar. Alternative medicines are several gargles, one of which contains strong vinegar and oil of roses. Or we may macerate onions in wine, and lay the mixture on the back of the neck, whilst bread is put into the mouth of the patient. Or we may pound aristolochia root, or dog's excrement, to a powder, and blow it on the neck with a reed.

Two gargles are prescribed for the benefit of the **tongue**; the first is mulberry juice boiled with sugar, and the second is goat's milk (p. 666).

Five prescriptions are given for ailments of the **teeth**. **Toothache** is caused by a worm in the tooth, and if the worm can be persuaded to come out the pain will stop. One of the ways of effecting this is to make hyoscyamus seed, leek seed, and onion seed, worked up with the fat of *kômê* birds (?), into pills, which are to be placed on hot coals, and the patient is to inhale the smoke. The fumes of the rhododaphne rose, if inhaled, will kill the tooth worm, and the fumes of aromatic resin, if inhaled, will draw the worm out. Pain caused by **holes in the teeth** is cured by placing alum, or garlic, or asafoetida in them. Another prescription tells us to lay the great vein of the placenta (?) on an aching tooth, or fat of figs, or small red roses. **Loose teeth** are made firm by applying to them a paste made of raisins boiled in olive oil or dill water. To whiten teeth use a **powder** made of horns of the goat and horns of the stag burnt to powder.

Seven prescriptions deal with ailments of the **mouth**. Warm wine and honey, or sumach, sour grapes, star of Bethlehem plant, and salt relieve pain in the mouth. To purify the breath suck tablets made of dried roses, cummin, barley, and honey. Gangrene in the mouth is to be treated with honey and spice applied to the wound with a myrtle wood stick; or use ṣôrê, pomegranates, leeks, salt, and aristolochia root. Excess of saliva is reduced by chewing cannabis Indica (ḥashîsh) and lettuce seed, or beetroot seed. **Cracked lips** are cured by an ointment made of raisins, terebinth gum, honey and oil, but alternative vegetable prescriptions are given. **Disease of the neck** (stiff-neck?) may be cured by a gargle made of honey, olive oil and water, or by powder made from the plant "live-for-ever" blown on the neck.

Chapping of the hands and **chilblains** are relieved by two kinds of ointment, the one containing the brains of a hare, the blood of a bird, the ashes of a crab, and a little ḥenna, and the other burnt olive leaves and sheep oil. "Throbbing" of the fingers is cured by applying to them the gall of a bull and the white of an egg.

Eight prescriptions deal with **ailments of the chest.** A plaster made of dates, figs, licorice root, the plant ox-tongue, and barley flour relieves pain. Aristolochia and olive leaves relieve tightness of the chest. A woman with pains in her breast may use one of three ointments made of sulphur and vinegar, dried herbs, peeled beans, and oil of roses, or thistle flour, mustard, barley flour and vinegar. To make the breasts have milk in them use an ointment made of burnt crab and sesame oil, or eat the testicles of a fox after fasting. Or cow's milk may be drunk, or an infusion of licorice root, dates, figs, and sweet (*i. e.*, new) milk. **A deep cough** will be lessened if the patient drinks syrup of pomegranates and dates for three days. Several alternative remedies are given, some of which are in the form of potions, and others in the form of liniments and ointments. Nut galls taken in eggs, or dates and arsenic, are good for **blood-spitting.** To children who cough give three figs containing fifty coriander seeds and fifty pomegranate seeds. **Hoarseness** is to be treated with boiled carob seeds (to be eaten), hot oil, licorice flour and root of vetches, roast onions, a gargle of oil and honey, or five figs, five dates, and licorice root.

For the **liver** several kinds of draughts are prescribed, all but one of them being of vegetable character, *e.g.*, chicory and anethum foeniculum; cypress gum, sumach, and barley; the plant "live-for-ever". Chicory seeds and pomegranate seeds; sugar and tincture of berberis; thorns, pomegranate seeds, roses, and sumach flowers. The exception is made of cows' gum and snow. **Jaundice** is cured by drinking an infusion of radish seed and mulberries, or by hanging the dog-tooth of a bitch over the patient. As a cure for **thirst** is prescribed an infusion of aromatic herbs mixed with white of egg, or a hot infusion of purslane. **Heart pains** are relieved by drinking infusions of pomegranates and chamomile; myrtle berries, juice of asparagus and juice of mustard; licorice root, wine and cypress gum; honey and yeast water. For **heart disease** the prescriptions are somewhat similar, and **wind over the heart** is removed by drinking vinegar and honey, or dill water and honey, or aristolochia sucked out of a hot egg, or cummin, terebinth gum, anethum foeniculum, and rock parsley, mixed with goat's milk.

Some children and grown up folk have a mania for **eating earth,** and even filth. To such administer

g

black cummin and lettuce seed three times a day, or crickets and bread, or licorice root, dates, and figs mixed with sweet milk, or feed them on the flesh of the yellow fox of the desert. To him that has no appetite administer pounded river crab and cabbage juice. The loss of tone in the stomach which follows the drinking of wine in excess is restored by a decoction of licorice root in wine, or by a decoction of rose leaves, oil of walnuts, and wine. Honey, vinegar, water and anethum foeniculum make a good **appetizer**; drink it on three consecutive mornings. **Palpitation** is cured by pomegranates, sumach, and honey.

Some prescriptions are loathsome. Thus the person who is very ill is ordered to drink the sweat of his feet mixed with his own faeces.

To **procure sleep** rub the face with pounded mandragora and vinegar, or with sour milk, lettuce seed, and oil of roses. **Excessive sleep** is prevented by drinking an infusion of ginger and cabbage seed. He who has **drunk poison** must take urine of children and wine, or the gall of a gazelle in goat's milk. Or let him suck blood from oxen. How this is to be done is not stated, but we may assume that a vein of an ox is to be opened in the neck, and that the patient applied his mouth to the wound, and sucked out the blood as some of the natives of the Sûdân do.

Pain in the **spleen** is relieved by taking macerated hare, or the spleen of a hare, or calf's gall, or the ashes of a burnt ram, or the head of a swallow burnt in the fire, or swallow's blood, or the spleen or liver of a fox, or burnt stag's horn, or infusions of willow leaves, bean plant and cabbage. Or hang a spleen over the patient for three days, and over the fireplace for one day, and as it dries up the patient's spleen will dry up.

Ten prescriptions deal with **colic,** flatulence, non-retention of food, indigestion, bile, and diarrhoea in children and adults. The remedies are chiefly of a vegetable character, *e. g.,* garlic, cummin, olive oil, peppercorns, thorns, mint, &c. The juice of a burning goat's horn drunk in wine is a remedy for diarrhoea. In the case of children suffering from diarrhoea rub the gall of a gazelle over the navel, and rub rue and honey into the anus, and the flow will cease.

Haemorrhage of the belly is stopped by eating boiled purslane and sumach, or by milk and honey, hare soup, mutton broth, or by cummin and vinegar. If these fail make a linen pledget and dip it in cummin, goose-fat, and the white of an egg, and plug the anus with it. Powdered ram's horn kills intestinal **worms.**

Five prescriptions deal with diseases of the **anus.** For scabies use an ointment made of alum, spices, and vinegar, or barley flour and wine, or insert pledgets of linen dipped in ox gall, garlic, and fat of the fat-tailed sheep. A protruding anus may be replaced by applying one or other of the following medicines: 1. Sheep oil, or cow oil. 2. Cummin, pine gum, ox-gall, and spice. 3. Garlic and bitumen. 4. Pounded sea-shells and pitch. 5. Smelter's dross, raisins (stoned), and oxide of copper. 6. Glass waste, iron rust, and lime. 7. Burn a spider's web and myrtle leaves in a fire over which the patient is to sit, and let him anoint himself with oil. An enlarged anus is to be anointed with leeks and oil of roses, and fissures in the anus are to be rubbed with a mixture of alkali, hyoscyamus,

g*

raisins, and the urine of children, or with pig's gall
mixed with the fat of a black serpent with a red neck,
or with a mixture of wild mint, merdasangh, myrtle
leaves, and oil. A closed anus is opened by sprinkling
on it gourd ashes, fennel ashes, plantain ashes, and
aloes. A mixture of black raisins, pomegranates, and
wine will heal wounds and close fissures in the private
parts. To restore vigour to a **wasted member** apply
ointment made of the oil of a red cow, sheep's oil,
and pounded bardîlâ leaves. **Wind in the stomach**
is removed by taking cummin and salt in wine, and
walnuts and vinegar.

A woman in difficult labour is helped by hanging
the *tâwlê* plants first over her right shoulder, and
then over the child when it emerges from the womb.
Decoctions of thorns, leeks and cummin are beneficial,
as also are burnt flax stalks in wine. In cases where
the child is dead in its mother, and will not come
down, burn either the hoof of a horse, or the hoof
of an ass, or thorn seed under her, or make her drink
a bitch's milk in wine, or a draught containing two
measures of ox-dung, or a decoction of shepherd's
staff.

In cases of **strangury** administer purslane seed in
wine, or rinds of pomegranates, or dove's dung in
wine, or onions and flax stalks, or the powder of a
roasted crab in wine, or cock-droppings in wine. For
strangury in a child administer an infusion of rue,
vinegar, and carthamus tinctorius both to the patient
and to his mother. To relieve continuous **micturation**
give the patient bread on which a crow has been
baked. **Retention** of **urine** is removed by rubbing
the patient's member with the juice of nacissus leaves,

or with a mixture of mice-dung and saliva. To remove **blood in the urine** administer a mixture of the gall of a horse, frankincense, caper berries and cummin, or cummin and gingers, or myrtle leaves and madder, or mix dove's dung with his urine, heat it, and let the patient drink it. To cure **involuntary micturation** in bed administer to the patient a burnt cock's comb in wine, or the testicles of a fox, or cock, or the bladder of a ram or pig, or the brain of a hare with the seed of dill and rock parsley, or the dried liver of a crow, or three drachms of thorns, or cummin, or raisins and sparrow's blood.

Four prescriptions deal with the **testicles**. For pain in the testicles apply a mixture of black raisins, cummin, and beans, or wine lees and sheep oil. Aching testicles are relieved by an application of a mixture of stacte, wheat, wax, ox-fat, pig-fat, and oil of roses, or of chopped lettuce leaves, or sulphur and vinegar. "Wind in the testicles" is removed by an application of wolf's gall in wine, or by a mixture of cypress gum, ox-tongue, wine and honey, or by the fat on the kidneys of a gazelle and unslaked lime. When this ailment appears in a boy apply a plaster made of lime, and gingers, and the urine of the patient. For pain in the knees and back apply a bran poultice, or rub the knees with naphtha for three days. For pain in the shoulders apply a poultice of barley meal. For "wind" in the legs burn cock's feathers and terebinth gum together, extinguish them in vinegar, stir the mixture well, and use it as a liniment. Two prescriptions are given for the "red wind", an ailment which I can neither describe nor explain. The principal ingredient is mountain leaves obtained from a mountain that has the shape of a fig-leaf, before sunrise on the fourteenth day of the month (p. 683).

The prescriptions for **ulcers**, wind, **pleurisy**, and pain in the loins closely resemble those given in the first section of the book. To cure **hiccoughs** shake the patient violently, or administer to him three mouthfuls of very strong vinegar, or of very hot or very cold water. To prevent **snoring** place a stallion's tooth under the head of the snorer, and he will snore no more. To **expel a devil** from a man tie the heart of an ass in the skin of a stag, and hang it over him, or burn the heart of a dove under him. If a man wishes to stand well with the governing powers let him cause the head of a crow to be hung up over him; and if he wishes to stand well with his neighbours let him cause the heart, eye, and skin of a wolf to be hung up over him.

Two prescriptions deal with the **complexion** and certain ailments. On the face that is like wax we must lay the warm lung of a camel, and the patient shall do well. To make a thin man fat give him pounded galbanum mixed with goat's milk to drink. To prevent **suppuration** in a burn, smear it with boiling water and the white of an egg. For **burns** and **scalds** apply white of egg and olive oil, or ashes of senna wood and paper, barley flour, yellow raisins, rinds of pomegranates, and whites of eggs, mixed to form an ointment. For the **bite** of a **dog** eight prescriptions are given, the principal ingredients being vegetable in character. One of the prescriptions orders that roasted hair of the dog that bit the patient be given to him! If you hang up over yourself (*i. e.*, wear) a dog's tooth, a **mad dog** will not bite you. Feed a dog on bread containing bitter almonds, and he will follow you everywhere; tie a weasel's tail above your loins, and dogs will not bark at you. For **snake-bite** administer cat's gall in wine, or cut out the flesh round the bite, and apply onions, salt, and vinegar to the wound. Or cut a frog in pieces, and lay them on it. Several vegetable remedies are added. For **scorpion sting** apply mustard and hot water, or unslaked lime and oil of roses, or tie a silver drachm over the wound.

To **expel serpents** and reptiles from a house, make a mixture of galbanum root, sapesta seed, a stag's horn and hoof, opium, and vinegar, set light to it, and the serpents will flee from the house. Burn the hair and hoofs of goats in a house and the **beetles** will flee; burn the white of an egg and the hair of a fox, and the **crickets** in a house will die; mix cummin, sulphur, and hyoscyamus (or, cannabis Indica) with water and pour the mixture on the nests of **ants,** and they will flee. Cypress leaves and cummin scattered about a house will kill **fleas. Moth** may be kept from bee-hives by sprinkling fresh milk and children's urine over the legs of the stands; pour wine and water on the hives, or fumigate them with burnt ass's dung, and the bees will not leave the hives. To **kill mice** fumigate the house with burnt parsley, or the dust of a burnt thigh of a camel. To drive away **gnats** fumigate your bedroom with burnt galbanum and sulphur, or colocasia, or place a little extract of hemp under your head. To cure **sickness in cattle** pound up a dried rose with salt, and give the mixture to the beasts in their food for seven days. To prevent the **escape of chickens** give them wheat soaked in yellow mint water. To **catch birds** throw out to them wheat which has been soaked in urine for one day and then boiled in sulphur; every bird that eats the wheat will die, or may be easily caught. To make **hens lay eggs** give them a paste made of wheat flour and wine.

Prescriptions for **pomades** for the **hair** are numerous. To make it grow luxuriantly use one of the following pomades. Fox fat and sesame oil, the brain of a hare and the marrow from a hare's leg, the ashes of a burnt swallow mixed with the urine of men who are virgins, wolf's fat and oak ashes, boiled bat's head and olive oil. Or use a plaster of olive leaves, or drink partridge gall mixed with fine wine when the moon is riding in the highest height of the sky. To make the **eyebrows grow** anoint them with burnt spice and sesame oil, or with an ointment made of pounded walnuts and date stones and oil of roses. Useful **depilatories** are: dried *kômê* flesh and ant's eggs pounded, and bats' blood. To prevent the **hair turning white** pour two drops of the gall of a swallow inside the mouth on the right side, and two on the left. Or use ointments made of bees roasted in olive oil, or the fat of a black raven and wet thorns, and ox gall and swallows' dung.

Four prescriptions deal with **fever**. For **summer fever** lay cloths soaked in hot water and oil of roses on the hands and feet of the patient, or anoint all his members with roasted mustard and oil of roses. **Fever caused by travel** is got rid of by sitting in hot water, and anointing yourself with sweet oil. For the fever that comes every second day burn the bone of a pig under the patient; for the fever that comes

every third day hang the shell of a crab on the patient,
or the upper lip of a mole. To make a boil burst,
poultice it with a poultice made of unleavened bread,
garlic, and vinegar.

Ten prescriptions deal with soft **sores**, watery sores,
obstinate sores, stinking sores, &c. and among the
remedies enumerated are: burnt ox horn and woman's
milk; sumach and honey; pounded shells and the white
of an egg; rose ashes, mulberry leaves, raisins and
antelope fat, alkali, honey, and old sheep oil; a human
bone pounded up; lead and oil; onion and oil of roses;
and figs and rue. **Leprosy** was treated with ointments made of vegetable substances and vinegar. After
an application of one of these the patient was made
to sit in the sun. The kind of **leprosy** called "Kharsâ"
was treated with ointments made of crab and ox-dung;
the blood of black doves and turtle doves, spice, and
salt; the dung of doves, human semen and spice; burnt
ox horn, olive oil, and old wine; gum of cedar and
terebinth, sulphur, and the white of an egg, and the
patient was to stand in the sun. In the prescriptions
for **gangrene** the principal ingredients are unslaked
lime (3 parts), arsenic (1 part), and vinegar, or arsenic,
alkali, and yellow hellebore, or alum, white silver, sâghâ,
and lentil flour. The man who gnaws his own flesh
is to be rubbed with wine lees, to stand in the sun
for a time, and to be washed in hot water.

The prescriptions for boils and itch in the back of
the head and neck resemble those in the first section
of the book, and the principal ingredients are alkali,
soap, henna, sulphur, garlic, and vinegar.

The next group of prescriptions (pp. 695, 696) deal
with the Edhrâ boil, ringworm, the **mulberry** boil,

which comes in man and beast, and sores in cattle. It is very difficult to identify these, but among them we appear to have the so-called "Baghdad boil", or the "Aleppo button", or the "date mark", of which I have seen many terrible examples in Mesopotamia. Each of these scourges begins with a small whitey-brown swelling, of which there are two kinds, hard and soft. The swelling may appear in any soft part of the body, the face and neck, the buttocks, the anus, &c. The hard swelling grows steadily until it literally looks like a date on the cheek or neck, and then its surface decays and the swelling turns into a large running sore, which even with careful attention takes weeks, and sometimes months to heal. It leaves behind it a large scar ("date mark") of a livid colour in people with fair complexions, and of a dark brown colour in people with dark complexions. When this sore appears in certain parts of the body it causes very great pain. The soft swelling also appears in various parts of the body, and soon turns into a running sore; unfortunately several swellings frequently appear at the same time and become confluent. I have seen nine on a man's face at once. Young children suffer greatly from this scourge, as do adults who are habitually over-worked, or under-fed.

Following the prescriptions for boils comes an appreciation of **bean flour** as a beautifier of the complexion, and our author says that the young men and women who sell bean flour frequently use it when they wash, and that their complexions are as "white as snow".

The next group of prescriptions provides remedies for wounds caused by swords, spears, arrows, &c. The wounds are first to be cleansed by applying onions to the lips of the wounds, and then an ointment containing several vegetable substances is to be applied. This ointment is called **Marham,** and was evidently very well known. There were several kinds of Marham, but the most important were the Red Marham, the White Marham, and the Black Marham. Honey, wax, soft soap, olive oil, madder, terebinth gum, gum ammoniac, red lead, and unsalted butter were the principal ingredients. The wound caused by the surgeon's knife is healed with soot and oil, or croton ashes and oil. Lead (a bullet?) may be extracted from the body by applying the flesh of the marten to the wound. To remove worms from wounds sprinkle over them a mixture of alkali, aristolochia, and lime.

Here the prescriptions come to an end. The remaining paragraphs contain a series of instructions that clearly represent a number of beliefs and superstitions about animals and birds, magical roots and plants, &c. Thus the stones in the crop of a **white cock** with a divided comb cure St. Vitus's dance and drive away devils. **Coriander seed** grown in the head of a black raven will enable a man to see whatever he pleases. Every member of the **mole** possesses medicinal properties that are beneficial for man, provided that it be prepared properly with quicksilver ointment. The claw of the right leg of a **cock** gives victory in the law court, the bones of the left wing keep away fever, and cock's gall and honey make the eyes bright. Eat the flesh of the **wild ass** in the summer and you shall suffer from no disease. Make a man drink the foam from the mouth of a **camel in must,** when the moon is directly overhead, and he will become a bold speaker.

Anoint your disciple's face with **wolf's gall**, and make him drink some, and you will marvel at his wisdom. The head of a **bat** placed in the hat makes a man invincible; bat's eyes keep away scorpions, and a bat's heart will drive away ants from their nests.

Next follows a series of sections describing the benefits that accrue to a man who carries about with him the heart or some member of the birds and animals, the names of which are here given: The Eagle, Vulture, Hawk, White Falcon, Falcon, Partridge, Hare, Crab, Frog, Ox, Fox, Bear, Weasel, Hedgehog, Pelican, and the Black Crow. A long paragraph is devoted to "Solomon's bird", perhaps the hoopoe, every part of which, if properly treated, produces a magical medicine, which has a marvellous effect. This is followed by a description of King Solomon's medicine for the eyes, which was composed of a mixture of mole's eyes and iron ore. This when applied to the eyes enabled a man to see every thing that he wished to see, and it caused the devils to flee from before the man whose eyes were anointed with it. Moreover, if he looked at his enemies he appeared to them to be fire, and, of course, they fled before him. The effect of this eye medicine lasted ten days, during which no pig's flesh was to be eaten by the man on whose eyes the eye-paint was. The last paragraphs of this section are devoted to descriptions of **roots** that possess magical qualities. Among these are the Kahînâ Root, the Live-for-ever Root, the Madhîn Sûhâr Root, the Semra Root, the Healer Root, the Sephînâytâ Root, the Dakartâ Root, the Abreshôm Root, the Edhlâ Root, &c.

The section of the Manuscript of the Book of

Medicines dealing with these subjects begins on Fol. 211 *b* (English translation, Vol. 11, p. 520), and the various Chapters may be thus summarized.

1. **Weather forecasts** for the week beginning on the 19th of Tammûz and ending on the twenty-sixth. The forecasts concern rain, and are based on the appearances of clouds at various times of the day and night (p. 520).

2. **Forecasts derived from shooting stars.** A star shooting from east to west portends the death of the king of Persia, and sickness and disease; shooting from west to east it portends wrath and temporary tribulation; shooting from north to south it portends evil to the king, barrenness of women and family feuds for three years; shooting from south to north, a red star portends pestilence (if it falls on the earth), war in a far country, sickness, and conquest; shooting from west to south or from north to south, if it breaks up, the star portends war, pestilence and slaughter; a star falling from heaven to earth portends peace; a star bursting up from the east portends fecundity among cattle, and general prosperity (p. 520).

3. A **second series of forecasts** derived from shooting stars; the general import of the series is similar to that of the preceding series. The portent concerning the waging of war by the king of Persia on Bêth Hûzâyê (Hûzistân) is said to be in accordance with the Book of Andronicus. This authority is probably Andronicus Cyrrhestes the astronomer, who flourished about B.C. 100, and who was the builder of the "tower of the winds" at Athens, which was also called the "horologium". He is also mentioned by Eusebius of Caesarea in his work on "the Star"; see Wright,

Journal of Sacred Literature, 1866, where the Syriac text is given (p. 521).

4. Forecasts concerning rain, derived from the appearance of the sun in the morning and evening (p. 522).

5. A **table shewing** what the **weather** will be throughout the year when the sun rises on New Year's Day in a certain Sign of the Zodiac, and stating what kind of crop may be expected. *E. g.*, "If the "year be born in the Sign of the Scales, the early "seed and the latter crop will not appear. If the "year be born in the sign of the Goat, the early seed "will lack rain sorely, but there will be an abundant "crop on the plain." This list seems to have been drawn up for the use of the farmers who lived in a particular district, and intended to guide them as to the time when seed was to be sown (p. 522).

6. Two sections dealing with the **Signs of the Zodiac** (Malwâshê) and the **Year,** which appear to be supplementary to No. 5. With the Ram are associated full rivers and streams. With the Crab plagues of locusts and mice, scarcity of fruit in India, good crops in Ṭibaryâ, a severe winter in Media, &c. If Bêl (Jupiter) be associated with this sign there will be peace, and if Mars, there will be blood and war (p. 524).

7. Of the **marrying of wives.** This paragraph is difficult to explain. At the head of it stand six letters, which have the numerical values of 2. 3. 4. 5. 6. 7 respectively. Each letter indicates a certain fact in some system of divination that was used by men in order to find out whether they ought to marry or not, and whether they will have any success with women (p. 524).

8. How to find out in which hand a man is hiding some object. Take the total of the numerical values of the letters that form the names of the man and his mother, and divide by six; if the remainder be an even number the object is in his right hand, and if it be an odd number, in his left. Suppose the man's name is "Joseph" and his mother's "Mary". Now Joseph,

$$\text{ܐܘܣܦ} = 10 + 6 + 60 + 80 \quad = 156$$

and Mary,

$$\text{ܡܪܝܡ} = 40 + 200 + 10 + 40 = 290$$

Adding 156 to 290 we have 446. Dividing 446 by 6 we have 73 and 5 over. This being an odd number, the object must be in the man's left hand (pp. 524, 525).

9. An elaborate system of **divination** whereby the thief may be detected (p. 525).

10. Here, curiously enough, comes **a list of the numerical signs** used in the prescriptions (p. 525), and this is followed by a table of the **weights and measures** used in medicine (p. 526).

11. Divination. To find out when called whether the caller wishes to do you a bad turn or a good one. Stand in the sun, and find out the number of paces in the length of your shadow, and add it to the totals of the numerical values of the letters in your own name and in that of your mother. Divide the total by two: if the remainder be naught, sit still; if one, go and see what is wanted. Taking the number of paces as 10, and the names of Joseph and Mary as before, by adding these together, 10 + 156 + 290 we get 456. Divide 456 by two, and there is no remainder; therefore sit still (p. 526).

12. **Divination.** To find out whether the husband or the wife will die first (p. 526).

13. **A table of lunar months** by which it is possible to find out on how many days of a solar month the moon will rise (p. 527—528). Appended is a list of the names of the months of the Arabs.

14. **Rule** for finding out which day of the week will be the first day of the new moon (p. 529).

15. **Rule** for finding out in which hour of the day or night the moon will appear (p. 530).

16. **Rule** for finding out how long the moon will shine each day, and at what hour it will set.

17. **Divination.** How to find out the cause of a person's sickness. Add up the totals of the numerical values of the letters in the name of the sick man and his mother, and divide them by nine. Assuming that the names are Joseph and Mary, as before, we have $156 + 290 = 446$; this number divided by nine gives 49 and 5 over. Turning to the paragraph we see that if the remainder after dividing the total by 9 be five, the illness is caused by a devil. Assuming that the names are Alexander and Sârâ we have

ܐܠܟܣܢܕܪܘܣ $= 1 + 30 + 20 + 60 + 50 + 4 + 200 + 6 + 60 = 431$
and ܣܪܐ $= 60 + 200 + 1$　　　　　　　　　$= 261$
_____
　　　　　　　　　　　　　　　　　Total 692

Dividing 692 by 9 we get 76 and 8 over. Turning to the paragraph we find that the remainder 8 indicates that the man's disease is caused by a blow of Satan (p. 532).

18. **Divination.** This section resembles the preceding, but is more elaborate, and the amount of the remainder after the division of the total by 9 is

coupled with one of the days of the week. Thus if
the remainder be three, and if the man was attacked
by his sickness on the third day of the week, the
day of Mars, his disease is hot and dry; he bathed
in water and the air smote him. Make three lamp
wicks, or torches, out of his garments. Set one at
his head, and one by his right side and one by
his left side, and he will be ill for fifteen days. When
we come to the remainder being eight, as there is
no eighth day of the weak, we are told that the
patient had a bad dream on the fourth day of the
week. When the total divides exactly by 9, it indicates
that the man sat on the ground on the Eve of the
Sabbath, and did not call upon the Name of the
Living God. Therefore he will continue to be sick
for eighteen days, and his illness is due to vengeance
(p. 533).

19. Divination. Another system in which, having
added together the numerical values of the letters
of the two names, and divided it by 9, we deduct
from the letters *Jôdh* to *Sâdhê* 1. 2. 3. 4. 5. 6. 7. 8.
9 respectively, and from the four letters *Kôph* to *Tâw*
1. 2. 3. 4 respectively (p. 533).

20. Divination. Similar to the preceding, but the
total of the numerical values of the letters of the two
names must be divided by 7.

21. Divination. To find out whether a sick man
will recover. Add together the numerical values of the
letters in his name, and those in the name of the day
on which he fell sick, and divide the total by 3. If
1 remains the man will live, if 2 remains his sickness
will be protracted, and if there be no remainder he
will die (p. 533).

h

22. Divination. Direction supplementary to the preceding about dividing the totals of the numerical values of the letters in the names of the days of the weeks (p. 534).

23. Divination. Another system whereby to find out whether a sick man will recover from his illness. In this case we add up the numerical values of the letters in the names of the man, and of his mother and of the day in which he visits the physician, and divide the total by 3. If the man comes on the first day of the week, and 1 remains he will recover rapidly; if two remains his sickness will last a long time; if there be no remainder he is nigh unto death (p. 534).

24. Divination. Four ways of finding out whether a sick man is going to live or die. Give a dog some bread that has been under a sick man's head for a night; if the dog eats it the man will live, if not, he will die. Or give a dog some of the water of which the sick man has drunk; if he drinks it the man will live, if he does not, the man will die. Or take a nail clipping from a toe of the sick man, and throw it into a cup of water; if the clipping sinks he will live, if not, he will die (p. 534).

25. Divination. How to find out whether a sick man will recover, and whether something that has been lost will be found, and if a slave who has run away will be brought back. Add up the number of days that have passed between the seventeenth day of the month Shebât and the day on which the man fell sick, or on which the thing was lost, or on which the slave ran away, and divide them by thirty-five. Then construct three tablets in which the numbers 1 to 35 are arranged in the following way:

Table I.   1. 4. 7. 13. 16. 19. 22. 23. 28. 31.
Table II.   5. 8. 11. 14. 17. 18. 26. 20.[1] 25. 29. 32. 35.
Table III.   3. 6. 2. 9. 10. 12. 15. 21. 24. 29. 30. 33. 34.
Then divide the number of days by 35, and if it does not divide exactly note which of the three tables contains the number of the days that are over. If it be in the first table the man will live, the lost thing will be found, and the runaway slave will be recaptured; if it be in· the third table, the exact opposite will happen in each case; if it be in the second, the man will recover slowly, the lost thing will be found after a long time, and the runaway slave will be caught eventually, and brought back (p. 535).

26. Divination concerning a sick man, based on the consideration of which day of the month his sickness attacked him. If he be attacked on the first day he will recover on the second. If he be attacked on the second day he will be ill for the whole month. If he be attacked on the fourth day, or the thirteenth day, or nineteenth, or twenty-third, or twenty-fourth, or twenty-sixth day he will die; if on the tenth day he will die in seven days, and if on the eighteenth he will die in twenty days (p. 536).

27. Divination concerning the recovery of a sick man, derived from the **days of the moon.** If he be attacked by his sickness on the twelfth, thirteenth, fifteenth, eighteenth, twenty-first, twenty-third, twenty-fifth, or thirtieth day, he will die (p. 537).

28. Another method. Find out on which day of the lunar month the man was taken ill. Add up the

---

[1] I have substituted 17 and 20 for a second 14 and a second 23, which are manifestly mistakes.

h*

numerical values of the letters that form the name of the man, and add to them the number of the days that have passed since the new moon, and then divide the total by nineteen. If the remainder be 1. 3. 5. 7. 9. 14. 16. 19 he will live; if it be 4. 6. 10. 12. 18, he may live or die, the chances being equal; and if it be 2. 8. 11. 13. 15. 17, he will die (p. 538).

29. Another method. Add together the numerical values of the letters of the man's name, add to them the number of the days of the moon's age, and twenty; divide the total by thirty. Then examine the two following Tables of numbers, and see in which of them is the number that remains after dividing the total by thirty. If it be in the first table he will live, and if in the second he will die.

Table No. 1.   Table No. 2.

1.	2.	3.
17.	8.	7.
9.	10.	11.
13.	14.	19.
16.	23.	20.
22.	26.	28.

5.	6.	12.
15.	18.	21.
27.	25.	
29.	30.	
4.		
24.		

(p. 539).

30. Another method, based on the light and dark days of the moon. The moon has light on the odd days, which are, 1. 3. 5. 7. 9. 11. 13. 15. 17. 19. 21. 23. 25. 27. 29; the moon is dark on the even days, which are, 2. 4. 6. 8. 10. 12. 14. 16. 18. 20. 22. 24. 26. 28. 30. If the man falls sick when the moon has light he will last in a weak state for a long time. Add together the numerical values of the letters of the man's name, and add to them the number of the day of the moon when he fell sick. The total is to be divided by some number, but as the scribe has omitted to say what that number is, we cannot complete the system. The result is, however, pretty clear. If the remainder be an odd number the man will live, if an even one he will die (p. 539).

31. To find out which of two men will win a duel. Add together the numerical values of the letters in the name of each warrior, and divide each total by nine. When Alexander the Great was about to fight Darius, he used this method. He found that the numerical values of the letters in the name Alexander, ܐܠܟܣܢܕܪܘܣ were $1 + 30 + 20 + 60 + 50 + 4 + 200 + 6 + 60 = 431$; and that the numerical values of the letters in the name Darius ܕܪܝܘܫ were $4 + 200 + 10 + 6 + 300 = 520$. If we divide the first total by 9 we get 47 and 8 over, and if we divide the second total by 7 we get 57 and 8 over. Since 9 is greater than 8 Alexander conquered Darius! This same method may be employed with success in cases of sickness, or in cases where things have been stolen, or slaves have run away (p. 541).

32. The **Regents of the days of the week.** First day, Hermes and the Sun. Second day, Zeus and the Moon. Third day, Aphrodite and Ares. Fourth day, Kronos and Hermes. Fifth day, the Sun and Zeus. Sixth day, the Moon and Aphrodite. The Sabbath, Arês and Kronos (p. 542).

33. **List of the remainders** after dividing the totals of the numerical values of the letters in the names of the stars and of the days of the week by nine (p. 542).

33. **Divinations** concerning the sick, based upon calculations as to the age of the moon when the man took to his bed, and the position of the moon among

the Signs of the Zodiac. If a man falls sick in the month of Îlûl when the moon is in the Sign Leo, he will recover. If a man falls sick during the month of the Second Teshrî, and the moon is in the Balance, he will be sick for ten days, and will then recover (p. 546).

34. Description of the course of the moon through the Zodiac (p. 546).

36. Description of the sicknesses that attack men under each Sign of the Zodiac. The following is a specimen paragraph. He who is smitten in Capricorn will be smitten severely; he will be smitten with heavy eyes. If he be smitten on the first day he will die; if on the second, his illness will be severe; if on the third, water will flow from his eyes in abundance and he will be healed (p. 548).

37. Another of the same kind. The man smitten under Aquarius and Pisces will not suffer severely if one propitious star be there. If Mars be there, he will go silly. If the Sun, he will go blind. If Venus, he will fall through love. If Mercury, he will be attacked by sickness. If the Moon, his ailment will be light (p. 549).

38. Divination concerning a sick man; from the Book of the Signs of the Zodiac. If a man, who was born under Arês and falls sick under Cancer, lingers for thirty days he will not die. This section contains twelve paragraphs, each referring to one of the Signs of the Zodiac (p. 550).

39. A List of the Signs of the Zodiac with **propitious stars.** The Ram with the Moon and Hermes. The Bull with the Sun, and Zeus and the Moon (p. 550).

40. The Eight Rulers and the days which they

rule. Venus rules the 1st and 9th days. Mars rules the 2nd and 11th days, &c. (p. 551).

41. The **diseases produced in men** by the Planets. Saturn produces headache, gout, chill, cough, shortness of breath, dropsy, &c. The Planets enumerated are Kronos, Zeus, Arês, the Sun, Aphrodite, Hermes, the Moon. Kronos kills by drowning, Zeus by earthquake and thunderbolt, Arês by metal and rheum, the Sun by rheum, Aphrodite by food and lust, Hermes by apoplexy, and the Moon by insanity (p. 551).

42. The positions of the **Seven Governors.** Kronos is in the zone of air. Zeus is in the zone of air. Arês is in a zone of fire. The Sun is in a zone of fire. Balti, or Zûhrah, or Aphrodite, is in a zone of water. Hermes, or Nâbô, or 'Ùṭrâdh, is in a zone of water. The moon is in a zone of water. The Sun and Moon occupy each one House, and the other five Governors two Houses (p. 552).

43. The **Revolutions of the Planets.**

2	Revolutions of	Kronos	= 60 years	
5	„	„ Zeus	= 60 „	
4	„	„ Arês	= 60 „	
60	„	„ the Sun	= 60 „	
120	„	„ Hermes	= 60 „	
720	„	„ the Moon	= 60 „	(p. 553)

44. The **Kingdoms of the Planets.**

PLANET.	KINGDOM.	SETTING PLACE.
Sun	Ram	Balance
Aphrodite	Fishes	Virgin
Hermes	Bull	Scorpion
Kronos	Balance	Ram
Zeus	Crab	Goat
Arês	Goat	Crab (p. 553).

45. Planets by which to travel. Eastwards. Aphrodite, the Moon, Arês. Westwards. The Sun, Zeus, and Hermes (p. 554).

46. Forecasts for the use of **travellers** based upon the position of the moon in the Zodiac, *e. g.*: If a man sets out on a journey, and he finds that the moon is in the Ram, let him return home immediately. If a man goes forth to war or to hunt when the moon is in the Fishes, he will be successful, but every other enterprise will fail. The Christian scribe betrays himself by the quotation, "Except the Lord build the "house, the builders thereof weary themselves vainly" (Psalm cxxvii. 1, p. 554).

47. Lucky and unlucky days. From the 27th day of the moon to the half of the 4th day do nothing at all; from the half of the 4th day to the 11th day do anything; from the 12th day to the 27th do everything. This is followed by a list of the times of the 15th day of the moon when, according to its place, certain things may be done, *e.g.*, When the moon is in the Ram on the 15th day, in the early hours it is good to travel (p. 555).

48. Lucky and unlucky days for travelling. The lucky days are: 1. 2. 4. 9. 13. 16. 17. 19. 20. 21. 22. 24. 25. 26. 27. 28. 29. 30. (p. 556).

49. The **Unlucky days of the months.**

Month	Days	Month	Days
Nîsân	3.	1st Teshrî	3. 6. 20.
Îyyâr	6. 20.	2nd „	3. 5. 11.
Khazîrân	3. 18.	1st Kânôn	3. 20.
Tammûz	6. 20.	2nd „	2. 3. 11. 14.
Âbh	1. 4. 15.	Shebâṭ	7. 11. 20. 21.
Îlûl	3. 10. 20.	Âdhâr	4. 5. 20. 21.

Any man who falls sick on these days will die. Any man who is born on these days will die. Nothing undertaken on these days will prosper, only God can alter their evil influence (p. 557).

50. The **Lucky days of the months**. The Lucky days are: 1. 2. 6. 7. 8. 9. 10. 11. 12. 14. 15. 16. 17. 18. 19. 20. 22. 23. 26. 27. 28. 29. 30. The 24th day was unlucky because Pharaoh was born on that day (p. 558).

51. **Omens for travellers**. If a pig meets a man, or a dog which is pleased, or two men, or two married women, or camels, or a man watering the road meet thee, it means good luck; go on thy way. A barking dog, or unmarried people, or bulls, or a black man, or a woman throwing out filth, are signs of bad luck; go back (p. 559).

52. **Advice for buyers and sellers** (p. 559).

53. A List of **Lucky and Unlucky Days**, with very elaborate details of what it is lucky to do, and what is not (p. 560).

Here follow one or more extracts from the Book of Basil on the heavens.

54. **Heaven is round** on all sides, and it travels from east to west. The earth is in the middle of it like an egg. The Circle of the Signs of the Zodiac lies across the sphere of heaven obliquely. Heaven is divided into 365 parts, and each part is one day. The Sun and the Planets move through the Signs of the Zodiac from west to east. The course of Kêwân, or Kronos, is 3 years, of Bel, or Zeus, 12 years, of Arês $1^1/_2$ years, of the Sun 1 year, of Aphrodite 10 months, of Hermes 6 years, of the Moon $29^1/_3$ days. The Sun is eighteen times as large as the earth. The darkness

of the moon (*i. e.*, eclipse) is caused by the earth pre-
venting the light of the sun from falling on the moon,
for the moon obtains its light from the sun. The moon
becomes dark twice each year. The moon must be
fifteen days old before it can suffer eclipse, and no
eclipse of the sun can take place unless the moon is
born (p. 568).

55. List of the **names of the Signs of the Zodiac**
and their corresponding months (p. 568). Besides the
Signs of the Zodiac there are in heaven the Giant
(Orion), and the "dog of the Giant" (Sirius), and the
Cross.

56. A **List of the luminaries** after which the
days of the week are called. The number of hours
in the year is 8766, *i. e.*, the year contains $365\frac{1}{4}$ days
(p. 568).

57. Of Eclipses. When the head or tail of the
Serpent is in the seventh space of the sun, and the
moon is in the same space as the sun, an eclipse
takes place on the fourteenth day of the moon. If
the colour of the moon is black there will be sickness
on earth, and gloom, and wind, and the flocks will
die. If reddish black, there will be pestilence, slaughter,
sickness, and famine. If yellow, there will be liver
disease, fruit and crops will perish, and birds and
animals will die. If it be like dust, there will be
want, and snow, and ice, and frost, and the cattle
will die. In partial eclipses only the part of the earth
facing the dark part of the moon will be affected.
The various coloured marks seen on the moon during
an eclipse are due to the ˌinfluences of the Planets
(p. 570).

58. Eclipses of the Sun and Moon. A series

of portents derived from the situation of the eclipse (p. 570).

59. Calculations concerning eclipses (p. 571).

60. Calculations concerning the upper and lower Signs of the Zodiac (p. 571).

61. Calculations concerning the Five Stars (Planets), and concerning the position of Zeus, Arês, Hermes, and Aphrodite (p. 572).

62. List of the Signs of the Zodiac (p. 573).

63. The temperatures of the Seven Zones, wherein are placed the planets and the Sun and Moon (p. 573).

64. The characteristics and dispositions of the Signs of the Zodiac. Kronos is evil always. Zeus is red and beneficent. Arês is little, loves blood, and destroys every thing. The Sun is red and beneficent. Aphrodite is beautiful, she destroys nothing, she is prone to adultery and harlotry. Hermes is swarthy, his character is neutral, he is addicted to trafficking and learning. The Moon is a creature of change, and is bald (p. 574).

65. The similitudes and colours of the Planets. Gold = the Sun, which is yellow. Silver = the Moon, which is light blue. Iron = Arês, which is a poor red. Brass = Aphrodite, which is white or a good green. Lead = Kêwân (Kronos), which is a deep black. Electrum = Hermes, which is green, like the emerald. Tin = Zeus, which is dark like beryl (p. 574).

66. List of the Courses of the Signs of the Zodiac (p. 574).

67. The same in a series of forty-eight diagrams (p. 578).

68. Another List (p. 602).

69. The Planets of the Day and Night (p. 603).

70. **Lists of the three kinds of the Signs of the Zodiac** (p. 603).

71. **The Duration of the Sun** in each of the Signs of the Zodiac (p. 603).

72. **The Colours and Symbols** of the Signs of the Zodiac (p. 604).

73. **The Countries ruled** by the Signs of the Zodiac

SIGN	COUNTRY	SIGN	COUNTRY
Ram	Persia	Balance	Libya
Bull	Babylon	Scorpion	Italy
Twins	Cappadocia	Bowman	Crete, Cilicia
Crab	Armenia	Goat	Syria
Lion	Asia	Water-carrier	Egypt
Virgin	Greece	Fishes	India (p. 604).

74. **List of the countries traversed** by the Planets in their courses (p. 604).

75. **Divination** based on the Signs of the Zodiac that are associated and those that are not (p. 605).

76. **Lists of the Signs of the Zodiac** that are, and that are not associated (p. 605).

77. **Another treatment** of the same subject (p. 606).

78. **The Equinoxes.** These take place on the 24th day of Âdhâr and on the 24th day of Îlûl (p. 606).

79. **The Number of the Degrees of the Sun.** Stand in the sun and measure the length of thy shadow. In the First Kânôn it will measure 28 feet, in the Second Kânôn 27 feet, in Shebât 26 feet, in Âdhâr 25 feet, in Nîsân 24 feet, in Îyyâr 23 feet, and in Khazîrân 22 feet. In Tammûz the length is 23 feet, and this increases month by month until the First Kânôn when it again becomes 28 feet. The

length of thy shadow naturally varies during each hour of the day.

80. Portents for each hour of the day and night of the seven days of the week, with lists of the Planets that rule the hours, *e. g.*:

NIGHT OF SECOND DAY	REGENT	PORTENT
1st Hour	Zeus	Good for ruling and sailing
2nd „	Arês	Bad for every thing
3rd „	Sun	Good
4th „	Aphrodite	Favourable
5th „	Hermes	Neither good nor bad
6th „	Moon	Favourable
7th „	Kronos	Troublesome
8th „	Zeus	Favourable for feasting and travel
9th „	Arês	Bad for every thing
10th „	Moon	Neither good nor bad
11th „	Aphrodite	Favourable
12th „	Hermes	Favourable for travelling (p. 609).

81. A List of the days of the week and their Regents by day and by night, with descriptions of the future of him that is born on each of them, and statements about the things that may be done on each with advantage to the doer. *E. g.*, The second day of the week is ruled by the Moon. It is good for travelling, buying, selling, sowing seed, and sexual intercourse. He who is born on this day will suffer from many sicknesses. The night of the fifth day of the week is ruled by the Sun. Good for sexual intercourse, and building, and setting out on a journey. He who is born therein will become a prince, and he will go into the presence of the Governor, and he will be an adulterer (p. 616).

82. The Character of a man, according to the Planet he is born under.

PLANET	CHARACTER
Moon	Crafty, wicked, untrustworthy, fornicator
	Of good external appearance
Arês	A lover of war and bloodshed
	Eyes red or blue
Hermes	Learned and wise, a lover of learning
	Of dark complexion, physically weak
Aphrodite	A lover of wine, women, and pleasure
	Ruddy and handsome to look upon
Zeus	Hard-hearted, of inflexible will, a lover of wine and women
	Of wretched appearance, mean-looking
Sun	Well disposed, amiable, affectionate
	Of dark complexion, with small eyes
Kronos	[Wanting] (p. 617).

83. The **natural characters of men** as deduced from the Signs of the Zodiac under which they are born, and the month of the year in which they are born (p. 617).

84. The **odours of men** as deduced from the Signs of the Zodiac under which they are born, and the month of the year in which they are born (p. 618).

85. The **members of a man** and their Regents among the Signs of the Zodiac and the Planets, *e. g.,* the head = the Ram, the neck and shoulders = the Bull, the arms = the Twins, the breast = the Crab, &c. (p. 619).

86. To find out under which Planet a man was born (p. 620).

87. Divination concerning a sick man, obtained by adding up the numerical values of the letters in his

124. The times of the Risings of the Pleiades, Orion, Sirius, Zûhrah (Venus), and the Little Waggon (p. 651).

125. Omens derived from appearances seen in the heavens. The form of a man portends pestilence; the form of a bull portends war, slaughter, and abundance; the form of a horse portends abundance, and pestilence among children, and barrenness among women; the form of a mule portends pestilence among children, and barrenness among women; the form of a lion portends famine; the form of a panther portends the death of the wild beasts of the mountain; the form of a wolf portends that father and son will kill each other; fire in the heavens portends famine, war, and pestilence (p. 652).

126. Concerning Comets, for which there are three names, Karîthâ, Bûbhâ, and Lakethâ. They cause wars and other things, which depend on their position in the sky, and the quarter of the heavens to which their tails are directed. When a comet appears in the east, and stays there a long time, the events that its advent portends happen quickly; when a comet appears in the west the events that it portends happen after some delay. There are three kinds of comets, *viz.*, the star with a tail, the star with a spear, and the star with a beard; the first kind portends famine and pestilence, the second wars and strifes, and the third a change of dynasty (p. 653).

127, 128. Two sets of **portents derived from shooting stars** (pp. 652, 653).

i

MEDICINE AMONG THE EGYPTIANS, GREEKS, AND SYRIANS.

THE oldest people of whose medical knowledge we have any exact idea are the Egyptians, and an examination of the medical literature of Egypt proves that they were the founders of the chief systems of medicine, which were, with modifications and improvements, in use throughout Greece, Arabia, Syria, and many other parts of Western Asia down to the Middle Ages. The principal sources of our knowledge of Egyptian medicine are the medical papyri preserved in the museums of London, Berlin, Leipzig, and Philadelphia. The London papyrus (XVIIIth dynasty) is in the British Museum (No. 10,059), and was first described by the late Dr. Birch,[1] who pointed out that some of the prescriptions contained in it are declared therein to date from the time of King Khufu, the builder of the First Pyramid. The two Berlin papyri (XVIIIth dynasty) were first discussed by Brugsch[2], who also published a facsimile of them in

---

[1] *Aeg. Zeitschrift*, 1871, pp. 59—79. A facsimile and transcription of the hieratic text were published by Wreszinski (*Der Londoner Medizinische Papyrus*, Leipzig, 1912). See also Lüring, *Die über die medicinischen Kenntnisse der alten Aegypter berichtenden Papyri*, Leipzig, 1888.

[2] *Monatsschrift für Wissenschaft und Litteratur*, 1853, pp. 44—56.

his *Recueil*[1]. The Leipzig papyrus, more commonly known as the "Ebers Papyrus"[2], was published in facsimile by Dr. Ebers, together with a very scholarly Vocabulary by Dr. L. Stern in 1875[3]. It contains 110 large quarto pages written in hieratic, and its value for the history of medicine is incalculable; this fact was first proved by Dr. J. Hirschberg in his work *Ueber die Augenheilkunde der alten Aegypter*, Leipzig and Berlin, 1888. The Philadelphia papyrus was acquired in Egypt by Dr. Reisner, and was published by him in facsimile with a Vocabulary in 1905.[4] There is no evidence whatsoever that would justify us in assuming that the prescriptions and the knowledge of elementary anatomy and therapeutics, exhibited in these papyri, are the product of the period when the above papyri were written, and it may therefore be said that a large portion of their contents is taken from older medical papyri, and is the work of physicians who flourished under the Ancient and Middle Empires. When the Egyptians learned to write, they began at once to commit to papyrus the medical formulas which had been in use among their "medicine men" for a very long time, just as they did the magical prayers which formed their earliest religious literature. It is impossible to

---

[1] *Recueil*, II. 85—107. See also the edition of Wreszinski, *Der Grosse Medizinische Papyrus des Berliner Museums*, Leipzig, 1909.

[2] It was purchased in Egypt from a native of Luxor by Dr. Georg Ebers in 1872.

[3] *Papyros Ebers, das hermetische Buch conservirt in der Universitäts-Bibliothek zu Leipzig*, Leipzig, 2 vols. fol. See also Wreszinski, *Der Papyrus Ebers*, Part I. Leipzig, 1913. See also a German translation of the whole papyrus by Dr. H. Joachim, Berlin, 1890.

[4] Under the title of the "Hearst Medical Papyrus".

assign a date to the period when this took place, but, if we accept Egyptian tradition on the subject, we may believe that the Egyptians possessed a tolerably good working knowledge of the use of plants medicinally in the earliest centuries of Dynastic civilization.

In all periods of Egyptian ·history we see that the practice of medicine went hand in hand with that of magic, and the belief was always common that the magic formula, which the physician recited as the patient took the medicine, was an important, if not the *most* important element in the cure. As long as men believed implicitly that pain and disease were sent into human beings by the gods and goddesses, and evil spirits, words of power that were able to placate the irate supernatural powers were as necessary for the patient as medicine. There must have been, however, here and there physicians who believed more in medicine than magic, and a proof of this is perhaps given by the Ebers Papyrus, where we find fewer spells and incantations, and words of power than in other papyri of the same kind. Be this as it may, it is quite certain that the position of the physician under the Ancient Empire was one of great importance and dignity. In proof of this statement reference may be made to the inscription of Ptaḥuash, the chief architect of king Neferárikará, which is found in the tomb of this official at Abû-Ṣîr.¹ The king was one day inspecting some works that had been executed for him by Ptaḥuash, when an accident happened to Ptaḥuash, the nature of which is not clear. The king and his court were much grieved at the accident, and the sick

---

¹ For the text see Sethe, *Urkunden* I. 40—45.

or injured man was taken at once to the palace for treatment. The chief officials and the king's own physicians, 〔hieroglyphs〕 *senu*, were summoned, and the king caused his case of medical papyri 〔hieroglyphs〕 〔hieroglyphs〕, to be brought, but when the chief physician had examined the sick man he told the king that his case was hopeless 〔hieroglyphs〕. At this news the king was greatly grieved and, giving orders that everything possible was to be done, he retired to his chamber and prayed to Rā. Ptaḥuash died, as the physicians said he would, and the king had the body oiled (embalmed?) and he provided an ebony coffin for his faithful servant.

As to the scientific value of the anatomical knowledge of the Egyptians opinions differ. Some think that they knew very little anatomy, because the practice of mummification taught them to deal only with the belly. In operative surgery the Egyptian physicians obtained good results. They bandaged suppurating ulcers, practised venesection and cupping by means of horns sawn off near the point.[1] They performed circumcision, using knives of flint, and castration, and lithotomy. They were skilled in treating diseases of the eye, and their reputation as oculists was great. According to Dr. Baas they were also acquainted with pterygium and the arcus senilis, with inflammation in the vascular parts of the eye, ophthalmic catarrh, lippitudo, smaragdus (glaucoma?), blood in the eye, fatty

---

[1] Baas, *Outlines of the History of Medicine*, New York, 1910, p. 18.

degeneration, granulations and whitening; with diseases of the heart, of the ears, of the skin (including leprosy, small-pox, acne of the face, eruptions upon the head, erysipelas, itching of the leg, sweating of the feet, &c.); with diseases of the hair, verminous diseases, haematuria, dysuria, too frequent urination, the urinary troubles of children; with diseases of the sexual organs, of the stomach, toothache, headache, &c.[1] The Egyptians knew that the heart was the most important organ of the body, and that the arteries radiating from it reached all the members; and they seem to have thought that all the diseases known to them arose from defect in one or more of the arteries.

The prescriptions in the medical papyri prove that the Egyptians were well acquainted with the medicinal properties of a large number of plants, and the number of vegetable medicines employed by them was very great. Among these may be mentioned absinthe, aloes, acanthus, coriander seed, crocus, cucumber, cyperus, dates, endive, or chicory, figs, gums of various kinds, onions, lotus, mastix, wild mint, myrrh, nasturtium, opium, pistacia terebinthus, sesame (seed?), saffron, hyoscyamus, mandragora, pomegranate, olive oil, styrax, sycamore, and tamarisk. Many of the ingredients in Egyptian medicine were of animal origin, *e.g.*, lizard's dung; the blood, fat, dung, semen and testicles of the ass, bat's blood; the blood, dung and vulva of a dog; the fat, dung and uterus of a cat; the dung of a crocodile; the dung of an antelope; the milk of a woman who has just been confined; and fly-dung from the walls. These ingredients are loathsome according

---

[1] *Ibid.* p. 19.

to modern ideas, but our ancestors used medicines that were equally nauseous, as the following examples will show.

1. Remedy for cataract, "The head of a cole black "cat being burnt to ashes in a new pot, and some of "the ashes being blown into the eye every day, helps "such as have a skin growing over their sight; if "there happen any inflammation, moisten an oak-leaf "in water and lay over the eye. Mizaldur saith (by "this one only medicine) he cured such as have been "blind a whole year."[1]

2. Aches in the spine may be avoided by taking "ane littel fatt dogg" and after stuffing him with cum-ing seed "rosting" him, and keeping the droppings, adding to him "a handful of earth worms boyled quhill "they be leik lie."[2]

3. The ashes of Hens-feathers or Hens-bones burnt, and applied to the place, is an excellent remedy to stop bleeding in any part of the body. Culpepper's *Last Legacy*, London, 1685, p. 75.

4. Toads, Spiders, and Frogs, or their spawn have the same effect, but they do so by antipathy, because the blood flies from its centre (*Ibid.* p. 75).

5. Against the Falling Sickness. Take a jay, pull off her feathers, and pull out her guts, then fill her belly full of Cummin seeds, then dry her in an oven, till she be converted into mummy, a dram of her being beaten into powder, seeds and all, is an excellent remedy for the Falling Sickness, being taken in any

---

[1] *Nineteenth Century*, October 1912. See Culpepper's *Last Legacy*, p. 249.

[2] Lecky, *History of the Eighteenth Century* (Caldwell Papers).

convenient liquor every morning, put in Piony-water (*Ibid.* p. 76).

6. To eat the liver of a mad dog (being first dried and beaten into powder, a dram at a time is sufficient,) is an excellent, yea the best of remedies for the biting of a mad dog (*Ibid.* p. 76).

7. Take an owl, pull off her feathers, and pull out her guts, salt her well for a week; then put her into a pot, and stop it close, and put her into an oven, that so she may be brought into mummy, which being beat into powder, and mixed with boar's grease is an excellent remedy for the Gout, anointing the grieved place by the fire. I fancy this receipt much, it standing to good reason that a bird of Luna should help a disease of Saturn, and therefore desire a dram of the powder may be taken inwardly every morning (*Ibid.* p. 77).

8. If any be bitten by a spider, take a great quantity of flies, and bruise them, and apply them to the place (*Ibid.* p. 83).

9. Three drops of a man's water put in his ear every morning warm, helps the noise there (*Ibid.* p. 86).

10. The blood of a hare dried and taken inwardly breaks the stone in the bladder (*Ibid.* p. 87).

11. Pidgeons dung[1] mixed with vinegar is excellent to anoint warts with, if you would be rid of them (*Ibid.* p. 87).

12. A decoction of earth-worms, sellendine and ivy-

---

[1] In one of his letters Martin Luther states that the medicine which his wife recommended him to take did him no good; this medicine was a mixture of garlic and horse-dung! See Preserved Smith, *Life and Letters of Martin Luther*, London, 1911, p. 312.

berries in white wine, take equal quantities of each, is an excellent remedy for the yellow jaundice (*Ibid.* p. 88).

13. The foot of a great living toad being cut off when the moon is void of course, and hastens to the conjunction of the sun, cures one of the King's evil, being hung about their neck (*Ibid.* p. 94).

14. The dung of a cat dried and mixed with vinegar till it be pretty soft, takes away hairs, and hinders their growing any more (*Ibid.* p. 98).

15. The grease of an eel, boiled a little with the juice of housleek, and a little of it dropped into a deaf ear recovers the hearing in a short space (*Ibid.* p. 98).

16. Take and bruise nine wood-lice, let them remain all night in eight or nine spoonfuls of drink; in the morning strain it, and let the woman drink it up at one draught, and lay to her breast a linnen cloth, warmed and doubled up three or four times; the next morning let her take eight of the said lice used as before, the next morning seven, still diminishing one every morning, till she comes to take but one (*Ibid.* p. 100).

17. A pint of *aqua composita*, a bullock's gall, and an ounce of pepper, beaten very small, and all boiled to a salve, cures any sciatica, ache or gout, being applied to it, and changed once in twelve hours (*Ibid.* p. 101).

18. The powder of earth-worms, of mice dung, and of a hare's tooth, put into the hole of a rotten tooth, it will drop out without any instrument (*Ibid.* p. 107).

19. To stop bleeding at the nose. Put a piece of hot hog's turd as it comes from the hog, up the nose (*Ibid.* p. 212).

20. The water of adders' tongues snuffed up the nose, is very good (*Ibid.* p. 212).

21. For a stitch in the side. Take the urine of him that is ill, and boil worm wood and cummin seeds, bruised very well in it, and anoint the sides going to bed with it (*Ibid.* p. 233).

22. For a swelling of the navel. Take cow's dung, and boil it in the milk of the same cow into a plaster and apply it to the navel (*Ibid.* p. 235).

23. A dead mouse dried and beaten into powder, and given at a time (*i. e.*, all at once), helps such as cannot hold their water or have a diabetes, if you do the like three days together (*Ibid.* p. 246).

24. The brains of a weazel dried and drunk in vinegar cures the falling sickness (*Ibid.* p. 263).

25. Mice dung, with the ashes of burnt wasps, and burnt hazel-nuts, made into an ointment with vinegar of roses, do trimly deck a bald head with hairs, being anointed with it (*Ibid.* p. 266).

Many metallic preparations were also employed in medicine by the Egyptians such as oxide of zinc, oxide of lead, oxide of copper, sulphate of iron, &c. Medicines were administered in the form of ointments, liniments, plasters, cataplasms, poultices, pills, tablets, boluses, tinctures, decoctions, extracts, juices, and powders. The use of the inhaler was well known, and patients were ordered to inhale the fumes of drugs, which were vaporized by sprinkling upon red hot coals. The beneficial effects of massage were well known, and the Westcar Papyrus states that when, under the IVth dynasty, a prince paid a visit to a famous magician, he found him having his feet and legs massaged. The form of the Egyptian prescription was copied by the Greeks, Syrians, Arabs, and others, and may be thus described. The first part contained a statement as to the effect which the medicine would produce; this was followed by a list of the ingredients, the quantities being given with directions for preparation. Thus we have:

1. **A medicine for a sick belly** (Ebers Pap. Pl. 2).

Cummin	$\frac{1}{64}$
Goose feet	$\frac{1}{8}$
Milk	1 *tena*

Boil, strain through a cloth and take.

2. **Another. To open the bowels** (Ebers Pap. Pl. 5).

Uam seed	1
Ānkh plant	1
Kesebt fruit	1
Honey	1
Sheneft grain	1

Work up together and take for four days.

3. **Another. To empty the belly** (Ebers Pap. Pl. 5).

Cow's milk	1
Dough	1
Honey	1

Crush, pound, boil, and take for four days.

4. **Another. To empty the belly** (Ebers Pap. Pll. 8 and 9).

Honey	1
Shasha seed	1
Absinthe	1
Juniper berries	1
Uān berries	1
Utchait seeds	1
Cummin	1
Āam seed	1

Thu seed	I
Sea salt	I

Make into a ball and insert in the anus.

5. **A salve for the Uha disease** (Ebers Pap. Pl. 26).

Warm dhura grain	
Warm àat-plant	
Warm dûm tree dates	[Presumably in equal quantities.]
Memphite stone	
Milk of a woman who has just given birth to a boy	
Fresh olive oil	
Oil	

Heat up together and anoint the patient seven days therewith.

6. **Another. To drive the aa sickness, which is sent by God, out of the body of a man** (Ebers Pap. Pl. 34).

The first fruit of the Cyperus	$^1/_8$
Fragments of shasha gum	$^1/_8$
Theḥui berries	$^1/_{64}$
Àbu plant	$^1/_8$

Rub down to a powder, and let the patient drink it in beer when he is going to bed.

7. **When thou makest a diagnosis of a woman who has a pain in the pit of her stomach, [and] her arm, and breasts, and the region of her stomach is sick, and she has been told that she has the Uatch sickness, thou shalt say concerning it, It is death that has entered into [thy] mouth, and is abiding there. Make her a remedy of the [following] plants** (Ebers Pap. Pl. 37).

Teḥua berries	I
Opium	I

Peppermint	1
Ȧnek plant	1
Red seed of *sekhet*	1

Boil up in oil and let the patient drink the mixture.

8. **Another. To make the muscle (or, tendon, or nerve) pliable** (Ebers Pap. Pl. 83).

Fruit of the dûm palm	1
Beans	1
Ȧmā grain	1
Garlic	1
Shavings of the cedar tree	1
Shavings of the mulberry tree	1
Shavings of the willow tree	1
Shavings of the Zizyphus Lotus	1
Shavings of the sycamore tree	1
Shavings of the Uān tree	1
Acanthus gum	1
Lotus gum	1
Ȧm-tree gum	1
Sycamore gum	1
Red grain	1
Ȧm-tree berries	1
White oil	1
Goose oil	1
Pig's dung	1
Juniper berries	1
Myrrh	1
Onions	1
Field plants	1
Cyperus *usha*	1
Water melons	1
Tȧu plant (barley?)	1
Besbes plant (fennel)	1

Abu plant from the Delta	I
Linen waste (flax waste)	I
Sea salt	I
Rock salt	I
Aneb plant	I
Oxide of red lead	—
Green lead ore	I
Natron	I
Ox-fat [as much as sufficeth]	—
Shasha gum	I

Work up together, and apply in the form of a plaster.

From what has been said above it is clear that the Egyptians possessed a certain limited, but quite definite, knowledge of the general anatomy of the human body, and that they were able to perform a number of surgical operations with success. That they possessed a good and wide knowledge of practical botany is certain, and it cannot be denied that their vegetable medicines were many and effective. The curious thing about Egyptian medicine is that it never developed into a scientific system. The physician seems to have been content to go on with prescriptions that had been in use for hundreds, if not thousands, of years, and to have been averse to experiment. Had he dissected human bodies systematically instead of confining his operations to eviscerations for purposes of mummification, it is probable that he would have arrived at the great results that were obtained by the Greeks. Towards the close of the New Empire, or say, in the period which followed the XXth dynasty, magic again asserted its influence on Egyptian Medicine, just as it did on the Egyptian Religion, and spells, charms, words of power, and incantations were generally employed by physicians because they were acceptable to patients. This state of affairs continued until the Greeks obtained a foothold in Egypt, and then the old system of Egyptian Medicine passed into oblivion until the Middle Ages, when it again became popular, both in Egypt itself and in the countries round about.

If we consider for a moment the intercourse that Egypt had with Syria and Arabia, and other parts of Western Asia, from a very early period, it will be found impossible not to conclude that the knowledge of Egyptian Medicine must have made its way into these countries along with Egyptian civilization. The fame of the Egyptian physician would be carried wherever merchant caravans marched, and, when the Pharaohs undertook military expeditions into Western Asia and fought battles there, the skill of the military doctor would be patent to natives. How far the knowledge of Egyptian Medicine penetrated into Eastern Mesopotamia cannot be said, but in the seventh century before Christ the prescriptions that are found inscribed in cuneiform on tablets from the Royal Library at Nineveh have many points of resemblance with those of Egypt. As examples of these we quote the following[1]:

1. If a man is sick of the colic, and his stomach will not retain his food, but returns it into his mouth, his . . . . is pierced, and it rends him, and his

---

[1] For the Assyrian texts and German translations, see Dr. F. Küchler, *Beiträge zur Kenntnis der Assyrisch-Babylonischen Medizin*, Leipzig, 1904.

flesh is loose, and wind is expelled through the anus, and his bowels open, thou shalt administer for his healing,

> Half a Ḳa of syrup of dates
> Half a Ḳa of Kasu and oil
> Wine and water
> 3 shekels of pure oil
> 2 shekels of honey
> 10 shekels of Ammi plant

rub down, mix together, and let the patient drink it without tasting it[1] at night before the rising of the Goat star. Then let him drink half a Ḳa of Shi Ḳa, and wash out his mouth and his anus, and sprinkle himself [with the same]; so shall he be healed (p. 5).

2. If the same is the case thou shalt rub down together rock-salt and Amanu-salt, and let him drink it in strong beer[2] without tasting it, and let him wash out his mouth and anus therewith, and sprinkle himself therewith; so shall he be healed (p. 5).

3. When the stomach of a man will not retain food, thou shalt rub down seed of tamarisk, and mix it with honey and butter, and the patient shall drink it without tasting it; so shall he be healed (p. 27).

4. When a man has drunk wine, and his head is confused (?), and he forgets his words, and his speech is wiped out, and his reasoning powers are wanting, and his eyes stare at those about him, thou shalt take for his healing

---

[1] *I. e.*, he shall gulp it.

[2] Or perhaps some kind of sweet beer or wine of an intoxicating or narcotic character.

Shi. Shi plant
Shi. Man plant
Tar. Mush plant
Ḫaldappānu plant
Il plant
Dil plant
"Sea-tooth"
Nuḫurtu plant
Seed of the Sha, Gan, Gan plant
Kam. Ka. Du plant
Muh. Gul. La plant.

Rub down these eleven vegetable substances together, and let the patient drink it mixed with oil and wine, before the rising of the goddess Gula in the morning, before the sun rises, and before any one has kissed him; and he shall be healed (p. 33).

5. If the eye of a man be filled with yellow pus, thou shalt rub down the **Ut** plant, and he shall drink it in strong beer, so shall he be healed.

[Or] thou shalt dry Shushu, and bake it, and moisten it with wine, and the patient shall drink it before Shamash; so shall he be healed.

[Or] thou shalt dry . . . . . of the plant "dog's tongue", and bake it, and moisten it, and the patient shall drink it; so shall he be healed.

[Or] thou shalt pound up one shekel of great Mush Dim Gurin Na plant, and let the patient drink it in strong beer and oil (p. 61).

6. When the body of a man is yellow, and his face is yellow and black, and the root of his tongue is also black, the name of [his] disease is "Aḫḫazu". Thou shalt bake Mush Dim Gurin Na, and the patient

k

shall drink it in wine; so shall the Ahhazu fly away from his interior (p. 61).

7. When a man is sick with the Ahhazu sickness, and his head, and his face, and his whole body, and the root of his tongue are blocked (?), the physician shall not lay his hand on such a man, for he will die and cannot be healed (p. 61).

If the reader will compare the forms of these and the other prescriptions translated by Dr. Küchler the similarity will at once be apparent. Both the Egyptian and the Assyrian physician thought that they could tell what was the matter with a man merely by looking at him, and both decided what medicine to administer on the testimony of external appearances only. Both were equipped with alternative prescriptions, and both prescribed medicines the bulk of the ingredients of which were of vegetable origin. Both wished their medicines to be swallowed in draughts of strong beer, and both thought that the efficacy of certain prescriptions was increased by the recital of magical formulae. Both ordered the ingredients to be crushed, or pounded very small, i. e., rubbed down, and sometimes strained through a rag. The fact that medical spells are added to some Assyrian prescriptions proves that the Assyrian physicians had not, in the seventh century before Christ, freed their system of medicine from the bonds of magic, and it suggests that such prescriptions had come down to them from older times. This is very probable, and there are very good grounds for thinking that the Assyrians derived a great deal of their medical knowledge from the Babylonians. Very few fragments of what may be properly called medical tablets have come down to us from, let us say, the time of Khammurabi (about B. C. 1950), but magical texts[1] of the period of Khammurabi have been unearthed in Lower Babylonia, and medical texts may come from the same place any day. It is impossible to think that a nation so highly civilized as the Babylonians had no system of medicine, and it is equally impossible to think that the Sumerians before them had none. In the prescriptions found on Assyrian tablets quoted above many of the names of plants, or medical preparations, are Sumerian, and these can only have come to the Assyrian scribes who copied the tablets for the Royal Library from Babylonian or Sumerian originals. In pointing out the similarities in the form and contents of the Egyptian and Assyrian (or Babylonian, or Sumerian) prescriptions, it must not be imagined that I suggest that their authors were influenced by the works of Egyptian physicians, three or four thousand years before Christ, for it is far more likely that the early Mesopotamian system of medicine was modified or improved under the influence of physicians from Elam, or perhaps from some country further to the east.

What has been said as to Ninevite prescriptions being derived from Babylonian originals applies also to the copies of the Lists of Plants and Trees that have been found at Nineveh. And if we had all the Lists of Plants from which the Assyrian scribes copied, we should probably find that many of them were accompanied by descriptions of their forms and uses similar

---

[1] Compare Nos. 22,446, and 22,447 published by L. W. King in *Cuneiform Texts*, Part III, Plate 2 ff.; and Part V, Plate 4 ff. In these texts the magician divines by means of oil thrown upon water, and his divination rests upon the form that the oil takes on the water.

to those that were drawn up in Greek by Galen and other medical authorities. That kings of Babylon and Nineveh devoted much attention to the cultivation of plants is proved by the cuneiform documents that have come down to us. Thus a tablet in the British Museum (No. 46,226) gives a list of the plants that were grown in the gardens of Merodach-Baladan II, King of Babylon B.C. 721—710, and 703—702, and on it is a statement to the effect that it was copied from an older list. And Sennacherib, King of Assyria, B.C. 705—681, says in the great Cylinder in the British Museum (No. 103,000, col. VIII, l. 16ff.), "I laid out gardens above and below "the city, I planted for my subjects products of the "mountains and of all lands, all the herbs of the land "of Khatti, and murru-plants, among which fruitfulness "increased more than in their own country, all moun- "tain vines, all the fruits of the nations, herbs and "sirdu-trees for my subjects."[1] Further on (l. 64), the king says, "They clipped the trees that bore wool, and they shredded it for garments," and it has been argued with considerable success that these "trees that bore wool" were nothing more nor less than cotton plants.[2]

The source from which all knowledge of Babylonian and Assyrian medicine must ultimately come is the great collection of cuneiform tablets known as the "Kuyûnjik Collection", preserved in the British Museum. The first to make known its contents as a whole was Prof. C. Bezold, who published for the Trustees of the British Museum a general description of each tablet and its contents in the five volumes of his monumental

---

[1] See the edition and translation of the cylinder in *Cuneiform Texts* by L. W. King, Part XXVI, p. 29 and 31.

[2] L. W. King in *Proc. Soc. Bibl. Arch.* vol. xxxi, p. 339 ff.

work, " The Kouyunjik Collection in the British Museum ",
London, 1887 to 1899. His elaborate classified Index
and lists of numbers are not the least valuable parts
of this great work. His Catalogue deals with all the
Kuyûnjik tablets acquired by the British Museum up
to 1891; descriptions of those which have been acquired later will appear in the Supplementary Volume,
which has been prepared by Mr. L. W. King of the
British Museum, and is now in the press.

Many Assyriologists have devoted much time and
labour to the study of certain sections of Babylonian and
Assyrian medicine, but up to the present no work has
appeared which gives in a connected form all that is
known on the subject. Thus Professor Sayce published "An
ancient Babylonian Work on Medicine." [1] The letter of
the physician Aradnanâ was first published by S. Alden
Smith,[2] and republished by Professor R. F. Harper,[3] and
discussed by others.[4] M. Boissier and Prof. M. Jastrow
jun[r] have published many important Liver texts and
translations,[5] and Dr. Zehnpfund has described the
" Zukakîpu ", or cupping instrument in use among the
Babylonians.[6] Of especial value are the numerous con-
tributions to medical journals on Babylonian and Assyrian medicine by Baron Oefele, for he is both a

---

[1] *Zeit. für Assyr.*, vol. ii, pp. 1—14.

[2] *Die Keilschrifttexte Assurbanipals*, Leipzig, 1887—1889
(Smith 1064).

[3] *Assyrian Letters*, London and Chicago, 1892—1913.

[4] Johnston, *The Epistolary Literature*, Baltimore 1898.

[5] See Boissier, *Note sur un monument Babylonien se rap-
portant à l'extispicine*, Genf, 1899, and Jastrow in his work on
the Babylonian Religion.

[6] *Beiträge zur Assyriologie*, Leipzig. Vol. iv.

skilled physician and an Assyriologist. His writings are published in a number of continental journals which are unfortunately difficult to obtain, and the list of them is considerable.[1] He is the first investigator who has been able to translate the words as well as the medical meaning of the cuneiform medical texts, and it is much to be wished that his articles could be published in a collected form.

We have seen that in the seventh century before Christ the systems of medicine in use in Egypt and Mesopotamia had many points in common, and we now pass on to consider briefly the changes and developments in the medicine of the period that are due to the Greeks. The first physicians among the Greeks who really founded Greek medicine were not the priests of Aesculapius, but the Asklepiadae, who "travelled about in the practice of their profession and

---

[1] In the *Allgemeinen medicinischen Central-Zeitung*, Berlin 1898, No. 96 ff. we have: 1. Eine Uroskopie aus altmesopotamischer Medicin. 2. Auch die Aderlasslehre ist wie die Uroskopie Altmesopotamiens humoralpathologisch. 3. Auch Amulette haben die Griechen aus assyrisch-babylonischer Medicin entlehnt. 4. Auf den Weg über Salerno kam in unsere Volksmedicin neben der altaegyptischen auch die altmesopotamische Medicin. In the same journal for 1889 we have no less than twenty articles on subjects of Mesopotamian medicine. In the *Zeitschrift für Assyriologie*, Vol. xv, p. 109, he discusses the Assyrian names for Aristolochia, Andropogon, Schoenanthus, and Pistacia. In the *Zeit. für diaetetische und physikalische Therapie*, Vol. iv, Part 7, we have *Diaetetisches Handbuch der Bibliothek Sardanapals*; in No. 7 of the *Deutschen medicinischen Presse*, 1901, we have *Das neueste Alte aus Keilschriftmedicin*; and in *Die Heilkunde*, Vol. v, 1901, he treats at some length the diseases of the *Mammae*.

to make their fortune",[1] the most famous of whom was Hippokrates. Their anatomical knowledge was not great, indeed it must be considered rudimentary, for it seems not to have been greater than that which could be acquired by the slaughter of animals. Homer was acquainted with Egyptian drugs, and it is clear that all the great Greek physicians owed much of their knowledge of medicinal plants to the Egyptians, all of whom were, according to Homer, physicians. The Ionic School of Medicine was founded by Thales of Miletus (B.C. 639—544), a pupil of the Egyptian priests. The Italian School, or School of Crotona, was founded by Pythagoras (B.C. 580—489) of Samos, whose teacher was one Un-Nefer of Heliopolis. The Eliatic School was founded by Xenophanes of Colophon, in Elia, about B.C. 450. Hippokrates, the "Father of Medicine", was the son of Heraklides, by Phaenarete his wife, and was born at Cos, B.C. 460. After the death of his parents he went to study at Athens, and then he travelled for twelve years. He lived for the greater part of his life at Larissa in Thessaly, where he died about B.C. 377. He is also said to have lived for a time in Egypt. He was, according to Dr. Baas, the creator of profane, as distinguished from sacerdotal or guild medicine, of public, in place of secret medicine, in fact, he was the great creator of scientific medicine and artistic practice. The Hippokratic writings are in the Ionic dialect, and are fifty-three in number, but only a few of them were written by Hippokrates himself.[2] The

---

[1] Baas, *History of Medicine*, p. 85.
[2] See the summary in Baas, p. 101.

undying importance of Hippokrates in medicine rests,
not so much upon his enrichment of science with new
material, as upon the creation of a scientific (lay-)
medicine and art, and upon the method and really
great principles that he introduced for all time into
science, and especially into practice. To his eminently
practical bent is to be ascribed also the fact that he
created, and specially cultivated, those branches of
medicine that are profitable in practice: semeiology,
prognostics, diagnostics, aetiology, symptomatology, and
therapeutics, and in a much less degree systematic
pathology, anatomy, &c. Yet he did not create one
single name for his pictures of disease!

The Alexandrian School was founded in the third
century before Christ, and owed its existence entirely
to the Ptolemies, who supplied the men of science
and philosophers of their day with the means, the
occasion, and the leisure for their labours. This is
especially true of human anatomy, for at this period,
in spite of the prejudice of both Egyptians and Greeks
against the dissection of the human body for medical
purposes, bodies were freely placed at the disposal of
the physicians. Indeed it seems certain that at this
time even vivisection was permitted, with the view
of finding out the seat of the soul and the origin of
disease. The disciples of this school devoted themselves to the explanation, criticism, and elaboration
of the works of Hippokrates.[1] They were partly
commentators and compilers, and partly independent
workers. The greatest physician of this School was
Herophilus of Chalcedon (B.C. 335—280) who dissected

---

[1] See Schlosser in Baas, *op. cit.* p. 119.

dead bodies, and vivisected convicts, and an almost equally distinguished man was his contemporary Erasistratus of Julis on the island of Ceos, who died about B.C. 280. The School of the former survived until about A.D. 100, and the School of the latter until about A.D. 200.

Since Alexandria became under the rule of the Ptolemies one of the greatest trading centres of the ancient world, and since its commercial influence was wide-spread and far-reaching, it follows of necessity that the fame of its School of Medicine penetrated into many parts of Western Asia, especially into Palestine, Syria, and the northern parts at least of Arabia. The data necessary for constructing an exact account of the transmission of the Greek system of medicine into Syria are not forthcoming, but we may presume that in all the Colleges of the Jews there were physicians who were well acquainted with the systems of anatomy, pathology, and therapeutics that had been formulated by the Greeks. From these centres the knowledge of scientific medicine would penetrate eastwards, and reach such literary cities as Edessa and Amid, the modern Diarbekîr. A medical college is thought to have been in existence at Edessa in the second century of our era, and it existed until the time of Nestorius, when, owing to quarrels, which arose there between the Jacobites and Nestorians, it was broken up. A few years later it was reformed by Bishop Ibas, and medical teaching was given in it until A.D. 489, when it was finally closed by the Emperor Zeno. Some of the leaders of the School fled to Persia, taking, of course, their medical knowledge with them, but others established themselves at Nisibis, and founded a public hospital similar to that which had existed at Edessa. Bar Hebraeus (born 1226), the famous Syrian writer, says in his "Chronicle"[1] that when the Sassanian king Sapor I. (240—273) founded the city of Gundhê shâbhûr (Bêṯh Lapâṭ) he brought thither Greek physicians, who introduced into the East the system of medicine of Hippokrates. His actual words are: "And Sapor built "for himself a city which was like unto Constantinople, "and its name was Gundhî Sâbhôr, and he settled "his Greek wife therein. And there came with her "skilful men from among the Greek physicians, and "they sowed Hippokratic medicine in the East. And "there were also excellent Syrian physicians, such as "Sergius of Rîsh 'Ainâ, who was the first to translate "the philosophical and medical works of the Greeks "into Syriac, and Aṭanôs of Âmid, and Philagrius, and "Simon the monk, who belonged to Taibûthah, and "Gregory the Bishop, and Theodosius the Patriarch, "and Ḥônain, the excellent man, the son of Isaac, "and many others after them until this day. These "were all Syrians, but Aaron the Elder was not a "Syrian. Gôsyôs, an Alexandrian, translated his book "from Greek into Syriac."

ܘܫܒܘܪ ܒܢܐ ܠܗ ܡܕܝܢܬܐ ܕܕܡܝܐ ܠܩܘܣܛܢܛܝܢܘܦܘܠܝܣ. ܗܝ ܕܫܡܗ
ܓܘܢܕܝܫܒܘܪ. ܘܐܫܪܝܗ̇ ܒܗ̇ ܠܐܢܬܬܗ ܝܘܢܝܬܐ. ܘܐܬܘ ܥܡܗ̇ ܐܢܫܐ ܡܗܝܪ̈ܐ ܡܢ ܐܣ̈ܘܬܐ
ܝܘ̈ܢܝܐ. ܘܗܢܘܢ ܙܪܥܘܗ̇ ܠܐܣܝܘܬܐ ܕܐܦܘܩܪܛܝܣ ܒܡܕܢܚܐ. ܘܗܘܘ ܐܦ ܣܘܪ̈ܝܝܐ ܐܣ̈ܘܬܐ
ܡܝܬܪ̈ܐ ܐܝܟ ܣܪܓܝܣ ܕܪܫܥܝܢܐ ܗܘ ܕܩܕܡܐܝܬ ܦܫܩ ܣܦܪ̈ܐ ܕܦܝܠܘܣܘܦܘܬܐ ܘܐܣܝܘܬܐ ܡܢ
ܝܘ̈ܢܝܐ ܠܣܘܪܝܝܐ. ܘܐܛܢܘܣ ܐܡܕܝܐ ܘܦܝܠܓܪܝܘܣ ܘܫܡܥܘܢ ܕܝܪܝܐ ܕܡܢ ܛܝܒܘܬܗ.
ܘܓܪܝܓܘܪܝܘܣ ܐܦܝܣܩܘܦܐ ܘܬܐܘܕܘܣܝܘܣ ܦܛܪܝܪܟܐ. ܘܚܢܝܢ ܡܝܬܪܐ ܒܪ ܐܝܣܚܩ
ܘܣܓܝ̈ܐܐ ܒܬܪܝܗܘܢ ܥܕܡܐ ܠܝܘܡܢܐ. ܗܠܝܢ ܣܘܪ̈ܝܝܐ. ܐܗܪܘܢ ܕܝܢ ܩܫܝܫܐ ܠܐ
ܣܘܪܝܝܐ ܗܘ. ܐܠܐ ܓܘܣܝܘܣ ܐܠܟܣܢܕܪܝܐ ܦܫܩ ܟܬܒܗ ܡܢ ܝܘܢܝܐ ܠܣܘܪܝܝܐ ܀

[1] Ed. Bruns and Kirsch, Leipzig 1789, p. 62.

This passage is of very great importance for us, because it shows that the system of Hippokrates was introduced into Mesopotamia before the close of the third century, and under royal patronage. Bar Hebracus was himself a physician and he practised medicine at Aleppo, and his literary knowledge was such that he is unlikely to have made a mistake on such an important point as this.

As to the Sergius of Rîsh 'Ainâ whom he mentions, a good deal is known[1]. He was a good Greek scholar, and was well versed in the philosophy of Aristotle; he was a priest and also archiator of Rîsh 'Ainâ, but according to some a man of loose morals and avaricious. He died soon after 536. He was the first to make known the writings of the Greeks to the Jacobite Syrians by means of translations and commentaries. In his capacity as physician Sergius translated parts of the works of Galen. In Brit. Mus. Ms. Add. 14,661 we have Books VI—VIII of the treatise "De Simplicium Medicamentorum Temperamentis ac Facultatibus"[2]; and in Brit. Mus. Ms. Add. 17,156 there are fragments of his Syriac translation of "Ars Medica", and "De Alimentorum Facultatibus"[3]. The translations of Sergius were revised by Ḥonain ibn Ishâk, a physician to the Khalîfah Al-Mutawakkil (died about A. D. 873), but his work is lost. Atanôs, Philagrius, and Gregory the bishop are

---

[1] See Wright, *Syriac Literature*, pp. 88—93; Duval, *La Litt. Syriaque*, p. 274.

[2] See Wright, *Catalogue of the Syriac Mss.*, p. 1187; and see Merx's article in *Z. D. M. G.*, XXXIX (1885), p. 237 ff.; and Gottheil, *Journ. American Oriental Society*, vol. XX. 1899, p. 186.

[3] See Wright, *Catalogue*, p. 1188; and Sachau, *Inedita Syr.*, pp. 88—94.

unknown as physicians; if they composed medical works they have perished. Simon the monk, who flourished in the seventh century, wrote a book of medicine[1], but it also is lost. Theodosius the Patriarch, who was elected in 887, is no other than Romanus the physician; he wrote a medical Syntagma, which was of some repute[2], but it also is lost. Gôsyôs, the translator of the Syntagma of Aaron of Alexandria is, according to M. Duval[3], Gesius Petraeus, who lived in the reign of the Emperor Zeno. The Syntagma of Aaron consisted of thirty books[4].

Among other famous Syrian physicians may be mentioned: 1. George, the son of Bôkht-Ishô‘, of Bêth Lâpât in Khûzistân, who was physician to Al-Manṣûr the Khalîfah, in the eighth century. 2. Gabriel, his grandson who was in practice in Baghdâd, and was court physician to Ar-Rashîd; he died in 828 A.D. 3. Ḥonain ibn Ishâk al-‘Ibâdî, of Al-Ḥirah, the disciple of Yaḥyâ ibn Mâsawaihi, who died 873, studied in Alexandria. When he returned to Baghdad he issued translations in Syriac and Arabic of the works of Dioscorides, Hippokrates, Galen, and Paul of Aegina. Isaac, the son of Ḥonain, and his nephew called Ḥobaish ibn al-Ḥasan al-A‘sam, also translated many Greek works into Syriac and Arabic, and several of these gained currency under the name of Ḥonain[5]. As a rule they translated the

---

[1] See Assemânî, *B. O.*, III, Part 1, p. 181.
[2] See Bar Hebraeus, *Chron. Eccles.*, i, 391, ii, 213.
[3] *La Litt. Syriaque*, p. 274.
[4] *Chron. Dynastiarum*, ed. Pococke, p. 158.
[5] These men were among the earliest and the ablest of the Nestorian Christians who during the ninth and tenth centuries, making Baghdâdh their headquarters, supplied Muḥammadan

Greek first into Syriac which they afterwards translated into Arabic; but their Syriac versions have unfortunately, as it would appear, perished without exception. 4. John, the son of Serapion, composed in Syriac two works, the one in twelve and the other in seven books; the former was a treatise on antidotes. John flourished about the end of the ninth century.

The last of the great Syrian physicians was Abu'l-Faraj Gregory, commonly known as Bar 'Ebhrâyâ, or Bar Hebraeus, because his father Aaron was of Jewish descent. He was born in 1226, and practised medicine under his father and other great physicians, and wrote many large and important books; he died at Marâghah on July 30, 1286. He was famous as a physician, and attended the Tatar "King of Kings" in 1263; his medical writings are very numerous. He translated into Syriac and abridged Dioskorides's treatise "De Medicamentis Simplicibus" and the Commentary on the "Questiones Medicae" of Ḥonain in Syriac, and he wrote translations and commentaries in Arabic on the works of Hippo-krates, Galen, and other physicians. Assemânî states that he also wrote in Syriac a kind of Compendium, in the form of a large book, which contained all the opinions of the physicians[1] ܟܠܗܘܢ ܕܝܠܗ ܡܢ ܕܡܠ̈ܦܢܐ ܕܒܐ ܕܒ ܟܐܒܐ ܟ̈ܐܒܐ ܕܐܣܝ̈ܐ. Unfortunately, however, he gives no detailed description of the work. The late Professor William Wright, the greatest of Syriac and Arabic scholars, said that Bar Hebraeus was one of the most learned

---

scholars with nearly everything that they knew of Greek science, whether medicine, mathematics, or philosophy. See Wright, *Syr. Lit.,* p. 213.

[1] *Bibliotheca Orientalis,* tom. ii, p. 272, No. 26.

and versatile men that Syria ever produced[1], and Gibbon's eulogy of him may well be quoted here. "In his life he was an elegant writer of the Syriac "and Arabic tongues, a poet, physician, and historian, "a subtle philosopher, and a moderate divine. In his "death his funeral was attended by his rival the Nes-"torian patriarch, with a train of Greeks and Arme-"nians, who forgot their disputes, and mingled their "tears over the grave of an enemy"[2].

From the facts stated above we see that the system of medicine formulated by Hippokrates was introduced into Syria and Mesopotamia in the early centuries of the Christian Era, and that it was studied with conspicuous success by many great Syrian physicians for more than a thousand years. We may now pass on to consider the system of medicine propounded in the Syriac Book of Medicines printed in this work, and attempt to arrive at some idea as to its author and its date.

---

[1] *Syriac Literature*, p. 265.
[2] *Decline and Fall*, ed. Smith, Vol. vi, p. 55.

THE AUTHOR OF THE BOOK OF MEDICINES, AND HIS SYSTEM OF MEDICINE.

Of the author, or perhaps, authors, of the Books of Medicines and of the astrological section given in the present work we know nothing, but here and there we find indications and allusions that suggest the sources of these works and the period when they were grouped together in a Syriac dress. The first section, which deals with what we may call scientific medicine, and the second section, which contains astrological-medical texts, seem undoubtedly to have been written originally in Greek. The third section, which deals with native medicine, and is nothing more nor less than a mere "Book of Recipes", many of them of a most loathsome character, may have been originally written in Arabic, or Persian, or in some Syrian dialect. The author of the Book of Medicines, whoever he was, studied medicine in Alexandria, as we may see from the following: "Now when I was in Alexandria, a certain villager was "bitten by an asp in one of the fingers of his hand "when he was at no very great distance from the "city. Immediately he tied round the lowest joint of "his finger, which was close to the palm of his hand, "a strong bandage, and ran straightway to a certain "physician whom he know at the gate of the city, and "entreated him to cut off his finger from the lowest

"joint, namely that which was in the palm of his hand.
"He expected that if this could be done he would
"suffer no [further] injury, and his expectation was
"fulfilled as he thought it would be, for he was saved,
"and lived, and this only did he seek" (translation,
p. 25). He mentions a case of the use of the "tour-
niquet", and another case of a man who was bitten by
a viper, and who was saved by cutting off the joint
that had been bitten, presumably in the neighbourhood
of Alexandria, and it seems that he made note of these
cases, as physicians do.

Another certain thing in connection with the Book
of Medicines is that each Chapter in it is in reality a
Lecture. The writer of these lectures often speaks in
the first person, and addresses his hearers as "you",
and does not hesitate to apply to the subject matter
under discussion the results of his own personal experience. Thus on p. 141 (translation) he remarks,
"Therefore what I said immediately after I began my
"discourse I will say again now, and then I will pass
"on to another matter". And on p. 143, "Now on
"many occasions I have made it clear to you that the
"roots of some of the nerves are in the marrow of
"the spinal column", &c. And on p. 198, "Since there-
"fore ye know what are the muscles that perform these
"operations, it is quite easy for you to calculate which
"of them is diseased, and in which of them the ope-
"rative faculty is injured. And if ye make a mistake
"in any particular in respect of those that have been
"demonstrated to you in dissections, the facts about
"the causes of respiration, and the facts about the
"voice should remind you". And on p. 207, "If ye
"see that they are all in motion, then ye must know

"that one of the upper muscles of which I have spoken
"is primarily the cause of motion of this kind, and
"afterwards ye must try to determine which is the
"cause in the place". And on p. 288, "To you, how-
"ever, this is not a matter of difficulty, for you are
"correctly acquainted with the fact", &c.

As instances of the incorporation of personal ex-
periences in the Lectures the following may be quoted:
On p. 8 the lecturer mentions the cases of two men
whom he had known, who had lost their memory, and
whose mental faculties were injured, the one through
overstudy at night, and the other through overwork
in a vineyard. On p. 23 he describes the case of a
boy of thirteen who suffered from pains, which began
in his leg, and asscended the whole body, and finally
reached his head. Our author adds what he believed
to be the true cause of the pain. On p. 131 he mentions
the case of a man who fell off his horse and hurt his
back, and says that two of his fingers were paralysed
and one half of a third. He adds, "I ordered them to
"put plasters on the place that was injured, and the
"man was cured in a few days". On p. 141 he speaks
of a man who had a large gathering of pus near one
of his toes, and when the flesh was cut away, the
nerves were visible. The wound healed, but the foot
was stiff afterwards. Our author applied ligatures, and
says, "I healed the man completely". On pp. 244, 245
he describes two cases of phthisis, and tells us how
he treated them. On p. 292 a case of palpitation of
the heart which the author treated by letting blood in
the spring of the year is mentioned. On p. 310 is
described a case of a certain sophist who suffered from
fits of the falling sickness. On p. 445 two cases of

jaundice are mentioned, and our author tells us his views about them and also those of other physicians. On pp. 448 and 449 more cases of diseases caused by the bile are enumerated, and our author proves the accuracy of his views about the value of hot baths.

Many passages in the Book of Medicines prove that the author of it attached great value to dissections of the human body and of the bodies of animals, and in this respect he was a follower of one of the most advanced Schools of Medicine that flourished in Alexandria in the second century before Christ. The following extracts make this clear.

1. "And the cause of this is well known to you "who have seen the exit of the nerves that descend "from the brain" (p. 106).

2. "Those who dissect call these 'soft', as you "know" (p. 107).

3. "If the skin be stripped off, and the muscle "which is beneath it be laid bare, it will be seen that "the nerve doth not move; but if a man toucheth it, "it becometh sensitive" (p. 125).

4. "This we can only learn from those who dissect "the body .... And we learn through the dissectors "that all the members that produce voluntary motions "in man, that is to say, the nerves that cause motion, "have their place of exit from the neck, and below "the marrow of the spinal column. And ye may learn "also and see by dissections of the nerves that work "the breast, that they go forth from the marrow that "is in the neck" (p. 129).

5. "If therefore a man knoweth from the making "of dissections the sources of the nerves that come "to each one of the members, this man, I say, can

"cure successfully the want of feeling and the want
"of motion of each one of the members, as I myself
"did in the case of a man who fell from a horse,
"and hurt the upper part of his back, and feeling
"was paralysed in two fingers and the half of a third"
(p. 131).

6. "We have learned from those who have dissected
"the nerves that the membranes of the breast have
"their exit from behind the fourth and fifth vertebrae"
(p. 133).

7. "Now those who dissect the body call the
"little tubes which are composed of cartilages 'little
"oesophagi', and they have a kind of circular form"
(p. 242).

8. "For behold, we often see in the animals which
"we dissect that there is a very considerable quantity
"of liquid in the membrane that surroundeth the heart"
(p. 291). He then goes on to speak of an ape, the
body of which was wasting away, which he wanted,
apparently to vivisect, but he was unable to do so.
After death the animal was dissected by our author.
He also dissected a cock.

9. "The stomach hath two parts, which are called
"'tunics', or 'coats', by the dissectors of the body"
(p. 306).

10. "I must first call to your memory the muscles
"that ye have seen in dissections" (p. 383).

11. ". . . . on the contrary, we see clearly, not only
"from those of the apes that have been dissected,
"but also from those of other animals, that the liver
"in some cases is joined to a rib, and not in others"
(p. 385).

Among the medical authorities quoted in the Book

I*

of Medicines, Hippokrates comes first. Thus on p. 6 the author quotes the Aphorism, "Both sleep and "wakefulness, if they be indulged in immoderately, "are bad". On p. 15 he quotes a passage from a work of Hippokrates, entitled "On the Comings of Sicknesses".[1] On p. 19 he quotes the statement of Hippokrates, "If a man shall remain for a long time "fast bound by fear and sorrow weakness of the in-"tellect will come to him", and then goes on to criticize it, asking why Hippokrates did not state the causes of this weakness. On p. 127 he quotes the Aphorism that rigidity is caused by fullness and emptiness of the body. On p. 197 he refers to a treatise of Hippokrates on the Voice; on p. 211 he refers to another treatise on Shortness of Breath; on p. 290 he quotes a statement of Hippokrates on heart disease; on p. 397 he refers to his writings on digestion; and on p. 518 he mentions some of his observations on urinary diseases.

The Archilogenes mentioned on p. 29 as the author of a book, entitled "Protracted Ailments", is probably Archigenes, who practised in Rome in the reign of Trajan, and wrote many medical works.

Diocles, who is quoted on pp. 18, 19, is probably Diocles of Carystus in Euboea, who flourished about the middle of the fourth century B.C. He wrote a great number of treatises on fevers, materia medica, poisons, dietetics, anatomical subjects, &c. and distinguished between pneumonia and pleurisy, as well as between dropsies depending upon disease of the liver and spleen, established the differential diagnosis between

---

[1] See also p. 136.

ileus, colic, &c. He belonged to the Dogmatic School. The Secundus who is mentioned on p. 139 I cannot identify with certainty.

The perusal of the Book of Medicines will convince the reader that he has before him in a Syriac dress a series of Lectures on Human Anatomy, Pathology, and Therapeutics, which were written in Greek by a learned and distinguished physician, who had studied at Alexandria, and who was an earnest and convinced follower of Hippokrates. Each lecture proves that its author had imbibed not only the letter but the spirit of the teaching of Hippokrates, and from first to last he shows that all disease is governed by what we should call natural laws. He uses neither spells nor incantations with his prescriptions, the name of God is not once invoked, and his explanations of the causes of diseases—whether true or not must be decided by medical experts—are simple and clear. There is no mystery, no assumption of superior knowledge, and no attempt to appeal to the ignorance or credulity of men. He says (p. 9), "It is right that the physician should be an enquirer and a painstaking worker", and he asks (p. 10), "Do not all physicians learn by experiment?" He followed Hippokrates closely in adopting the "Humoral Theory." He thought that the body contained various kinds of bile, red, yellow, reddish-yellow, and black; if any one of these was present in excess some disease was the result. Many diseases were caused by phlegm and chyme, and many by some change in the composition of one or more members, and many by the presence of abscesses in the members. Sympathetic affections of the members are also carefully described. The duty of the physician was to let Nature work in its own way. Diseases ran a regular course, and the fluids that caused them would, if Nature were not interfered with, be expelled from the body by natural processes. The physician might assist Nature, and it was his duty to calculate when the "crisis" would happen, and to help nature at the right moment. The skill of the physician was shown by his ability to calculate the times of crises. Every circumstance in a man's life must be considered, whether the house in which he lived had a north or south aspect, his age and habits and manner of life, his physique and constitution, the place and country in which he lived, the climate, the season of the year, &c. The symptoms of every case were noted, and in making a diagnosis the appearance of every part of the patient's body was to be taken into consideration, and the experience so gained applied to other cases. In prognosis our author was greatly skilled, as many of his Chapters show. He believed greatly in blood-letting, and in the use of purges, but all such remedies were only to be applied after a careful consideration of the symptoms of each patient's case. If the physician cannot make a correct diagnosis he cannot cure the disease (p. 143).

In the treatment of disease special attention must be given to the diet of patients. Many diseases are caused by over-fullness and gluttony, which cause pain in the bones, and thicken the legs, and destroy the members. One philosopher went so far as to "wonder how a man who does not over-eat himself can die" (p. 645). Water from pools and wells and slow-moving rivers is injurious, but rain water, and spring water, and water from swiftly flowing rivers is healthful, light, and beneficial. How important to the patient was judicious dieting is proved by the dietary for the year given on p. 642 ff. Our author prescribes moderation in all things, moderate food, moderate work, moderate exercise, moderate pleasure, and moderate sleep.

All these considerations make it unlikely that he was a believer in the astrological forecasts, &c. which are given in the second portion of the manuscript, or that he had anything to do with the compilation of the third section of the book, which contains some four hundred "native" prescriptions, representing the most degraded form of the art of medicine, and appealing only to the most ignorant and credulous among men. Many of the astrological forecasts must be derived from the Babylonians and Assyrians, and it is to be hoped that an Assyriologist will some day find the time to identify them with their originals in cuneiform. The presence of these forecasts and the four hundred "native" prescriptions in the manuscript is probably due to the person who had the archetype copied, and who wished to include in one volume the various systems of medicine that had been in vogue in his native land.

Finally, a few words may be added as to the prescriptions in the Book of Medicines which were approved by the author of the Lectures on the Hippokratic system of medicine. The form of the prescription is, to all intents and purposes, that which we find in the Ebers Papyrus. First we have a statement as to what the medicine is intended to effect, then a list of the ingredients, the quantity following the name of the drug, and last of all directions for preparing and administering it. Thus Hiera Pikra is to be used for the stomach, and for all ailments of the head that are caused by its sympathy with the stomach. It contains mastic, crocus, bearded grain, cassia, hazelwort, balsam berries, cinnamon, 6 drachms of each, and aloes 12 drachms; these are to be pounded together and made into a powder, of which four drachms in a draught of honey and hot water form a dose (p. 50). In this case a little more than sufficient for 13 doses was made at a time. Sometimes the draught was flavoured with thyme instead of honey. In cases where the medicine was to be administered in liquid form it sometimes had to be purified and strained (p. 52). Solids were to be administered in the form of pills, boluses, and tablets, the dose being as large as a sheep's dropping, or a chick-pea, or a bean, or a gall nut, or a chestnut. Liquids were administered by the spoonful. Solids were to be crushed, or crushed and pounded, or reduced to a powder. Pills were to be swallowed or placed under the tongue. Minute directions are given as to the beating and boiling of medicines, as to pouring them into metal, or glass, or earthen ware receptacles, and as to preserving them for a time before use[1]. In some prescriptions the quantities of the ingredients are given with great care (p. 275), but in others the ingredients were to be measured by the handful (p. 279). One of the most elaborate prescriptions, that for making wine of lilies will be found on pp. 339, 340. Tables of the numerical signs and the weights and measures will be found on pp. 525, 526.

Of the greater number of the prescriptions the authors are unknown, but there are several with which the names of distinguished physicians are associated. Thus we have the Hiera of Galen (p. 48) which contains

---

[1] In some cases the period was six months (pp. 340, 345, &c.).

30 ingredients and, when mixed with some vegetable direction, was administered for urinary diseases. And we have:

1. The Pills of Galen (p. 52), which were called "Kûbbâyê", and were intended to expel noxious chymes from the head and to make the eyes bright. These pills were in reality boluses, as large as grapes, and seven to ten of them were to be administered in a draught before and after meals.

2. The Hiera of Archigenes, a vegetable medicine which was said to cure exhaustion, vertigo, diseases of the eyes, phlegm, leprosy, elephantiasis, dementia, the falling sickness, delirium, scrofula, cancers, scabies, ulceration of the kidneys, nettle-rash, hoarseness, tightness in breathing, the bites of mad dogs, the bites of deadly reptiles, disease of the womb, lumbago and sciatica! (p. 49).

3. The Hiera of Theodoretus (Theodore, the pupil of Agathinus?), for which duplicate prescriptions exist (pp. 50, 51).

4. A prescription for headache by Ptolemy (about B.C. 30), the contemporary of Hicesius and Menodorus (p. 54).

5. A prescription for coughs and asthma by Dioskorides (Phacas), who wrote twenty-four books on medicine (p. 219).

6. A prescription for haemorrhage by Andronius (?). It contains vitriol, flowers of copper, smelters' dross, &c. (p. 237).

7. 8. A prescription for ointment by Democritus of Abdera (B.C. 460—360), and a recipe for making a wine which was very comforting to the stomach (pp. 160, 341). Six measures of strong smelling wine

were poured on irises, seed of rock-parsley, cinnamon, bark of cinnamon, laurus malabathrum, myrrh, and twigs of the absinthe plant, all reduced to powder, and then placed in a glass bottle which was sealed with lime. Dose one cupful before and after a meal.

9. A prescription of Sòrînùs (Soranus of Ephesus (?), who flourished in the reigns of Trajan and Hadrian A.D. 98—138) (p. 94).

10. A prescription of Menestius (?), for pleurisy (p. 271).

11—13. Prescriptions for a diarrhoea plaster by "Esklepiades" (p. 370), a diarrhoea mixture (p. 487), and diarrhoea pills (p. 488).

14. A prescription for a plaster for hardness of the liver by Solon (p. 371).

15. A prescription for pains in the stomach for a medicine made of Beldor (root of the thorny caper?), "which has been handed down from Solomon". The author states that there were two ways of preparing this medicine (p. 345).

17. A prescription for a plaster by Piàgaryas (Philagrius, A.D. 360?) (p. 367).

18. A prescription for an ointment called after the "Twelve Apostles", it contained in addition to vegetable substances protoxide of lead, and cinnabar, and was very good for wounds (p. 165).

Next we have a series of "Antidotes", some of which bear the names of eminent physicians, kings, &c., e. g., the Antidote of Esklepiades (p. 427), the Antidote of Galen (p. 428), the Antidote of Caesar (p. 297) and the Antidote of Pêrôz, a Persian king (p. 502). Besides these we have the Ariston Antidote (p. 409), the Coralium Antidote (p. 303), the Gold Antidote (p. 298), the Musk

Antidote (p. 300), the Pearl Antidote (p. 301), the Wild rue Antidote (p. 300).

Foreign prescriptions are also included, *e. g.*, one for the "Egyptian" medicine (p. 100), one for the "Heliopolitan" medicine (p. 343), one for the "Persian" medicine (p. 151), and one for the "Indian" medicine (p. 150). A number of medicines are known by special names, which are generally derived from the names of the ingredients which give them their special powers, *e. g.*, "Baḳîḳôn", the calf's foot medicine for coughs (p. 268); "Parânosh", which contains iron oxide (p. 352); "Tarsûm", a medicine for colic (p. 353); the "Great Tripal" (p. 362); the "Little Tripal" (p. 354), both medicines for the stomach and beautifiers of the complexion; "Barnagsar", a medicine for the stomach containing Panid, an Indian substance containing sugar (p. 354); "Angabhdîgh", a liver medicine (p. 355); "Pelpelmôr", or "Pepper-root", which was a stomachic and an aphrodisiac (p. 355); the "Royal Pôlkariôn" (p. 361); "Miliṭis", a corruption of "Melilotus", for ulceration of the stomach (p. 367); "Adîsparmaṭôn" (= 'adh isparmaṭôn), which was compounded of several kinds of seeds of plants (p. 367); "Gughershân Shahrîrân" (p. 415); and the "Splendid" medicine (p. 229).

Among the medicines which were greatly prized were the "Great Atenasya" (ἀθανασία) *i. e.*, "Immortality" (p. 407); the "Little Atenasya" (p. 408); the "Atenasya of Asclepiades" (p. 408); "Têryâḳê" (ἡ θηριακή), a simple medicine containing gentian, laurel berries, aristolochia, myrrh and honey, and taken in doses of one drachm in hot water (p. 409); and "Metdôrîṭôs", which contained forty-nine ingredients!

(p. 375). Finally may be mentioned the "Great Hiera-Leghûdhâyâ" (p. 47) which contained pith of colocynth, roasted sea-onion, agarikon fungus, scamony (convolvulus), black hellebore, ammoniac, flowers of thyme, bdellium, chamaedrys, aloes, thyme, malabathrum, haprîkôn, horehound, teucrium polium, cassia, peppercorns (black, white and long), crocus, cinnamon, jackal's fat, polypodium, sagapenum, betonica, myrrh, rock parsley, long aristolochia, juice of artemisia Pontica, euphorbium, bearded grain, amomum gingiber, khemâmâ balsam, strychnus, gentian and honey. This medicine was supposed to cure all diseases caused by the chymes, disease in the hemicranium, idiocy, dementia, stupidity, delirium, vertigo, deafness. the falling sickness, asthma, pains in the kidneys, sciatica, pains in the tendons, gout, palsy, elephantiasis, leprosy, scabies, tumours, running sores, pig-sores, cancer, and diseases caused by black bile and collections of phlegm. It was said to re-establish the impaired constitution, to heal pains in the eyes and ears, to bring on the menstrual flow, and to cure protracted fevers and the fevers which come on daily and the fevers which come on every third day!

THE FEES OF THE PHYSICIAN.

THE Book of Medicines is silent about a good many
details of the life of the physician. Its author does not
say whether patients came to him generally or whether
he went to them, and from one end of the book to
the other nothing is said about the fees of the physician. In Syria his fees may have been paid in money
or in kind. The medicines were probably supplied by
what we should call a herbalist, or by an inferior class
of general practitioner.

In ancient Babylonia the physicians' fees were in
certain cases regulated by the king, as we see from
the Code of Laws that is found on the Stele of Khammurabi.[1] Here we are told:

"If a physician has performed a serious operation
"upon a man with the copper lancet, and has brought
"the man back to life, or if he has opened the cataract
"of a man with the copper lancet, and has brought
"the eye back to life, he shall receive a fee of ten
"shekels.

"If it be an officer of the Administration [on whom
"he has operated] the physician shall receive a fee of
"five shekels.

---

[1] See V. Scheil, Textes *Élamites-Sémitiques,* tom. IV, 1902,
p. 155.

"If it be a man's slave [on whom he has operated],
"the owner of the slave shall pay the physician two
"shekels.

"If a physician has performed a serious operation
"upon a man with the copper lancet, and has brought
"the man to death, or if he has opened the cataract
"of a man with the copper lancet, and has destroyed
"the man's eye,

"If a physician has performed a serious operation
"upon the slave of an officer of the Administration
"with the copper lancet, and has brought him to death,
"he shall replace the slave by another.

"If he has opened his cataract with the copper
"lancet, and has destroyed his eye, he shall pay to
"the owner of the slave one half of his value.

"If the physician heals a broken bone of a [free]
"man, or cures a disease in a limb, the patient shall
"give to the physician five shekels of gold.

"In the case of an officer of the Administration, he
"shall give the physician three shekels of gold.[1]

"In the case of the slave of a man the owner of
"the slave shall give the physician two shekels of
"gold."

In Egypt the physician was usually paid out of the
revenues of the temple to which he was attached. He
was obliged to treat a patient according to the rules
laid down in the books of his order; and if the patient
died, and it was found that he had been treated in
an unauthorized manner, the physician was liable to
be put to death. Military physicians on active service

---

[1] See Kohler and Peiser, *Hammurabi's Gesetz*, Bd. 1, p. 61,
§§ 215—223.

received no fees, and the poor who were satisfied with the treatment provided by physicians from the temples received their services gratis. Herodotus states (ii. 84) that Egypt was full of medical specialists, some undertaking to cure diseases of the eyes, others of the head, others again of the teeth, others of the intestines, and some those which are not local. The patient who employed a physician privately of course made his own contract with him. Egyptian physicians who were summoned to remote countries to treat foreign kings (see Herodotus iii, §§ 1 and 132) were, presumably, handsomely rewarded.

Among the early Greek physicians there was no fixed fee; the patient gave what his gratitude suggested. Later the patient made an agreement with the physician about the price to be paid before his treatment began. According to Herodotus (iii. 131) Democedes received as his annual salary from the treasury one talent, *i. e.*, about £344; the Athenians raised this salary to £406, and subsequently he was persuaded to go to Samos, where Polycrates gave him two talents annually, *i. e.*, about £487.10.0. (Attic standard). Pliny states (H. N. xxix. 3; vii. 37) that the fee which Erasistratus received for curing Antiochus was 100 talents, (Attic) *i. e.*, about £24.375. If the standard was the Ptolemaïc talent, the sum was £39.375.

Under the Romans the ordinary fee for the physician was about fifteen pence, but many eminent physicians made a very large income annually. Thus Albutius, Arruntius, Calpetanus, Cassius and Rubrius made each 250,000 sesterces per annum, *i. e.*, about £1953. The income of Stertinius, derived from private practice, was 600,000 sesterces, *i. e.*, about £4687. He agreed to treat the

Emperor for 500,000 sesterces, *i. e.,* about £3906, per annum as a favour. His brother received the same annual income from Claudius. Though the two brothers spent 30,000,000 sesterces, *i. e.,* about £234.375 in decorating Naples, they left a large fortune behind them at their death (Pliny, N. H. xxix. 5). The Roman "archiatri" were paid by the state, and they attended the poor gratis. Galen received for one case a fee of £ 420. Family physicians received large salaries, which were paid annually on the first of January.

CHAPTER III.

On all the diseases which take place in the head, and first on injuries and mental functions.

Now it hath already been made manifest that all the operations of the body may be divided into two divisions, namely
those which appertain to the spirit (or, soul) and those which
appertain to nature. Those which appertain to the spirit may
be divided into [three classes, namely], those which appertain
to the mind, and those which appertain to the senses, and
those which appertain to motion. It hath, moreover, been said
that it is the brain itself which performeth the operations of
the mind, some by means of other members, and some, that
is to say, those which belong to the mind, it performeth itself.
Now in this Chapter we are about to teach concerning the
injuries which happen to the operations of the mind, that is
to say, those which are caused by the three cavities of the
brain. The first is that which imagineth things and is called
"phantasy"[1] (or, "imagination"); the second is that which
thinketh out things and is named "intelligence" (or, "understanding"); and the third is that which remembereth and is
called "memory". Now the brain itself was not constituted by
nature to be an organ of perception only, but it hath been
the chief of the senses from the beginning. The brain itself by
means of nerves sendeth the power to feel[2] into all the members
of the body. And this is exceedingly well known from the
fact that, if any nerve whatsoever be severed, straightway that

---

* These numbers and the following refer to the origional Syriac text which is not reproduced here.
1 The gloss reads "vision of the mind".
2 *I. e.*, perceptive faculty.

member which it serveth becometh without feeling, because
that power, which came down into it from the brain and was
distributed throughout it, hath been withdrawn from it.   And
this is also exceedingly well known from the fact that in sleep
the senses either are wholly idle, or they work obscurely; in
the latter case it is well known that the power which cometh
Page 2 down from the head to the members descendeth in | a small
quantity.   It is customary to describe these two kinds of sleep
by the words "deep sleep" and "light sleep".   The soundness
of sleep varieth in proportion to the amount of power which
cometh down [from the head].   And thus it happeneth that
according as the power which cometh down and descendeth
[is great or little] even so is the depth of the sleep.   Therefore it seemeth that during the whole period of sleep the
power of the spirit (or, soul) is at rest, and that the natural
Fol. 2 a. force worketh | vigorously.   And this fact is well known.   When
this power is weary so soon as a man lieth down it is straightway made strong again, more especially if the man lieth down
after [partaking of] a moderate portion of food.   Moreover,
during the time of sleep, digestion taketh place satisfactorily
throughout the whole body, and not in the belly only; and,
moreover, very properly, that member also in which [is seated]
the principle of the rational soul is rested.   Now it seemeth
that the heart worketh slowly[1] [during sleep], since it hath not
need of a lengthy period of rest; with the brain, however, such
is not the case, for it worketh without cessation always during
sleep.   For this reason deep sleep falleth upon those who toil
excessively, inasmuch as power hath descended in abundance
and flowed from the head into them whilst they were toiling.
Because, then, of the emptying out (or, exhaustion) of that
power which was emitted by the brain, and because of the
fatigue caused by [his] excessive exertions, he who is tired
hath need both of rest and strength.   Therefore men fall asleep
easily after exertion and sleep heavily, and they do so also
after they have partaken of food whensoever it hath [much]
moisture in its composition.   And they likewise sleep much

---

[1] Literally, "little by little",

when they drink large quantities of wine, and they also sleep
to excess when they bathe in very hot water, which they pour
on their heads. For all these things appear to fill the brain,
and the brain hath need of the fullness whensoever | it is over- Page 3
worked and dried up through over-exertion. And from all
these things it is well known that, when the brain hath worked
very much and is disposed to rest, it causeth itself to fall
asleep naturally, and especially if it hath within itself the power
which feedeth it. If it hath anointed itself with moisture, or
hath become cold through excessive cold, then [its] sleep is
that which ariseth from absolute stupor and deathlike senselessness. And thus is it with all the [other] senses.

For the causes of all [these things] are the | moisture and Fol. 2 b.
the coldness which come into being in the brain, either by
themselves or in conjunction with some humour that possesseth
some quality which resembleth their qualities. Moreover, all
the medicines which do indeed cause sleep, and those which
do not cause real sleep at all, but only stupor, and silence,[1]
and torpor of the whole body, resemble these. Therefore all
those medicines which only make moist are rightly called
sleeping medicines; but those which cause coldness cannot
correctly be said to produce sleep or refreshment. For instead
of sleep they cause insensibility and stupor, and instead of
refreshment they cause difficulty of perception, and they bring
on insensibility. And we have also said above that this feeling
of torpor is a difficulty of perception and a difficulty in respect
of motions in the nervous bodies; and it ariseth, even as we
have already said, through other causes. Thus it ariseth from
a change in the composition [of the body], as well as from
medicines which are cold, and thence come sleep, and silence,
and stupor, and torpor.

Moreover, the dryness which cannot be measured, as well
as what is bitter, or a pungent and hot humour, produceth the
wakefulness[2] which doth not arise from grief (or, affliction) or
anxiety. Therefore as happenings like unto these are common

---

[1] *I. e.*, insensibility.
[2] *I. e.*, sleeplessness.

to the whole body because the head feeleth [them], so also do others happen because of voluntary motions in all their

Page 4 operations, | the head feeling them [at the same time]. Now those stupors which are called "apoplexy", and that disease which [causeth] falling down, and is called "*apôlômsîâ*",[1] are caused by the brain; but apoplexy is a disease which [also] ariseth through voluntary operatibns, such as very deep sleep. In those operations which are connected with the perceptions of the senses, there is another disease, that is to say, watching,[2] and there is that which causeth falling down; now both of these are [caused by] defective motions of the brain, and thus are made defective the other members of the body also.

And there are also two other happenings which take place,

Fol. 3 a. that is to say, torpor | and cessation of its (*i. e.*, the brain's) operations throughout the whole body when spasms which cause rigidity take place, in addition to delirium and in addition to stupor. When the sickness is that of the brain which is in the neck, it is as if a hand, or a foot, or some organ of the body, whichever it may be, is forcibly extended and contracted. In such a case it is the one nerve only which moveth the member of the body that hath the injury. In the case ot diseases in general, whatsoever their kinds, it is through the nature of the disease itself that the members of the body are contracted. Such [contractions] are difficult to describe, but they resemble trembling, and the fluttering of the eyelids, and twitching of the body. I call "twitching" in this case not the feeling of severe cold, but the tremor and motion which arise in the whole body, and all these appear to be produced by the nerves and organs.

Now there are cases when operations which are connected with the mental powers take place wherein each of them singly is affected, and there are cases wherein two are affected, and there are even cases wherein three are affected. It hath already been said in the First Chapter in dealing with the subject of

---

[1] A corruption of ܐܘܦܠܘܡܣܝܐ = ἀποπληξία. Epilepsy, however, seems to be referred to.

[2] *I. e.*, sleeplessness.

the difference between the things which happen, that when the imagination alone is injured phantasms appear, as in the case of the man who thought he felt | musicians tiring him out, and Page 5 when thought (or, reasoning power) is injured, as in the case of the man who threw vessels out of the window and broke them, and when, moreover, the memory is injured, as in the case of those who are attacked by the disease of ravenous hunger (βούλιμος), who forget not only themselves but also their acquaintances, and who do not remember even their own names.

Now the memory changeth in this life, for it appeareth very often that the disease of memory ariseth through an injury to the thinking power, just as frequently also the injury of the thinking power cometh into being at the same time as an injury to the memory, for the very same disease ariseth both in the thinking power and in the memory; but it becometh specially severe when the memory is destroyed together with the thinking power, and in that case insanity | ariseth.   Fol. 3 b.

Moreover, both the thinking power and the memory perish in the diseases of forgetfulness and stupidity, and the variation in all these diseases is of the self-same origin, that is to say, they all arise from the nature of the variation of the composition (or, constitution). Now this sheweth us that this disease is of a general character, and that all the members of the body which possess primarily the power of action are formed of similar constituents. According to this, then, we are compelled to recognize that this change (or, variation) in the composition of the body is necessarily one of coldness. For it seemeth that coldness produceth stupor in the mental functions, a fact which animals demonstrate to us very clearly, inasmuch as they are compelled to eat because of the cold, as do also produce stupor all the various things which help to make [the body] cool, and also all those foods which cause coolness, as, for example, the lettuce which produceth heavy sleep if a man eateth too freely thereof, and also heaviness of the head. All these things which subsist, with the exception of | severe sicknesses (or, pains), all of them, I say, are stupefy- Page 6 ing and cause heavy sleep, but [the effects of] very many of them appear to be lightened by the use of gargles. Besides these the heat and the cold which exist in the head also make this manifest.

Now warmth produceth wakefulness, but cold produceth deep sleep; moreover, the feelings which are due to the bile and to heat appear to produce wakefulness (or, alertness of mind), and talkativeness, and versatility. Now the feelings which arise from coldness and phlegm cause the opposites of these qualities, that is to say, heaviness in the members, and torpid actions. Therefore the strength of the feelings of stupor and of alertness is due first to the change in the composition of the body owing to cold and heat, and secondly to moistness and dryness. Now bathings of all kinds induce the head to sleep because they make it tender (or, moist), as do also the drinking of mixed wine, and all kinds of foods which cause softness [of the body]. And the period of life also influenceth this matter, for in the case of children their tender youth causeth sleep, whilst the dryness of old men produceth wake-

Fol. 4a. fulness. | Now therefore all these signs of the body happen to us so that they may make [us] to know that the moisture which is unnatural occupieth the second place in respect of its care for the soul, and that coldness occupieth the first. Therefore when thou increasest moisture by itself, it produceth deep and heavy sleep and stupor, just as when thou increasest dryness by itself it produceth wakefulness, according to the words which were spoken of these feelings by Hippokrates, who said, "Both sleep and wakefulness, if they be indulged in "immoderately, are bad."

Now if coldness be mingled to a sufficient degree with moistness it produceth the feeling of stupor and silence (or, inanition); for there are very many varieties of superfluity (or, excess)

Page 7 and of insufficiency, not | only in moistness and in coldness, but also in dryness and in heat. And there existeth another kind of change which is the natural cause of the operations that appertain to the soul (or, mind). As, then, sleep and wakefulness, when they are excessive [are connected], sleep being caused through the moisture of the composition of the body, and wakefulness through the dryness of the same, so

also in the excess and insufficiency of wakefulness and sleep
are the [varieties] connected, namely, the excess and the insufficiency which are in moisture and in dryness. Because,
therefore, he sheweth us that variations of the component parts
of the body such as these take place in two forms, it is found
also that the form of each one of the feelings (or, senses) is
of two kinds, the one arising from the cold and moist chyme,
and the other existing in the nature of the particles of the
body. When to the variation in the composition itself there
come the particles (or, bodies) of the chyme, with these are
the variations in the composition of the body which have been
described, and they are the opposites of each other. Now
there existeth another form which is composed of both, and
it taketh place in those [feelings] which are called stupor and
torpor because in them there are much bile and phlegm [mixed]
together. And these same kinds of changes in the composition
of the body are simple, and they subsist through that other
form which is composed | of both, and also through the an- Fol. 4*b*.
tagonism of cold and heat.

Now the bile is a reddish substance which is mixed with
phlegm, and it produceth a pain which is mingled with cold
and heat. But if it be admitted that a pain which is composed of two things which are opposite in character can exist
in the nature of the members of the body, it must also be
admitted that there are three changes (or, variations) in composition in each one of the things that are opposed each to
the other. Therefore all such pains (or, feelings) arise in the
brain, but they are different from each other, not only because
there are varieties in their compositions, and because they are
more or less | simple and compound in their mixtures, but also Page 8
because there subsists in the cavities of the brain a change in
composition at one time, and at another in the veins and arteries which are in it, or in that moisture which is implanted
in the body (or, matter) of the brain. Now besides these there
is a fourth kind, that is, when the composition of the matter
of the brain itself changeth; in such a case it is right to know
how those sleep whose memory and understanding have perished.
For the destruction of their understanding causeth imbecility.

And we must enquire carefully whether the sleep of those who are afflicted is heavy, or whether it is slumber of an ordinary kind, or whether their slumber is not different from ordinary sleep. For from these [indications] thou wilt be able [to learn] what the change in the composition is.

It is right also to observe whether anything is ejected through the nostrils, or whether anything cometh out through the mouth and descendeth from the head, or whether these parts appear to be dry. For from these [indications] also thou art able to make a calculation about the pain (or, disease), and also about that ailment which is called catarrh, and that which is the result of witchcraft. And together with these things also we must observe the kind of quantity of the substances which are expelled [from the nostrils and mouth], and enquire into the causes which preceded [them], which will make known to us what is the pain in the head, and whether it is hot like that caused by a flame, or whether it is cold like that caused by coldness. Unless a man is able to define all these things it is impossible [for him] to contrive a means of healing which will

Fol. 5 *a.* be suitable | for each of these pains. Now therefore in a case where the memory is destroyed, or seriously injured, if this hath arisen from a change in the composition of the brain due to cold, it is most certainly right to have recourse to warmth, and if it be due to warmth, recourse must most certainly be had to coolness, but [the brain] must not be dried up nor over-wetted. If the brain existeth with moisture it must be moistened, but if it be neither dry nor wet, then it will be preserved by this.

Page 9   Now I once knew a man | who had almost entirely lost his memory, and whose power of thought was injured through the vigil and toil of studying. And I once saw a man who was a vineyard labourer who suffered in the same way as he did through the exhaustion caused by labour in the vineyard, and by too little food. And both of these were injured in a manner which was apparent, and each suffered, the one through all those things which parch the brain and inflame it, and the other through those things which make it over-wet and cool it. Moreover, injuries take place in the operations of the mind through fevers, and the effect resembleth that which ariseth in delirium and forgetfulness. And besides the injuries which arise through fevers there are those which are due to dementia and madness, and there are others also which are due to passibility, and to the faculty of passibility which is possessed by the head. But these injuries which exist in the faculty of passibility are better known than the happenings which are closely connected with them one by one and at all times, and they do not progress and become things which are different. Those which exist in connection with passibility are not known thus, neither do they abide at all times and exist through other happenings. Let us, then, keep in mind, even as it hath been said above, that such is also the case with the capacity for suffering. The suffering which only taketh place together with the causes which produce it is useless and can be dissipated, but that which hath taken root already hath acquired the pain of that member which is [connected] with it, and although we can demonstrate the causes which have produced it, it is certainly permanent.

Therefore, that all the diseases connected with the working of the mind are situated in the brain is admitted by all those physicians who do not for the sake of combating opinions think one thing inwardly and | say something quite different Fol. 5b. with their mouths. Now, for us to find out the change in the composition [of the brain], and what it exactly is, is no small task. For in this matter it is right that the physician should be an enquirer and a painstaking worker, and he must not consider how he ought to speak against (i. e., contradict) the things which have been stated by the ancients in fine speech about | the mind (or, understanding) of the soul, or believe, as Page 10 do the ignorant, that it is situated in the brain. For in what other way can any skilled physician heal those who are suddenly stricken into silence, and those who fall down smitten with epileptic fits, and those who have convulsions in their backs, and those whose bodies have become rigid like logs of wood? Moreover, what physician hath, except in this manner, cured those who were paralytics and the entire halves of the bodies which were useless and motionless? Is the case, then, different in any way in respect of those sufferings which arise from convulsions? Do not all physicians learn by experiment? All the ancients make their works throughout treat of the healing of the demented, and of those also whose bodies are in a state of convulsion. And besides, they warmed the head, and with this treatment also they effected the cure of those who were suddenly stricken into silence, just as they cured those who fell into epileptic fits, and became ill. And when this disease arose from the stomach, or through one of the other members, they took especial pains to heal that member first of all, and they were most careful to work on the brain which doth not easily become affected.

On epilepsy, which is the sickness of falling down.

Now also the disease of those who fall down in that sickness which is called "epilepsy" is a rigidity of all the members of the body. It is not always thus like the rigidity which cometh from behind, or like that which cometh from the front, or like the other kind of rigidity which cometh from both these and is of the mind and is in all the body, but it lasteth for a short time only. And it is not different in this Fol. 6a. respect only from the other kinds of rigidity | (or, stiffening) of which we have spoken, but also in respect of injuries of the Page 11 understanding and of the senses[1] (or, feelings), | and from this it is known that the source of this disease is in the brain. But, inasmuch as it is speedily relieved in the cavities of the brain, it is meet to declare emphatically that it is the thick chyme which blocketh up the exits of the breath (or, spirit) and causeth the disease, at the same time shaking and making to twitch the very beginnings of the nerves. Therefore thou must eject therefrom that which afflicteth it. Perhaps also each of the ends of the nerves being submerged, after the manner of those which go forth from the back, there then cometh into being the disease (or, pain) of those who fall into epileptic fits and become ill. Now the fact that the origin of

---

[1] The scribe has repeated these words inadvertently.

the disease cometh suddenly into being, and that a man is straightway thrown into a fit, sheweth that this disease never ariseth through dryness or through lack of chyme, but always through fullness of the same. Now the obstruction in the tubes by the thick and viscous chyme taketh place suddenly because it is due to the fact that it is impossible without much time for the brain or its membrane to attain to such dryness as this.

And very closely connected with this disease is the disease of those who can neither see nor perceive with their senses, and who are quite unable to make any use whatsoever of any one of their senses, and also the disease of him who is stricken, and who knoweth nothing whatsoever of the things which are taking place, and who is injured both in his understanding faculties and in his power of memory.

From all these facts one is absolutely bound to think that the origin of the disease is in the brain, and that it is the chyme which maketh useless and blocketh up the exits of the soul-spirit, which is restrained within the cavities of the brain. Now why this spirit is called "soul-spirit", and what its power is, hath been shewn in other places. For when we examine those things which appear in the openings, it seemeth to us that it [cometh] from the soul, and that it dwelleth in the body | of the brain, | and belongeth to the thinking power, and the force of the senses, and the memory. Now its prime instrument whereby it performeth all the operations of the senses and of volition is that spirit which is in the cavities of the brain, and especially that which is in the posterior cavity. But it is not fitting for us to decide definitely concerning the medial cavity of the brain that it is not entirely the principal instrument. For there are very many convincing circumstances which also bring us to this conclusion, just as also [there are others] which lead us to think that the two cavities which are in front are the principal instrument. As regardeth the discovery of the means of healing, an exact knowledge of these things doth not benefit us in any way whatsoever. In respect of healing it is sufficient for us to make a good cure, and to know that the part which is affected is the brain, and that it is the thick and viscous chyme which is collected in its cavities. There-

Fol. 6b.
Page 12

fore, inasmuch as these things are useful to us for healing
purposes, we must enquire concerning the region which is
affected and the disease which is in it. Similarly also, the
difference between the thick chymes is useful to us that we
may knòw whether they originate in the phlegm or in the
black bile. Moreover, we must remember this also, that here
we give the appellation of "phlegmatic", wheresoever we speak
of them in an ordinary manner, to all those chymes which
hold in their composition moisture and coldness; but to those
which hold in their composition dryness and coldness we give
the appellation of "black bile". Now there are also great
differences among the phlegmatic chymes, and also between
those which originate in black bile. Many men daily vomit,
and spit up, and expel in breathing, being filled with vaporous
wind, chymes which originate in the phlegm, the component
parts of which, as may be perceived by the senses, do not
resemble each other. And of some of them the component
parts seem to resemble each other, whilst in reality they do
not do so.

Page 13   Of this class is | the living chyme which existeth in the
Fol. 7 a urine, and that also which is said to be like unto | glass; and
again, that kind of spittle which is not very wet, or watery,
appeareth to belong to this clàss. And it may also be perceived by a man's sense of taste, for though it may not appear
to him of himself that he possesseth it, yet he may be altogether
phlegm. For how many times do we perceive that the spittle
in our mouth is saltish, or acid, or tasteless, just as it is
watery and without any taste whatsoever which we can perceive, so long as we are in a healthy condition and are not
sick? Similarly also, that chyme of black bile possesseth in
its constitution many varieties of matter. Some portion of it
appeareth to be like the sediment of blood, and is sufficiently
dense like the dregs of wine, but one portion of it is very much
thinner in its consistency than these. To those who vomit it,
and also to those who smell it, it appeareth to be very acid,
and it dissolveth the earth which toucheth it and maketh
holes in it, and it produceth swellings (or, bubbles) like those
which appear in yeast when it is working. Now that portion

of the black bile which I have said resembleth a thick sediment doth not produce swellings (or, bubbles) when it is poured out upon the ground, unless it be met with on some occasion when it is exceedingly hot indeed in a burning fever. And that which I am in the habit of calling "chyme of bile", or "blood of black bile" is not at all acid in its nature, and I am not yet justified in calling chyme of this kind "black bile". Now this chyme is produced, and is abundant in men, through the constitution which they have had from the beginning [of their lives], or through that constitution which cometh to them through their habits of living, or through their various kinds of food which are changed into this chyme, which is present in digestion, and is that which is in the veins. Therefore also in proportion as the chyme of the | phlegm is thick, so doth Page 14 it produce the disease of those who fall into epileptic fits, and it blocketh up the vents of the cavities of the brain, that is to say, of the medial cavity, or of the posterior cavity. Now when this chyme existeth in excess in the matter | of the Fol. 7b. brain it produceth madness, just as when the black bile in-creaseth, which ariseth from the burning of that reddish-yellow substance, it setteth up violent dementia when it is in excess in the matter of the brain, whether it be with or whether it be without fever. Therefore among the insane, he whose mad-ness ariseth from the yellow bile is only mad to a certain degree, but he who is violent and cruel is the offspring of the reddish-yellow bile. There is, moreover, another kind of mad-ness, namely, that which is savage and ferocious, and this madness ariseth from the burning of the reddish-yellow bile. Now those who have become fools through the effects of violent fevers or through suffering caused by a diseased brain, must be placed in another class, and not with those whose special characteristic is the suffering of pain. Therefore those who lack their wits and do not understand properly, and who lose their memories during their illnesses, are merely called "idiots" by the physicians, and they do not give them the appellation of madmen, and they do not experience any relief from the violent effects of the fevers [during the passage of] the years. As therefore the fever which cometh with years

is one of the incidents of the disease which existeth in the
brain, so also the delirium which cometh on during a burning
fever is an incident due to the many hot vapours which ascend
to the brain, and it resembleth this disease. And also that
which taketh place is, as it were, an incident of the mistiness
(or, darkness) which produceth it, and it ariseth from some
pain (or, disease) in the belly. For the belly is a partner with
the head, and the head is a partner with the belly, and they
suffer together, because of the great size of the nerves which
descend from the head to the mouth of the belly. And because
of these nerves there is an intensity of feeling [in the belly]
Page 15 which is far greater than that which is in any other member | of
the body [when there is] a cancer in this member. Therefore
with all injuries of the head, which ascend to the milk of the
brain, are closely connected bilious vomitings; and also with
all pains of the head are closely connected nausea and pain
of the stomach. Very often severe pain and insanity are
Fol. 8 a. closely connected with the diseases which are called | "epi-
gastric" and with flatulence. Now association of diseases
which followeth in this case resembleth the delirium which
clingeth to severe fevers, and to certain diseases of the mouth
of the belly, and also to diseases in which arise the symptoms
of blindness. Similarly, delirium is more readily associated
with a hot tumour when it existeth in nerve substances than
when it is in any of the other members of the body. Sometimes this is due to the fact that the tumour possesseth heat
which ascendeth to the head, and sometimes to the fact that
it possesseth some breath which is vapoury or smoky.

On the madness which ariseth from black bile.

Now therefore as no small change taketh place in the
diseases which attack the head through sympathy, so also
the change which taketh place in them through the peculiar
property of suffering is not small. The thick chymes which
are gathered together and increase in the natural matter of
the brain affect it sometimes like an organic disease, and sometimes like substances made up of homogeneous particles. Now

when they block up the vents (or, openings) and the exits [of
the brain], the disease is organic, but when they work on the
brain itself the disease is [due to] a change in the substance
made up of homogeneous particles. The author of this observation is Hippokrates, who maketh it in the Sixth Chapter
of his treatise on the Comings of Sicknesses. Moreover, those
who are mad fall by these things. Now each one of these things
happeneth particularly in the region to which the sickness
inclineth, if it be in those bodies which fall, but if not, in the
mind of those who are mad. Therefore, by this | observation Page 16
he maketh known primarily that not always, but in the majority
of cases, there taketh place a conversion of these diseases one
into the other. For not through the chyme of black bile only
doth the madness of those who fall take place, but also through
phlegm, and therefore that which taketh place through the
chymes of black bile sometimes changeth the madness into
disease. But that | which happeneth through phlegm [is changed] Fol. 8b.
into another [kind of] disease; how this change is brought
about I will after a little space describe, but through it madness never happeneth. And following this remark of Hippokrates which hath been quoted there are [given] in his discourse other opinions [on the matter] not a few. For the soul
is a mingling of the operating stomachs, and also from their
mingling he speaketh of them as being changed. And because
of this he calleth that bile the brain, even as he doth organic
diseases, and saith that it inclineth towards the substance of
the brain. Now this taketh place in stoppages [of the vents
and exits] as we have said, but he calleth a mixture that
which resembleth a disease caused by substances made up of
homogeneous particles, and saith that it inclineth towards the
mind. Now in my own opinion, the fact that all physicians
have abandoned that view maketh it necessary [for him] first
of all to lay down an exact definition of the subject. For as
it is in the members of the body which are visible, [even so]
is it with them all, and the mingling itself is apparent, even as
that book of Hippokrates teacheth. The bile wandereth about
in all the skin, even in the disease called elephantiasis, and in
hydropisis, and in phthisis, and besides these diseases also in

the "departure of colours", and in diseases of the liver and of the spleen. Now sometimes it is one member of the body only which receiveth the reddish-yellow bile, or the phlegm, or the black bile; in these cases the composition of that member only is changed, and thus also is it with the brain. At one time all the blood in the veins becometh charged with black Page 17 bile, and then it smiteth (or, attacketh) also | the brain; but at another, all the blood of a man remaineth unchanged, and it happeneth that the blood which is in the brain only is changed. Now this taketh place in two ways: either because the chyme of the black bile is injected into it from another region, or because it is produced in that place itself. And the chyme itself is produced by the excessive heat of the place which burneth and is fiery hot, or from the reddish-yellow bile, or Fol. 9a from thick, black blood. This definition | helpeth towards healing in no small degree. For when this blood charged with black bile is in all the body, it is right to begin to effect a cure by making a letting out of the blood. But when blood of this kind is produced in the brain, the patient hath no need of blood-letting, as in the case of pain; but for any other thing it hath been found that he needeth blood-letting. Let the following indication of these things be with thee (*i.e.*, in thy mind). If there be black, bilious chyme in all the body, or if it be only collected in the brain, then before doing anything else enquire into the constitution of the body, and find out what it is like. Only remember that those whose bodies are soft, and white, and fat, never contain this kind of chyme. But in the folk who are spare of body, and black, and hairy, and in those who possess dilated (*i.e.*, varicose) veins chyme of this kind is readily produced in great abundance. There are cases, too, in which those also who possess a reddish (*i.e.*, fair) complexion change quite suddenly to the composition of black bile.

Now in both varieties there is a rule of this kind: those who are exceedingly red (*i.e.*, fair), especially when they find themselves contending with watching (or, vigil), and work, and anxiety, and with business affairs of a very delicate character, are kinsmen of these. Moreover, [thou must] ask questions on

such points lest peradventure | there be restrained some burst- Page 18
ing forth of blood, or some other process of emptying blood
which is proceeding forcibly, or the courses of women. After
these things [thou must enquire] also concerning the various
kinds of foods which the patient useth (*i.e.*, eateth), and find out
whether it is from these that the black bile is produced, or
by things which are of opposite natures. Now if he followeth
a method of life and maketh use of foods which produce black
bile, or if he leadeth a course of life and liveth on foods of
which the chyme is good, then it is right to theorize; and it
is right also to investigate his habits and to find out if he
liveth a life of toil to which are added sorrow, and insufficient
sleep, and anxiety. And, as we have said above, in other
cases pains (*i.e.*, attacks) of fever | produce the black bile. Fol. 9*b*.
Now in arriving at an exact diagnosis a knowledge of the
following is helpful in no small degree, namely, the time of the
year, and the composition of the atmosphere (*i.e.*, the weather)
before or during the attack, and the place in which the patient
is, and also his bodily stature.

Having therefore made investigations of all these things, and
having become convinced that the bilious blood is in the veins
of the whole body, thou wilt at length be able to make an
exact indication by the cutting of the vein which is in the arm.
It is best that the middle vein be cut, because it is the more
central of two, that is to say, the one that cometh from
the shoulder and that passeth through the upper arm and
cometh to the hand. Now, if the blood which floweth from
this member is not seen to contain black bile, stop the flow
immediately, but if it be seen that it doth contain black bile,
then draw out from it as much as thou thinkest should be
drawn out, according to the bodily constitution of the patient.

Of the sympathetic weakness of the epigastric region.

Now there is a third species of weakness which happeneth
when the head taketh [food] from the belly, and which
resembleth the disease of those who fall into epileptic fits; and
certain of the ancient physicians call | a disease of this kind Page 19

"epigastric", or "appertaining to the breathing organs". It will be sufficient to describe it and those incidents which cling to it, for Dîâḳlôs wrote about it in the work which is entitled "Of diseases and the causes why physicians exist". And in this book Dîâḳlôs writeth about the disease thus:—Now the other disease which attacketh the regions about the belly, and which resembleth the diseases which have been mentioned above, certain men call "inflation", and others "flatulence". To this disease there cling after food, and especially when men eat what is hot and is difficult to digest, eructations which have a bitter taste, and saliva which is wet and abundant, and which is burning hot in the epigastric region, and a distension of the

Fol. 10 a. bowels which abideth | and is not relieved immediately. Sometimes there accompany this disease also violent pains in the belly which ascend to the region between the shoulders, but these are relieved after the digestion of the food [which causeth them], though these self-same pains are felt again after men eat. Oft-times these pains cause them trouble both after they have fasted and after they have eaten. When they vomit, they vomit the food without difficulty. Such patients vomit substances which appertain to the phlegm and which are so bitter, and hot, and which are so acid that they make blunt the teeth of those who vomit them. Now these other things happen to them immediately after their childhood, and the attacks increase in duration and frequency so long as they remain with them.

And after Dîâḳlôs hath said these things, he addeth the causes next and writeth thus:—As regardeth those who are held fast in the pain of flatulence, the physician must think that they have excessive heat in those veins which receive the food from the belly, and that the blood in them is thick. Now this is an indication that the disease is in those veins, because the body will not receive the food, which remaineth

Page 20 in the belly | undigested. For first of all those veins take the food and then they generally convey it to the lower belly. If after a day hath passed they vomit it this is an indication also that the food doth not ascend into all the body, and then one must understand that there is in them great heat, which ariseth from the inflammations that take place in them through their foods. Now they are to be benefited by such meats as are cold and are wont to cool the heat and to reduce it.

And after the above observations Dîâḳlôs also wrote the following:—Certain men say that, in ailments of this kind, the mouth of the belly which is closed on the intestines is affected by a boil (or, tumour), and that through this boil the mouth is shut and preventeth the food from going down into the bowels at the seasons which are appointed to it; because of this | the food remaineth in the belly for a long time and, in Fol. 10b consequence, produceth the flatulence and inflammations which have been mentioned. These things, then, did Dîâḳlôs write, but he made short his observations in the paragraph on the symptoms of the ailment, which it would have been of far greater importance to have, and which he should have set forth in all the paragraph. I mean to say, he ought to have described the symptoms which indicate disease in the epigastric region; but it seemeth to me that he omitted to do so because from the name of the disease itself what these are becometh clearly apparent.

As concerning what we have learned from Hippokrates, who said, "If a man shall remain for a long time held fast bound "by fear and sorrow weakness of the intellect will come to "him," why, seeing that he wrote in his work on the "Causes" a description of all the causes of other incidents [of the disease], but wrote nothing whatsoever about the cause of the mind being injured, is a subject worth enquiring into. For supposing there be very great heat in those veins which are at the side of the belly, that is to say, those which hold fast this disease, and supposing there be an abscess (or, boil) [on] the lower door of the belly, why weakness of the intellect should cling to these things is what he hath failed | to say. That the belly Page 21 should be filled with the wind of flatulence, and that patients are relieved after the eructations and vomitings which are mentioned by Dîâḳlôs, are matters which are very well known, even though he mentioneth them not. But it is very difficult to connect the special symptoms of weakness of the intellect with that disease which is said to be in the belly.

2*

We, then, ourselves will add the following remarks and will explain the matter clearly. Whatsoever may be the disease which existeth in the belly, it hath characteristic pains. Now it appeareth that there is in it (*i. e.*, the belly) some disease which may be thus described: when the disease is imprisoned in that member in which the abscess (or, boil) is, the member becometh thick through (?) black bile. As there ascend from the belly to the eyes a certain smoky fume and dense vapours which produce in them symptoms that resemble those of darkness, even so is it in this case, for when there goeth up to the brain the fume of black bile, which is in the form of Fol. 11*a.* vapour or smoke, there arise the symptoms of | weakness of intellect in the mind. For, behold, we see always that headache ariseth through the black bile which is imprisoned in the belly, just as it happeneth that so soon as the bile is vomited by a man he becometh without pain immediately. And those pains which resemble these are griping and gnawing, even as some appear to be associated with a weight of some kind, and others with tension and stupor. Now all skilled physicians agree in stating that the pains of those who fall down in epileptic fits and become ill do not attack the head from the belly only, but also from the disease.

Of the varieties of fear which appear in those who are weak in intellect.

Now fear clingeth at all times to those who are weak in intellect, but the phantasm of fear doth not always come to them in exactly the same form of pain (or, sickness). For, behold, one member of this class imagineth that he hath Page 22 become an earthenware pot, and therefore he fleeth | from all those who meet him lest peradventure he become broken. And another of the class, on [hearing] the cocks crow and seeing them flap their wings, will also strike his arms against his sides, and imitate the sounds which the birds make. And another is timorous and is terribly afraid lest Atlas, who beareth up the heavens, will become tired out and remove himself, and so crush himself and destroy us also along with him. And a myriad other things like unto these will the weak in intellect imagine vainly in their disease. Now those who are weak in intellect differ each from the other. Some of them suffer from fear, and are sorrowful, and they curse their lives, and hate all the children of men; but they are not all of them eager to die, for there are some of them with whom the fear of death is constantly present throughout the years of their life. On the other hand, there are some who seem to be different from these for they are not afraid of death.

Therefore Hippokrates appeareth to me to be correct when he summeth up in two classes the symptoms of them all, that is to say, those which are the result of fear, and those which are the result of sorrow (or, grief). For, in the case of sorrow, the weak in intellect hate all | whom they see, and Fol. 11b. they are afflicted with grief always, and they are afraid of children. And there are also other men who will not travel in the thick darkness of night. For as this darkness causeth all the children of men to fear, with the exception of those who have been trained to travel in the dark, or those who are travellers by profession, so also doth the colour of the black bile, like a kind of darkness, obscure the region of thought, and produce most abject fear and terror of this kind, which change the operations of the soul, that is to say, the chymes, and the constitution of the soul, of whatever kind it may be. And this is openly admitted both by physicians and by skilled philosophers.

Now, so far as I myself am concerned, I have proved this fact in the Chapter wherein I have shewn that the powers of the mind are very intimately connected with the | compositions Page 23 of the substances of the body, and with the chymes and [their] compositions. Therefore, those who have not apprehended the power of the chymes have not dared to write anything at all on the disease. If, then, the things which happen in the belly first of all become quiet and then increase, and that disease of weakness of mind clingeth to them, a man will be relieved when his belly is opened, or when he vomiteth, or when he digesteth his food properly, or when eructations come to him. In such a case [physicians] may call the disease "epigastric" or "flatuous", or may say that the attack [is caused by] sorrow and fear. Now when the symptoms of dementia appear in a severe form, no incident at all, or if there be any a small one, taking place in the belly, then it is right for us to think that it is the brain itself which is the cause of the suffering, just as when black bile collecteth in it. Now from these considerations it is right for us to lay down a rule and arrive at a conclusion as to whether this chyme is in the brain only, or whether it is in the whole body. We have mentioned these already a little way back, but I again call them to the minds of my fellow practitioners who know them, and who have seen me on several occasions make a cure of dementia of this kind by means of frequent bathings, and by a gentle regimen which is easily carried out, not to Fol. 12a. mention other things which are beneficial. | This I do when it hath not become difficult through the length of time which hath expired to scatter the chyme which is deep; but when this disease is of old standing, it needeth to be treated according to the methods which have been described, and other and stronger remedies are required.

Now this kind of dementia taketh place when hot pains have been felt in the head immediately before the attack, and when these have arisen either from inflammation, or from some pain caused by a hot (*i.e.,* angry) sore, or from a fever which Page 24 produceth delirium. And it also taketh place | from excessive anxiety, and grief, and overwork. These remarks on this kind of dementia will suffice.

Of the falling sickness (epilepsy) which taketh place in sympathy with other members of the body which fall thereby and become sick.

Now we are going to speak about that sickness of those who fall and become sick, and shake, for these things take place also sometimes when the brain itself is sick, and sometimes when it is sick at the same time as the other members are sick. Let us also define [the sickness] and distinguish [these things] very carefully, for the physicians have with ex-

ceedingly great carelessness omitted to define it. Now therefore as three kinds of dementia are to be distinguished by their
characteristics, so also doth the sickness of those who fall
possess three varieties; that the brain suffereth is common to
all three. Now the brain suffereth either when there is disease
in it, a thing which happeneth in many of those who fall, or
when something goeth up in sympathy to it from the mouth
of the belly, which the physicians are accustomed to call the
stomach, and giveth the brain trouble, like those happenings
which take place through (or, from) the stomach in the eyes,
which resemble suffusion (or, cataract), one becoming so some
time [after the fall], and the other at the fall, for it taketh
the characteristic of some member in which the disease beginneth, and thus it is felt by the patient that it goeth up to
the head.

Now first of all consider the following [story]. There was
a certain youth who was about thirteen years old at the time
when I myself was a young man, and I saw him | with the Fol. 12 _b_.
physicians who were held in honour among us, and who were
gathered together in order to find out some means of healing
him. Now I heard that boy relate that the beginning of the
pain came to him from his leg, and that from this point it
proceeded directly upwards | through his thigh, and through Page 25
the epigastric region above it, and through his side, and so on
towards the neck up to his head, and he said that immediately
the pain approached his head he was sensible of it no longer.
And when he was asked by the physicians to describe what
kind of thing it was that went up to his head, the boy himself was not able to do so. Finally, however, a certain young
man who was not a person of learning, but who knew enough
to perceive what had taken place, was able to explain the
matter very clearly indeed to the others. He said that that
which went up [to the head] was like some cold blast, and
that a change took place in the unity of the members, or that
it was some spirit-substance, which was not worth wonder-
ing at.

And this was what he said. That chyme which was produced
in the member which was suffering possessed some strong power,

which was like unto that venom (or, bile) which malicious reptiles eject. But who would believe us unless they saw the things [of this kind] which take place frequently? For through the sting of the scorpion, and through small tarantula spiders a great change taketh place in the whole body, although only a very small and unknowable quantity of some substance is injected into it by the stings of these reptiles. For as concerning the little spider (or, louse) which biteth, which is a reptile, though a very small one, we do not think that it ejecteth any venom [from its mouth into that body which is bitten by it. Now the blow of the spike in the tail of the roach of the sea, and also that of [the tail] of the scorpion on land, are seen manifestly to be exceedingly sharp and severe, though it is not apparent that there is any hole whatsoever in their heads through which they can eject venom. Nevertheless, we are

Fol. 13a. bound | to understand concerning these creatures that they do eject into the body some windy or watery substance, which is

Page 26 very minute in its composition but exceedingly powerful | in its effect. For, behold, when a man is struck by a scorpion in our presence he says that he is smitten by hail-stones, and all his body becometh cold, and he breaketh out into a cold sweat, and though he may be healed it is only with great difficulty that he can be saved. Now, therefore, it is not beyond the bounds of possibility that there may be also in the body itself some substance of this kind, and that it may be produced in it, besides that which is external. And this substance, being more especially in one nerve-member, may transmit its power by means of unity of substance to the head of the nerves, either by change, as I have already said, or by transmission of a blast of spirit-substance. Instances of this also appear frequently very plainly, that is to say, when the scorpion driveth its sting into the nerves, or veins, or arteries, those who are stung are seized by severe pains. Now the sting of the scorpion is able to penetrate into the body itself, and to pass through the whole of the thickness of the skin. The sting of the wasp is small, and it affecteth the outside of the skin only, which also is well known from the following fact. There are cases when the power of the venom cometh through the

skin only into the whole body, because all the skin is united
to all the nerves. Therefore it is easy, seeing that it is diffused
in all the power of the venom which hath been ejected, for it
by means of the skin to pass over to each of the members
over which it is spread and to which it is closely attached.
From these again it can spread itself to the others because of
their unity, and again from these which are affected to the
others. When it hath reached one of the principal members,
then doth the danger of death cleave to the man. Now the
bandages on the members which are above their fellow-mem-
bers operate in respect of these, | provided they be strong, and Fol. 13*b*.
they are beneficial | in no small degree, and helpful. Page 27

For we have gained experience of these matters through
[treating] the bites of vipers, and the stings of scorpions, and
the darting blow of the asp; about the latter one cannot be
absolutely certain, for death [usually] overtaketh immediately
those who are struck by it. Now when I was in Alexandria,
a certain villager was bitten by an asp in one of the fingers
of his hand, when he was at no very great distance from the
city. Immediately he tied round the lowest joint of his finger,
which was close to the palm of his hand, a strong bandage,
and ran straightway to a certain physician whom he knew at
the gate of the city, and entreated him to cut off his finger
from the lowest joint, namely, that which was in the palm of
the hand. He expected that if this could be done he would
suffer no [further] injury, and his expectation was fulfilled as
he thought it would be, for he was saved, and lived, and this
[treatment] only did he seek. And I saw another to whom an
antidote which was compounded of the flesh of vipers had
been administered, and after his finger had been cut off he
was healed. And I also saw another villager who had been
bitten by a viper into his finger, which he cut off straightway
with a reaping knife, from the joint which was below the spot
in which he had been bitten. Now he was by trade a vineyard
man, and he also lived without the application of an antidote,
and the wound in his finger was made whole by the use of
the customary bandage.

Moreover, as regards that youth who fell down and was

sick through the disease which went up into his body from
his leg, whilst those physicians who were gathered together
to find out some means of healing him were wishing to make
certain about its cure, it seemed to them that, first of all, they
were bound to make clean his whole body, and then to apply
to the member which was diseased a plaster compounded of
the yellow thapsos plant, or of black mustard. And whilst
these things were happening they tied a cord round the leg,
Page 28 above that spot | whence the pain originated, and straightway
Fol. 14a. restrained the flow which was taking place before these things
were done. I merely mention these things in passing so that
one may not wonder how it is possible that such an ailment
as this can be produced from any one of the lower members.

Now it remaineth for us to seek out the cause of the pain
which cometh on during the spasms which make stiff the
bodies of those who fall into fits sympathetically. I myself
have seen sometimes a falling take place sympathetically which
was of this kind, as well as other severe spasms accompanied
by little motions which resembled twitchings. It seemeth to
me that there happeneth something of the kind which resembleth
that which is sometimes seen to take place in the stomach
during an attack of the ailment which is called "gurgling".
And I have also seen very many other patients in which this
hath happened, and in the mouths of whose stomachs the
pain hath increased. And I have also seen the falling of those
who fell sympathetically and not through suffering of the brain
itself, and the movement resembled the tottering motion which
taketh place during short lapses of sensibility, and not during
the ailments in which spasms make the body rigid. And from
this I have seen (i.e., concluded) that this movement which
taketh place in the brain also resembleth that which taketh
place in the stomach, and that it is caused by the chymes
which trouble it. There are times, too, when it becometh
heavy and is injured (or, bitten) through a superfluity of meats,
and the kind of "gurgling" which it maketh is in accordance
with the kind of food which is eaten. But I have also
seen on several occasions the rigidity caused by spasms
which taketh place in the whole body, and that when

this acid chyme is vomited all these ailments immediately
sink to rest.

Now it is not a thing to wonder at that the initial portion
of the nerves should produce such a movement as this, seeing
that it is anxious to expel from it that which cometh | to it Page 29
from that member wherein is the source of the ailment. For
in this wise all the other happenings | which move the nature Fol. 14*b*.
of the nerves in a trembling manner appear to take place,
that is to say, they produce therein a falling which can be
felt, besides a quaking and trembling movement, and they
come into existence through great cold. Of this kind also is
oblivion (*i. e.*, forgetfulness). Now that state of insensibility,
which is like death because it taketh place suddenly, indicateth
that some thick, or viscous, or cold chyme filleth and blocketh
up the cavities of the brain. And this doth not arise through
a change in the composition of the substance, as when happen
stupor, and dementia, and raving madness, and insanity, and
the destruction of the memory, and the blunting of the per-
ceptions also, and wrong action of the movements [of the
members of the body]. Concerning all the ailments which are
like these, whatever may be the state of insensibility, I compare the greatness of the danger to the greatness of the
nature of the respiration. For as respiration taketh place in
those that are asleep, who, although they do not possess any
other voluntary operation, yet are seen to be motionless in
their sleep, even so also is it in the other soporific ailments;
the body is insensible and is without motion, and the only
function that is maintained is the breathing, which is performed by the organs that work the chest. And we have
of all these things an understanding which is founded on the
law of proofs, just as we have also concerning the fact that
the beginning of motions taketh place in the organs of volition
through the nerves which come to them. For that knowledge
of the openings [mentioned] in the First Chapter proveth to
us that [the collection of] all the nerves is the brain. I did
not say of the head only, but I added that which is in front
of the brain, the brain of the spinal cord. For from this also
there appear to go forth many | nerves, but they serve Page 30

the brain, which sendeth [impulses] to the spinal cord. When, then, thou seest that the breathing is obstructed overmuch, and that it is only maintained with the greatest difficulty, in the case of ailments which produce stupor, understand that the brain is suffering no small pain.

Fol. 15 a.   ## Of the ailment which is called vertigo, or dizziness.

All the above-mentioned ailments take place in the head, and with them must also be mentioned the ailment called vertigo, for it is well known, and what manner of ailment it is is known from its appellation. Now those who suffer from vertigo become dizzy through small causes, and they often fall on the ground, more especially when they run round in a circle. And that dizziness which only cometh to others after going round and round several times, taketh place in them through going round once only. Moreover, if they only look upon something which is going round straightway do they fall down; and although it be only a wheel or something else of the same kind which revolveth, or if they look at the eddies or whirl-pools in the rivers, straightway do they become dizzy. Now this happeneth to them more especially when their head be-cometh hot through the sun or some other hot body. There-fore it seemeth that that which happeneth to ordinary people through going round and round in a circle excessively, happeneth to those who are liable to dizziness without their going round and round in a circle. It is well known to every man that there cometh to those who go round and round many times a disturbing, disordered, and irregular motion of the chymes and breath. Therefore there must happen to those also who have a tendency to the ailment of vertigo something of the same kind. And for this reason men are cured thereof by the cutting of those ligaments which are behind their ears; and it is well known that those tendons which are behind their ears are cut to their lowest depth (*i.e.*, to their roots), so that there may be a wound between the two sides of the cutting. Not all, however, are cured by this remedy.

This also is known: this ailment happeneth through the

other tendons which are greater than these and which | go up Page 31 [to the head] through that ligature which is called "like a net", because there ascendeth from them some spirit which is hot and vaporous, and it filleth the brain. It is possible that in the brain itself some irregular change of its composition taketh place, and produceth therein this kind of spirit; but it is well known also that this ailment is of the brain through the perception itself | of those who suffer from vertigo. Sometimes Fol. 15 b. this happeneth through the brain itself being affected, and sometimes through its sympathy with the mouth of the belly. And this is admitted by Archilogenes, who writeth in his book on "Protracted Ailments" thus: "Now this ailment of vertigo "is also caused by two things, by the brain and by the other "parts of the epigastric region." And he trieth to distinguish the vertigo which ariseth through the head from the vertigo which ariseth through the epigastric region, and he saith concerning vertigo that it is a sickness of the head. Now his words are: "Before it cometh on there are ringings in the ears, "and headache, and the feeling of heaviness, and the breath- "ing is affected, or some one of the other functions which take "place here." Now he addeth in his discourse [the word] "here", wishing, at least such is my opinion, to prove that that ailment taketh place through the head. And before an attack of the vertigo which taketh place through the mouth of the stomach, he saith that a pain in the heart and vomitings come on. But, as I have said above very many times, although sometimes the head may suffer in sympathy with the other members, still it is right to think that that suffering which taketh place in the head is due to the head itself.

On the violent pain which taketh place in the head.

Now as concerning that ailment which is accurately called by the physicians "headache", there is no man who will dispute the fact that it is an ailment of the head. For this ailment, to state the matter very briefly, is a sickness of | the Page 32 head of long standing which it is exceedingly difficult to dispel, which becometh intense and very severe, and which is also brought about by such small causes that the sufferer is unable to imagine what hath caused it. There is no [sound of] a hammer, and no loud noise, and no blindingly strong light, and no motion whatsoever; but [the headache cometh], and the patient wisheth to lie down in a quiet place and in the dark, by reason of the intensity of the pains. Some sufferers imagine that they are being struck by some hammer (or, club); and again some feel as if their head was being torn asunder or being crushed in pieces; and very many of them feel that Fol. 16a. the pain cometh down | to them to the roots of their eyes. And there are times when the pains are stilled for a short period, after which they return, just as it happeneth in the cases of those who have an attack of the falling sickness, and there is a middle period wherein they do not suffer pain.

Now it is well known that this ailment of vertigo ariseth from weakness of the head, just as do the other pains of the head, but vertigo cometh more readily than the other pains in the members of those whose heads are weak. Yet there is also in the pains of the head a certain change (or, variation). For the heads of very many men are easily filled, and, more-over, the whole constitution of their bodies is exceedingly ready to fill them. Others, however, possess also those mem-bers which are said to receive this ailment, and they are the membranes of the brain, and that other membrane which is beneath the skin, and these are weak, and they become affected easily. When men who have constitutions like these are attacked by severe pains they collapse quickly with this sickness of headache. And we do not think that we shall go beyond the truth if we say that some of them are made ill by those membranes which surround (or, envelop) the brain, and others by that membrane which is beneath the skin.

Now the distinguishing characteristic of these kinds of vertigo is whether the pains go down into the roots of the eyes or not. Page 33 For it is absolutely essential | that this pain should come to the roots of the eyes of those who have the ailment located between the two bones of the skull, because the ends of the nerves come to them from the brain, and also from its two membranes there come to it the arteries and veins which are in the membranes.

Of the pain of one half of the head.

Besides these pains there is the pain which affecteth men in one half of their heads. In some this feeling of pain ariseth outside the bone of the skull, and in some cases they feel it right down to the base of the head. It is felt distinctly in one of its two sides, and either to the right or left of that cord which is stretched along its length, and is below it, and inside the Fol. 16*b*. bones of the skull. And that line which divideth the brain in the middle being subtended coincideth with it, and formeth the boundary of the two cavities of the brain which are in front. Now the substances (or, bodies) which fill the head easily are those which have in their compositions an abundance of hot and vaporous wind, or a superfluity of the bile which is collected in the mouth of the belly. And those pains which arise through this wind are protracted, and to those to which the feeling of duration is coupled this description is also given. Now those pains which arise from a great superfluity [of bile] possess a feeling of oppression. When they exist with a certain amount of redness of colour, they arise from a superfluity of hot chymes. To another man it happeneth that he hath headache continually after drinking wine, whether it be much or whether it be little; this happeneth more especially if the natural disposition of the sufferer be hot (or, excitable), and it is also caused by hot (or, pungent) scents, such as styrax (στύραξ), and *ḳôpîôn*,[1] and all kinds of substances which when burnt produce pungent odours. Some men cannot bear even the smell of frankincense. In other men headache is caused Page 34 by some intense feeling, as, for example, that which ariseth in the mouth of the belly. Many men possess such a sensitive stomach that they cannot bear any sharp pressure or touch on it, or any thing else of a similar kind; on the other hand, there are some who have no sensitiveness at all in their stomachs. Again, there are others who are seen to eject or to vomit whatsoever is very nauseous; this they do if they merely smell what is nauseous, for they are not able to bear its smell, al-

---

[1] A word of doubtful meaning; see PAYNE SMITH, *Thes.*, col. 3553.

though they do not perceive things which pain them when they are present in a large quantity.

Now it is possible that changes such as these take place in the brains of men, according to the odours themselves. Some odours they can smell without feeling any discomfort, just as if they had not been brought near them, whilst others they cannot bear to smell at all. Thus it appeareth quite clearly Fol. 17 a. that all ailments of this kind belong to the head. | It may be said that in my opinion all the sicknesses which take place in the head, whether through the head itself or whether through its sympathy [with other members of the body], are accompanied by exact indications. Therefore we may gather together all the sicknesses of the head, and the causes which produce them, into one small group.

Now headaches can arise from external causes, as, for example, through the heat of the sun, or through cold, or through the heaviness of the atmosphere. And they can also arise from blows [on the head], or from a fall, or from the drinking of wine, whether in large or small quantities. These are the external causes which produce headaches, and they are called "primary", or "predisposing causes". And the head can also suffer through internal causes, which are called "initial causes", either through some ailment in itself, or through sympathy with the body generally, or with one of its members. When the head suffereth through some ailment in itself there arise in it Page 35 pains | of three kinds, *viz.*, those which are due to substances of like composition to itself, those which are organic, and those which are due to the solution of its integrity. If it be attacked by any one of the class which is caused by a substance of like composition to itself, it is attacked by one of those eight noxious constituents which have been mentioned above. Some of these have substance in them, and some are without. Now this substance appertaineth either to the chymes or to wind. Moreover, there are times when these change the matter of the brain into that kind of matter of which they themselves are composed, and there are others when they subsist in the vacant spaces in the head. When the head is attacked by some organic disease the pains are caused either by tumours or by stoppages, and these are caused either by thick and viscous chymes, or by chymes which are hot and fluent, or by chymes which are cold. Those chymes which block up the exits [of the head] are difficult to describe. The solution of the integrity of the head is caused by boils, and by wounds, and by the smashing and cleavage [of the bones].

And certain sicknesses arise in the head through [its] sympathy [with the body] generally or with one of its members. And sometimes, through the fullness of the whole body, the chymes or wind go up and do it an injury. There are occasions, too, when it is injured through | the chymes, that is to say, Fol. 17b. there is either thick and impure blood imprisoned in the veins, or the black, or the red bile, or phlegm oppresseth it. And the wind is either the vapours of the hot, moist chymes, or the fumes of the hot and dry chymes, or the dense emanations which ascend from the body generally or from one of its members. And the head also suffereth, with the stomach, from the fume of the bile, or from the phlegm which is imprisoned within it, when it ascendeth to the head and filleth it with vapour. Moreover, the head suffereth pain through places in the epigastric region, or some other member, and also, as hath already been said, through dementia, and through the falling sickness, and through vertigo. These are, very briefly, all the causes | which produce headaches. These, then, shall Page 36 be to us as it were a canon throughout this treatise, both as regardeth the head and all the members which suffer pain.

On the natural formation of the head and on the indications of the natural compositions of the brain.

Now therefore it is meet that we should, before everything else, write concerning the cure of headaches. And we must speak a little concerning the natural compositions (or, constituents) of the head, for this also, necessarily, is required of us. A small head is an indication that the construction of the brain is bad, though it doth not follow necessarily that that of the brain in a large head is very good; but if, through the firmness of the power of the womb which hath found material in abundance, the head is small, it is an indication of bad [construction]. Now it is right to distinguish all the various kinds of heads. If they are properly formed they indicate moral excellence, and also, if they are round, and their veins are firm, and their apertures delicate, and their eyes are steady and sparkling and sharp, these are signs of moral excellence. But when heads bend forwards, and have loose veins, or when they are immoderately | long, these characteristics are, in the majority of cases, signs of wickedness. Therefore as we attribute large heads to a bad brain even so do we attribute the above characteristics to the [bad] brain. And just as moral excellence is sometimes associated with a large head, so also a few men with such physical characteristics of the head possess moral excellence, according as whether the formative power of the womb was potent or not. The brain which is well constituted, and moderate in its spiritual movement, and is not over strenuous in its action is not quickly injured by the things which happen externally. Folk who possess these characteristics have reddish hair in their babyhood, and it remaineth red during their childhood; in their early manhood it becometh very red indeed, being neither crisp nor lank, and they do not easily become bald.

Fol. 18*a*.

Page 37 **On the simple compositions (or, constituents) of the brain.**

Now if the compositions of the brain become immoderately hot, the head becometh red and hot, and the arteries which are in the eyes become large (*i. e.,* swelled). The hair of men with such a brain is abundant in their early childhood, and black and strong and crisp, according to the measure of heat [which is in the brain]. There are cases in which it is first red and afterwards black; later, when the body hath reached its full stature, [the head] becometh bald; and the roof of the mouth and the nostrils are clean (*i. e.,* free from hair?). Now the head of such a man is quickly filled with foods and drinks, and with odours (or, scents) which inflame, and with things which take place externally; and his sleep is little and not deep. As for the brain which is cold, the superfluity thereof cometh down on to the palate and through the nostrils; his hair is lank and red, and he doth not quickly become bald. He is speedily injured by a cold wind, and he is attacked easily by sicknesses which are due to cold, and by coughs, and his cheeks feel cold to the touch, and the arteries of his eyes are deep and sluggish [in action]. Now the indications of the brain which is dry are: a lack of superfluous fluid from the brain on the palate and in the nostrils, and a readiness to keep awake, and hair which is strongly fixed in the head and which groweth but slightly. The hair of such a man is crisp in early childhood, but he becometh bald quickly.

The brain which is moist in its | composition produceth hair Fol. 18*b*. which is long and abundant, and the head in which it is never becometh bald. And it maketh the eyes to be dark, and causeth the mucus to be abundant, and produceth excessive and deep sleep.

On the compound constituents of the brain.

Now if the brain be hot and dry it doth not produce mucus, and it endoweth the senses with the power to become exact and certain; and the head in which is such a brain becometh bald easily, and the first growth of hair is scanty, and the hair itself is glossy and black and crisp. And the face feeleth hot to the touch, and the cheeks are red until the time of early manhood. If both heat and moisture are predominant in the constituents of the brain, though they exceed the Page 38 normal but to a slight degree, the body becometh hot and red, and the arteries of the eyes grow large, and the mucus is abundant though somewhat dried up, and the hair is lank and red. Heads with brains of this kind do not become bald easily, but they are quickly filled and become heavy through the things which inflame. And when the composition of the brain is over-heated and over-moist, it causeth the head to ache, and the moisture in it becometh abundant, and it is easily affected by the things which make it hot and moist, and its senses are not vigilant for any length of time. And when the composition of the brain is cold and dry, it causeth the

3*

head to become cold and without colour, and the arteries which are in the eyes become contracted, and it is easily injured by any accidental cause. And for this reason men with brains of this kind are easily cured. In their early manhood their senses are not acute, and with advancing years they shrivel up quickly, in fact, to speak briefly, they age very quickly, and become white-haired. In their childhood their hair scarcely groweth at all, and what doth grow is fine and red. A brain which is cold and moist in its composition filleth the head, and predisposeth to sleep, and maketh feeble the senses (or, feelings), and produceth mucus, and noises, and wheezings, and throweth a man into fits of coughing and shivering. Now the head wherein is such a brain never becometh bald.

Fol. 19a.   ## On the hot pains which arise in the head.

Now there are also hot pains which come into the head in the season of summer, and in a hot country, and [especially] into [the heads of] young men who are ardent in temperament, and of those who live in a squalid manner, and of those who labour overmuch. And in their constitutions red bile and blood predominate, and [one] of them, or both, when gathered together [will cause headaches]. The head suffereth pain through some small cause, as, for example, [the heat of] the sun or dryness [of the air], especially when through its own weakness, or that of its original constituents, it is easily suscep-
Page 39 tible of suffering. Now there are attached to the head | certain symptoms(?) which are indicative of its suffering, [as, for example,] the feeling of excessive heat. And there are times when there accompany it also fever, and thirst, and inflammatory pains, and urine, fiery-hot, which is passed in very small quantities, and a terrible throbbing in the arteries. Now the primary cause of the pain in the head must be learned from the patients themselves.

Of the cure of headaches.

Now the cure of headaches may be effected by the following things: An airy house with a northern aspect, in which water is sprinkled about frequently; and branches and leaves of trees which give refreshment and coolness are to be dipped in water and then laid upon the beds of the sick and all about the house. And cool water is to be poured over the heads of the sick, and tinctures of vinegar and oil of roses applied. And some refreshing juice must be administered, such as the juice of grapes, or the juice of the "shepherd's staff",[1] or the juice of river-weed,[2] or the juice of unripe pomegranates, or the juice of purslane,[3] or the juice of water lentils, or the juice of chicory (or, endive), or the juice of mallows (μαλάχη), or the juice of flax seed mixed with oil of roses, and laid (i.e., rubbed) all over the head or upon the temples, or mixed with one of the other juices. Or again, the whole head and temples may be covered with a plaster made of vinegar and oil of roses, or oil of violets, and a little fine wheaten flour. And we bathe the nostrils with oil of violets mixed with the milk of a woman. And we administer to the patient food which hath little stimulating power and is easily digested, | such as Fol. 19b. the soup made of orach or spinach, mixed with the sweet oil of sesame seed, and tender vegetables and endive, and cold water to drink. And there are times when I give oxymel (ὀξύμελι, i.e., vinegar mixed with honey). Let us remember this example when we have to come to a decision as to the cause which produceth the sickness, and to alleviate the results thereof.

Of the pains in the head which are caused by cold.

Now pains caused by the cold arise in the head in the season of winter and in a cold country, more especially in the heads of old men, and of those who are cold constitutionally.

---

[1] Arab. بطباط, Gr. πολύγονον ἄρρεν. See PAYNE SMITH, col. 1250.

[2] Gr. ποταμογείτων, or pond-weed.

[3] ܦܲܪܦܚܝܼܢܐ. Chald. פַּרְפְּחִינָא, Arab. فَرْفَخ, Pers. پُرْپَهَن, Gr. ἀνδράχνη, portulaca.

Page 40 In these the phlegm is predominant, and they live | useless lives, and they eat habitually cold foods, which give birth to phlegm, and there are produced in them chymes which live when all these are gathered together in one place, and especially when the head is cold and weak. In such cases the head is easily attacked by the slightest cold or chill. Now to all these pains are attached results (or, symptoms) by which we are specially able to recognize them, and these are as follows:— Men who have pains in the head feel the cold very severely; they are never attacked by thirst, and their sleep is mingled with heavy stupor and accompanied by fantastic dreams, and they crave for warm air, and their sense of touch (or, feeling) is defective, their urine is pale and thick, and the beating in their arteries is feeble. Besides these there is the investigation concerning the predisposing cause.

Of the cure of pains in the head which are caused by cold.

Now the means to be used in curing these pains are as follows:—A warm house, with an eastern aspect, and a good fire placed inside it, or baths, if they be available, and anointing with unguents which produce warmth, and the pouring over the head of water in which have been steeped chamonile flowers, or aniseed, or laurel leaves, or any other plant or shrub of this kind, and the laying on the head of oil of nard, or of narcissus flowers, or of aniseed, or of marjoram, and the administering of foods which produce heat, and of drinks of honey water. And if there be no fever let the patient drink wine mixed with hot water containing mastic or aniseed, and let him wear warm clothing, and there are times when an application of oil made from the euphorbium plant, or with the

Fol. 20a. balsam of the behen-nut, is beneficial. In cases where the pain is caused by dryness a wet treatment must be employed, and in cases where the pain is caused by damp a dry treatment must be employed. And in respect of the other systems [of colds], systems of treatment which are opposite in their nature to them must be employed. Thus the system which is hot and dry we cool and moisten, and the system which is cold and dry we warm and moisten, and the system which is hot and moist we cool and make dry, and the system which is cold and dry we warm and moisten. In all these cases we employ as remedies things which are of opposite natures to the diseases. | Page 41

On the headache which is caused by the drinking of wine.

Now for this kind of headache it is unnecessary for definitions to be laid down, for the drinkers of wine are found at all times, and in every place, and among people of all ages and conditions. The predisposing cause maketh known the pains, and the initial cause, which is within the body, showeth it, and its intensity is in proportion to the size of the flask of wine which hath been drunk. As concerning the predisposing cause we are able to obtain information from the sick man or from his kinsfolk. And we can find out concerning the initial cause [by enquiring] if his head is weak, and if it is susceptible to the fumes of wine, or if fullness of any kind collecteth in [his] head. When two of these things are associated they produce headache. And in respect of wine we can enquire whether it is strength (or, newness) which hath injured the patient, or its excessive quantity. Now there are effects which are associated with the headache caused by [the drinking of] wine, namely, drunkenness, and falling about, and lack of control of the mental faculties, and stupor, and sleep, and the complete helplessness and relaxation of the whole head, and great heaviness, and urine which is white and cloudy but is sometimes thick, and that full throbbing in the arteries which is called "rotatory".

Of the cure of the headache which is caused by drinking wine.

For this sickness quietness and sleep are necessary until the patient hath got rid of his drunkenness, and then it is proper to make him take a bath. He must be fed on foods which contain good juices, and are light and do not inflame [the head], such as soup made of barley flour, and dainty

Fol. 20*b.* broths | cooked with sweet chick-peas, and leeks, and anise, and on tender vegetables and endive. And on his head must be poured an infusion of chamomile flowers and oil of roses or violets. And thou must take very great care to clear out his whole body, and to wash out from his head the fumes of the wine, and to have him washed and anointed with oil, and to make arrangements for him to have quietness and sleep.

Page 42 **On the pain in the head caused by a blow or by a fall.**

Now it is meet for us to know how the man hath been struck on the head, and from what height the man hath fallen on his head; [either of these may] produce a serious illness. Sometimes both the head and the brain within it are very violently shaken, and then stupor and delirium supervene; this is a very difficult illness [to cure], for in the majority of cases it bringeth on death. And at other times the head is so seriously contused, together with the skin and the membranes which are beneath it, that very often the bones of the skull are broken, and they enter the brain and inflict such injuries upon it that delirium is the result. Sometimes also the skin of the head is lacerated, little or much, and it happeneth that blood floweth freely; the effects of an injury of this kind are burning fevers, and there are cases in which delirium or insensibility followeth, and also sores which fester and cause acute pains. Now at times the whole body becometh rigid, or convulsed with spasms, and the urine is bloody and thick, and there is a distracting throbbing in the arteries.

Of the cure of the pain caused by a blow on the head, or a fall.

Now if in addition to the primary cause of the pain there is another which is associated with the interior organs of the body itself, that is to say, if there be excessive fullness of the body, and the vigour of health, to such a degree that when

pain attacketh the head, the matter floweth thither and produceth a sore, it is right for us to begin our cure with the
letting of blood from the upper artery of the arm, and with
an effective clearing out of the body. If possible, the latter
must be done by means of a draught, and if this faileth then
a strong purge must be administered, so that it | may draw Fol. 21 a.
away the matter from the member which suffereth. At the
same time we must not neglect to be careful about the head,
and we must bathe it with warm wine and oil, provided that
there be no fever. If there be fever we must bathe it with
oil of roses [mixed] with vinegar. And if there be a sore (or,
wound), we must first of all apply to it fomentations and
plasters, whereby it may be kept clean | and in a healthy Page 43
state, and then with the medicaments used we may mix, little
by little, small quantities of the medicines which quiet and
alleviate the pains. And when the sore is in a quieter state
[we may use] medicines which have greater power to alleviate
pain. And if with the bruise on the head there is breaking
of the skin, it may be healed by means of warm wine and oil,
and by plasters in which are mixed myrrh, aloes, and frankincense. And if there be copious bleeding it may be stopped
by the application of powdered aloes. Take one part of aloes
and two parts of frankincense, and apply the mixture sometimes dry, and sometimes mixed with the white of an egg, or
with vinegar, or with wine. And if there be many cracks (or,
openings) in the skin of the head, or in the openings which
are beneath it, so many in fact that the bone of the head is
laid bare, then we must wash the wound with wine and oil
warmed, or we must bathe the wound with warm wine only,
and bind it up with soft sponges. And if the bone of the
head is seen to be broken, in such a way that it cannot unite,
we must remove the fractured piece, and pour hot oil of roses
on the place where it was, and then we must bring the lips
of the wound close together, and sew them with a thread of
silk, if such be available, and if not, with a thread made of
flax. And after the sewing up, moist plasters made of aloes
and frankincense and mixed with the white of an egg or wine
must be made and smeared on the bandages, which must be

laid on the wound after it hath been sewn up. And above the wound other bandages of many folds soaked in wine Fol. 21 *b*. and oil must be placed and fastened in position | quickly. And once daily these must be undone, and the wound dressed and bound up again until the sore is quieted and the [lips of the] wound unite. And we must take care about the whole body, and feed it with light foods which do not produce excessive over-fulness. And the patient must abstain from wine and flesh, and he must avoid the sexual embrace and other labours, Page 44 and we must secure for him | rest and sleep.

Of the headache which cometh in sympathy with the whole body.

Now when there is a superabundance of effusions in all the body, and it becometh soaked therewith, and fever burneth therein, and matter inclineth towards the head, headache is produced. Thou wilt be able to recognize this condition from the fullness of all the body, and the inflation of its veins, and from the redness of the face, and from the heavy movements of the eyelids, and from the cloudy and clotted urine, and from the terrible throbbing of the arteries.

Of the cure of the same.

It is meet that we should begin our cure by first of all letting blood from the vein which hath already been mentioned, or by cuppings on the neck, and by clearing out the whole body. Then we should bathe and wash the head with the medicines which relieve it of its vapours, and administer foods which are easily digested.

Of the headaches which arise through the stomach or through other parts of the abdomen.

Now the symptoms of these headaches have already been described above at great length, and it is therefore unnecessary for us to consider them again here overmuch, and we only mention them as a little reminder of what hath been said above. Now the great nerves descend from the head to the belly, and thus it happeneth frequently that when a man is struck on the head, dizziness and vomitings come to him. And if the bile and phlegm be gathered together in the stomach they produce severe pains | in the head, and dementia, and Fol. 22 a. raving, and vertigo; but if red bile be collected in the stomach it produceth gripings, and smoky eructations, and a bitter taste in the mouth, and thirst. If phlegm be collected in the stomach it produceth acid eructations, and an excess of saliva in the mouth, and vomitings in which crude phlegm is mingled, and the failure in the power of tasting whether a thing be acid or sweet. Now the effects which take place in the head as the result of its | sympathy with the stomach, or abdomen, or any Page 45 of the members of the lower part of the body, are well known. The regular throbbing of the head, which is not, however, continuous, and the fact that the patient is relieved by the digestion in the lower part of the body in old age, and the dullness which cometh over the eyes, and the bilious or watery and cloudy urine, and the terrible throbbing of the arteries will inform thee concerning the fullness of the stomach. Now it hath been said above that the sickness of dementia, and the falling sickness, and vertigo take place in the head when there is sickness in the stomach or in the other members of the body.

On the cure of the pains which arise in the head through the stomach, or through other parts of the abdomen.

These pains are cured when the chymes which are imprisoned in the stomach and which are the cause of the pain are purged away. Now they can be be purged away first by making the patient vomit and then by means of a purge administered through the anus. The means used in producing vomiting dependeth upon the age of the patient. Sometimes it is produced by radishes and honey, or by oxymel, and sometimes by means of some [other] mixed medicine. And after we have purged the stomach, we administer food which is easy to digest and doth not produce watery vapours. If bile be found to be in the stomach, let the physician administer to the patient an emetic of ἱερὰ πικρά[1] with honey-water, for this is the chief medicine used by physicians for this ailment. If there be

Fol. 22 b. phlegm in the stomach let the physician purge him with medicines which will bring down the phlegm, such as cakes in which aloes, or some other stomachic, have been mixed. And after his purging let the physician give him honey-water to drink, or the spice cummin, or three peppercorns. If there be red bile in his stomach after the vomiting caused by ἱερὰ πικρά, let the physician give him a decoction of absinthe (*artemisia Pontica*), and when the stomach hath been purged let him

Page 46 apply to the head fomentations and plasters which will subdue | the vapours in him, and which will strengthen him, so that he may not again receive other vapours. And we keep the sick man far from the foods which fill the head, such as leeks, onions, and garlic, and from wine in large quantities, and from dates, and from all those things which beget evil humours, and make vapours to ascend into the head. We first of all administer those foods which soften the belly, and afterwards those which are of an astringent character, and which make the mouth of the belly to contract so that watery vapour may not ascend to the head.

Of the ailments which are proper to the head.

It hath already been said very many times that all the ailments which arise in the head are of two distinct kinds, *i. e.*, they either arise in sympathy with the other members of the body, or they are due to suffering in the head itself. As regardeth those which take place sympathetically, we have already described them so far as it was possible, and also the remedies which are suitable for them; but there is still lacking some mention of the means to be employed in curing those pains which arise in the head through its own proper ailments. These are the pains: some of them are acute, [and last but a short time], and others are protracted. The acute pains are

---

[1] This is some kind of bitter herb.

those which arise through the weakness of the head, and they produce a strong and severe throbbing, sometimes in the whole head, sometimes in one half of it, and sometimes in a certain portion of it. Now these usually come on in the time of fever, but sometimes when there is no fever; they are caused sometimes through the fullness, and sometimes through the emptiness, of the body. Now the protracted ailments (or, pains) are dementia, and the falling sickness, and vertigo; | and Fol. 23a. all their symptoms have already been described clearly in many paragraphs.

Of the cures of the pains proper to the head.

Now the cure for these pains dependeth upon the kind of ailment and what it teacheth us, and upon the natural constitution of him that is to be healed, and whether the patient suffereth from fullness or from emptiness, from heat or from Page 47 cold, from moisture or from dryness; and we must employ in each case the means of cure which the law of healing demandeth. In the case of fullness we empty the head, and in the case of emptiness we fill it, in the case of cold we warm it, in the case of heat we cool it, in the case of dryness we make it moist, and in the case of moisture we dry it. Thus also do we in other cases. If the bile be predominant in the body we first of all let blood from the upper vein; if other chymes are seen to be predominant in the blood we eject them by means of the purgative which is employed in clearing out from the body that chyme which is present in it in excess. Now we are able to identify the chyme which is present in the body in excess from the constitution of the sick man, and from his physical stature, and from his manner of life, and from the season of the year. If it appeareth to us that phlegm is present in excess we administer the medicine which cleareth it away, and if red bile be present in excess we administer the medicine which cleareth away the bile, and similarly if black bile be present in excess we administer the medicine which cleareth this away. And when we have emptied the whole body, and the pain in the head still lurketh there, then we make use of tinctures and plasters on the head, and injections in the nostrils, and lotions, beginning with those which reduce the mental operations of the minds of the patients (?), and then we come to those which are stronger and which are employed in protracted and difficult illnesses, and to cutting off the hair, and to painting the head with tinctures of *aparḳôn*, and thapsia (θαψία), and natron (?), and mustard, and a tincture of wild barley, and cuppings and incisions in the flesh, and drawing [the flesh] together by means of clysters, and poundings (*i. e.*, massage) of the hands and

Fol. 23*b.* feet, | and the cutting of the veins and arteries. Now it is well known that we cut the arteries which are in the temples when there are much rheum and heat and swelling in the head, or in the eyes. And we cut the arteries which are behind the ears when there are vertigo and

Page 48 those protracted ailments in the head | which produce heat and swelling. And there are occasions when we cauterize them in order that there may be a wound in the midst and that cut portions may not unite again. And we also cut the veins for headache as near that portion of the head which is affected as possible. And more especially do we cut that vein which runneth down straight between the eyes, when the portions of the head which are near (or, towards) the forehead are affected. We also cut those veins that go up by the side of the ears, keeping in mind the fact that to relieve the pain which ariseth in the inside of the bones (?) of the head the cutting of the veins and arteries is superfluous. These things are, however, beneficial in the case where the pain existeth above the bones of the head.

Now the art of curing all the pains (or, ailments) which occur in the head hath been very fully described, and it is necessary that we should also set down in this Chapter all the medicines, both simple and compound, which have been tested by use for a very long time and which experience testifieth that they are exceedingly beneficial in all the protracted and obstinate ailments which occur in the head. First of all we write down the draughts which drive out all bad ailments from the body.

Great Hiera-lĕghudhâyâ. This expelleth from the body all the manifold forms of ailments which arise from the chymes without producing in it weakness, and it cureth all the sicknesses of the head, namely, disease in the hemicranium, and idiocy, and dementia, and stupidity, and raving, and vertigo, and deafness, and the falling sickness, and asthma of long standing, and pains in the kidneys and in the sciatic nerve, and pains in the tendons, and gout, and those whose limbs shake and are palsied, and elephantiasis, and leprosy, and scabies, and tumours, and running sores, and pig-sores, and cancers, and all the sicknesses | which are begotten of black bile, or of crude Fol. 24a. phlegm which is not distributed. And it re-establisheth the constitution which is impaired, and it healeth pains in the eyes and pains in the ears, and it bringeth on the menstrual flow, and it cureth protracted fevers, and fevers which come on for | a day, and those which come on every third day.   Page 49

Pith of colocynth	5	drachms
Roasted sea-onion	$2^1/_2$	„
Agarikon fungus	$2^1/_2$	„
Skamônia (convolvulus)	$2^1/_2$	„
Black hellebore	$2^1/_2$	„
Ammoniac	$2^1/_2$	„
Flowers of thyme	3	„
Bdellium	3	„
Chamadrâôs (chamaedrys)	3	„
Aloes	3	„
Thyme	2	„
Malabathrum (betelnut?)	2	„
Haprikôn	2	„
Parsîôn (horehound)	2	„
Teucrium polium	2	„
Cassia	2	„
Peppers, of the three kinds[1]	2	„
Crocus	2	„
Cinnamon	2	„
Jackal's fat	2	„

---

[1] *I. e.*, black pepper, white pepper and "long pepper".

Polypodium	2	drachms
Sagapenum (fennel)	2	„
Betonica	2	„
Myrrh	2	„
Petroselinum	2	„
Aristolochia makra	2	„
Juice of the artemisia Pontica	2	„
Euphorbium	2	„
Bearded grain	2	„
Amomum gingiber	2	„
Khĕmâmâ (ἄμωμον) balsam	2	„
Strychnus	$1^1/_2$	„
Gentian	$1^1/_2$	„

Honey, as much as is necessary.

Take in a draught of three drachms of warm water and honey, or in an infusion of flowers of thyme.

Another Hiera of Galen. This is to be used in all sicknesses connected with the retention and excessive flow of the urine, and it is to be drunk mixed with celery water (or, parsley water), or wild spikenard (hazelwort), or pastinaca (*i. e.*, parsnip).

Pith of colocynth	4	drachms	
Chamaedrys	3	„	and 2 *denkê*
Roasted sea-onion	3	„	„
Agarikon fungus	3	„	„
Skamônia (convolvulus)	3	„	„
Black hellebore	3	„	„
Strychnus	3	„	„
Ammoniac	3	„	„
Hôpríḵôn	3	„	„
Flowers of thyme	$1^1/_2$	„	
Teucrium polium	$1^1/_2$	„	
Bdellium (mûḵlâ)	$1^1/_2$	„	
Chamaepitys azuga	$1^1/_2$	„	
Aloes	$1^1/_2$	„	
Cassia	$1^1/_2$	„	
Polypodium	$1^1/_2$	„	
Myrrh	5	*denkê*	

Peppers, three kinds of	5 *denḳê*
Cinnamon	5 ,,
Crocus	5 ,,
Opopanax	5 ,,
Fennel	5 ,,
Betonica	5 ,,
Castoreum	5 ,,
Round aristolochia	5 ,,
Gentian	5 ,,
Euphorbium	5 ,,
Petroselinum	4 ,,
Draught	2 drachms

Honey as much as is necessary.

To be used like the preceding prescription.

Another Hiera of Archigenes. This is to be used in all cases where the ailments are protracted, | and for exhaustion, and Fol. 24*b*. vertigo, and obstruction(?) of the humours of the eyes, and for all ailments which are produced by red bile or black bile, for phlegm, and for leprosy, and elephantiasis, and dementia, | and Page 50 the falling sickness, and all kinds of delirium, and scrofula, and cancers, and scabies, and sores on the kidneys, and nettle-rash(?), and hoarseness, and tightness in breathing, whatsoever may be the cause of it, and the bites of a mad dog, and the bites of vicious reptiles which eject deadly venom, and disease of the womb, and diseases of the loins and sciatic nerve.

Colocynth	12 drachms
Horehound	2 ounces
Strychnus	2 ,,
Black hellebore	2 ,,
Saḳmônârîn (convolvulus)	2 ,,
Pepper, white	2 ,,
Pepper, long	2 ,,
Roasted sea-onion (squills)	1 ounce
Euphorbium	1 ,,
Aloes	1 ,,
Crocus	1 ,,
Gentian	1 ,,
Petroselinum	1 ,,

4

Ammoniac	1 ounce
Opopanax	1 „
Teucrium polium	2 drachms
Cinnamon, or Cassia	Double quantity of all the [other] drugs.
Fennel	2 drachms
Myrrh	2 „
Bearded grain	2 „
Flowers of the pistacia lentiscus tree	2 „
Wild marjoram	2 „
Aristolochia makra	2 „

Honey as much as is necessary.

Take four drachms in a draught of water in which thyme hath been infused.

Another simple Hiera, which is called "Pikra" (ἱερὰ πικρὰ) and is to be used for the stomach, and for all the ailments of the head which occur sympathetically.

Mastic	6 drachms
Crocus	6 „
Bearded grain	6 „
Cassia	6 „
Wild spikenard (hazelwort)	6 „
Berries of the balsam tree	6 „
Cinnamon	6 „
Aloes	12 „

Pound and administer in a dry powder. Take four drachms in a draught with honey and hot water.

Another Hiera which is named after Theodoretus [administered] with the nuts (or, fruit) of incense. It is to be used in all sicknesses which are protracted, and for the clouding over of the eyesight, and for dementia, and delirium, and the falling sickness, and vertigo, and elephantiasis, and leprosy, and all the sicknesses which are the products of black bile, and obstinate fevers, and disease of the liver, and disease of the spleen and the kidneys and the colon, and gout, and disease of the excrementary organs, and the purification of women, and the tightening up of the womb. And it looseneth the bowels and Fol. 25a. produceth | painless stools.

Aloes	60	drachms
Agarikon fungus	24	,,
Crocus	6	,,
Rhubarb	3	,,
Wild spikenard(?)	4	,,
Iris	4	,,
Bitter colocynth	4	,,
Wood-oil	4	,,
Fruit of balsam trees	4	,,
Ḳûshtâ (κόστος)	8	,,
Eḳrôn (ἄκορος)	6	,,
Mastic	6	,,
Wild thyme	6	,,
Cinnamon	6	,,
Cassia (horse's tail)	12	,,
Flowers of thyme	8	,,
Bearded grain	6	,,
Chamaedrys	8	,,
Meum	2	,,
Peppers, black, white and long,	4	,,
Euphorbium	4	,,
Flowers of the pistacia len-tiscus tree	2	,,
Gentian	3	,,
Khemâmâ (ἄμωμον) balsam	2	,,
Saḳmônârîn (scamony)	14	,,
Skimmed (*i. e.*, run) honey	Double quantity of all the other drugs.	

Page 51

Take four drachms in a draught of water wherein flowers of thyme have been infused.

A Prescription which was tried (*i. e.*, used) by Theodoretus, and is to be used in the cases described in the preceding [paragraph].

Aloes	60	drachms
Agarikon fungus	16	,,
Crocus	6	,,
Aḳrôn (ἄκορος)	6	,,
Mastic	6	,,

4*

Cinnamon	6 drachms
Bearded grain	6 ,,
Rhubarb	4 ,,
Oil of balsam	4 ,,
Balsam berries	4 ,,
Euphorbium	4 ,,
Pepper, black, white and long,	4 ,,
Gentian	4 ,,
Chamaedrys	8 ,,
Ḳûshtâ (κόστος)	12 ,,
Aprîmôn	12 ,,
Meum	2 ,,
Flowers of the pistacia	2 ,,
Ἄμωμον balsam	2 ,,
Saḳmônârîn (scamony)	. . . . . ,,

To be used like the preceding prescription.

An infusion of Apṭar fungus which is to be drunk after the above-mentioned medicines, or by itself, and which is given in all the sicknesses which arise from black bile.

Flowers of thyme	12 drachms
Myrobalsamus chebula	12 ,,
Herbs with the outer layers removed	12 ,,
Large dried grapes from which the stones have been removed	12 ,,
Agarikon fungus	11 *liṭrê*

Shake all these up in water until only one *liṭrâ* remaineth, and then purify and strain the water by itself, or infuse in it one of the medicines mentioned above.

The pills of Galen which are called "Ḳûbâyé", and which expel all the noxious chymes from the head and make the eyes bright. | Take equal quantities of aloes, skamônia, colocynth and absinthe, or a tincture thereof, and mastic, pound them up and macerate them | in water of fox-grapes, and then make of the mixture boluses as large as grapes, and administer from seven to ten of them in a draught both before and after meals, and they will expel the chymes from the head with great efficiency.

Fol. 25 *b*.

Page 52

Purgatives of aloes and mastic which are used for head-ache and which cleanse the stomach.

Aloes	3 parts
Mastic	1 part

Pound up and mix with the juice of cabbage (κράμβη), and make boluses and administer one drachm in a draught.

Stomach pills, which are used for headache, and the stomach, and fever, and pains in the back, and gout, and ailments of the excrementary organs, and which expel black bile and phlegm.

Flowers of thyme	16	drachms
Colocynth	16	,,
Agarikon fungus	10	,,
Aloes	48	,,
Bearded grain	4	,,
Ḳôshtâ (Costus)	4	,,
Fruit of incense plants	4	,,
Flowers of the pistacia	4	,,
Crocus	4	,,
Cassia	6	,,
and some add Scamony	4	,,

Pound up and mix with the juice of fox-grapes, and make into pills, and administer one drachm in hot water.

Another stomachic.

Aloes	10	drachms
Colocynth	5	,,
Saḳmônârîn (scamony)	5	,,
Myrobalsamus chebula	7	,,
Herbs	7	,,
Convolvulus turpethum, fine, white,	8	,,
Parsley seed	2	,,
Mastic	2	,,
Absinthe	2	,,
Agrimonia eupatorium	2	,,

Make pills of these mixed with the juice of cabbage, or night-shade, and administer two drachms in a draught before and after a meal.

The above medicines are used for all the protracted sick-nesses which are begotten of the thick and viscous chyme, and

especially for those which are sympathetic and produce those
pains in the head which have been mentioned above.

**Of the medicines which are injected into the nostrils, and
are used for obstinate (or, protracted) pains in the head,**
Fol. 26 a. **and for dementia, and delirium, and the falling sickness, |**
Page 53 **and for vertigo, | and for all the pains in the head which
are due to the thick and viscous chymes, and to black bile
and phlegm.**

A medicine which is injected into the nostrils, and is used
for pains in the head of long standing, and is administered to
those who need it because of cysts in the eyes, and aberration
of the mind. It removeth obstructions, and expelleth excessive
moisture.

Nigella	8 drachms
Salt	1 drachm
Ammoniac	1 „
Juice of the wild cucumber	1 „

Pound finely and mix with oil of lilies, or oil of pistacia terebinthus, or oil of bitter almonds, and place in a glass vessel.
When this is to be taken sprinkle some of the powder in the
nostrils, and tell the patient to draw in the breath, or mix the
powder with oil of bitter almonds and inject into the nostrils.

Or [take] the juice of beet, the purest possible, pound it and
press it and inject it into the nostrils. It will open the closed
passages, and will cure an obstinate headache. Or take the
root of the medicago sativa and treat it and administer it in
a similar way. Or take the plant ears of mice (μυὸς ὦτα,
myosotis), and treat it and administer it in a similar way. Or
take viscous elecampane (or, fleabane), and treat it and administer it in a similar way. Or dissolve the finest myrrh in
honey and hot water and inject.

Another prescription of Ptolemy, which is used for those
who suffer from headaches of long standing, and clouding
over of the eyes.

White hellebore	4 drachms
Ḳîndôsh	2 „

Red poppy	2 drachms
Castoreum (beaver secretion)	2 ,,

Pound, and pour or inject into the nostrils. It will cause the patient to sneeze violently and will clear the head.

Another medicine which is used in sicknesses of long standing caused by delirium and the falling sickness. Take nigella, and aloes, and poppy, and weigh equal quantities in a pair of scales, and macerate them in old oil, and inject into the nostrils.

Another. Take myrrh, four drachms, and fifty peppers, pound them' in oil of celandine, and inject into the nostrils.

Another medicine which is used for the protracted sicknesses of those who suffer from the king's evil. Now thou must not be afraid if it causeth much inflammation, for thou shalt feel relief in a very short time.

Mountain grapes	1 drachm
Nigella	2 drachms

Pound in vinegar to a thick consistency, | and inject into the patient's nostrils as many days as he is able to bear it, until Page 54 Fol. 26*b*. the scabs be cleared away from his eyes. Or, take some juice of the wild cucumber, equal in amount to a chick-pea, and pound it in the milk of women, and when the peeling process beginneth, let the patient wash his face in hot water, and the king's evil will be cured, and his headache greatly relieved.

Another to be used when there is prolonged and severe throbbing in the head. Take in equal quantities

Castoreum (beaver secretion)

Opopanax

Crocus

Wolf's gall

Rub down, and pour into juice of beet, or fleabane. This is also used for pain in the hemicranium (or, for hemiplegia).

Another. Take nigella and poppy in equal quantities, and rub them down in water, or in one of the juices which have been mentioned, and inject it into the nostrils, and the mucus will be well cleared out.

Another which is used for hemiplegia, palsy and ague.

Juice of cucumbers	1 spoonful
Beetroot	1 „
Root of medicago sativa	1 „
Nigella	2 drachms
Rue seed	2 „

Pound these and rub them down with the juices and inject into the nostrils. It will open the closed passages and eject the thick and cold chymes.

[The above medicines] are used for those pains in the head which arise through the thick and viscous chymes, and for the stoppages which are caused by winds. Now among all the galls of birds and animals, some are thin and some are thick, according to the nature of the creatures from which they are taken. The galls of birds are hotter and thinner than those of animals, and the galls of animals of the mountain are hotter than those of the animals of the plain (*i. e.,* domestic animals). And according as the chyme which thou wishest to clear out appeareth to thee to be thick or not, choose thou the gall which is suitable, then rub it down and dissolve it in one of the juices which are written down above, or with the milk of a woman, or with oil of violets, and inject. Or dissolve honey in hot water and inject it. Also dissolve a little new euphorbium in some oil which it seemeth to thee | may be used, and inject. |

Page 55
Fol. 27 a.

Another, which is to be used for headache of long standing.

Glycyrrhiza (licorice) pounded and washed	2 drachms
Goats' fat	5 measures
Sweet oil	1 measure

Heat over a slow fire until the fat riseth to the top and the oil remaineth, then, if necessary, put some of the oil into the nostrils of the patient.

Another, which is to be injected into the nostrils, and which is to be used for headache of long standing.

Opium	1 drachm
Castoreum	1 „
Fat of a jackal	1 „
Fennel	1 „

| Thistle juice | 1 drachm |
| Ķindôsh | like all [the others] |

Pound up, mix together, and blow into the nostrils, and it will relieve obstinate pains.

Another, which is to be used for headache of long standing.

Opopanax	1 drachm
Castoreum	1 „
Opium	1 „
Fennel	1 „
Musk	1 *denķâ*

Pound and mix and dissolve in the milk of a woman, and inject two drops. This is to be used for pain in the head and hemicranium, and is a certain [remedy].

Another for headache, and dementia, and the falling sickness.

| Fennel | 2 drachms |
| Natron | 2 „ |

Rub down and pound them well, and mix them up with the stale urine of a man, and smear the mixture on nut-fibre, or on lye-ashes, and dry it in the sun, and then shake it off the fibre, and pound it up, and add thereto two drachms of ķindôsh, and two drachms of castoreum, and mix them together, and use either as a powder which is snuffed up the nose, or as an injection when mixed with one or other of the juices which are used in such cases. And it will relieve the phlegm and vertigo, and those who suffer from rheum in the eyes.

Another, which is used for these pains.

Ķôshtâ (κόστος)	2 drachms
Root of pyrethrum	2 „
Peppers	1 drachm
Ruta sylvestris (wild mint)	1 „
Nigella	1 „

Pound and mix together, and blow into the nostrils.

Another, which is used for those who suffer from torpor in the head and delirium.

| Crocus | 2 drachms |
| Opopanax | 2 „ |

Fennel	2 drachms
Opium	1 drachm

Pound each of these drugs separately, and then mix them together, and work them up with the extract of "ears of mice" (myosotis), or the extract of | fleabane, and make them into pills the size of peppercorns. For use dissolve in some liquid which appeareth to thee to be suitable, and add a small quantity of oil of cows, and inject into the nostrils.

Page 56
Fol. 27b.

Another, which is to be used for headache of long standing, and for clouding over of the eyes, and for curing the effusion of water in the eyes.

Musk	$2^{1}/_{6}$ drachms
Camphor	$2^{1}/_{6}$,,
Castoreum	1 drachm

Pound and rub down and blow into the nostrils, or rub down with oil of jessamine and the milk of a woman, or wine, and a little spirit, and inject into the nostrils.

Another, which is used for hemiplegia, and ague and obstinate pains in the head.

Fat of a jackal	2 drachms
Fennel	2 ,,
Pure mômîâ (bitumen)	2 ,,
Ruta sylvestris (wild mint)	2 ,,
Fat made from horns	2 ,,
Peppers, round, and long,	2 ,,
Gum ammoniac	2 ,,
Castoreum	2 ,,
Euphorbium	2 ,,
Urine of a camel	4 ,,

Pound the dry ingredients, and then mix them with the urine, and put the mixture into a glass vessel, and add to it strong vinegar, and let it stand for seven or fourteen days. Pour into each nostril daily five drops.

Another, which hath been well-tried and is a sure remedy, and is used for headache of long standing.

Mûmîâ (bitumen)	1 drachm
Nut of the incense tree	1 ,,
Amber	1 ,,

Camphor	1 drachm
Musk	1 ,,

Powder together, and mix with oil of jessamine and of balsam, and inject into the nostrils. This is a sure remedy.

Another.

Castoreum	3 drachms
Myrrh	3 ,,
White hellebore	3 ,,
Ḳindôsh	3 ,,
Euphorbium	1 drachm

Pound, and mix together, and blow into the nostrils, and it will relieve the pain in the head. Now others add to this mixture Nigella, according to the first weight.

Pills which are beneficial for those who suffer from the falling sickness, and for those whose heads are in a dazed condition, and for those who suffer from the spirit which the Persians call "Badsîsâjân" (*i. e.*, convulsions).

Take of red sulphur a piece about the size of a lentil and dissolve it in oil of jessamine and the milk of a woman and inject. Or take of yellow sulphur which is pure and very sparkling a piece of the size of the bitter vetch, and rub it down in oil of jessamine, | or with old oil, | and inject. Or take the gall of a weazel and dissolve it in oil of jessamine, and inject. Or take oil of violets and inject, and it will be beneficial. Or take camel-brain, and dry it, and pound it and put in the nostrils of those who suffer from the falling sickness, with extract of beet, once a month. Or use marzangûsh (origanum majorana), or fleabane, thrice a month. And let the patient drink the rennet of a hare, a mithḳâl and half a drachm in quantity, and vinegar three times a month.

And observe, when a white she-ass droppeth a foal, there is found on the face of the foal a membrane which resembleth red flesh. Take this off and place it on a tree until it is dry, and then take it and keep it carefully. And when it is necessary take a piece of it about the size of a lentil, and pound it finely and inject it into the nostrils. And twist a

Page 57
Fol. 28a.

somewhat thick thong of this membrane, and hang it on the neck of a child or a man who suffereth from attacks of the falling sickness, and he shall be cured. Or dissolve fennel in honey and hot water and inject, or use opopanax in a similar way. Or take of yellow sulphur which sparkleth greatly a piece as large as a peppercorn, and rub it down in the fermented urine of a man, and inject it three times a month. Now all these medicines which have been described are beneficial in all the diseases which arise from the thick, viscous, and cold chymes, and from stoppages and excessive moisture. If, however, they be employed in the sicknesses which are hot and dry, that is to say, those which are of an opposite nature, they cause danger, especially if the sicknesses be accompanied by fever.

Of the medicines which are used for the headache which is due to heat and dryness, and for that which is due to dryness and emptiness.

Take oil of violets, and the milk of a woman, mix them together and inject, and the mixture will prove beneficial. Or Page 58 take oil | of the seeds of the bitter gourd, and mix with the white of an egg and inject. Or take oil of roses with extract Fol. 28b. of chicory, or by itself, and inject. Or | take oil of violets and extract of lettuces, mix them together and administer to those who are sleepless. These medicines are beneficial in the case of headache which ariseth from heat and dryness, and is accompanied by fever. [Besides these there may be mentioned] the sprinkling (or, bathing) of the head with infusions of matricaria chamomilla, oil of violets, origanum mara or wild marjoram, and pearl-barley, or barley-groats; and cooling (or, refreshing) cataplasms such as extract of purslane and oil of roses laid upon the crown of the head and on the cheeks; or extract of flowers of grapes used in a similar manner; or some refreshing tincture mixed with the flour of wheat; or guimauve, and oil of roses, or oil of violets, mixed together and applied to the whole head, and on the temples. [All these] are refreshing.

Of the other medicines which are used in cases of head-ache which is caused by cold and dryness.

Take fresh cows' oil, which has not been boiled, and heat and keep covered for from two to seven days, and sesame oil, and put them warmed into the nostrils three days. Or melt down oil made from the marrow of leg-bones of animals and pour into the nostrils. Or mix sweet sesame oil with the extract of glycyrrhiza and pour into the nostrils. Or take fresh goose-fat, or pig-fat, or the fat of hens, and melt it down without letting any salt come near it, and pour it into the nostrils for seven days. Let the fats be melted down in a double vessel so that the smoke may not be able to reach them. And also, let them make decoctions of the fat of goats, or of the *espadhbĕḳâ* (*i.e.*, the fatty parts) of the head and feet of cattle, and tinctures of substances which moisten and warm the head, and plasters made of unguents which warm, and moisten, and strengthen the head. | Page 59

Of the gargles which are to be used for protracted pains in the head, and which expel the exudations of sores in the throat, and lighten the head, and draw down the light phlegm.

Mix the root of pyrethrum, and mountain grapes and mastic together, and let the patient swallow the mixture; and he must open | his mouth and spit out the saliva which will Fol. 29a. fill it.

Or take in equal parts marjoram, long peppers and pome-granate seeds, pound them together and mix them well, and let the patient take the mixture and gargle his throat therewith.

Another gargle which will expel freely exudations of the brain.

Marjoram	2 sticks	
Long [peppers]	2 drachms	
Pyrethrum	2	„
Sumach	2	„

Ginger	1 drachm
Gelîlâthâ (peppers)	1 „
Natron	1 „
Wild grapes	1 „
Pomegranate seeds	3 measures.

Pound these and mix well with honey and administer.

And if thou art dealing with some ailment of the mouth mix the [above] with mulberries and let the patient gargle his throat therewith.

Another. Take marjoram, leaves of thyme, pomegranate seeds, and aloes in equal quantities, tie up in a cloth, and soak in wine for drinking or in old spirit for a whole night; in the morning squeeze out the juice carefully, and give it to the patient and let him gargle therewith.

Another.

Peppermint	4 drachms
Marjoram	4 „
Wild grapes	4 „
Mustard	3 „
Pyrethrum	3 „
Bark of camphor root	6 „
Stoned grapes, or figs, or grapes,	
with sweet new wine, or vinegar.	

Rub down, throw medicinal herbs on them, mix together, roll, and make into pills about the size of *bendâķê* (beads?), and dry in the shade. And when the case requireth it, let the patient swallow one at a time and roll it about in the throat, for it will bring down the thick rheum and relieve the head.

Or pound up mustard, and work it up with honey or vinegar, and let it be swallowed.

Or pound the bark of the root of the camphor tree, and work up the powder in honey, and let the patient roll it about in the throat. Now if the camphor-bark powder be given with vinegar it is also beneficial.

Or pound up ginger, and work it up with vinegar, or honey, and keep it in the throat.

Now the above gargles are beneficial not only for ailments Page 60 in the head, | but also for [any] ailment of the mouth, and for the running of the eyes and ears, and for the rheum which is transferred to the nerves and which maketh them to become unstrung. They are very beneficial when they are used as gargles.

Of the medicines which are made into an ointment and which are used for protracted ailments of the head, and are beneficial when smeared on the face and on the temples. | Fol. 29*b*.

An ointment which is to be used for a pain in the head, and which is wonderfully beneficial when it is smeared on the face and temples.

Crocus	16	drachms
Lîḳîṭrîn (electrum?)	10	,,
Alum of the furnace	13	,,
Green oil (omphacium)	3	,,
Copper-ore (χαλκῖτις)	3	,,
Myrrh	4	,,
Opium	4	,,
Gum Arabic	16	,,

Rub down into a powder and work up with fine wine, and make into cakes (or, balls), and dry in the shade and keep. When need demandeth dissolve in vinegar, or wine, and use as an ointment.

Another which is to be used for the [same] ailments.

Verdigris	3	drachms
Copper oxide	4	,,
Iron scrapings	4	,,
Arsenic	3	,,
Red earth	4	,,
Copper-ore	4	,,
Juice of terebinth	1	stick
Wax	45	drachms
Old oil	2	measures

Rub down the dry constituents, melt the moist ones, mix them and use [when required].

Another which is good for the hemicranium, and for the whole head, and for the ailment of the anus.

Thapsia	3 drachms
Euphorbium	4 „
Opium	4 „
Myrrh	2 „
Fat of the jackal	2 „

Rub these down together and work up with wine, and make into cakes (or, pills), and when occasion ariseth for use dissolve in vinegar and use as an ointment.

Another which is to be used for cold and protracted ailments of the head.

Wax	3 drachms
Euphorbium	1 drachm
Oil of nard, or of jessamine, or of balsam	6 drachms

Melt down the wax and let it go cold, and then powder the euphorbium and mix with it, and use as an ointment and bandage.

Page 61   **Another which is to be used for headache caused by |** **heat and excessive fever.** Take frankincense, myrrh, Persian gum (σαρκοκόλλα), and crocus, rub them down together and make into a powder, and work them up with the white of an egg, or with extract of lettuce, or with extract of coriander seed, or with extract of hyoscyamus, and smear on the temples.

Fol. 30a.   **Another which is to be used | for excessive heat of the head.** Take two measures of white sandal oil, and one measure of Persian gum, and rub down and work up into a paste with the white of an egg, and use as an ointment.

Another which is to be used for violent inflammation.

Myrrh	1 drachm
Opium	1 „
Red worms which are found in clay soil	2 drachms
Flour of wheat	2 „

Rub down with strong vinegar and use as an ointment.

Litharge (protoxide of lead), large and washed,	1 measure
A little cold water	—

The yolk of an egg   —
Oil of roses   —

Work into a thick paste and smear over the head.

Another, which hath been well tested, and which is to be used for him that hath the sensation of excessive heat. Take aloes, mastic, celandine, and crocus root in equal quantities, and rub them down to a powder, and work up with vinegar, and use as an ointment.

Another, which is good for the head from which hot rheum floweth down.

Aloes	2 drachms
Persian gum	2 ,,
Frankincense	2 ,,
Crocus	1 drachm
Myrrh	1 ,,
Opium	1 drachm
Nut galls	4 drachms
Bôlasṭîôn (βαλαύστιον)	4 ,,

Rub down with vinegar and a very little [white of] egg, and spread on a piece of cloth or on a bandage(?), and lay it on the places which are affected and it will do them good.

Since therefore we have described all the healing medicines which are good for all kinds of headaches, these [remarks] must suffice.

HERE ENDETH THE THIRD CHAPTER.

5

CHAPTER IV.

Of the injuries which happen to the organs of breathing, and of all the ailments which affect the nostrils and the symptoms thereof, and of their cure, and of the medicines which are good for them.

NOW a demonstration hath been given in the preceding Chapter concerning all the rational powers of the soul which are produced by the cavities of the brain, and concerning all Page 62 the symptoms of the | injuries to which they are liable, and concerning the art of curing them, and concerning the simple and compound medicines which are suitable for them. And Fol. 30*b*. it hath been described | therein how the brain is the source and centre of all the senses (or, feelings), and how the faculty of feeling is derived therefrom and conducted to all the members which possess the sense of feeling. Now the faculty of breathing is performed by the front cavities of the brain and by the nostrils. The nostrils serve two purposes, for they are the organ of the faculty of smell, and of breathing and respiration. When the faculty of breathing is impaired, or when the front chambers of the brain are obstructed by any noxious substance, or by a boil, or by a fleshy growth, it closeth the nostrils, or the interior openings [in the nose] which are like cisterns. Or there may be in them that sore which is called a "foetid nexus", or that pustule which on account of its form is called "polypodium", [a word] which [means] "many feet", or there may be a flow of blood which produceth foetor. And frequently [the stoppage] may be the result of a flow of acid (or, pungent) or salt or foetid matter which runneth down from the head by the sides of the nostrils; and this also may be combated by means of the medicines which are injected into the nostrils. Now an injury to the faculty of breathing may be also caused by some ailment of the head.

Of the means of curing of the same.

Now if the injury should be in the brain itself, and nothing appeareth in the nostrils and no smell of foetid matter cometh from them, and if when the patient speaketh there be no symptom of any injury in his speech that would indicate that there is any stoppage in his nose, or that any foetid or acid blood is coming down from his head, or from any of the medicines which have been described above, and there is no apparent cause for [the injury], then all thy care must be devoted to the top of the head. And thou must employ bathings and spongings, and plasters which are | antidotes to Page 63 the ailment, and which strengthen the head, and draughts which will clear the head, and the letting of blood if it be necessary, and the medicines that are applied to the mouth and throat, and that expel the liquid (or, moisture) which is collected there. But if the passages be blocked up by haemor- Fol. 31 a. rhoids, or fleshy growths, then cut these away with an iron knife, or cauterize them, and cleanse the places with astringent medicines. If the obstruction be caused by the ulcer which is called the "foetid nexus", then medicaments which clear away the foetid matter and dry up the ulcer are required. Now the medicaments are these.

Of the medicaments which are good for the nostrils.

The medicine, that is good for that disease which is called "foetid nexus." Take moist calaminth (καλαμίνθη), and pound it, and squeeze out the juice, and inject it into the nostrils. Or dry calaminth, and rub it down into a powder, and blow it up the nostrils. Or take white hellebore and pepperwort, and rub them down to a powder, and blow some of it into the nostrils each day. And pour over the head warm sweet water; this is beneficial in cases of the pustule which is called "many feet" (polypodium).

Another medicine for the "foetid nexus", and for fleshy growth, and for haemorrhoids.

Litharge (protoxide of lead)	2 drachms
Myrrh	4 ,,
Arsenic	4 ,,
Nut galls	1 drachm
Copper acide	1 ,,

Rub down to a powder and use. First wash the nostrils with fragrant wine, and then insert into them with a spatula some of the medicine.

Another, for the foetid nexus and for haemorrhoids. Take fullers' herbs and myrrh in equal quantities, rub them down into a powder; make a plug of linen and dip it in wine, and then roll it in the medicine and place it in the nostrils.

Another, for the [same] ailments. Take equal quantities of ḵĕlîḵaṭrîn (i. e., قلقطار, copper-ore), chalcitis, alkali, cinnabar, vitriol, and protoxide of lead, and rub them down to a powder, and apply either by means of a plug or by blowing.

Another. Take fullers' herbs and pith of colocynth, and rub them down to a powder, and apply by means of a plug in the manner written above. Or rub down nigella with vinegar Page 64 and inject into the nostrils, and it will remove the obstructions | Or take saponaria and pepper in equal quantities and rub down and blow into the nostrils. Or mix cinnabar and honey together and lay on a plug [and apply]. Or fullers' herbs and chalkstone laid upon cotton wool; this will consume the fleshy Fol. 31b. growth, and will drive out the haemorrhoids. | If there be a flow of blood from the nostrils apply the above-mentioned medicines.

Medicines which are to be used for a flow of blood from the nostrils.

Take one part of aloes and two parts of frankincense, and rub them down well; make a plug the size of the nostrils, dip it in vinegar and squeeze it, and roll it in the medicines, and insert it in the nostrils from which blood is flowing, and press with thy fingers the upper portion of the nostrils on the outside.

Another, which is to be used for a flow of blood from any part of the body. Take in equal parts: Acacia, frankincense, the plant goat's beard (ὑποκιστίς), chalcitis, and burnt paper, rub down into a powder and blow into the nostrils. Or apply it by means of a plug and vinegar according to the direction given above.

Another. Take the Greek juice of leeks and let the wet droppings of an ass stand in it, and then rub it down and press out the liquid. Dip a plug into it and insert in the nostrils. Or inject some of the liquid into the nostrils and it will stop the bleeding.

Another, which is good for stopping inordinate bleeding in any part of the body whatsoever. Take two drachms of burnt vitriol and one drachm of opium, and rub down to a powder and apply by means of a plug and vinegar as already described.

Another, which is good for stopping bleeding in any part of the body.

Protoxide of lead	4 drachms
Nut galls	6 ,,
Vitriol	3 ,,
Burnt copper	2 ,,

Rub down to a powder, and blow into the nostrils, or lay on a plug.

Another, which is called "acid saliva", and which stoppeth bleeding in a wonderful manner.

Burnt copper	6 drachms
Vitriol	6 ,,
Samtĕrìn	4 ,,
Nut galls	16 ,,
Myrrh	10 ,,
Chalcitis	10 ,,
Burnt paper	10 ,,

If paper be not available then burn the seeds of the pomegranate instead.

Rub these down to a powder and apply | by means of a plug Page 65 and vinegar, or blow into the nostrils, and the bleeding will stop.

Another, which stoppeth bleeding in any part of the body whatsoever.

Acacia	1 drachm
Frankincense	1 „
Samtĕrîn	1 „
Persian gum	1 „
Copper-ore	1 „
Protoxide of lead	1 „
Vitriol	1 „
Burnt paper	2 drachms
Sandarax	1 1/3 „

Fol. 32 a.

Rub down to a powder and apply, according to the directions given above, with extract of polygonum, or extract of the plantain.

Now of all these medicines which stop bleeding, some are composed of ingredients which check and stop it, and others of ingredients which burn it up and cause obstructions. And we apply both kinds to the face and to the temples, and we make of them cataplasms which burn up and check the flow of blood. If it be necessary to treat the whole head, we smear it over with clay mixed with vinegar, and apply the same to the face and temples. Or we make a paste of lime and vinegar, and smear it over the head in the same way. And we may lay the sick man on his back, and pour cold water or vinegar water over his face, and insert in his nostrils wool soaked in strong vinegar. And if the flow of blood increase, we tie tight bands round his hands, and feet, and testicles, and we put cupping vessels over the liver if the blood cometh from the right side, and over the spleen if it cometh from the left side. And we let blood from the band of the head, and we apply sharp clysters, so that [the blood] may be drawn down to the parts of the body lower down.

Of cold in the head and of the rheum which runneth down therefrom upon the cheek.

Now the cause of the cold, that is to say, of the rheum which runneth down the nostrils, and of that also which runneth down into the mouth, is the brain. But in cold and in heat, like a body the parts of which are homogeneous, the

brain suffereth when it undergoeth a change in its composition,
and like an organic member it becometh full. As, then, diar-
rhoea is an ailment | which taketh place in the belly, on account Page 66
of some defect in the digestion, so also doth each of the ailments which have been mentioned [take place on account of
some defect] in the brain. But we only call "rheum" that
superfluity which floweth down from the brain and cometh
in the mouth, whilst to that which descendeth from the brain
into the nostrils we give the name of "catarrh". | Now deaf- Fol. 32*b*.
ness ariseth from rheum if it soaketh into the throat, and if
rheum oozeth through and descendeth to the gums and palate
then a boil or an abscess is formed there. And if the rheum
descendeth to the two sides of the cavity of the mouth, on
this side and on that, the swellings of mumps arise therein.
And moreover, the rheum descendeth to the belly and to the
lower part of the body. Now the injuries that arise from such
diseases (?) as these are discussed in the Chapters [of this work]
which are devoted to them.

Inasmuch as the subjects of our remarks at this moment are
the nostrils, it is right that we should discuss the catarrh which
descendeth to the nostrils. Now when this descendeth into
the nostrils, it sometimes produceth a feeling of inflammation,
and sometimes a coldness and swelling of the veins, and
tension in the head. And as regards the kind of matter which
is excreted, sometimes it is bitter, sometimes it is salt or acid,
and sometimes it is tasteless.

Of its cure.

Him that hath the feeling of heat do we cure by means of
the medicaments which cause coolness, whilst as for the other
kind of cold, that is to say, for him that hath a true catarrh,
we cure this by making use of the medicaments which cause
warmth and diminish the catarrh at the same time. For both,
however, we employ the medicaments that break through the
obstructions [in the nostrils], and that bring down the chymes
which cause the trouble in the head and strengthen it.

Medicines which are good for catarrh of the nostrils.

Take equal quantities of aromatic costus and nigella sativa, pound them and tie them in a rag, and make the patient inhale the smell.

Or take equal quantities of the flour of wheat and wild thyme, mix them together and heat them over hot coals, and let the smoke from them be inhaled by the patient through his nostrils. Or take nigella sativa by itself, burn it on a Page 67 metal plate (or, in a frying pan), and whilst it is | smouldering tie up in a rag and give it to the patient to smell.

Or take a millstone, and heat it thoroughly in a fire, clean away the ashes and sprinkle vinegar upon it, if the patient hath a feeling of heat, and wine if he feeleth catarrh. And let him bend his head over it, and wrap himself up well in his clothes, and let him inhale the fumes through his nostrils. And Fol. 33a. let him fumigate himself also | with frankincense, or with styrax or with wood of aloes, and let him take baths continually, and let him wash his hands and feet in hot water and anoint them with oil of narcissus, or oil of nard, or oil of aniseed, or oil of chamomile, and let him anoint also his anus and the skin of his genitals, and let him anoint his head with the same kind of oil. And let him abstain from foods of flesh, and from the drinking of wine, and let him partake of light and easily digested food, and drink hot water, and cover up his head against the cold, and if he feel the cold let him put on his head some warmth-producing oil, and if he feel the heat let him use oil of violets. Now we have in this Fourth Chapter treated at sufficient length the ailments of the respiratory organs and of catarrh of the nostrils.

HERE ENDETH THE FOURTH CHAPTER.

CHAPTER V.

Of the composition and placing of the eyes, and of the injuries which happen to them, and the symptoms of the same, and of the means to be employed in healing them.

NOW a sufficiently full description hath been written above concerning the injuries which happen to the brain, which is the primary source of the senses (or, feelings), which are effected by the front cavities of the brain, and concerning the ailments of the nostrils which are its instrument, and concerning the symptoms of the same, and concerning the means that are suitable for effecting the cure of each one of them. It is therefore right for us to teach here concerning the sense (or, faculty) of sight, which is the second [in importance] after breathing (now the sense of sight is effected through the brain by means of the first exit of the nerves | which come Page 68 from the brain, over which the power of sight passeth, and by means of the composition and placing of the various parts of the eyes), and concerning all the injuries which happen to them, and concerning the symptoms of the same, and concerning the means to be employed in effecting their cure.

Now this pair of nerves cometh from the first exit from the brain, and the nerves mingle with each other and then separate, the one [going] to the right and the other to the left, | and Fol. 33 b. by them the power of sight is transmitted from the brain. At the termination of these nerves is placed that liquid which is called "glass", because it is like glass which hath been liquefied and poured out. And above this liquid is placed the tunic which is like unto a grape, because it resembleth the skin of a grape in its fineness of texture and in its colour. Now there is hollowed out a round hole in the "cavity" (*i.e.*, pupil) thereof, for by this name is called the middle of the black portion of the eye, where sight is situated. And above this tunic is that second liquid which resembleth crystal, but which, according to others, resembleth the white of an egg. Now in Greek this liquid is called "*k̂rôs̱talîdôs*" (κρυσταλλοειδής), and it is described [by us] as being like unto crystal because in its transparency and fineness of texture it is exactly similar to ice made of clean water which hath been frozen. Moreover, there is spread out above it the second tunic which is called "horny" (*i.e.*, cornea) because of its similarity to those horny substances which are cut and planed down until they become thin, white [tablets]. These men place in the lanterns wherein lighted lamps are set instead of thin slices of limestone (?) when they wish the lamps to light up the space around the lanterns, the flame of the lamp being thus uninjured by any gust of

Page 69 wind or rain. Now this tunic of the eye resembleth this horn in its composition and solidity.

And above this tunic of the eye is spread out that membrane which is like a net; now this membrane is of a fleshy nature, and there are in it very many small veins that become clearly visible in any disease of the eyes, whensoever they are overcharged with blood. And outside this membrane are the eyelids and the eyelashes, and the corners which are formed by them. Now the eyes possess naturally for their proper movement from the second exit six muscles that appertain to

Fol. 34 *a.* the nerves which come | from the brain, besides the other muscles that surround the root (*i.e.*, base) of the opening which is in the nerves, which root cometh to them from the first exit of the brain. Such is the construction of the eyes, and it is well known that their operations appertain to the feelings.

Now the injuries which affect the eyes are of three kinds. The first kind is produced when the first organ of sense is itself affected; the second kind is produced when the power of the sense of sight is itself affected; and the third kind is produced when one of the members which minister to the first organ of sense is affected. The first organ of perception in the eyes is that which is called "*k̂rôs̱talîdôs*" (κρυσταλλοειδής), that is to say, "crystalline", for it hath been demonstrated in other books that this alone is changed by colours and that it is sensitive to them. Now the power of sensitiveness (or, feeling) is that which cometh to this liquid from the brain by means of the nerve which cometh thereto. The members, then, which exist for the service of this organ are all the other tunics, and muscles, and parts of the eyes. If therefore any one of these members that have been mentioned be injured, either the sight becometh impaired, or it absolutely ceaseth to exist. Now the eyes diminish in strength, sometimes both are affected, and sometimes only one of them; and they lose the power of movement and of sensitiveness. Sometimes | they Page 70 lose the power of movement only, and sometimes the faculty of perception only. And there are times when the eyelids only are affected, and sometimes when they are said to be suffering it would be more correct to say that the injury was in the eye, either as regardeth its movement or its power of perception. When therefore, even if one cannot see anything wrong in the eye, it happeneth that the sense of sight hath perished, it is the nerve that cometh from the brain which possesseth the cause of the disease, either because there is some sore (or, tumid growth) in it, or because its body (*i. e.*, substance) hath become hard, or because it hath become injured through some flow of liquid, | or because in addition to Fol. 34*b*. these causes, the cavity (or, hole) that is in it hath become blocked up.

Now all these things happen to the eye, just as they happen in organic members, and there take place in it, even as in the members that are similarly constituted, eight changes in composition. And besides these also [there are times] when that luminous spirit that cometh [to the eye] from the brain is transmitted only in a very small quantity or not at all. Now when the power of movement only is lost in one eye, it followeth of necessity that the nerve that cometh to it from the second exit of the brain is suffering from some one of the diseases, which I will mention here, in those nerves that come to it from the first exit. For, according to what we have learned from the openings, the muscles which move the eye are six in number.

And there are others, that is to say, those that surround the root (*i.e.*, base) of the hole which is in that nerve that cometh thereto from the first exit of the brain. Now it happeneth oft-times that even when this itself is in no wise affected, the muscles suffer from one of these diseases, which I will now mention, either in their own nature, or when that nerve that cometh to each one of them is injured.

For to each one of these muscles a certain part cometh from the nerve that is separated from that second exit [of the Page 71 brain], just as there doth also to those | muscles, that surround the root (*i.e.*, base) of the hole [in the nerve]; but whether it is right for us to say they are two, or three, or one, mattereth nothing in this place. For we know that the functions of the muscles are these: to raise the eyes upwards, to lower them, to move them together in every direction, and to keep them motionless. In order that we may not incline to any side out of the straight path that nerve is pliant (or, soft), and we give unto it the name of "seer of the hole". Now the muscles which move the eye are six in number. If that muscle which raiseth the eye upwards be affected, then the whole eye, it Fol. 35 a. seemeth, turneth itself downwards, | and hath rest. If that muscle which directeth the eye downwards be affected, then the eye is strained upwards. Similarly also, if that muscle whereby the eye is moved to the little angle (*i.e.*, towards the side of the face) be affected, straightway the eye inclineth to the great angle (*i.e.*, towards the nose). And if the muscle which turneth the eye to the great angle be affected, the eye moveth itself towards the little angle. And if one of the two muscles which make the eye revolve be affected, the eye turneth round in an oblique direction.

Inasmuch as I have said that there are other muscles which surround the root of the optic nerve, it is right to know that a defective condition of these affecteth the whole eye, but only outwardly, for many who suffer in this respect enjoy vision which is unimpaired. The optic nerve may be stretched out (or, strained) without any disease whatsoever being therein, but if it happen to be diseased those who are afflicted in this manner see very badly. Now if their disease be severe (or, acute) it is well known that they will never see any more. Also as concerneth the oblique vision (squinting?) of the eyes, it is right to know that that inclination of the sight towards one angle is a protector of a natural operation of vision. The straining of the eye upwards, and the bending of the eye downwards, as well as those other inclinings which take place in the eyes obliquely, | all these inclinings which take place Page 72 cause the eyes to see double.

Now the upper eyelid possesseth large muscles that make it to move; but the lower eyelid doth not move at all, because there are only small muscles in it, which can only be set in motion with great difficulty. Moreover, as concerning the things which were said a little way back about the muscles that move the eyes, one must understand that they apply also to those muscles that move the eyelids downwards. These therefore are the diseases that affect the parts of the eyes, that is to say, those whose situations are not visible, besides those others that take place in the eyes in sympathy with other members of the body. For phantasms, which resemble those causes by mistiness, appear in | the eye although there is no disease in Fol. 35 b. it, but these are due to the fact that it hath relationship with the mouth of the belly and with the brain. Now it is very right that we should distinguish between those that are caused by the mouth of the belly and the brain, and those that are caused solely by disease in the eye itself. If the disease be in the eyes, the phantasms appear first in one eye, and not in both at the same time. For, in the majority of cases, the phantasms that do appear are due to a defect in the chyme of the belly, and then they appear in both eyes. Now those that are due to mistiness (cataract?) do not begin to appear in both eyes, and similarly they are not visible at all times. But afterwards, when a certain time hath elapsed, we are able to come to a decision about the disease thus:—If after three or four months the phantasms of mistiness still appear, and if having examined the eyes most carefully thou art unable to find any film closing over them like darkness, then thou art able to declare that the eyes are suffering through the "mouth of the belly." But if it be only a short time since the disease

began to manifest itself, enquire first of all if the phantasms
Page 73 have been visible always, | from the first day when the patient
began to suffer, and if there hath not passed a single day in
which they did not appear to him as before, or if there have
been certain days which formed an interval when he was well,
and wherein they did not appear.

When thou learnest that the phantasms do not appear [to
the patient] at all times, thou mayest assume that they are
due to the mouth of the belly, especially so when, his digestion
being perfect, the patient saith that no phantasms have appeared
to him. And more than ever thou mayest assume this to be
the case if the patient feeleth, at the time when the phantasms
begin to appear, a certain biting (or, griping) that taketh place
in the mouth of the belly. Moreover, thou mayest be the more
certain if, in addition to these symptoms, when the chyme that
caused the griping pain hath been vomited, the appearances
of the phantasms cease altogether. These things, then, thou
Fol. 36 a. wilt learn immediately from an investigation | of this kind which
must begin on the first day wherein thou seest the patient,
if, as I have already said, thou dost see that the eyes are
perfectly healthy. When, however, thou seest that there is
over the pupil of one eye something which resembleth a film,
or that one eye itself is more bilious than the other, or it
seemeth to thee that it is not perfectly clean (or, clear), then
thou must recognize that that which resembleth a film is the
beginning of mistiness (cataract?). But if patients possess
naturally pupils of the eyes that are cloudy, then examine
closely and see whether the two pupils appear to be exactly
alike; and in addition to this enquire if it is not a long time
since the phantasms have begun to appear. And if such be
the case, then order the patient to partake of food in smaller
quantities than is usual with him, and let there be nothing
injurious in the chyme of such food. And on the following
day, when such food hath been digested perfectly, ask the
patient questions about the phantasms that have been wont to
appear to him. And if he confesseth that they have not appeared
at all, or that they have appeared to him with less frequency,
Page 74 then thou wilt know of a certainty | that the disease ariseth

through the stomach. But if they have remained with him as usual, even after this treatment, then thou wilt know that their appearances are due not to the sympathy of the eyes with other parts of the body, but to disease in the eyes themselves. And thou mayest be the more convinced that such is the case if, when the patient hath taken a draught of the extract of aloes which is called "*pîkrâ*" (ἱερὰ πίκρα), he continueth in the same state as before. For if the disease be due to the stomach, it is easily cured by a draught of this healing medicine, especially when coupled with a good digestion, for then the two things are brought together, namely, the symptom of the part that is affected and the cure of the disease.

And there are also phantasms which are like unto those produced by mistiness (cataract?), and which appear when the brain itself is diseased; these are due to the effects of the disease called "delirium", and appear in men who are smitten with severe fevers. Take, for example, the case of the man who imagined that he saw musicians standing in the corner of his house, or again the case of those who thought that they were plucking ragged bands from garments and chopped straw from the walls. It is well known, then, that that which begetteth these happenings is derived from a | cause that is [always] in Fol. 36*b*. the [same] form, and that they do not arise from the actual place which is affected beforehand; this applieth to the phantasms that are due to the brain, or to obstructions [therein], as we say, sympathetically. For when bilious chyme becometh collected in the brain, and is accompanied by a burning fever, there is some disease therein. And it is like the things that burn in a fire, and that therefore naturally produce some smoke; and the smoke of the chyme resembleth that which is produced by oil [when burning] in a lamp. When this smoke falleth into the veins and arteries that come to the eyes, it becometh in them the cause of the phantasms. For ye have heard [already] that veins and arteries come from the nerves to the eyes, and the membrane that is spread above and is called "net-like" causeth them to mingle together.

Now the eyes also become clouded through gazing too much at the sphere | of the sun, and through imprisonment in a Page 75

place which is totally dark. And there are to be seen eyes
which, although uninjured, are unable to see when those nerves
which transmit the power of vision from the brain are blocked
up by a boil or any other obstruction. Now the part that
is affected by disease may be distinguished in this manner.
When we close one of our eyes the other becometh larger
than it was previously, provided that the nerves be not obstructed. But if when we close one of our eyes the other
becometh dilated, it is well known that the nerves thereof are
obstructed, and that the eye itself is not affected by disease.

Now there is another kind of mistiness (cataract?) of the
eye that is called "suffusion of the eye by a blinding humour",
and that subsisteth between the horny tunic of the eye and
the crystalline fluid, and its source is some liquid that con-
gealeth in front of the little pupil of the eye which transmitteth
the sight. Now when this disease beginneth to come into
being, thou mayest recognize it by the following symptoms:—
The patient imagineth that there is something that is passing
Fol. 37 a. before his eyes like water, or like gnats, or like flies, or | like
rays of light, or like darkness, or like smoke. And patients
see other kinds of phantasms (or, appearances) before their
eyes, and more particularly when they are awakened out of
sleep; and when this evil hath run its full course the patient
is no longer able to see anything at all. Now the forms (or,
changes) of the colour of that humour are very many. Some
are like air (hazy?), some resemble glass, some are white, some
are dusky, some are purple, and some are bluish in colour.
Now when they are in a perfect (or, complete) state they are
in a rounded form, and when those which are coagulated are
heaped up they are like the white and like the dusky-coloured.
And thus thou mayest recognize those that are to be removed.
Page 76 If, having made the patient to stand upright, | and commanded
him to straighten his eye, and pressed with thy thumb upon
the eye above the eyelashes, and taken it quickly and seen the
humour, if I say [having done these things], it doth not move
from side to side, then most certainly it may be removed. But if
it be scattered by the pressure of thy finger and then returneth
to its [former] position, that humour is of a malignant character.

And the eyes are also made misty (or, covered with a film) by the fume of the malignant humour which is secreted between the tunics of the eyes, and which darkeneth the visual power; those that are afflicted with this ailment see that which is some distance from them better than they see that which is near. Now the reason why they see more clearly that which is afar off is in looking at a distant object the humour is reduced in denseness, and they see better in consequence. But they are unable to see that which is near to them because of the denseness of the humour which darkeneth the vision. And, in the cases of those who can see in the day-time and cannot see at night-time, the cause is the same. And as concerneth those who can see clearly an object which is close to them, and who can only see dimly that which is afar off this result ariseth through the weakness of the visual power.

And mistiness also | taketh place in the eyes when the tunics Fol. 37 *b.* and ·the humour are dried up and constricted, for then doth the power of vision become weakened. And this disease is very serious indeed, for it is much more difficult to make moist dryness of the eye than to dry up the humour. And there arise trouble (or, disturbance) and cloudiness in the eyes, when that membrane which is above the horny tunic of the eye is struck, or when the tunic of the eye itself receiveth some blow externally, or [when they are touched] by | smoke, Page 77 or by fat, or by marrow, or by [sour] milk, or by anything which falleth into them. And frequently even blood is formed in the eyes by reason of something that striketh them externally, and such causes very often give rise to very severe disease of the eyes when the eyes are constitutionally weak, or when these causes are very great (or, serious).

Up to this point we have spoken about two kinds only of the diseases which affect the eyes, the one kind being due to their homogeneousness, and the other to organic disturbance. There now remaineth to us to speak about that kind of disease which is common to the two kinds already mentioned, that is to say, the solution of unity; now this taketh place through wounds (or, sores) which destroy the component parts of the eyes. There are some cases in which the sores eat away the

6

eyes and cloud over the sight until at length they destroy the
eyes from the outside, and there are others in which the sores
produce pus, and they burst out through the eyes, and destroy
the tunics of the eyes and the humour that is between them.
Such sores, provided that they do not happen to be en-
countered immediately opposite to the pupil of the eye, may
be healed, without doing any damage at all to the sight, except perhaps a very little. Now if they be situated exactly
opposite the pupil of the eye, and if they extend as far as
that grape-like tunic, so that only a very small portion of it
be visible, the sore is called the "fly's head". But, if the sore
Fol. 38 a. be so widely extended that it goeth | outside the grape-like
tunic, it is called the "daughter of the grape", because of its
resemblance to a grape. And this is the case also with the
abscesses which come inside the eye, and the pustules and
cysts which generate pus, and which burst outwardly and cause
far greater pains and sufferings than before they burst. And
these also, if they happen to come exactly opposite the pupil
of the eye, bring a man to total blindness. If, however, they
Page 78 come on one side of the pupil, the evil is lessened. |

Now all the wounds, or abscesses, which come into the
eyes, whether external or internal, produce alike in the eyes
films and scabs in the various places where the sores are.
And there are also the hot ulcers which come in the membrane that envelopeth the eyes and containeth nerves and
veins, and they cause severe diseases. These diseases are
called "diarrhoea" and "inflammation" of the eye, and they
produce such intense agony that people who suffer from them
are driven to commit suicide. Now the rheum of the boil
(or, abscess) appeareth quite plainly, whether it be much or
whether it be little, and whether it causeth pain (or, disease),
or whether it doth not; it is light-coloured, or red, according
as the veins are seen to be full of blood. And also we can
learn from the patients themselves in what way they suffer.
And the pains caused by these diseases are different in character.
Some of them cause the patient to feel stabs, made as it were
by a pointed instrument, and some cause a feeling of tension,
and others a feeling of weight, and others produce a gnawing

sensation. And according to the effects which each produceth so is the kind of rheum that is transmitted to the ulcer (or, abscess). These diseases also produce severe pains in the membrane that hath been mentioned.

And as concerning the rheum that is transmitted to the eyes from the head, and that resembleth in kind that which is called "running at the nose", when this descendeth into the eyes it produceth in the eyes themselves and on the insides of the eyelids granulation, and very small swellings, and pimples, and watery humour, and pus, and frequently it causeth boils (or, styes) in the corners of the eyes. Sometimes it descendeth by means of the veins | and arteries that are inside the bowls Fol. 38*b*. of the head, and sometimes by means of those that are outside them. Now we may recognize the presence of this rheum, which is transmitted from within, by the sores that appear round about the eyes, and by the sneezings that accompany feeble flows of tears, and by the cloudiness as of gnats [flying before the eyes], and by the falling out of the eyelashes; also we hear patients crying out repeatedly that they feel as if their eyes were being torn out by their roots.

And we are able to recognize that rheum which is in the veins | and arteries, and which descendeth externally, by the Page 79 swelling of the veins, and by the red flush on the cheeks, and by the terrible throbbing of the arteries, and by the terrible feeling of pain which is in them, and by the heat of the whole face. The suitable treatment for this disease is to cut the veins of the temples, and those that are behind the ears, and, if it be necessary, the *kônê*[1] also. Now the disease which is inside the bowls of the head is very difficult to cure.

There is also another disease of the eyes, which is called *ṭĕpherâ* (or, *ṭephrâ*, *i.e.*, πτερύγιον[2]) and is produced by the membrane of the vein that is described as being "like a net". It beginneth to grow from the great corner of the eye and cometh as far as the dark portion of the eyes which it covereth

---

[1] Some funnel-shaped portions of the head. The Lexicons fail to help us here.

[2] This disease is caused by a membrane that grows over the eye from the inner corner.

over. It can be cured sometimes by medicines, and sometimes by cutting. Sometimes, moreover, the eyelashes bend over and grow inwards towards the eyes and become like wires and pierce the eye. And sometimes the eyelids fall down and stick tightly together at the corners of the eyes. And there is also above the eyelashes the tumid sore which is called "*shûrnâḳâ*" (*i.e.*, sarcoma), and because of its weight patients are unable to raise their eyes, and it causeth grave injuries to them. And there groweth also in the bones of the nostrils, close to the great corner (*i.e.*, the inner corner) of the eyes, the running sore which is called "*sôrîghâ*", or "*nâsôrâ*". Sometimes it dischargeth inside the eye and destroyeth it by the watery fluid that runneth from it; sometimes it dischargeth inside the nostrils, and pus runneth from the nose; and sometimes it dischargeth outside it. Now this may be cured by opening the

Fol. 39a. sore | and cutting it out down to the bone.

Now among healthy eyes we esteem most highly those which are small naturally, and we also praise most those which are naturally slightly sunken. The eyes that are large and that protrude are always found to be delicate (or, weak), especially if they are not naturally so. Now therefore, I think

Page 80 that as far as | it is possible, sufficient hath been written concerning the diseases which attack the eyes, and concerning their symptoms. It is now right to describe the means that are to be used in healing them, and the means of healing used must be in accordance with the symptoms.

Of the healing of the diseases of the eyes.

Now in respect of the diseases of the eyes which arise through an abscess or through rheum, first of all are beneficial the sparing use of food and drink, abstinence from coition, and rest of the body. And if it seemeth to us that the blood is in excess, we must let blood from the upper vein of the arm; and if the pains in the eyes or the mistiness be due to sympathy of the eyes with the stomach, we must empty the stomach by means of *piḳrâ* (ἱερὰ πίκρα), or by means of vomiting. And if the pains in the eyes be due to the [over-]fullness

of the whole body, we must expel the [over-]fullness by means
of a cleansing aperient which will expel the chyme that
appeareth to us to be in excess. And if the pain be due to
the condensation caused by chill, we must employ baths, or
use fomentations with water in which chamomile hath been
infused. If the patients are suffering in their eyes through
cold and damp, we must administer new wine. And if, after
the emptying of the whole body and the use of medicines,
the pains still continue in the eyes, we must pay especial
attention also to the head, by cutting the veins and arteries,
according to the plan which is described in the section dealing
with headaches, and by [cutting] the veins that empty the
head, and by smearing the temples' with liniment, and by
drawing away [the rheum] by means of violent drenches, and
by tying bandages tightly round the hands and feet, and by
fomentations with water in which have been infused chamomile
and aniseed and trigonella. In cases when the pains are | of Fol. 39b.
long standing, foment also with water containing an infusion
of origanum (mentha pulegium), and marjoram (σάμψυχον), and
mugwort (ἀρτεμισία). And if it be necessary to allay the
irritation and cool the eyes, foment with oil of roses and oil
of violets and with tincture of the plant "live for ever"; and
use the plasters that check the rheum in its beginning and
further progress, and those that give relief and freedom from
pain, both when the pain increaseth | and diminisheth. And Page 81
we must also wash the eyes under the eyelids with an infusion
of trigonella and the white of an egg, or with the milk of a
woman, or with the milk of an ass slightly warmed, either
with these by themselves, or with the addition of one of the
medicines that are beneficial in these diseases dissolved in them.

The medicines that are useful in cases of the severe pains in the eyes which are called "dissolution"[1], and the abscess that is caused by inflammatory rheum.

Ointment that is to be used for severe pains in the eyes and in cases where attacks of rheum are frequent.

Opium	6 drachms
Lycium tincture (λύκιον)	6 ,,
White lead (ψιμύθιον)	8 ,,
Gum Arabic	12 ,,

Rub down to a powder and work up with an infusion of "king's crown" (melilotus officinalis), and make into ointment and use. Thou shalt make the infusion of "king's crown" thus:—Take of "king's crown" one lîṭrâ and of rain water two measures (kesṭîn), and boil them until only one half of the two measures remaineth; then clarify and use.

Another, which is to be used for pains in the eyes.

Glaucium (γλαύκιον, or blue celandine)	48 drachms
Opium	12 ,,
Indian lycium	6 ,,
Hyoscyamus tincture	6 ,,
Fresh roses from which the thick stalks with the leaves have been cut	48 ,,

Macerate in rain water or in an infusion of "king's crown", and make ointment and use.

Another, which is to be used for sores in the eyes caused by rheum, and for severe pains, and for eyes that are wholly covered by sores.

Cadmin (καδμεία), washed	40 drachms
White lead (ψιμύθιον)	80 ,,
Opium	10 ,,
Gum Arabic	6 ,,

Add sufficient rain water to work up into a paste.

Another ointment that is called "star", and is to be used Fol. 40a. for ulcers in the eyes, and pus in the eyes, | and abscesses which go deep in them.

---

[1] Literally "diarrhoea".

Cadmin (καδμεία) burnt and washed	16 drachms	
Deposit of furnace smoke (πομφόλυξ), washed	16 „	
White lead, washed	16 „	
Lead, burnt and washed	8 „	Page 82
Stibium (antimony)	10 „	
Starch (ἄμυλον)	18 „	
Frankincense	10 „	
Myrrh	8 „	
Opium	8 „	

Mix with rain water sufficient to form a paste, and make into ointment and use in the manner described above.

Another kind of ointment which is called "Lîbânôn", because it containeth frankincense, and is used for inflammation of the eyes, and for sores which are deep-seated in them.

Frankincense	10 drachms
Cadmin (καδμεία)	10 „
White lead	40 „
Opium	6 „
Gum arabic	6 „

And rain water [as much as is necessary].

Another, which is to be used for all pains in the eyes and especially for white, filmy spots. Take a bunch of grapes and squeeze out the juicy matter. Dry that juicy matter and keep it by thee, and when occasion requireth rub it down into a powder, and mix it with the gum that is used by the physicians and with the milk of a woman, or with the fine white of an egg; warm this and bathe the eyes with it. Begin this treatment by administering to the patient pills that are suitable for the ailment, and at the same time bathe the eyes [with the above medicine], and thou wilt be astonished at its efficacy.

Another kind of ointment, which is called "ḳûḳnôs" (i. e., "swan", κώκνος) because of its white colour, and which is used for sores and wounds in the eyes.

Cadmin (καδμεία) washed and burnt	40 drachms
White lead (ψιμύθιον), washed	44 „
Starch (ἄμυλον)	44 „

[And add] as much rain water as is necessary.

[Another kind of] ointment which is called "rose" because of the large quantity of roses which are in it; it is used for inflammation of the eyes, and for the stinging caused by rheum.

Moist roses from which the stalks have been cut	52 drachms
Crocus	15 ,,
Indian nard	4 ,,
Cinnabar	2 ,,
Brass shavings	2 ,,
Cadmin, washed	44 ,,
Opium	3 ,,
Kohl (antimony)	3 ,,
Myrrh	3 ,,
Gum Arabic	20 ,,

[Add] rain water, or the juice of thorn-berries as much as is Fol. 40*b*. necessary. This ointment is very useful for severe pains, | and for inflammation, and for tumours, and for rheum of long Page 83 standing, | and blearedness of the eyes.

Another [kind of ointment] which is called "myrrh" because of the myrrh that is in it; it checketh and relieveth pains.

Fresh roses	11 drachms
Crocus	50 ,,
Opium	4 ,,
Myrrh	2 ,,
Gum Arabic	50 ,,

Rub down and mix with wine having a bouquet, and apply with a very little white of egg.

Another [kind of ointment] which is called "Dĕyâla", because it is mixed with myrrh; it is used for severe pains in the eyes, and for blearedness, and it removeth wounds from the eyes and filleth them up.

Cadmin	4 drachms
Burnt brass	2 ,,
Acacia	12 ,,
Opium	1 drachm
Myrrh	1 ,,
Gum Arabic	12 drachms

These must be stirred up together with rain water for forty days, in the season of summer, once or twice each day, and the mortar must be well covered up. Then make the ointment and apply it.

Another [kind] which is called "Îdinôn", which I use on myself; it is beneficial for disturbances and twitchings of the eyes, and acute pains; and it filleth up cracks and checketh the rheum in the eyes.

Cadmin	40	drachms
White lead	40	,,
Tragacanth	5	,,
Gum Arabic	5	,,
Opium	7	,,
Starch	2	,,

Pound and work these up together in rain water, and apply with fine white of an egg, or with the milk of a woman.

Another which is called "Ḳedrôn"; it checketh tears, and cleanseth wounds in the eyes, and filleth them.

Pure myrrh	1	drachm
Crocus	2	drachms
Burnt brass	2	,,
Seed of roses	2	,,
Gum Arabic	2	,,
Cadmin	2	,,
Opium	2	,,

Pound and mix well together in wine with a bouquet, and make the ointment and apply.

Another which is called "Esṭapṭiḳân" (*i. e.*, "Styptic"), and is to be used for weak eyes which water and are cloudy, and for filminess and for watering.

Crocus	4	drachms
Opium	4	,,
Round peppers	4	,,
Ninkâ	2	,,
Gum Arabic	8	,,
Celandine	8	,,
Persian gum, σαρκοκόλλα	8	,,
Indian salt	1	drachm

Page 84

Armenian borax	12 drachms
Yellow arsenic	2 „

Mix into an ointment with wine | and apply.

Of the dry medicines which are to be used for the eyes, and which are applied in cases of disease of the eyes.

The medicine which is called "Pramṭiḵôn" (stibium), and which is to be used for the running of the eyes, and rheum, and pains, and swollen sores (styes?). It is to be smeared inside and outside the eyelids, and in the corners of the eyes.

Crocus	2 drachms
Celandine	5 „
Aloes	2 „
Opium	$\frac{1}{2}$ drachm
Rose seed	1 „
Tamarix, μυρίκη	1 „
Persian gum	20 drachms

Rub down to a powder and use dry.

Another "Pramṭiḵôn" (stibium) which I myself use, and it is most certainly beneficial in cases of pains in the eyes.

Crocus	6 drachms
Aloes	6 „
Opium	4 „
Celandine	12 „
Gum Arabic	12 „

Mix well together and smear the eyelids with the dry mixture.

Another. Red elixir (ξήριον) which is to be used for sores in the eyes, and ulcers and boils, and it cleareth away and cleanseth any unnatural flow of humour in the eyes.

Opium	8 drachms
Haematite شادنج	8 „
Burnt copper	8 „
Starch	8 „
Gum Arabic	25 „
White lead	84 „
Cadmin	40 „

Pound well the burnt copper with water until it becometh a

dry powder, and treat the haematite in the same manner, and rub down the other ingredients and mix them together and apply as a dry powder.

Another elixir, a white one, which is used for rheum, and for excessive tears, and for filmy spots.

White lead	8 drachms
Gum Arabic	4 ,,
Starch	4 ,,
Cadmin	4 ,,
Persian gum	1 drachm
Tragacanth	1 ,,
Opium	2¹/₂ drachms

Pound, wash, and use.

Another, which is to be used for a severe pain | and for sores. Page 85

Celandine	1 part
Persian gum	2 parts

Rub down to a powder and use.

Another, which is to be used for sores and ulcers in the eyes, and for severe pains.

Persian gum	1 *manyâ*

Pound and rub down in the milk of an ass for seven days. When the milk diminisheth | add more to it and stir seven Fol. 41 *b*. times a day. Then dry, and rub down to a powder, and smear [the eyelids]. This is a well-tried remedy.

The above are the medicaments which are to be used for sores and wounds which come in the eyes and cause pain, and they cool, and relieve, and check their growth without [causing] pain, and they alleviate pains.

Medical preparations which are laid on the eyes, and which are used in cases of severe pains in the eyes, and for ulcers which come through excessive rheum.

Lôḳiôn (πυξάκανθα) Indian	
Fresh roses	
Crocus	in equal quantities(?)
Opium	

Rub down and work up in wine, and make into pills (or, tablets) and dry in a shady place. At the time of using dissolve in grape juice (must) or in vinegar, and smear the forehead and temples.

Another.

Crocus	
Myrrh	
Opium	in equal quantities (?)
Frankincense	
Gum Arabic	

Rub down into a powder and make into pills (or, tablets) in wine, and when about to use dissolve in tincture of hyoscyamus or mandragora, and paint [over the eyes and temples].

Another, which causeth pain but stoppeth the rheum immediately; and it is beneficial also in cases of scabies of the eyebrows.

Burnt brass	4	drachms
Opium	2	„
Crocus	4	„
White peppers	2	„
Gum Arabic	4	„

Pound and make up with rain water, and smear the outsides of the eyelids therewith, but be careful that it doth not touch the eyes.

Another, for pains in the eyes.

Seed of hyoscyamus	2	drachms
Opium	1	drachm
Myrrh	4	drachms
Fine white flour	4	„
Castoreum	1	drachm
Crocus	1	„

The yelks of two roasted eggs.

Work up together and apply according to the directions given above.

Another, which is good for severe pains caused by excessive rheum; it produceth sleep immediately.

Glôḳîôn (γλαύκιον)	12	drachms
Haematite stone |	3	„
Opium	3	„

Page 86

Extract of mandragora	4 drachms
Gum Tragacanth	4 „

Work up with rain water, and make into tablets, and dry in the shade, | and at time of use dissolve in vinegar, or in the Fol. 42*a*. white of an egg, or in fresh extract of coriander seed, and paint [round the eyes].

Another. Take rinds of pomegranates, lentils, and dried roses, steep them in water and macerate them and pound them together, mix with oil of roses and lay at the side of the eyes. Or, pound up chicory and mix with the fine flour of barley and oil of roses, and put a plaster on sores that are inflamed.

Another, which is good for sores that are hard. Chop and crush and boil cabbage leaves with cow-oil, and apply [round the eyes].

Or, pound up wild marjoram with fine barley flour and oil of roses, and apply [round the eyes].

Or, take the inner part of a piece of meat, pound it up with wine and yelks of eggs and oil of roses, and apply [round the eyes].

Another, for inflammatory pains.

Crocus	
Frankincense	
Aloes	
Opium	in equal quantities (?)
Myrrh	
Persian gum	

Rub down into a powder, and dissolve in extract of coriander, or in extract of hyoscyamus, and smear [about the eyes]. This preparation is also good for great weakness of the eyes and for watering of the eyes.

Another.

Fine barley flour	4 drachms
Myrrh	2 „
Opium	1 drachm

Pound and mix with vinegar and oil of roses, and apply.

Or, pound purslane, and mix it with the fine flour of barley and oil of roses, and apply.

The medicines which are good for dimness of vision, and for a flow of water from the eyes, and for the disease of not being able to see in the twilight, and which are to be smeared on the eyelids moist.

A preparation which is good for flow of water from the eyes. Empty the whole of the gall of a bull into a vessel of brass, and let it stand ten days. Then take myrrh, and crocus, and balsam, and the gall of a tortoise, of each of these two drachms, and ten peppers, and of the finest honey double the quantity of | the gall of the tortoise. Mix these together in a brass vessel, and use as *kohl* moist.

Page 87

A preparation of Sôrînôs which is exceedingly good for water in the eyes.

Gall of a hyena	4 drachms
Foreign gall	4 　„
Oil of balsam |	4 　„
Sal Ammoniac	1 drachm
Crocus	1 　„
Finest honey	half a pot
Tincture of aniseed	half a pot

Fol. 42*b*.

Mix together and apply.

Another preparation, which is incomparable, and which is good for a flow of water from the eyes, and for the sores which spread over the whites of the eyes.

Cadmin	4 drachms	
Fresh chalcitis	4 　„	
White lead	4 　„	
White peppers	3 　„	
Juice of *ḳewrînê*	2 　„	
Crocus	1 drachm	
Opium	1 　„	
Sagapenum (σαγάπηνον)	1 　„	
Butter of the olive	1 　„	
Arsenic	1 　„	
Myrrh	1 　„	and 2 *denḳê*
Gum Arabic	2 drachms	
Hellebore	2 　„	

Incense, of ammoniac	1 drachm
Balsam	1 „ and 2 *denk̲ê*
Gall of a bull	¹/₂ „
Pulp of colocynth	¹/₂ „

Pound together and then work up with tincture of aniseed, and apply.

Another, for dimness of vision and stoppage of the flow of water.

Wolf's gall	4 drachms
Opopanax	3 „
Round peppercorns	3 „
Old olive oil	2 „
Oil of balsam	2 „
Tincture of aniseed	2 „
Cadmin	1 drachm
Run honey	1 measure

Rub down and mix well together, and apply.

Now it must be well understood that all the above medicines, which are used for running of the eyes, are to be smeared on the eyelids when moist, and they must be kept in vessels of brass which are made for the purpose.

Another. The gall of the *shabbûṭ* (*i.e.*, cephalus) fish, the gall of a partridge, and the gall of a wolf, and tincture of aniseed, pound and mix them together, and smear on the eyelids.

Now for this disease every kind of gall, no matter from what beast or bird, when mixed with run honey and tincture of aniseed and tincture of mint, is good, because of the natural fineness which they possess, and they attract and keep the moisture | which is collected between the tunics of the eyes. Page 88

Another, which is good for [these] ailments. Burn the heads of young swallows (or, bats), and pound them, and work up together with honey, and smear on the eyelids.

Another, well-tried medicine for [these] ailments.

Ginger	1 drachm	
Bone		1 „	Fol. 43 *a*.
Dried mustard leaves	2 drachms	

To these add as much tincture of aniseed as is necessary, pound, and mix them, and smear the eyelids with them.

Another, which is good for those who see as it were flies, and whose eyes are dim. Wood lettuce (مَرمَاحُوزٍ) and caper-root (κάππαρις, كَبَر), bruise and pound, and smear on the eyelids.

Another, for those who do not see in the night, and for those who do not see in the day-time what is near to them. Take *dĕyâlâ* tablets, and *ḳaḳnadîn* tablets, and seven long peppercorns, and a liver of a black goat. Crush the tablets and peppercorns and put them inside the liver, sew it up, and roast it over hot coals until it is half cooked. Then take out the tablets and the peppercorns and dry [them] in the shade, and give the liver to the man who is suffering, and let him eat it, standing behind the door. Then pound the tablets and the peppercorns into a fine powder and smear the eyelids with it at the ninth hour each day, three smearings. Let the patient keep his head covered, and he shall become well.

Another, for these ailments.

Gall of a partridge	4 drachms
Extract of aniseed	2 ,,
Run honey	2 ,,

Mix together and apply.

Another, for those who cannot see in the twilight, and for those who do not see things near them, and for running of the eyes.

Gall of a raven	4 drachms
Gall of a partridge	4 ,,
Gall of a vulture	4 ,,
Gall of a cock	4 ,,
Gall of a hyena	4 ,,
Gall of a goat	4 ,,
Balsam	$^1/_2$ drachm
Run honey	3 drachms

Mix together and make into an ointment, and apply to the eyes in water or vinegar.

Another, which is to be used for effusion of the eyes. Take the gall of a tortoise, one portion, and run honey, two portions, mix together and apply to the eyes as an ointment.

Another, which is to be used for those who cannot see in

the twilight. Take the liver of a goat (or, stag), roast it until it is half-cooked, squeeze out the juice and catch the liquid which floweth from it | in a glass vessel, and smear the eyes with it, Page 89 and let the patient eat the liver. Now others stew the liver in a cooking pot, and order the patient to hold his head over the pot and to keep his head covered on all sides.

The above medicines | constitute the means which are to be Fol. 43*b*. used for dimness of sight, and effusion of the eyes, and nocturnal blindness.

The medicines which are to be used for the rheum which cometh down into the eyes and eyelids, and causeth in them the pains which have been described above, that is to say, the medicines which are to be used for dimness of sight, and swellings in eyelids, and the formation of pus, and *kenpâ*, and erosion of the corners of the eyes, and filmy spots in the eyes, and the falling out of the eyelashes, and the manifold "cords" which spring up in them.

First, the medicine which is called "Bâsîlîḳôn", *i.e.*, "Royal", which is to be used for the diseases that have been described, and also for keeping the eyes in a healthy state. It is a sure cure and has been well tried by many.

Dross of brass	10	drachms
"Sea spume", *i.e.*, Bastard-sponge (ἀλκυόνειον)	10	,,
Burnt brass	5	,,
White lead	3	,,
Mountain salt (rock-salt)	3	,,
Sal ammoniac	2	,,
Long peppercorns	2	,,
Ḳarpallôn (caryophyllus aromaticus) *i.e.*, aromatic basil (?)	1	drachm
Lichen	1	,,
Camphor	$^1/_2$	,,

Pound these together with the greatest care, and smear the eyes therewith.

Another "Royal" medicine which is to be used for the diseases that cause "keenness of sight".

Dross of brass	7	drachms
Pallon (φύλλον) betel, or base cinnamon	2	,,

7

Burnt brass	4 drachms
Sal ammoniac	2 „
Long peppercorns	2 „
Round peppercorns	4 „
White lead	4 „
Rock salt	4 „
Indian salt	1 drachm
Aromatic basil	1 „
Lichen	1 „
Flour	1 „
Sea spume (bastard-sponge)	10 drachms

Pound these together very carefully and apply.

Another, which is called "Banushag" *(i. e., "Violet"),* or "Banushag Rakîkâ", and which is to be used for swellings of Page 90 the eyelids, and the gathering of pus, and | red (*i. e.*, inflamed) eyelids, and the falling out of the eyelashes, and abundant scabies.

Armenian natron	6 drachms
Cinnabar	8 „
Round peppercorns	5 „
Aloes[1]	4 „

Pound together and apply.

Another, which is called "Banushag Kharîfâ", and which is used for the diseases that have been described.

Dross of brass	8 drachms
Aloes	2 „
Haematite	5 „
Burnt brass	5 „
Natron	8 „
Opium	1 drachm
Cinnabar	3 *estîrê*	

Fol. 44a.

Pound and smear on the eyelids.

Another, which is called "Shakhîkâ", and which is beneficial for moist and watery eyes, and for redness of the eyelids.

---

[1] After "aloes" comes the word ܙܘܙܐ the use of which here is difficult to explain, unless we omit the word *zûzê*, and read "Aloes 4 *estîrê*".

Acacia	40 drachms
Opium	10 „
Haematite	8 „
Aloes	3 „ and .... *estîrê*
Myrrh	6 „

Macerate and apply.

Another, which is used for swelling and scabies of the eyelids.

| Dross of brass | 4 drachms |
| Burnt brass | 2 „ |

Break up with vinegar in the sun, and dry, then make into a powder and smear the eyelids with it.

Another.

Dross of brass	1 drachm
Burnt brass	1 „
Indian nard	1 „
Burnt pepper	$^1/_2$ „

Pound together and smear the eyelids with it.

Another.

| Flower of brass (χαλκοῦ ἄνθος) | 2 parts |
| Ḳlimîyâ (cici, or croton) | 1 part |

Rub down and use as an ointment. Or, rub down with wine, and dry in the sun until it setteth, and then use as an ointment.

Another, which is good for dry blearedness of the eyes, and for the diseases that have been enumerated, and for the "claw" (or, "nail") of the eyes, and for scabies. This is an exceedingly beneficial ointment.

Dross of brass	10 drachms
Chalcitis	10 „
Peppercorns	10 in number
Ear of wheat [nard]	1 drachm

Crush the dross of brass and the chalcitis in wine, and when they are dry add to them the nard and peppercorns, first having pounded them, and rub them all down together and use like *koḥl*.

Another, a moist one, which is to be used for wet blearedness | of the eyes, and for the disease which is called "Tetâ", Page 91 and for scars, and for an overgrowth of the skin, and for "claws", and for the eating away of the corner of the eye.

7*

Vitriol (μίσυ) which is burnt	8	drachms
Myrrh	4	,,
Finest honey	1	measure

Mix up together and use.

Another, which is called the "Egyptian", and which is used for scars, and which removeth blearedness.

Cinnabar	6	drachms
Gum ammoniac	6	,,
Sal ammoniac	6	,,
Colocynth pulp	2	,,
Bull's gall	2	,,
Armenian salt	1¹/₂	,,
White peppercorns	40	,,

Run honey, as much as is necessary.

Pound, mix together, and put into a brass vessel, and use.

Another, which is used for the disease called "Bar
Fol. 44b. Ṭôṭithâ", that is to say, "Gûglâ" (σταφύλωμα?). | Take wild beetles (scarabaei), pound them, press out the juice, and smear on the eyes with a kohl stick, and the eyes will be opened and become healed.

Another, which is used for filminess and blearedness, and which healed the eyes in a marvellous manner. Take fine natron, which is called "pill salt", and rub it down with old olive oil, and smear the eyes with it, and it will cleanse them greatly.

Another, which is called the "Egyptian", and is used for wounds (or, scars), and for filminess and blearedness of long standing, and for "claws".

Cinnabar	4	drachms
Gum ammoniac	2	,,
Sal ammoniac	1	drachm

Rub down and pound, and then rub down again with honey, and smear on the eyes.

Or take burnt brass and crush it carefully in wine and dry it; crush it once more and then use it as kohl.

Or crush brass shavings in wine in the sun for seven days, and then dry them, and rub down and use as kohl.

Another, which is used for filminess of long standing, and for redness of the eyes, and for scabies. The dung of a female lizard when rubbed down to a powder and used as *kohl* cleanseth the eyes of filminess.

Also, rub down to a powder the skeleton of the *sepia* fish (σηπία) and mix it with honey, and smear on the part affected.

Another, which is used for the disease that is called "Bar Tetâ", and for scabies and running sores.

Burnt brass	4 drachms
Sal | ammoniac	4 „
Gum Arabic	2 „

Page 92

Rub down to a powder in vinegar, and use as *kohl*.

Another, which cleanseth filminess, and filleth the cracks, and removeth scabies of long standing.

Sugar of the axe (*i. e.*, rock sugar)	1 drachm
Sepia	2 drachms
Dross of brass	2 „
White lead	2 „
Fine flour	3 „
Unpierced pearls	3 „
Cassia	1 drachm
Alexandrian glass	4 drachms
Castoreum (κάστορος ἄρχεις)	2 „
Bats' dung	2 „
Gum Arabic	1 drachm
Tragacanth	1 „
Persian Gum (sarcocolla)	1 „
Lizard's dung	1 „
Starch	1 „
Ointment of lye ashes	1 „
Burnt brass	4 drachms

Rub down to a powder very carefully, and use dry as *kohl*.

Another, which is used for scabies of long standing. Take the scales of the *sepia* fish, that is to say, "gûfrâ of the sea", and rock sugar, and ashes of a [burnt] jawbone of an ass, crush them, pound them, and beat to a powder, and use as *kohl*.

Now scabies may be cleared away quickly if thou takest | Fol. 45 *a*. pepperwort plants (λεπίδιον) uncrushed, and dost lay them inside the eyelids close to the scabies, and dost leave them until the sores disappear. After this use the medicine mentioned above as *kohl*.

Another, which is called in Persian "Absĕrîshân" (*i.e.,* "refreshment"), and which is used to preserve the health of sound eyes, and for blearedness, and for eyes which run with water.

Kohl	12 drachms
Pyrites	7 „
Unpierced pearls	2 „
Camphor	1 *denķâ*
Musk	2 *denķê*
Laurus malabathrum	1 drachm
Crocus	1 „

Crush very carefully into a powder and apply.

Another, which is used for eyelashes that fall off and for strengthening the cords of the eye.

Burnt brass	2 drachms
Acacia	2 „
Dry ink of the scribe	5 „
Juice of leeks	2 spoonfuls.

Rub down together into a thick paste, and put it in a glass vessel and use as *kohl*.

Another, which is used for scabies, and hairlessness of the eyes, and for eyelashes that fall off.

Brass powder	1 drachm
Incense, ammoniacal	1 „

Rub together and use as *kohl*.

Page 93   The above are the | medicinal preparations which are used for all the diseases that take place in the eyes.

Now the claw-shaped growths in the corners of the eyes can usually be healed by means of the medicines which burn up scabies and films; but, if they cannot be healed by medicines, we cut them away with an iron instrument, and then we dry up the wounds by means of medicines.

Of the matting together of the hairs of the eyes.

Now, as concerning the superfluous hairs which grow inside the eyelids and pierce the eyes, root them out with a pair of

tweezers, and paint the places whence they have been plucked
with the blood of ticks [1](?), and they will not grow again. Or
paint the places whence they have been plucked with the
blood of green frogs, or with incense mixed with burnt shells
of the marine creature *khalzûnâ*, or with powdered steel shavings mixed with spittle. Now for about half an hour ap-
plications of these will cause intense pain, but the hairs will
not grow again.

Or take psyllium (ψύλλιον), or lye-ashes (κώνειον), and crush
in water |, and smear on the places, and leave it there until Fol. 45*b*.
it drieth them up.

Or take a goat's gall and cabbage juice, rub down well, and
apply as an ointment.

Or mix sea spume with psyllium (?), use as an ointment, and
they will cause no further pain.

(Make these to cling above the eyelids by means of a plaster,
or by means of Persian gum and spittle.)

Or [take] aloes, or ammoniac, or frankincense, dissolved in
the white of an egg, and lift the mixture up and make it stick
above [the eyelids].

Now others perforate the eyelids, down to the roots of the
hairs, and pass through them a fine needle through which is
threaded a small portion of the hair of the head, and they
draw out the hair of the eyelid by means of this filament, and
make it come out after the needle, and thus the eyes remain
as they were originally.

As concerning the disturbance of the eyes which taketh
place through blows, and the formation of blood in them, for
these things are good waters wherein white frankincense hath
been dissolved, or aloes, or myrrh, or crocus, or tamarix
(μυρίκη), after they have been pounded up and dissolved in
water, and they may be injected into the eyes, | having been Page 94
slightly warmed. Or let the blood of a dove, or of a cock,
or of a partridge, be dropped warm from the bird into
the eyes. Also sponges dipped in warm wine and laid
on the outside of the eyelids are beneficial. Or treat them

---

[1] Or, perhaps, with the juice of the croton plant.

with infusions of the herbs matricaria chamomilla and "king's crown" mixed together. Or lay on the eyes strips of linen on which have been smeared pounded flax seed mixed with goats' butter.

Now as concerning the eyelids which are doubled up inwards, a disease which is called "agânâ", we turn the eyelids back into their proper shape, and treat them, and the skin groweth straight again.

And as concerning the eyes the corners of which stick together, we treat them, and heal them by means of medicines.

And as concerning the watery cyst which cometh on the eyelids, and which hath already been described, we cut it and Fol. 46a. remove the superfluous and tumid flesh which is inside it, and we bring the edges together and bind them with bandages smeared with myrrh and aloes.

And as concerning that haemorrhoïd which cometh by the side of the nose, near the great corner of the eye, we cut it off, and root it out down to the bone, and in this way it is healed.

This is the way in which the diseases which come in the eyes are healed, that is to say, all those diseases that have already been described together with their characteristic features. Now what we have written in this Fifth Chapter concerning all the injuries which attack the eyes, and concerning their symptoms, and the manner in which they are cured, is sufficient.

HERE ENDETH THE FIFTH CHAPTER WHICH IS ON THE EYES.

CHAPTER VI.

On the composition and construction of the tongue, and on the injuries which attack it, and on their symptoms.

Now we have said above that from the brain, which is the sense of senses (*i. e.*, the source of all sensation), the first exit of the nerves cometh to the eyes and causeth in them the sense of sight, and that the second exit is divided between those muscles that move the eyes. The third exit of the nerves cometh to the tongue, and causeth therein the sense Page 95 of taste and feeling; now the nerves that make it move are those which come to it from the seventh exit, which is by the side of the beginning of the spinal cord. The tongue is an instrument of two-fold use, viz., taste and speech, and its auxiliaries in this are the lips, the teeth, and the nose. Now as it is the construction that is in the eyes which sheweth us the causes of the things that happen, so also is it the tongue which sheweth us, if we understand well, which is the primary instrument of feeling in it, and which are the members which perform its service. For we shew in other places that because the tongue serveth for two powers, that is to say, for one which belongeth to sensation and for the other which belongeth to the will, it possesseth two operations. Now we see sometimes that the motion of the tongue is injured, | and sometimes Fol. 46*b.* its sense of taste; and at other times besides this we see that its sense of touch is injured. But the nerves which belong to the sense of touch (or, feeling) are not one group and those which belong to the sense of taste another, as are those which appertain to its motions. And those which come to it from the third exit are not only those which make known the things that are to be felt, but also those that are to be tasted. For the sense of taste is injured far oftener than the sense of touch (or, feeling), although they both belong to the [same] nerves, as if it had need of the sense of exact knowledge. For that sense appeareth to be much more dull of perception than the others, as, for example, that of sight, which is well known to be the most delicate of them all. Now after sight the sense most delicate is that of hearing, and similarly also is it in the case of dullness of perception (or, touch), for after that sense cometh next the sense of taste. In the middle of the four of these cometh the sense of drawing the breath.

Page 96   Now the motion of the tongue is derived from the exit | of the seven nerves which go forth from the brain, therefore when the two sides of the brain, on the right and on the left, are affected, in that spot where the exit cometh forth, there is a danger of cessation of speech. But when only one side of the brain is affected, then the condition which is called "half speech" superveneth. Sometimes it happeneth that the motion of the tongue is slightly affected only, and sometimes that a part of it, or the whole, is rendered powerless. Sometimes, though in a totally different manner, [this] happeneth in the members which are below the head; but it happeneth sometimes that the whole of the half of the body up to the head is rigid. Now very often the speech alone appeareth to us in this injured state, besides those other functions which are in the person, the sense of taste therein, however, being in no wise injured, and the voluntary operation thereof being Fol. 47a. unimpaired. And the cause of this | is well known to you who have seen the exit of the nerves that descend from the brain; now you may decide concerning the person from the first part of the brain wherefrom descend the nerves. All the members which are below the person (or, face) descend from the side which is behind the brain, for there cometh from this side also the exit of the seven nerves which descend to the muscle of the tongue. And from these nerves arise the voluntary motions of the tongue.

For this reason it is well known that when the front part of the brain is affected by itself, the motion of the tongue only remaineth uninjured; but all the other members of the person (or, face) lose their voluntary motions and feelings of sensation on one side, either right or left. Therefore the tongue is associated closely with the brain. As the motion of the tongue is injured through that exit of the seven nerves, so also is its sense of feeling (or, perception) impeded through the exit of the three nerves. Those who cut (*i.e.*, those who dissect?) call these | "soft", as you well know. It descendeth Page 97 and is divided in that tunic which embraceth the tongue, just as also those nerves which come from the exit of the seven are divided among those muscles that give motion to the tongue.

Now the diseases which arise through the tongue are not difficult to discover, and they are mentioned in what I have said a little way back about the diseases of the mind. For the change of constitution of the muscles of the tongue like homogeneous diseases, impedeth its movements, just as the change in the constitution of the membrane that envelopeth it outside abrogateth its two powers of sense, that is to say, the sense of touch and the sense of taste.

Now organic diseases of the tongue are inflamed boils, and hard, tumid, persistent swellings [containing] pus. And if in the body proper of the tongue, and in that tunic which en- Fol. 47b. velopeth it with those soft nerves, and in that portion of the brain wherefrom these nerves go forth, there be collected any moisture (or, fluid) of a bad kind which is bitter and saltish, now the sense of taste in the tongue always acteth in this manner, [it is because] the various kinds of foods that it tasteth, whatsoever they may be, are changed into that constitution which it possesseth. And it is like that which happeneth in the eyes of those whose sight becometh dim because of a pain in the stomach; or like that which happeneth in the ears and maketh in them a great noise, although it is no external sound which causeth it, but the murmuring of the internal membranes which maketh an imaginary sound of this kind; or like the occurrence in the nose of a smell that ariseth from the foetid chyme which is there.

Now there are also times when the nerves become slack or

reduced to a rigid, helpless condition, either those which move the tongue, or those which give it the faculty of taste, or all Page 98 of them together, and in that state it is without motion and without the faculty of perception.

Of the healing of the tongue.

Now the healing of the interior. is like the healing of all nervous bodies which become slack, or are reduced to a help- less condition, or become dense, or in which are sores and pus. First of all, if it appeareth to us that the body is [over] full, we let blood from it, and then we purify the chyme which is in excess by means of cathartics. And, if it be necessary, we also cut those veins which are under the tongue, after clearing out the whole body. And we also clear them out by means of gargles and drenches in the nose, and we put wet compresses on the head and neck, and hot bandages are placed on the nape of the neck and on all the head; and we employ those means that are good for the nerves which are written down in the Chapter on nervous diseases. And if there be an inflamed sore (or, boil) on the tongue, we give the pátient a gargle made of extract of fox-grapes, or of chicory, or lettuce, or polygonum, or sour milk, or an effusion of lentils, and dried Fol. 48a. roses, and the rinds of pomegranates, | and worm-root, and a preparation of sweet wheat flour. Let these be steeped in water, and let the patient gargle with the infusion. If the boil be hard, let him gargle with ass's milk, or with goat's milk, or with honey-water, or with an infusion of worm-root, or tri- gonella, or with water in which dates or figs have been steeped. Or let him gargle with sweet wine three times a day. And we reduce his belly by means of ἱερὰ πίκρα, and we give him food which is easily digested. And if there be pustules and ulcers on the tongue we make him hold in his mouth an effusion of some astringent or drying substance, such as water mixed with tincture of mulberries, or beetroot water, or rhubarb water, or water of myrtle, or water of olives, or water of lentils, or rose water, or some "dry" wine, or water wherein rhubarb hath been steeped, or water of burnt palm fibre, or we make

him gargle with any fluid which will reduce the | sores and Page 99
alleviate the pain in the boils.

A medicine for a relaxed tongue.

Cinnamon	3 drachms
Aromatic herbs	3 „
Aromatic amomum	3 „
Nard	3 „
Laurus malabathrum	1 drachm
Aristolochia (ἀριστολοχία)	1 „
Parsley seed	1 „
Mint	1 „
Seed of dill	1 „
Stacte	3 drachms
Ligusticum	1 drachm
Seseli officinale (meadow saxifrage	1 „
Dûkôn (daucus carota)	1 „
Cummin	1 „
Aniseed	1 „
Unpierced pearls	2 drachms
Electrum	2 „

Crush and pound small and work up with honey, and give as
a draught one *gûmâ* measure in a *menkithâ* measure of wine
and water.

Another. Take gum of the pistacia terebinthus, two drachms,
and Cyrenaean fat, one drachm, work up together, and make a
stiff paste of them, and administer. Let the patient who is
suffering from a relaxed tongue hold this under his tongue. Or
let him use frequently a gargle made of mustard and honey,
and let him take baths frequently. Now they prescribe means
of healing (or, medicine), according to the nature of the disease
and according to the symptoms which are described above,
and according to the medicine (or, remedy) which hath been
laid down in [cases of] diseases in the head, and according to
that which hath been laid down in [cases of] diseases of the
nerves. Concerning, however, the tongue and its healing these Fol. 48 b.
things must | suffice for the sixth Chapter.

HERE ENDETH THE SIXTH CHAPTER.

CHAPTER VII.

Of diseases of the ears and of the hearing.

Now therefore concerning the sense of hearing, we say that it also is carried out by means of that fourth pair of nerves which come from the brain, and are expanded, and form the ears, by means of which the sense of hearing is effected. Now we do not require symptoms from all the diseases which arise in the parts of the ears which are visible because they are [readily] understood, but those which become agents that do harm to the hearing, even though no injury is visible in the cavities of the ears, must be thoroughly and clearly diagnosed. If the operation of hearing be affected, then we think that the Page 100 aural nerve | is diseased; but, if it be affected together with the other members which are in the person, we say that the cause, which manifesteth itself externally, is in the brain, and that, in that case, the brain itself is affected, either as a homogeneous or as an organic member.

Hardness of hearing and deafness come to a man either through some member of the ears, or through that nerve which cometh to them. Now those diseases that arise in this nerve and in the brain are precisely identical with those which we have described as arising in that nerve which cometh to the eyes. And those also that arise in the members (or, parts) of the ears are of the same composition as those that arise in the tunics of the eyes, but not in everything, and like those they do harm. For as there is in the eyes the crystalline fluid, so also is there in the ears that internal limit in the tubes of hearing where they touch those nerves which come from the brain. And all those parts which are outside this limit in the curvings of the tube of the ear resemble, with the exception of the crystalline fluid, those tunics which are in the eyes. Therefore the same is the case with the diseases of both ears and eyes, and it is meet for them to be comprehended from what hath been said above. | Therefore all those diseases that Fol. 49*a*. arise in both ears and eyes are alike, since they take place in bodies that are homogeneous, and are due to change of composition and solution of unity. Moreover, the others which take place in the tube of the ear through obstructions, or through some unnatural swellings, are caused by the abscesses of inflamed boils, and by the hard swelling which is in them. For in the tube of the ear there is often some boil, or pustule, or some unnatural growth of flesh, and often the wax which is produced in the tube of the ear blocketh it up. And there arise inside the ears | sounds, and singings, and whistlings, Page 101 which are caused by excessive sensitiveness of perception in the brain, or by some wind which bloweth into them; and through heat, and even through cold the ears ofttimes suffer pain. Moreover, sores appear externally, by the side of the ears, and sometimes they make a hole into the ear, and a discharge of matter cometh from the ear which causeth it pain. And besides these, diseases of the ear arise from external causes, [chiefly] through things falling into them as, for example, water when bathing (or, washing). Or perhaps there is some poisonous thing in them, or some noxious insect entereth the ear, or perhaps a stone chipping droppeth into it, or some other small substance. These are the causes which produce diseases of the ears.

Of the healing of the diseases of the ears, and of the medicines which may be applied to them with advantage.

Now if the difficulty of hearing be due to overfullness of the brain or of the whole body, or if there be noises and singings in the ears, and it seemeth to us that blood is in excess, we must let blood from the upper vein of the arm, or empty the whole body by means of some cleansing agent which is suitable for that chyme which is in excess, and by means of [*hiera*] *piḳrâ*. And then we must empty the brain by means of gargles, and excite sneezings, and then we must put into

Fol. 49*b*. the ears the medicines which | penetrate the obstructions, and cleanse the tubes of hearing. The following are beneficial in cases of singings in the ears and for the sounds which arise through a blowing wind that hath no exit through which to escape:

Warm oil of balsam and pour into the ears. Or dip the woolly fibre of plants into white naphtha, and put plugs of it into the ears. Or warm extract of absinthe and pour into the ears. Or mix the tincture of the outside of a radish with oil of roses and pour it into the ears. Or warm [some] oil of almonds and pour into the ears. Or pound up nigella with

Page 102 oil of terebinth, | and pour into the ears. Or rub down black or white hellebore, and mix with honey and pour into the ears, or work up the hellebore with boiled honey, and make little balls of it and place in the entrance of the ear. Do this once a day and it will open the hearing. Or warm up the extract of the leaves of wild bitter herbs and pour into the ears. Or the extract of wild cucumbers and pour into the ears. Or heat together castoreum with oil of laurel and pour into the ears. Or bull's gall, or goat's gall, or any other kind of gall, dissolve in some kind of oil, or in oil of roses and pour into the ears. Or oil of myrtle, or oil of narcissus, or oil of jasmine, or fine oil of balsam, or of nard, or of chamomile, or of dill, or of one of the fine oils which have been boiled with perfumes, warm it and pour into the ears, and dip into it plugs of the light-coloured woolly fibre and place plugs of it in the ears. Or take some refined fat, such as goose fat, or the fat of hens, rub it smooth and mix with it one of the different kinds of gall which have been mentioned above, and pour into the ears. The medicines which have been described [above] are also beneficial in the disease of the ears that ariseth from causes of cold. Now the following medicines are also good in

Fol. 50*a*. cases of disease of the ears which | arise from cold. Squeeze out the inside of an onion and fill it with oil and pour it, having been warmed, into the ears. Or pound up garlic with oil and pour into the ears. Or dissolve a little euphorbium in oil of roses and pour into the ears. Or pound some peppercorns in oil and pour into the ears. Or pound up rue with oil and pour into the ears. All the medicines here described are good for "cold" and for prolonged diseases, and for removing the thick chyme which causeth stoppages [in the ears]; but in the cases of the diseases which are dry, and which carry with them inflammation, and for the difficulty of hearing which ariseth from emptiness, they are injurious.

Now if there be in the ears a sore which is inflamed, or an angry boil, | and it causeth pain, the following medicines are Page 103 beneficial. Take vinegar and oil of roses, heat them and pour into the ears; or dip the woolly fibre of the woad plant (ἴσατις) plant [into the mixture] and put into the ears. Or take two drachms each of spleenwort (σπλήνιον) and *basalikòn* [nut], and rub down with oil of nard or oil of roses, and, if the patient have the feeling of inflammation, pour the mixture, when heated, into the ears, a little at a time. Or warm oil of violets and pour into the ears. Or take one of the various kinds of fats which have been mentioned, and melt it down in a double vessel over hot coals, and having warmed it a little pour it into the ears. If the patient hath intense pain due to the extreme sensitiveness of feeling which appertaineth to the sick man, [add] a little opium. Or pour into the ears one of the medicines which have been described. Or dissolve a portion of the Philo drug, which is good for pain in the colon, in the milk of a woman, or in fine white of an egg, and make warm and pour into the ears. Or treat the boils in the ears with an infusion of chamomile, or of dill, or of cotton seed, or of trigonella, and sponge the ears with these infusions. Or infuse chamomile, and dill, and mugwort, and laurel leaves, and "king's crown", and mint, and thyme in water in a vessel, the mouth of which is well plugged up, place in the mouth of the vessel a tube made of reed, and let the place where the reed is inserted be well covered up so that the steam | of the water Fol. 50b. may not escape, and then let the patient place his ear over the tube so that the vapour of the infusion may rise into his ear. And the physician shall warm one of the soothing oils which have been mentioned and pour it into the ear, and he

8

shall dip a piece of woolly fibre (lint?) and place it in it, and above this he shall place a small layer of wool which hath been lightly dipped in the oil. And such a wad of wool the patient shall lay over his ears frequently, and especially in the season of winter. And if the gatherings in the ears be not checked, but produce pus and become deep holes, and there is a discharge of pus from the ears, the physician shall employ **Page 104** such remedies as are good | [for stopping] the discharge of pus. First of all let him employ the remedies which alleviate the pain, and subsequently, when there is no pain, let him make use of those which cleanse [the ears], and remove the pus and dry it up.

Medicines which are good for ears that discharge pus. Pound iron scoriae with vinegar in the sun until the mixture assumeth a thick consistency, and heat it, and pour it into the ears once a day until the sores are dried up.

A medicine which is to be used for ears that discharge pus. Take one drachm of burnt alum and one drachm of myrrh, and rub them down with honey, and pour into the ear; smear a wad of woolly fibre with the mixture and place in the ear.

Another, which is good for a boil in the ears which is due to rheum. Take two drachms each of myrrh, aloes, frankincense, opium, and fresh shoemakers' vitriol, work up with vinegar and dissolve in a vessel, and apply (or, use).

For ears which are swollen and stink, and for the excessive growth of flesh which is called "Egyptian".

Cinnabar	8 drachms
Brass shavings	3　"
Honey	1 measure

Rub down with honey, and heat over coals until the mixture becometh red, and then place in a vessel and apply with a wad (or, wick). Or dissolve in vinegar and pour into the ears.

[Another medicine] which is to be used for pains of long standing in the ears.

White hellebore	2 drachms
Myrrh	2　"
Frankincense	2　"

Crocus	2 drachms
Opium	2 „
Castoreum	3 „
Cinnabar	26(?) „
Peppercorns	1 drachm

Rub down and mix thoroughly the myrrh, and opium, and Fol. 51 a.
frankincense and castoreum with vinegar wherein the rind of
pomegranates hath been infused until they are dissolved. Then
place with them, in powdered form, the hellebore, and crocus,
and peppercorns, and cinnabar, rub them all down together
and pour on them honey wine as much as is necessary, and
let the mixture be as thick in consistency as honey. At the
time for application dissolve the mixture in vinegar or wine,
and warm, and pour into the ears. Dip a wad of woolly fibre
into it, and place it in the entrance to the ear.

[Another medicine] which is to be used for ears that discharge stinking pus and water, and for gatherings behind
the ears, and for boils of the nose, | and for the eating Page 105
away which taketh place in the gums, and for the "pegs"
which arise in the pudenda, and for scabies, and for the
itch which is called "Egyptian".

Cinnabar	3 (?) drachms
Honey	8 „
Vinegar	4 „

Boil until the mixture becometh thick, and apply with lint; or
dissolve in vinegar, and pour into the ears.

[Another medicine] which is to be used for ears that
discharge pus, and have worms in them. Take alum, and
Cappadocian salt, break up, pound, and rub down in old wine,
and warm, and pour into the ears. Then wash the ears and
pour in oil of roses seven days; or wash them in asses' milk
and honey and hot water.

[Another medicine] which is to be used for ears that
discharge matter but have no pain in them. Take a little
salt and dissolve in women's milk, and pour into the ears. Or
dissolve glaucium with a little natron in vinegar, and pour into
the ears. Or warm up old salt water in which olives have
been soaked, or water in which fish have been kept, and pour

8*

it into the ears—this is good for old wounds—or warm up extract of leeks, and pour into the ears. Or warm up myrrh water, and pour into the ears.

[Another medicine] which is good for ears that have sores in them and that give pain.

Myrrh	1	drachm
Aloes	4	drachms
Frankincense	3	,,
Nitre pills	3	,,
Galbanum	2	,,
Crocus	4	,,
Opium	3	,,
Bitter | almonds (peeled)	24	,,

Fol. 51 b.

Crush these and work them up in vinegar, and when the occasion calleth, provided that the pain be severe, dissolve the mixture in oil of roses, and pour [into the ear]. If there be hardness of hearing dissolve in vinegar, and pour [into the ear], and if there be sores in the ear, dissolve in wine, and pour [into the ear].

[The following] are to be used for ears from which blood runneth, so that they may not drip with blood continually. Mix extract of leeks and vinegar together, and heat, and pour into the ears. Or take extract of pomegranates which have been soaked in vinegar; warm this and pour it into the ears. Or dissolve the rennet of a hare in extract of leeks, or in vinegar, and warm and pour into the ears. Or boil it in vinegar, and strain, and pour into the ears when warm. |

Page 106

Another [which is good] for ears which run with pus. Dissolve honey in hot water, and pour into the ears. Steep myrtle leaves in wine, and pour into the ears. Or dissolve alum in honey water and pour into the ears. Or mix urine and wine together, and warm, and pour into the ears. Or dissolve natron in wine and make hot, and pour into the ears.

Another [which is good] for ears in which there are swellings of pus, and which run with pus. Dissolve celandine in weak vinegar and make hot, and pour into the ears. Or use aloes and treat and apply in the same way.

For ears which have worms in them. Rub down white hellebore in vinegar and honey, and make hot, and pour into the ears. Or dip a wad of linen into this mixture and plug the ears therewith. Or blow into the ears hellebore in the form of a powder. Or drop hot extract of brambles into the ears. Or mix extract of onions with honey, and make it hot, and pour it into the ears. Or mix extract of mountain-thyme, or garden-thyme, with honey, and make it hot, and pour into the ears. Or mix extract of thyme with wine and oil of narcissus, and heat it, and pour it into the ears. Or warm the gall of any animal whatsoever when fresh in the skin of a pomegranate, and pour into the ears. Or rub down aromatic amômum and mix it with honey, and warm, and pour into the Fol. 52*a.* ears. Or mix the gall of any kind of animal which is | available with the milk of a woman, and warm, and pour into the ears. Or rub down absinthium and blow into the ears, or infuse it in wine and pour into them. And if there be worms in the ears they will die, and if there be any insect in them of any kind whatsoever, it will die.

Another [which is good] for ears into which insects crawl. Dissolve scamonia (σκαμωνία) in vinegar, or in the extract of absinthium, and pour into the ear. Or rub down long aristolochia, or round aristolochia, and sprinkle or blow it into the ear. Or rub down alum with the root of bitter rue, and mix with oil of cedar, and pour into the ear. Or mix together onion juice with the urine of young men, and pour into the ear. Or rub down fresh ḳaḵlîdîs, and mix with honey, and Page 107 warm the mixture, and | pour it into the ear.

Another [which is good] for ears that have worms in them, or in which any kind of insect is wont to crawl. Take the juice which runneth from the flesh of the bull that is being roasted, and strain when warm, and pour into the ear. Or warm extract of origanum (mint), or extract of absinthium, or extract of brambles, and pour into the ears, and the worms will come out, and every other insect, of no matter what kind. Or fill the ear with old oil, and let the patient lie down in the sun, and the worms will come out. Or bruise the leaves of the plum tree, and squeeze out the juice, and heat it, and pour it into the ears. Or warm the juice of the root of the caper tree, and pour it into the ears. Or squeeze out the juice of pungent onions, and mix it with stale (or, foetid) urine, and pour into the ears.

For ears which are hard of hearing. Mix goat's gall with goat's urine, and heat the mixture, and pour it into the ears. Or mix extract of rue with goat's gall, and pour into the ears. Or mix castoreum with oil of fennel, and pour into the ears. Or soak laurel berries or laurel leaves in water, and pour the water into the ears.

Another, which is to be used for noises which are heard in the ears through sickness. For these warm vinegar is
Fol. 52*b.* beneficial. Or juice of absinthium | and oil of roses. Or black hellebore [mixed] with oil of roses or with vinegar.

Another, for wax in the ears. Rub down burnt natron and sprinkle in the ears, and afterwards drop in vinegar. And lay wool on them, and leave them [covered] the whole night, and on the morrow wash them with warm water and oil. Or crush pepperwort and natron (one drachm each) and work up with figs, and make a plaster of this, and lay it over the entrance to the ear and leave it there for a whole night. On the morning following take the plaster off, cleanse the ear, and it will cause much wax to come out and will open the hearing.

Another, which is to be used in cases where something has fallen into the ears. If water, or a splinter of stone, or the seed of some plant hath fallen into the ears, wrap a piece
Page 108 of wool | round the top of a koḥl needle, and dip it into the gum of terebinth, or the gum of the fig-tree, or the gum [of any kind], or pitch, or any kind of sticky substance, and insert the needle carefully into the tube of the ear, so that that which hath fallen into the ear may stick to it, and then take it out. Or, if it be possible, take hold of it with a small pair of tweezers, and bring it out. And if it cometh not out, blow into the ear some greasy substance, and close the mouth and nostrils of the patient, and let him incline his ear over that side which is affected, having laid the palm of his hand over the ear, and then leap up. And they must do this also when

water falleth into the ears. Now for all the diseases of the ears which are written down in this Chapter, the above are the means that are to be employed in healing them, and the medicines that are to be used. This must suffice concerning the ears in this Seventh Chapter.

HERE ENDETH THE SEVENTH CHAPTER.

CHAPTER VIII.

Of the faculty of feeling and of all the injuries which occur to the nerves.

AFTER the discourse on those four senses and on the mem-
Fol. 53a. bers by which they are effected, | and on the injuries which
occur to each one of them, and on the means to be used in
healing them, the order of this treatise demandeth of us
to speak about the fifth property which is common to all
the members, that is to say, that by which the sense of
touch is effected by means of the nerves and muscles which
concern them and which are in all the body. Therefore we
have decided to treat of the nerves in the present Chapter.
Now the nerves do not appear to be of one kind only, as a
large number of men think, but each of them is entirely different
in its form. And first of all it is meet that we should distinguish
between each, and should describe what is the name and form
(or, kind) of each one of them, so that a man may be able
to know them, and to discern the aids which are suitable to
each. There are five kinds of body (or, composition) belong-
ing to the nerves which are stretched in minute filaments
through the whole body. | The five kinds of nerves possess
five appellations which distinguish each from the other, and
these are they: *Warîdê*, the ARTERIAL VEINS, *gĕyâdhê*, the
NERVES, *sheryânê*, the ARTERIES, *asârê*, the LIGAMENTS, and
yathrê, the TENDONS. Now these five kinds are distinct each
from the other, not only in respect of their natural composition,
and their formation, and their setting, but also in respect of
their beginnings and the places whence they derive their source.
Now we are in the habit of calling ARTERIAL VEINS those

which have the beginning of their exit in the liver, and their natural formation is like that of a pipe, and they are filled with the blood which goeth up from the liver to the whole body. And we call ARTERIES those which have their source in the heart, and they are divided and radiate from thence through the whole body. They also are of tubular formation, and they are filled with thin, refined blood, and they are moved at all times with a motion of dilatation and contraction by the heart. And we call NERVES those which have many exits from the spinal cord, and they also split up and radiate in all directions throughout the body. In them is [located] the whole power of sensation. Now these (*i. e.*, the LIGAMENTS) are simple, after the manner of the nerves, but they are different from them, because their nature is harder than that of the nerves, and because | also there is no sense of feeling in them, Fol. 53*b*. and because their places of exit are not from the brain (*mûkhâ*), nor from the brain of the head, nor from the spinal cord, but they are in the joints, and they go forth from the bones, and they bind together the joints, and they also support that other bone which is below them. Therefore are they called LIGA-MENTS, because they tie together the joints. Now the TENDONS also partake of the nature of the ligaments and the nerves, and they are in the majority of cases closely attached to the muscles, and they extend in all directions and are like very fine fila-ments. Therefore, since they have in their composition a portion of the nature of the nerves, there is in them the sense of feeling; | but inasmuch as, on the other hand, they partake Page 110 of the nature of the ligaments, the sense of feeling in them is slight.

Now there are in the body other kinds of members besides these five which have been described, even as hath been said in the beginning of [this] treatise, and there are also other differences in these other members which have been [already] described; and concerning these also it is right to say a little of the very much which might be said. The arterial veins and the arteries, and all the other members of the body, possess natural powers whereby they are nourished, but the nerves, besides these natural powers which they possess, have also rational powers (or, soul-powers) which give feeling (or, perception) and motion to the whole body. And they are the servants of the wish of the soul, for which reason they are also called "voluntary powers", even as I have said. Now the natural powers, I say, are the power of attraction, the power of grasping tightly, the power of digesting (or, rubbing down food), and the power of expulsion. These effect the service of the body in all the members of the body, great as well as small. When food entereth the body, if they are in a healthy condition, they perform their operations with avidity, whether it be in the belly, or in the liver, or in any one of the members. When food cometh to the member, it draweth it in, and graspeth it tightly, and is fed thereby, and it propelleth and casteth forth from it the superfluity which remaineth behind of the food that is unsuitable for it. This taketh place when Fol. 54a. the natural powers are in a healthy condition, | whether we wish it or whether we do not.

But since the use of the word demandeth it, we must also say, in connection with these natural powers which we have mentioned, under what condition each one of them performeth its functions. From the fact that the power of attraction is wont continually to draw to itself heat it is well known that it operateth in coldness. And the power of grasping tightly operateth in dryness, and the power of digestion in warmth and moistness, and the power of expulsion in coldness and Page 111 moistness. Now these powers are the ministers of the | natural powers: for the nerves, besides the natural powers which they possess, possess also the power of performing the wish of the soul, which they effect by means of the various motions of the body. And because of this the injuries that happen to them sometimes bring to nought the voluntary functions and sometimes the natural functions also. Now concerning the functions which appertain to the sensations that take place in certain senses in the body we have spoken sufficiently. Moreover, it is time to speak about the faculty of feeling, which is in all the members of the body that have the sense of perception which we call "feeling". And this also is in proportion to the other [sensations], and, like them, is capable of perceiving (or, feeling) the injuries which happen to it. Now it suffereth injury when there is numbness in the whole body, especially in the hands and feet, and it produceth difficulty of motion and difficulty of touch. And it appeareth clearly that this numbness ariseth through cold and through the compression of the nerve bodies, and besides these causes of numbness there is the numbness that ariseth when the marine creature which is called *"nârḳâ"* (*i.e.*, νάρκη, the torpedo fish) toucheth a man. And also when there is no external cause for it numbness ariseth from the body through a wasteful and useless course of life, and from overeating and overdrinking and from the overthickness of the chyme that most certainly proceedeth from the restriction of the purgings which are wont to take place. Now that is the restricting and immediate | cause, by Fol. 54*b.* whatsoever name one may wish to call it. For this kind of numbness is the disease which ariseth in the nerve bodies, and it impedeth the working of the power which cometh to the member from the head. And the working of this power is impeded if it happen that there be a fissure in the nerve, even as we see clearly is the case in those nerves which come to the eyes because of stoppages or because of constrictions. If there be a fissure in the nerves for the breath | of the soul, Page 112 which is as it were a kind of open road from the brain to the members to which the nerves lead, it is well known to every man that when this road is blocked up, this breath is also prevented from coming down; and because the nerve becometh constricted externally the orifice thereof is also narrowed. This also is evident. If there are no fissures in the nerves, even as the rays of the sun pass through water or air, in exactly the same manner doth the power which cometh from the head pass through the nerve bodies and descend [into the members], for it doth not escape through any fissure. And this descent [of the power] is also impeded if the nerves be changed [in composition] and become unnaturally thick, which also, according to my own opinion, needeth not many proofs, if a man remembereth the things which take place in water and in the air. Now darkness (or, fog), and smoke, and cloud subsist in the air, and prevent and impede the light of the sun from

passing through it and descending through them in a pure
state. In like manner therefore if the nerves be impeded [in
their operations] they become unnaturally hard, and impede
(or, injure) the descent of the power. Now the nerves are
injured either through being nourished upon the thick and
viscous chyme, or through being subjected forcibly to the influence of cold, or through the pressure of something which is
hard, and which constricteth them externally, or through being
Fol. 55 a. pressed by the hands, | or through being compressed by some
hard swelling, or through a rupture of the joints, or through
the pressure of a broken bone; for all these produce at first
numbness in the nerves, and finally, in some cases, cause them
to be without motion and without the power of feeling. Now
Page 113 this disease of the nerves is called | "dissolution", which is a
kind of numbness, suddenly only doth a man suffer from it,
and if it taketh place in all the nerves, it suddenly causeth the
whole body to be motionless and senseless, and it bringeth on
death immediately through the body's being deprived of the
power of breathing. If now the top of the spinal column be
injured, those members which are in the head, and those only,
perceive it, and they move as long as they are alive. If the
injury be in the brain, straightway all the members become
without motion and without the power to feel. Now these two
[classes of sufferers] are able to live, in so far as they are able
to live at all, *viz.*, those who are seized with choking (or, suffocation) only, and those in whom the marrow of the spinal
column is diseased in the upper part of the back and the two
shoulders, in that place wherefrom go forth those nerves which
come to the breast, one of the vertebrae having escaped [in-
jury], or there being there some other kind of disease; in these
also, I say, all those members which are below that vertebra
become incapable of feeling and motionless.

Now in minor diseases all the lower members become numb,
but because breath is preserved in them these do not die.
Feeling in the hands perisheth entirely, together with motion,
in those in whom the fifth vertebra of the spinal column is
affected. But in those in whom the sixth vertebra is affected
the motion of their hands doth not entirely perish, because the

front portions of the arm are preserved uninjured; and the power of feeling is not wholly destroyed when the seventh Fol. 55 *b*. vertebra of the spinal column is diseased. If the disease be in the eighth vertebra then the sense of feeling is only slightly injured; and if the ninth vertebra is diseased, the sense of feeling in the hands is not injured in any way. Now the faculty of speech | is destroyed in all those in whom that part Page 114 of the spinal column which is in the neck is injured; but when all the vertebrae of the spinal column are uninjured the faculty of speech is not always injured. Speaking generally, I should say that every perception in the body is injured when those nerves from which it ariseth are diseased. And as concerning all the members which are moved by voluntary motion through the nerves, to which also sensation (or, feeling) cometh from it to them, it is necessary that injuries in them should be of two kinds, when the nerve is diseased; one is the injury to the sense of feeling and the other is the injury to the faculty of motion. Sometimes members are seen to move, though they have no feeling in them; and others possess the power to feel without being moved. To the tongue and to the eyes no question of this kind appertaineth, because these members possess two kinds of nerves. Concerning those members which have in them hard nerves, to these it is very right to apply this question. If the skin be stripped off, and the muscle which is beneath it be laid bare, it will be seen that the nerve doth not move; but if a man toucheth it, it becometh sensitive. In this case it is right for us to think that some great injury hath come upon it, so that it may receive some portion, but only as much as it is able to feel, of the power of the soul. To set in motion the muscle is not sufficient, because sensation, though affected, still existeth in the muscle, though it effecteth nothing; therefore this feeling is able to exist, though its power is little. Now the motion of the muscle existeth when it is active and ceaseth in disease. The whole body or the whole member is entirely changed by this, and therefore hath need of the power of the soul, which is very great. The opposite | thou wilt never be able to find, that is to say, that Fol. 56 *a*. the muscle can be moved (or, set in motion), the skin thereof

being removed, and the muscle laid bare, and it will never
Page 115 have the power | of feeling.   But if the skin which is over it
lose the power of feeling it is not a matter for wonder that
the muscle itself can be set in motion.   Similarly also it is not
a matter for wonder if there be two muscles, one with motion
and the other without motion, or one which is sensitive and
the other which is not.   For as in respect of these it is possible
that the nerve of the one may be injured, whilst the other is
without injury, so it is possible that that nerve which moveth
in the skin may be injured on an occasion, whilst that which
cometh to the muscle may be uninjured.   In like manner, on
the other hand (or, contrarywise), this may happen, that is to
say, that the nerve which belongeth to the skin may be uninjured, whilst that which belongeth to the muscle may be
injured.

Of the spasm which maketh the whole body rigid.

Now when the whole body is seen to be rigid, straightway
it is known to every man that that portion of the body, which
is to it what the root is to the tree which beareth its branches,
is diseased.   For the common exit of all the nerves resembleth
a certain part of the root of a tree, and it is not like the
branch which hath a little foliage.   Now it doth resemble the
branch in respect of one of the members, that is to say, either
the hand or the foot, when it happeneth that it becometh
rigid, or that one of them becometh numb.   Now a single
member, when it becometh rigid or numb by itself, indicateth
that the source of those nerves which come to it is affected
(or, diseased), just as any branch of a tree [which withereth]
indicateth that the root thereof is diseased.   When, then, the
whole body is seized with a disease of this kind, it sheweth
that the common source of all the nerves which are below
Fol. 56b. the face | is diseased in the same way as doth that part of
the tree which is above the ground.   Now the source of this
Page 116 is | the foremost portions of the marrow of the spinal column,
and therefore all the skilled physicians apply [their] remedies
to this, even though there be no depth in the diseases such as

those which have been mentioned. For if the face [alone] of the whole body is seen to be rigid, they at once apply remedies to the marrow of the spinal column, and not only to the place where the marrow of the spinal cord emergeth. Now we often see that the lips are seized with spasm, and also the eyes, and the skin of the cheeks, and the root of the tongue. Moreover, since we have learned from the dissectors of the body that all these members are moved by muscles which receive nerves from the brain, we understand from their rigidity that the marrow of the spinal column is diseased; just as, when we see that these are unaffected whilst all the other members are rigid, we know straightway that the top of the marrow of the spinal column is diseased.

Now Hippokrates said that the spasm which produceth rigidity is due to [over] fullness and [over] emptiness. This remark of his is true, though only wise men understand it; but I myself am persuaded that the spasm which causeth rigidity of the body is due to these causes. And if every voluntary motion is seen to take place when the muscles in the members whereto they are attached are lifted up, and it is impossible for this lifting up to take place without the contraction of the muscle towards its source, it is obvious; even from this very fact, that rigidity of the body taketh place involuntarily, and that the motion of members which are in a spasm is changed from what it is naturally. As therefore in the time of con-valescence there is that effort of the will which is in the begin-ning of the nerves, now in the brain it giveth motion first of all to the nerves and so through them to the muscles, so also is it without this cause, if we find that it is possible for the nerves to be stretched from any other cause, and we are able | Page 117 to understand what are the things which beget spasms of rigidity | in the body. Now for him who hath seen the nerve Fol. 57 a. substances, that is to say the strings of the harp, and how they are frequently snapped asunder by the immeasurable condition of the air wherein they become exceedingly weak, this will not be difficult to understand. And this very thing happeneth in the body also. Now the conditions of the air in which the strings of a harp are stretched tight and are

snapped asunder are well known, for these things take place
either when the air is exceedingly dry or when it is exceedingly moist. For moisture, because it wetteth them, maketh
them swell to an unnatural degree, and they therefore snap,
and dryness hath a similar effect, as also the sun drieth up
skins and maketh them contract, and the sun being elevated
above natural objects stretcheth them tightly. And when
leather straps are dried by the fire they also are seen to
become stretched and to snap asunder.

Now from all these things which we have said it cannot
henceforth be difficult to understand the facts which concern
the members that become rigid during spasms; for this disease
ariseth in them either through dryness, which is an emptiness
or an insufficiency of moisture, or through an excess of moisture,
which is an insufficiency of dryness. And this [form of] the
disease is called by Hippokrates [over] fullness. When therefore rigidity of the body ariseth from toil, and vigil, and
poverty, and anxiety, and a dry and burning fever, such as
existeth in the insane, it is proper to assign as its cause dryness and emptiness. In the case of the man who is always
drunk, and who overeateth greatly, and who leadeth an evil
and debauched life, it is proper to think that disease which
ariseth from the opposites of [toil and want] causeth the spasms
of rigidity. Now [over] fullness is the opposite of [over] emptiness. Very often also the strings of a harp which are frequently
overstretched snap asunder. And it is for this reason that
players on the harp, when they have finished using [their]
harps | and are going to lay them aside, unstring | the strings
that are on them. Well, therefore, did Hippokrates say that
the spasms of rigidity in the body arise through [over] fullness and
[over] emptiness, for such spasms are [the results of] excessive
stretchings of the nerves.

Page 118
Fol. 57 *b.*

On Convulsion of the whole body (paralysis).

Now the disease of convulsion of the whole body is not
made known to the sense like these [above-mentioned], but it
needeth a word which will indicate when the whole body is

injured (or, affected), and when the operative sense of the nerves is affected. This we can only learn from those who dissect the body. When therefore all the nerves lose their faculty of sense and motion this disease is called "Shûtâḳâ" (apoplexy); but if this happeneth in one side of the body, either the right or the left, then it is called "Mĕrashshalûthâ" (paralysis). And it is well known that it belongeth to that side in which it taketh place, and is described as paralysis of the right or left side, just as when paralysis taketh place in one member only, it belongeth to that member only, for sometimes the whole foot only is paralysed, and at other times the whole hand only is paralysed. And the sole of the foot also can by itself be paralysed, and the whole region below the knees, and a portion of the hand only can also be paralysed. And we learn through the dissectors that all the members which produce voluntary motions in man, that is to say, the nerves which cause motion, have their place of exit from the neck and below the marrow of the spinal column. And ye may learn also and see by dissections of the nerves which work the breast, that they go forth from the marrow which is in the neck. And besides these things ye may see from the malignant ulcers which arise therein, and which are cut out from every part of it, that there is the sense of feeling in all the parts of the body which are below it, and that they (*i. e.,* the ulcers) do not produce it. And ye may see also | that Page 119 the spinal column receiveth the power of feeling and voluntary motions from the brain. And ye may also see in dissections of the | oblique sections of the spinal column, which extend to Fol. 58*a*. one half of its length, that not all the members which are below are paralysed, but only those which are placed below the side in which they exist, whether they be on the right side or on the left. Therefore it is well known that when the disease is in the place of exit from the spinal column, the power which cometh to it from the marrow is held back, and all the members which are below, with the exception of the face, are without motion and without feeling. When one half of the place of exit is affected paralysis of all the members which are below doth not follow, but only of those which are

9

on the right side or the left. Now if thou wilt observe paralysis of this kind thou wilt see that there are times when some of the members of the face also are affected, and that that side of the face which is paralysed is twisted towards the other side of the face. Because thou hast learned from dissections that the nerves are sent forth from the spinal narrow into the whole face, thou must know that in the case where certain of these, together with the whole body, are paralysed, the disease is in the spinal narrow. Now this happeneth to others who have only the face affected, as also to those who have only one member of it affected, that is to say the tongue, or the eye, or the palate, or the lip, because they have not their source in one place but in divers parts of the brain where the nerves have [their origin], for this is clearly apparent in [making] dissections. Now apoplexy, because all the operations of the soul are injured, sheweth us plainly that it is the brain itself which is affected. And the indication (or, symptom) of the seriousness (or, greatness) of the disease is the extent of the injury which the respiration hath received, for in those in whom respiration remaineth the injury to the brain is little. Page 120 Now the worst form of respiration | possible is that which barely existeth and is only carried on with the greatest difficulty, for then those who are stricken down with apoplexy die because of the destruction of their power of breathing. The fact that the members of the body do not move sheweth a Fol. 58b. man | that he is useless as regardeth the works of the world, but this state doth not produce sudden death. We have actually seen a man, all of whose other members were paralysed, cause all the members of his face to work in a natural manner. It must be well understood that his power to breathe still existed, and it would not have been possible for him to live long had this power been destroyed. Therefore we thought that this exit of the nerves, which cometh forth from the spinal column, a little way down, and which cometh to the fatty part of the breast, was seized with the disease. Now it is well known that he involuntarily voided urine and excrement.

And we have also seen other men all of whose lower members, except the hands, were paralysed through some fall. And paralysis, when it existeth in the whole body, the face being preserved free from injury, sheweth that the disease is located in the top of the spinal column; similarly also if the spasm of rigidity be in the whole body, this fact maketh known that the same spot in the spinal column is diseased, as well as the fact that no disease remaineth in the face. If the face also be affected, it sheweth that the disease is in the brain; but if the rigidity be in one member then it followeth of necessity that either the nerve which moveth that member, or the muscles thereof, must be affected. If therefore a man knoweth from the making of dissections the sources of the nerves which come to each one of the members, this man, I say, can cure successfully the want of feeling and the want of motion of each one of the members, even as I myself did in the case of a man who fell from a horse, and | hurt the upper part of his back, Page 121 and feeling was paralysed in two [of his] fingers and the half of a third. I ordered them to put plasters on the place which was injured, and the man was cured in a few days. And with the following fact physicians are not at all acquainted: that the roots of the nerves which are divided (*i.e.*, radiate) in the skin Fol. 59 a. of the hand, wherefrom cometh the power to feel, are of one kind, and that those which move the muscles are of another.

Of the diseases of the face.

Now as concerning each of the other parts of the face, there are times when not only their power of motion but also their power of feeling is entirely destroyed, and there are times when these powers are injured only. And sometimes the parts of which the operations are injured have in them some pain, and sometimes the nerves which come to them from the brain are injured, or even the brain itself may be affected; but all these are clearly defined each from the other by means of the sensations which cleave to them. Now when it happeneth that one member only is injured in its feeling or motion, or in both, it is that member itself which possesseth in it the cause of the disease, and the parts thereof suffer pain (or, are diseased) either organically or through change of condition. When,

9*

however, many members are affected at the same time we
must enquire whether they possess feeling or motion from one
nerve source, or whether from many nerve sources, when they
are in a natural state. Now ye learn that it is the third exit
[of the nerves] which giveth sensation and motion to the muscles
that are in the cheeks, and to those that are in the palate,
and to the lips, and to the ends of the nostrils (or, tip of the
nose); now to the tongue it only giveth sensation in the same
proportion as to the other parts of the mouth. Ye have seen
already that it is the second place of exit of the nerves which
setteth in motion the muscles of the eyes, and behold, it also
giveth to them the visual sensation.

Page 122 | If then all those members which receive nerves from that
third exit appear sometimes to be injured on one side, [the
physician is] bound to think that the disease is in the nerves,
but the place in the brain wherefrom go forth the nerves is
affected by the disease first of all. When, then, the brain is

Fol. 59b. diseased | on both sides, on the right and on the left, in that
spot in which the third place of exit of the nerves is situated,
it is well known that the other places that are near it become
diseased sympathetically. Therefore the nerves of the first
exit and those of the second are injured, and with the injury
to them is associated injury to the members that are in the
eyes. If, then, a muscle or a nerve be diseased in any way
whatsoever, either of itself or sympathetically, then the member
is deflected to the muscle which is opposite; for if the muscle
which moveth the right [side of the] lip becometh paralysed, the
portion of the lip which is there is not deflected on the left
side. Now when the left side [of the lip] becometh paralysed
it is deflected to that of the right side. And it is exactly the
same also with the whole jaw, and the ends of the nostrils,
but the cheeks, which, as we have learned, are set in motion
by a minute muscle, are not deflected to the side which is
opposite to that which is paralysed. Now this minute muscle
hath not nerves issuing from the third exit, but from the
vertebrae which are in the neck come the nerves which are
in its every part, except to one small portion of it which doth
receive nerves from that exit which is well known, where that small

[muscle] hath its place, which is very high therein. Now it is time for us to treat of the spinal column.

Of disease of the marrow which is in the spinal column.

Now the disease of the marrow of the spinal column teacheth us liltle, although the actual knowledge of the same is not by any means | a small matter. For if a man doth not re- Page 123 member each of the exits wherefrom the nerves issue, and to which of the members of the body the nerve cometh, he cannot know in which vertebra the marrow is diseased; but to the man who remembereth these things, the knowledge of | the place which is diseased is easy. Moreover, Fol. 60a. in respect of the symptoms [shewn] by all the other members, the remembering of the nerves that go forth from the spinal column is necessary. And behold, we can shew straightway when the first vertebrae are diseased that there ariseth correspondingly the disease which is called "suffocation" (angina); this happeneth more frequently to children than to grown men. Now it is well known that the upper portions of the spinal column are far more important than the lower; but a disease of this kind happeneth much more frequently below the first two vertebrae and the injury which ariseth from it is very much less .severe than [if it were caused by the first two vertebrae]. We have learned from those who have dissected the nerves that the membranes of the breast have their exit from behind the fourth and fifth vertebrae, and we have also learned that the breathing which is unforced taketh place through this membrane; and also the membrane which cometh from the muscle between the ribs helpeth when we have need to take a great (*i.e.*, deep) breath, just as that upper muscle which is by the neck also helpeth when we have very great need of breath. Now we do not help ourselves by means of those muscles when we need deep and violent respiration only, but also for other causes. According, then, to the knowledge of the place which is diseased it is sufficient for us to take only that which is necessary from the word[s] of Hippokrates, not only [concerning] the changes which are inside the vertebrae,

but also their curvatures externally which cling to their ex-
tensions and which take place towards the inner region; for
the nerve bodies which are there are attracted (or, drawn aside)
Page 124 by the | unnatural sores (or, ulcers) which arise there. Now
when one of the vertebrae only is drawn inwards, its position
sinketh down correspondingly in the spinal column, and when
two or three vertebrae which are placed one following the
Fol. 60b. other are drawn inwards, their | position also sinketh down.
When those which are drawn in are situated in the midst of
the vertebrae, and there remaineth one of the many in which
there is no disease, they bend outwards. Now when the drawing is to one side, the right or the left, the spinal column is
twisted towards that side. These two [facts] doth Hippokrates
mention in that Discourse which we have quoted a little way
back; for in that he said that the sore (or, ulcer) was a direct
[cause] (?) he maketh known the drawing inwards [of the verte-
brae], and inasmuch as he said that some of them incline to
one side, he indicated clearly the twisting of the spinal column.
And most accurately and excellently did he write down for us
the things which we read in his words, saying:—"In the draw-
"ing aside which is within division doth not take place in a
man." And as concerning the twisting of the spinal column
he saith that "it taketh place towards the hand", that is to
say, not downwards to the ribs, but towards the lumbar regions
or towards the feet.

Now it is meet that we should first of all know that other
things happen when the marrow of the spinal column is diseased.
A certain kind of disease occurreth in it through a change
of its constitution only, or through the chyme which is trans-
muted in it when there are present running sores, and ulcers,
and boils. And there are also other kinds of diseases which
happen to the marrow of the spinal cord when it is com-
pressed in any way, either by one vertebra or by several
vertebrae which are twisted. When the spinal cord becometh
affected by any disease whatsoever, on the right side or the
left, with the exception of the twisting of the vertebrae, then,
Page 125 I say, when the disease is in one side of it, all | the members
of the body that are placed directly below the side which is

diseased are injured in respect of their motion and power of sensation. And similarly, when the whole of the marrow of the spinal column becometh diseased, every one of the members, on the right side and on the left, which are below the region which is diseased, becometh paralysed. Now when a vertebra is deflected outwards | or inwards, it is possible that Fol. 61 a. the lower members may be injured in no wise in respect of their motion and powers of sensation; but it is possible sometimes that they may be injured according to the limit which is assigned to the disease. When then the marrow doth not become curved little by little in one of its parts, but suddenly, as, for instance, through some smashing blow, then all the lower members become injured of necessity; but if it becometh curved little by little, in a circular fashion, then no injury whatsoever taketh place in those members which are below the vertebra that hath become changed in position.

Now the twisting of the spinal column to one side doth necessarily injure those members which come to the vertebrae that have been twisted; this happeneth more particularly in the case of the vertebrae of the neck, and to a less degree in the case of the lumbar regions. As concerning the vertebrae which are joined closely together in the neck, each one of them equally formeth the orifice (or, hole) from out of which goeth the nerve; but in the case of those vertebrae which are opposite the breast, in the upper vertebra the portion of the orifice is especially spacious, and in the lower it is so to a much less degree. In the case of the vertebrae which are near the lumbar regions, the whole of the nerve goeth forth from the first vertebra. Therefore in this case (or, here) the twisting of the vertebrae to one side carrieth the marrow of the spinal column and the nerves with it, and without pressure preserveth its natural exit. When twistings arise in the neck, because the nerves go forth from the region that is between the two vertebrae, they are compressed in those portions near which cometh the twisting of the spinal column. Now the other parts which are opposite to them are distended | by these Page 126 things, but the distensions, when a large pustule accompanieth them, are also accompanied by rigidity. In the case where

there is pressure on the nerves, paralysis followeth in those
members into which the nerve on which there is pressure
entereth. Rightly, then, doth Hippokrates say that the suf-
focations that we have described in the extract from his Dis-
Fol. 61*b*. course, which hath been copied above, | follow division as far
as (or, towards) the hand, inasmuch as the hands receive [the
powers of motion and sensation] from the heel (or, end) of the
neck, but the nerves come to the muscles which are between
the ribs from the vertebrae that are placed in it, with the exception of some small [power] that cometh to them from the
upper [vertebrae]. And all the other members that are in the
face remain entirely without injury, their powers of sensation
and movement being unimpeded by the twistings of those
vertebrae which are here mentioned. Now the temples only,
as well as the cheeks, are injured, since they receive motion
from the flat (or, wide) muscle, and from those other muscles
which possess nerves that are distributed in them, and that
begin in the marrow which is in the neck, and that also are
injured through diseases of the spinal column, or of the nerves
which come forth from it. How these muscles are [placed]
and what they are, ye may learn from the nerves of the fissures which come from the spinal column, just as ye have
also learned the things that take place—which it is right ye
should know—in connection with the vertebrae that are in the
spinal column, and about its curvatures outwardly, and its
drawings in inwardly, and its twistings to one side or the other.

In addition to these it is right for you to know also that the
vertebrae depart from their positions sometimes, either through
some fall, or through some blow, or through some unnatural
swollen sore which contracteth or looseneth the ligatures that
Page 127 hold fast the vertebrae | and the nerves of the spinal column.
Now these are of two kinds generically; some of them are in
the members, inwardly, and some of them arise in an unnatural
manner, that is to say, the swollen sores, which Hippokrates
groupeth under a single name, and he calleth them "unopened
abscesses". Now it is well known that, in the section which
he wrote in the Second Discourse of the "Comings of Sicknesses", [he saith that when] the muscles which are situated

in the neck are diseased there is, perhaps, some abscess in them at a certain place, and that | therefore the physicians are in the Fol. 62a. habit of lancing any soft abscess when it is ripe. And why did he give to them the name of "the strangles" unless there was some connection between them, besides [saying that] the places which are in the breast and lungs are diseased in some particular? In this matter only is the feverish trembling of men with the strangles different from pleurisy of the ribs and from disease of the lungs, that is, they feel a contraction in their throats. Now all these diseases cause some difficulty in respiration. There is no danger of dying to those who have the strangles, but they can only swallow anything with difficulty, and they suffer to an intense degree when their power to drink is cut off, and the drink cometh from their nose. And there are cases when the abscess moveth about, and when the parts which are near the palate and the tongue are swollen, as also [Hippokrates] wrote. Now the destruction of the other functions taketh place when the tongue is diseased. Therefore it is laid upon us now to discover and to lay down the following proposition, which is based upon a large number of cases: when the places which are diseased are very numerous injury to one function taketh place: in addition to the proposition that when one of the organs of respiration is diseased, whether it be diseased itself, or whether it be affected sympathetically, it is impossible for the faculty of breathing to be injured, for the organs | of respiration are many, and with them also are very Page 128 many other members which must become diseased with them. For this reason many definitions have been sought for by us, so that when a man maketh use of them he may be able to discover the places which are diseased first of all, and also those which are associated with them in pain. A man can learn this immediately from the manner of respiration alone, and it is possible to arrive at some kind of decision concerning the place which is diseased and the disease thereof. It may happen that a patient may be seen whose whole chest is moved in breathing to such an extent that the motion passeth up from both [his] sides as far as the clavicle of his shoulder, and Fol. 62b. passeth through the sides of his breast from behind, and the

motion may also be seen going up as far as his collar-bones.
Respiration of this kind affordeth a demonstration of three
diseases. The first is when there is excessive heat in the lungs
and in the heart; the second, when there is a certain amount
of constriction in one of the organs of respiration; and the
third, when, in addition to these, the muscles which set in
motion the chest are infirm and weak in power.

It is therefore right when a man seeth this kind of respiration
to make an examination of it. First of all, he must touch the
arteries (*i. e.,* feel the pulse), for these shew us the excess of
heat, which is written in the foreknowledge that is from them.
Secondly, we must examine the breathing and see whether it
be abundant, and sudden, and accompanied by great puffing.
Thirdly, in addition to these, we must feel the breast outside.
When we find that it is hot, then we possess all the symptoms
of an excess of heat. And these symptoms are accompanied
by transient flushes of heat in the face, and glittering of the
eyes, and fiery heat in the head, and intense thirst, and dryness of the tongue and roughness, and the patient will confess
Page 129 (*i. e.,* complain) that he is being burnt up with fire. | Now, if
the symptoms of inflammation are not very numerous, and the
breast is moved outwards to a great degree, there is some
constriction (or, pressure) on the organs of breathing, and
therefore some part in the region of the throat is diseased, or
a superfluity of moisture filleth the lungs, or chest, or some
abscess hath burst in one of them. Of this kind [of abscess]
is that which is called *apesṭîmâ* (αποστημα). Now I have already enumerated all those diseases of the throat which take
Fol. 63a. place in it either directly or sympathetically, and those | which
take place in the lungs and chest we shall define most carefully a little way further on. At present we have decided to
speak concerning the diseases of the spinal column, and concerning diseases in general, and concerning that disease which
is called "weakness" (*i. e.,* debility); for sometimes a man
suffereth also from weakness. Now weakness taketh place in
him sometimes through the brain, since he receiveth from it
two powers, namely, one which produceth the sense of feeling
and one which produceth the faculty of motion, and sometimes

also through the spinal column, in which case he suffereth
through change of its constituent parts, either in some part of
it or the whole. Now this [fact] doth not even enter the
mind of many physicians, although they often see men so
greatly prostrated by sicknesses that they can only move with
the greatest difficulty the fingers of their hands, though mean-
while they breathe with all parts of the chest, and have not
the slightest heat of inflammation which would compel them
to gasp. Now this happeneth to them through the following
cause: that breathing which taketh place without force (*i. e.*,
effort, or straining) is effected by the membrane, which is
called "parnôs", by itself during those times when it is quiescent
and healthy; but when this membrane is [not] in a healthy
state it is unable by itself to supply the needs of respiration,
and those muscles which are between the ribs must help it. | Page 130
And when a man is in the most urgent need of breath, those
upper muscles also render assistance, for it is obviously they
which set in motion the collar-bones. When therefore one
seeth a man with all the parts of his breast in motion, though
he is not hurriedly drawing his breath, then let him examine
the amount of the motion of his breast, and of his nostrils;
and if he findeth that his nostrils are distended, and that he
is moving his breast in an outward direction, the patient is not
like those who, through the excess | of heat in inflammatory Fol. 63*b*.
fevers, have their power of respiration in an unhealthy (?) condition. Now these very often move the whole of their breasts
in an outward direction, but again not like those patients who
do so because of the constriction of one of the organs of
respiration, such as is wont to happen in the strangles, and in
the sudden discharge of rheum. In both these cases there
ariseth shortness of breath, even as it is also [said] in that
Book which is called simply "Of the lungs", for these are they
who exhale their breath in great puffs and hastily. As therefore through the unhealthy (?) condition of the power [of breath-
ing] all the parts of the breast are compelled to move in an
outward direction, so also through some small sickness one
class of the parts of the chest only moves.

And behold, that skilled physician Secundus, as one who

was able to devote himself closely to those who were sick,
used to say that he could perceive (or, feel) distinctly the
weakness in the power [of breathing] which was diffused [in
the breast], and for this reason he was obliged to know of
[the movements of] those muscles which were between the
ribs, and sometimes also of those upper muscles when [the
patient] drew a breath. And when he fastened some belts
(or, bandages) round the parts of the patient's body which
were in the direction of his epigastric region, the respiration
of the membrane [of the breast] (or, diaphragm?) was by it-
Page 131 self sufficient for him | during the time in which he rested
therein. Now whether the weakness arose from that muscle
which is in the diaphragm, or from those nerves which come
to it, or from both, the matter is one which is too difficult for
me to explain, but the weakness must have been due to the
respiratory [organs]. And [Secundus giveth another instance]
of a man who was once kicked so severely on his collar-bone
by a horse that there was danger of death overtaking him,
because there was an abscess on the membrane, which they
[sometimes] call "parnôs", and sometimes "diaphragm", and
sometimes the "girdle". Now although this man was delivered
from the danger of death, the weakness of this membrane
remained with him always. And in the case of another man
Fol. 64a. who was cured | of a pulmonary complaint, and who was not
robust, his arm, both the front and the back, as far as the
elbow (ἀγκῶν), and also no small portion of the forearm (?)
below his elbow, and the part which reached to the top of
his fingers, remained without feeling. His fingers also were
slightly injured, and they could not move. Therefore in the
case of this man the nerves of the first and second muscles,
and of those between the ribs were injured, for from these
cometh the first part, which is of no small depth, and it is
mixed with the part which is in front of it, and then it is also
spread through many parts, which ye see in dissections; for
from these they come to the top of the fingers, within the inside of the arm. Now this nerve is the second, and it is also
very fine, and it is separate from the other nerves; and it
cometh through an opening to the arm, under the skin, and

it is divided in the skin on the front and back of the arm.
Now this man | was quickly cured when a suitable ligature Page 132
was placed on the place whence the nerves issue over the
first and second muscles, which are between the ribs, just as
other men whose legs were slightly paralysed have been cured
by means of the ligatures which were placed over the girdle
of their backs, above that portion of the spinal column from
which issue the nerves that descend to the legs. We do not
apply any kind of ligature to the legs which are paralysed,
because the disease is not located in them but in the marrow
of the spinal cord.

Another man had a great quantity of pus at a place which
was near one of his toes, and when the flesh in front of the
bone which was there was cut away, the nerves of his foot
were visible; and when the place was healed he found the
movement of his foot to be stiff (or, heavy). From this fact
I decided (or, argued) that some portion of the abscess remained
there, and that it had become hard | in one of these nerves. Fol. 64b.
Wishing to lighten for him the pain in his diseased limb, I put
ligatures on it, and I healed the man completely.

Now what we have stated in respect of the members which
are in the face is also true in respect of all the other members. When any of the numerous operative faculties becometh
injured, provided they be in one place, it is possible that all
its muscles are diseased through some internal cause; now it
is also possible that some internal nerve of all these muscles
may itself be the thing which is diseased. For, behold, a
certain man was once catching fish in a river, and he became
so chilled in those parts of the body which were round about
his anus and his bladder that he evacuated and made water
involuntarily. This man was cured at once by means of hot
bandages which were placed over the muscles | that were Page 133
diseased. And another man, through some unknown cause,
fell prostrate on the ground in agony. [I placed] many bandages on him, and only with the greatest difficulty, and only
after a very long time was he made whole, because these
nerves that were in the bone above his anus were diseased.
Therefore what I said immediately after I began my dis-

course I say again now, and then I will pass on to another matter.

Now when a man knoweth each of the fissures and each of the nerves which issue from the spinal column, and to which member they come, this man will be able to recognize accurately the places that are diseased. And ye have actually had experience in testing these statements during the practice of [your] profession, because ye have often seen the obvious benefit which cometh through such knowledge to those who are afflicted. For many physicians not only vexed the feet and hands vainly and uselessly with the hot bandages which they placed upon them night and day, imagining that the place which was diseased was in the spinal column, or in one Fol. 65a. of the nerves which issue from it, but they, | only a very short time ago, stripped the skin off the head of a man by placing on it bandages which caused very great heat in it, thinking that by means of them they were bringing back the faculty of feeling which had been greatly injured. This man we ourselves healed, for we found out the place which was diseased by asking him questions concerning each one of the primary causes of the illness and other matters, among which were rain and a strong wind. And he said that when he was travelling along the road he was so much wetted about the region of his neck by the skin covering which was placed on his head that he could feel quite distinctly the bitter cold which attacked him. Now inasmuch as I was persuaded that four nerves went Page 134 up to his head from the first vertebrae of the neck | and that it was from these that the skin of his head acquired the faculty of sensation, I understood easily which part was diseased. And when I had laid bandages upon this part the nerves of the head and the diseased part of the neck were healed immediately, for the nerves themselves were in no wise diseased. Yet because the physicians had only very little knowledge of the nerves which were in each of the parts of his body, and it was easy for them to apply things which gave him relief in some small portion of the beginning of the nerve, for this reason [I say] they were vexing those members which were not diseased.

Now on many occasions I have made it clear to you that the roots of some of the nerves are in the marrow of the spinal column, and that some of them, like as it were branches which have their exit from the spinal column, are divided [in the members]; and these again, being split up and divided, come to the great and little parts of the skin. Therefore it is not right for us to wonder at the men who dissect the apertures of the body, and to blame them because they are not able to gain touch of them; for some of them give the faculty of sensation only to the members, and some of them give the power of movement only. Because of this in all the kinds of paralysis which take place in the members, when the | nerves Fol. 65*b.* which give the power to feel are injured, the member is prevented from feeling, and if the nerves which set it in motion are injured, its power of movement is paralysed. And if both kinds of nerves be injured, the power of movement, as well as the power to feel, perisheth. Now sufficient hath been said concerning the nerves and concerning the matters which take place in them, and it is therefore time for us to speak about the means to be employed in healing their diseases.

Now the system laid down in this treatise demandeth that we should first of all diagnose the case and then apply means for healing it; for if a man is not able to make | a diagnosis Page 135 he will not be able to cure the disease properly.

Of the healing of rigidity of the body or of paralysis, which taketh place in the nerves of the whole body or in one of the members thereof.

Now if rigidity or paralysis taketh place in the whole body or in one of the members, we recognize that it is due to fullness. If it seemeth to us that there is blood to excess in the body, we must first of all let blood from the middle vein of the arm, or from the upper vein, and then we must empty the whole body by means of a purgative which shall clear up the chyme that hath caused the sickness. And we must administer strong clysters of the plant centuary (κενταύριον), and the plant paganum (πήγανον), and anoint with hot oil, and bathe and foment the head and neck, and apply tinctures made of the finest scents, and plasters which are indeed strong, and medicines which will break through the obstructions of the head and nerves (now these are to be thrust up the nose), and which are written down in the Chapter on headaches, and those gargles which are also written down there, and which expel and bring down [moisture] from the mouth, and relieve the head and nerves, and washings (or, baths) which settle the chyme and strengthen the nerves. And this also must thou remember: rigidity of the body is the tension of the nerves, and it is necessary to make them loose and to soften them; and paralysis is a slackening and looseness of the nerves, and

Fol. 66 a. it is necessary to tighten them. | If these things happen through fullness, they need emptying; and if they happen through emptiness and weakness, it is necessary to fill and to strengthen. If that rule be correct that that which is opposite healeth that which is its opposite, we must always and in every sickness accept it. In cases of fullness we must empty out, and in cases of emptiness we must fill, and what is tense we must slacken, and what is slack we must tighten, and what is warm we must cool, and what is cool we must warm, and what is

Page 136 dry we must make wet, | and what is wet we must make dry, and we must cure also that which ariseth in the composition of the mingling elements by means of a composition of an opposite character. We must also be convinced of this fact, that it is easier by far to empty the body than to fill it. This also doth Hippokrates hint concerning the obstinacy of paralysis, when he speaketh about that which taketh place through excessive emptiness, saying, "It is better for rigidity of the "body to take place through fullness than through emptiness." Which is the better? that the nerves should be feeble naturally, as is the case in old men, or that this effect should be produced by some preparatory course of treatment? The fullness [caused] by medicines which are administered in prolonged sickness we obtain through the cumulative effect of simple medicines. Sometimes we employ simple medicines, and sometimes compound medicines, according to the obstinacy (or, difficult characters) of the diseases. And fullness is also obtained by means of delicate foods which do not produce harmful superfluity, and by abstention from copulation, and by sleep, and by the wakefulness which is immeasurable [*i. e.*, insomnia].

Of the medicines which are to be used for nervous diseases. First of all as concerneth POTIONS. Now we employ in cases of rigidity and paralysis which arise through fullness those well known medicines which have been selected as the result of experience covering a long period of time, viz., *hiera lĕgûrîâ*, or the electuary of Theodoretus, or *hiera pîḳrâ*, or other [similar] medicines, which because of their constituent parts have been prescribed in protracted sicknesses caused by the head. These are the principal medicines for nervous paralysis:—

| **Purgatives of euphorbium which are used for paralysis Fol. 66*b*. and for the active chymes which are secreted by the nerves and in hemiplegia.**

Agrîḳôn (ἀγαρικόν, or fungus)	3	drachms
Pulp of colocynth	3	,,
Euphorbium	3	,,
Sagapenum (σαγάπηνον)	3	,,
Bdellium	3	,,
Aloes	2	,,

Rub down and work with cabbage juice into a paste. Administer in a draught one gramma (γράμμα), or half a drachm, or if a large quantity is necessary, administer one drachm.

Another kind of purgative (or, pill) which is used for rigidity, and paralysis, and pain in the back, and gout, | for Page 137 it expelleth thick phlegm.

Ammoniac	4	drachms
Bdellium	4	,,
Sagapenum	4	,,
Aloes	4	,,
Opopanax	4	,,
Castoreum	4	,,
Seed of rue	4	,,
Euphorbium	3	,,
Pulp of colocynth	7	,,

Rub down and work into a paste with the juice of fox grapes, or of cabbage. Administer as a draught one drachm in hot water.

Another [medicine] which containeth pepperwort and which physicians use in palsy and hemiplegia, and in diseases of the excretory organs and the tendons, and for pain in the belly, and colon, and fundament, and it expelleth the active chymes from the body.

Green myrobalsum chebula	10 drachms
Aloes	20 „
Ginger	2 „
Peppercorns, round	1 drachm
Peppercorns, long	1 „
Mustard	3 drachms
Indian salt	4 „
Pepperwort	4 „
Pulp of colocynth	4 „
Pĕnîd (?)	4 „

Make a paste mixed with cabbage juice, or the juice of fox grapes, and administer two drachms in a draught of hot wine.

Another [medicine] which is used for rigidity of the body.

Oil of mulberries boiled and mixed with *hiera pîkrâ*	2 drachms
Infusion of thorns	2 „
Rue	2 „
Spica	2 „
Seed of rue	2 „
Dried grapes	2 „

Make an infusion of these, and take one measure of it, and four drachms of mulberry oil, and one or two drachms of *hiera pîkrâ*. Let him drink the *hiera pîkrâ* one day, and on another day the mulberry oil in the infusion which hath been described, and let him do thus for seven days. At the ninth Fol. 67a. hour | let him eat bread with broth made of olive oil and chickpeas and leeks and aniseed and parsley and rue; and if he be feeble let them put into the broth francolins and turtledoves. At eventide let him gargle his throat with mustard and honey, or with a decoction of camphor root and honey, or with extract of aloes and thyme and sour grapes, which have been steeped in old sweet wine from evening till morning, or with a decoction of pyrethrum and origanum majorana (*i. e.,* marjoram) and mountain grapes and thyme | and long Page 138 peppercorns and mint, or with those excellent gargles which have been written down in [the Chapter on] headaches. And at first let him vomit after food, and later before food, either with the help of radishes and honey, or oxymel, which we are in the habit of using to cleanse the stomach.

It is also useful to them to drink mulberry oil prepared in the following manner.

Oil expressed from mulberries	40	*estîrê*
Myrobalsum chebula	4	„
Belilkê	4	„
Amlag	4	„
Peppercorns (round and long)	4	„
Ginger	4	„
Rue	4	„
Thorns	4	„

Pound these [drugs] together and add to them four measures (*dûlkê*) of water, pour the mixture into a vessel and place it on a soft (*i. e.,* slow) fire, and boil until it is reduced to one half. Strain the liquid off the drugs, and pour into it oil of mulberries, and then boil the mixture until the water hath evaporated. Take [the residue] and administer in draughts with the infusions which have been mentioned above; and anoint with the oil the members which are relaxed or palsied, for it will be of benefit to them.

[Another medicine].

Mustard	5	*estîrê*
Mint	5	„
Nigella Sativa	10	drachms
Sesame seed, husked	40	*estîrê*

Mix these together, and grind them, and squeeze out the oil from them, and administer in a draught of infusion of fennel five drachms, or seven drachms, and anoint the whole body with this oil also. And if the members of the face also are

paralysed, and the cause thereof is the brain, we must put
Fol. 67*b.* into the nostrils of the patient | crane gall, or hawk gall, or
wolf gall, or turbot (?) gall, or any other kind of gall which is
available, [mixed] with the juice of beet, or the juice of marjoram, or the juice of conyza (κόνυζα, *i. e.*, elecampane, or
fleabane), or with the juice of that root which is called "mouse's
ears", or with a little euphorbium mixed with the milk of a
woman. Of this last we must put into the nostrils of the
patient about one spoonful. Or dissolve sagapenum in *sôsan*
Page 139 oil, and pour it into them. Or dissolve a little myrrh | in honey
and the milk of a woman, and pour into the nostrils. In short,
all the medicines which have been prescribed for prolonged
pains in the head are also very useful for prolonged pains (or,
sicknesses) of the nerves.

Another medicine which is to be used for rigidity. Put
some sweet oil of olives, or narcissus oil, or fennel oil, or
chamomile oil, warm into a bull's bladder, or a bag made of
some soft substance, and place it on the neck of the sick man,
or on any other place which is diseased, and let him wear it
continually. And anoint the spot which is diseased from time
to time with oil of rue which hath been prepared in the following manner:—Take fresh leaves of the rue plant one handful,
and pour on them olive oil sufficient to float over them, and
boil it over coals which are not flaming. When the mixture
hath boiled, strain off the oil and throw other leaves into it,
and boil the oil and strain it [as before]. Throw leaves into
it three times, and strain the oil. Then rub down castoreum
and throw into it, and anoint the parts of the body which are
diseased. Now others mix a little euphorbium in it with the
castoreum, and make it very hot. And place the patient in
an infusion of cotton seed and trigonella and "king's crown"
and "live for ever" and guimauve (marshmallow), and anoint
him with that oil before and after he batheth.

**Another [medicine] which is useful for those who are
seized with spasms, or with pains in the excretory organs,
or with pains in the kidneys; this also is to be used when
bathing.** Infuse a large quantity of fennel in water and sweet
Fol. 68*a.* oil. [Take] | a hyena, or a fox, or a little puppy, and soak in

this water until the flesh and skin are stripped off and only
the bones remain. Pour off the water and let the patient sit
in it twice a day. And apply to the patient outwardly plasters
and bandages soaked in fats of all kinds which possess the
power to build up and to warm, such as calf's fat, or asses'
fat, or the fat of the wild ass, or | the fat of the stag, or the Page 140
fat of the lion, or the fat of the wolf, or the fat of the hyena,
or the fat of some bird. Spread the fat when mixed with
medicines on bandages, and lay these on the diseased parts.
And let there be unguents on the head also, of the kind which
give warmth, provided there be no fever or inflammation or
excessive thirst, such as oil of chamomile, or of fennel, or of
narcissus, or of nard, or of pine, or of myrtle. If there be
fever and inflammation and thirst let oil of mulberries or oil
of violets be placed on the head. And let them make for him
purges of oil of rue which hath been mentioned above, mixed
with a little castoreum, and tinctures of fennel (or, dill) and
centaureum. And let them place in his anus the following
pills after he hath ejected the purge.

Black hellebore	2 drachms
Natron	1 drachm
Doves' dung	1 „

Rub down and make into a paste with boiled honey until it
acquireth a thick consistence, and make into boluses the size
of acorns, and insert in his anus. And after he hath ejected
them inject asses' milk and oil of sheep, or of cows, or of the
fat-tailed sheep, or melt down one of the oils which have been
enumerated, and inject as a purge.

Again, the ointments which are to be used for rigidity,
and for paralysis of the nerves, and for pain in the excretory organs and in the kidneys, and in hemiplegia.

Rub down pyrethrum by itself with oil and anoint [the body].

Mix euphorbium with oil and anoint the body.

Or let them make for the patient the following unguent
which is called "coriander" (or, "plantain").

Estûmkê (?)[1]	1 ounce

---

[1] The text seems to be corrupt here; read *estûrkâ* "styrax"?

Stacte	2	ounces
Wax	3	„
Oil of balsam	4	„
Oil of nard	5	„

Fol. 68 *b*.

Rub down smooth, mix together, and anoint [the body], or smear on a linen rag and moisten the places which are diseased.

And [1] **the unguents also of a warming kind which are to be put on the head if there be not fever, or inflammation,** Page 141 **or much | thirst,** such as oil of chamomile, or oil of narcissus, or oil of nard, or oil of pine, or oil of laurel. And if there be fever and inflammation and thirst, oil of roses, or oil of violets, let them put on his head. And let them make for him purgings of oil of rue, which have been described above, with a little castoreum, and an infusion of fennel (or, dill) and centaureum. And let them place in his anus, after he hath ejected the purge the following pills: two drachms of black hellebore, and one drachm of natron, and one drachm of doves' dung. Rub these down and work them up with boiled honey until they obtain a stiff consistency; make the boluses as large as acorns and place them in his anus. And after he hath ejected these inject into him asses' milk and sheep's milk, or barley water and sheep's oil, or cows' oil, or oil of the fat-tailed sheep, or one of the unguents which have been described above. Rub it down and put it into him as an injection.

And the unguents which are good for rigidity and paralysis of the nerves, and for pain in the excretory organs and kidneys, and for hemiplegia.

Another unguent, [called] "Indian" and the "peppery", **which is good for paralysis and palsy, and for attacks of** Fol. 69*a*. **wind which twist | the nerves, and for pain in the back,** **and sciatica, and for pains in the loins, womb, and bladder.**

Root of the lily (*shar*, or, *shal*)	10	drachms
Par	10	„

---

[1] The following prescriptions have been repeated inadvertently by the scribe.

Bal (caper berry)[1]	10	drachms
Hazelwort (or spikenard, or male nard)	10	„
Aromatic kostos (κόστος)	10	„
Elecampane (inula helenium)	10	„
Pepperwort	10	„
The nut of vomiting	10	„
Zarûnbâd (*not* zarûnbar)	10	„
Doronicum scorpioides	10	„
Peppercorns, long,	10	„
Iris florentina	10	„
Dârudâd		10	„
Or cedar wood,	10	„
Or cypress wood,	10	„
Goats' milk	2	„
Water	2	measures (*dûlķê*)
Sesame oil	1	*lîṭrâ*

Page 142

Pound the drugs thoroughly well, and then pour on them the milk and water and boil the mixture until it is reduced to one half. Then pour in oil, and boil over the fire until the water and the milk have evaporated. Strain the oil and anoint therewith the nerves which are paralysed, and mix it with the plasters which are used. Pour some of it also into the injection for those pains which have been enumerated above. And inject it also into women whose wombs have contracted a chill, and who suffer from stoppages of old duration.

Another unguent, called "Persian", which is beneficial in cases of paralysis of the nerves.

Resina (ῥητίνη, راتينج)	45	drachms
Camphor oil	20	„
Wax	16	„
Opopanax	4	„
Galbanum	4	„

Pound finely the resina, dissolve the opopanax in oil, rub down the wax with the camphor oil, then pour on the galbanum and mix all these together, and anoint the body therewith, or smear

---

[1] Shar, Par, and Bal are Indian names of drugs.

it on a strip of linen and lay it on the parts which are diseased.

Another [unguent] which is to be used for paralysis and apoplexy.

Aromatic costos	40 drachms
Cassia	16 „
Leaves of the marmahôz (مرماحوز)	12 „
Laurel berries	12 „
Castoreum	10 „
Styrax of honey	16 „
Sesame oil	4 *lîṭrê*

Pound these very finely, all except the styrax and the castoreum, and macerate the mixture in wine for a whole night and day, pour on it the oil, and boil it over a smokeless fire until the wine is evaporated, then, having put in the styrax and castoreum, lift it off the fire and strain the oil, and use.

Fol. 69 *b*.   **Another.** Steep | the root of bitter herbs in oil and rub the
Page 143 body with it. Or | dissolve styrax in iris oil and use as an unguent.

Another [unguent] made with pine berries, which is good for all the pains and shiverings which occur during fevers.

Pine berries, viscous, fresh and small	2 *lîṭrê*
White styrax	1 *lîṭrâ*
Sweet olive oil	3 *lîṭrê*

Pound the pine berries and mix with the styrax, and work them up well together, then dissolve the mixture in oil, put it in a glass vessel, and hang it up in the sun for forty days. Then take it down and use the unguent.

An Indian unguent called "Leshîd", which is good for paralysis and rigidity of the nerves and muscles, and for every pain which is due to cold. It may be drunk(?), or used as an unguent, or may be mixed with the medicaments which are used in these sicknesses. And it softeneth hard abscesses.

Alînôn, that is to say, resin,	10 *estîrê*
Long peppercorns	20 „

Indian salt	5	*estîrê*
Rock salt	5	"
Sweet Lâsarpìcium (silphium)	15	"
Red grapes (raisins)	2	handfuls
Dry white sugar	1	*estîrâ*
Peeled garlic, uncrushed,	5	*ḳapîdê*
Hot milk of cows and goats	12	*liṭrê*
Water	2	measures (*ḳôlâ*)

Put the dry drugs and the garlic and the milk and the water
into a saucepan, and boil until only one fourth of the mixture
remaineth.   Then strain, and squeeze out the [solid] stuff, and
throw it away.   Into the juice which remaineth pour forty-two
measures (*liṭrê*) of pure cows' oil.   Then throw in the drugs,
having pounded them very finely, and boil the mixture until
the oil showeth clear above the medicaments.   Shake gently
and boil again until all the water hath evaporated.   Take the
saucepan off the fire, and let it cool, and then skim off the
oil from | the liquid and the medicaments, and pour it into a Page 144
glass phial, and squeeze the medicaments dry and lay | them Fol. 70a.
aside by themselves.   Use the oil in all the sicknesses which
have been mentioned above.   It may be given as a draught
mixed with an infusion of thorns, and it may be applied to
the body externally as an ointment, and it may be poured
into injections, as well as on the plasters which are employed
to soften hard abscesses.   This oil is to be drunk as follows:—
During the first week one *ḳaisâ*, in husked-barley water, or in
a decoction of thorns.   During the second week two *ḳaisê* each
day; and let the patient drink from the time of morning.   And
in the evening let him anoint the whole of his body with the
oil, and he must keep himself away from everything which is
acid or salt, and from everything which is cold.   His meal
shall take place at the ninth hour, and it shall consist of broth
made from white meat, or from young doves, or from a chicken;
and let him drink old wine with a bouquet.   Apply the coarser
parts of this oil to all kinds of hard abscesses, for it will arrest
scrofula and cancers, and it will open tumours of the womb,
and it will soften dried nerves which are rigid and will make
them supple (or, pliant) again.   And it may be laid upon

diseased parts of the body by itself in the form of plasters, and it is a medicine of the first and most important kind.

Another [unguent] which is to be used for paralysis.

Peppercorns ⎫
Pyrethrum ⎪
⎬ [in equal quantities]
Euphorbium ⎪
Castoreum ⎭

Pound and mix with oil and use as an ointment. Let the patient drink the pure tinctures of aniseed and of irises and of rock parsley, and the dregs which remain from these medicaments. And from time to time let him take one gramma of castoreum mixed with an infusion of rue and honey; or let him take aristolochia, and rub it down, and take as a dose one drachm mixed with hot water and honey. Let him drink antidotes (θηριακός) in wine mixed with water, and let him relieve himself by the application of some warming unguent, and anoint himself therewith, and let him inject an injection which shall be made as follows:

Fennel seed	1 handful
Thorns	1 bundle
Rue	1 bunch

Page 145

Boil these well in water, and then of this water take three measures (*menḳeyân*), and oil of the rue which hath been pre-
Fol. 70*b*. scribed above, and one drachm of castoreum, | mix them together and inject at the time of going to bed for three nights. Or let the physician make the following injection:

Oil of rue	6 drachms
Goose fat which hath been well rubbed down	4 „

Mix them together with a decoction of fennel and inject. Or sprinkle on the patient's hands and feet an infusion of pistacia lentiscus. Or inject cows' oil mixed with an infusion of fennel and thorns, and sprinkle on his hands and feet an infusion of gentian, or of *esnîd espîd* (root of white sandal), or of the twigs of the cypress, or of acacia, or of blackberry. Infuse these in water and pour it on his hands and feet, and let the physician pour it into a large vessel and let the patient sit in the water.

Of other medicaments (or, plasters) which are to be used for paralysis and rigidity.

Wax	5	*estîrê*
Euphorbium	2	drachms
Round peppercorns	2	„
Moist zôpâ[1]	2	„
Ammoniac	2	„
Crocus	2	„
Oil of balsam	2	„
Castoreum	2	„
Juice of terebinth	2	„
Opopanax	2	„
Galbanum	2	„
Styrax	1	drachm
Cow-oil	1	box

Rub down smoothly the wax with oil, pour on the mixture the other liquid substances, and rub down, and pound them into a dry mass; then throw on them the [dry] substances, and work them all up together, and apply, or spread a plaster of it on a strip of linen and fasten it on the neck, or on the places which are paralysed.

Another [unguent] which is good for pains.

Opopanax	1	drachm
Myrrh	1	„
Alkali of the fuller	1	„
Cypress	1	„
Castoreum	1	„
Gum	2	drachms
Mountain rue	2	„
Frankincense	2	„
Seed of agîs	1	drachm
Pyrethrum	1	„
Galbanum	1	„
Euphorbium	1	„
Wax	7	drachms
Cedar gum	7	„

---

[1] = *zôftâ* (?) a Persian medicine.

Page 146

Oil of the wild cucumber	3 drachms
Oil of gall nuts (oak-apples)	1 drachm
Oil of myrtle	3 drachms

Apply by smearing on all the members which are paralysed.

Another plaster which is to be used with fine flour, and which is good for those whose members are paralysed, and for those whose nerves are dried up.

Greek bread	2 pieces (?)
Ammoniac	3 ounces
Bdellium	3 „
Resin of shells (shell gum?)	3 „
Cotton seed	1 ounce
Trigonella	1 „
Dates	1 „
Figs	1 „
Mallows	1 „
Chamomile	1 „
Wine, with a bouquet, sufficient for macerating all these drugs in it	
Gum of cedar	1 ounce
Gum of terebinth	1 „
Styrax	1 „
Oil of nard, or oil of camphor	1 „
Calf's fat	1 „
Wax	4 ounces

Fol. 71 a.

Prepare in the proper manner and apply by smearing over the members which are dried or paralysed.

Another, from Ḳatganres, which is to be used for fatigue, and for tension of the nerves, and for spasms in the back; and for convulsions.

Oil of roses	3 ounces
Camphorated oil	3 „
Oil of glaucium	3 „
Anrînôn oil	3 „
Wax	3 „

| Gum of terebinth | 3 ounces |
| Fine honey | 3 „ |

Boil the honey by itself. Rub down smooth the moist in-gredients, and pour them into a mortar and pound them all together, and take and apply.

Another [unguent] which is good for every kind of pain, and for rigidity and for paralysis of the face, and for the whole body.

Gum of terebinth	56 drachms
Wax	24 „
Ammoniacal incense	36 „
Oil of the fat-tailed sheep	44 „
Boiled honey	24 „
Old oil	2 *kûṭlôs*[1]

Dissolve the ammoniac in vinegar, and spread into a paste the liquid drugs, mix together and smear on the parts which are paralysed.

Another [unguent] which is used for fatigue and spasms from behind, and for pain in the loins, and in cases of palsy and tremblings.

Myrrh	4 drachms	
Incense	4 „	Page 147
Ammoniac	4 „	
Frankincense	4 „	
Natron	4 „	
The plant artemisia	4 „	
Opopanax	3 „	
Galbanum	4 „	
Gum of terebinth	4 *lîṭrê*	
Wax	4 „	
Oil of malabathrum	$^1/_2$ *kesṭâ*	
Oil of cypress	$^1/_2$ „	
Vinegar as much as is necessary.		

Dissolve the ammoniac and myrrh and artemisia and natron in vinegar, and throw into the mixture the opopanax and the

---

[1] The cotula = half a sextarius.

galbanum in powder, and mix together the liquid drugs, and work up, and take and use as an ointment

Another [unguent] .... **castoreum, which is used in**
Fol. 71*b*. **cases of paralysis and .....**

Moist hyssop	3	ounces
Gum of terebinth	3	„
Galbanum	3	„
Ammoniac	3	„
White peppercorns	3	„
Castoreum	3	„
Euphorbium	3	„
Alkali of the fuller	3	„
Aphronitrum	3	„
Opopanax	3	„
Brain of the fat-tailed sheep	4	„
Oil of glycyrrhiza	2	„
Old oil	3	*lîtrê*
Wax	1	*lîtrâ*

Work up in a suitable manner, and use as an ointment on the parts which are diseased.

Another. If a man who hath the palsy useth this unguent he will find relief from his sickness.

Oil of myrtle	4	ounces
Oil of amîlînôn	4	„
Oil of cypress	4	„
Old oil	1/2 a	*lîtrâ*
Gum of cypress	1	*lîtrâ*
Wax	1	„
Galbanum (or, styrax)	3	*lîtrê*
Opopanax	2	„
Artemisia	2	„
Bryony	4	„
Aphronitrum	4	„
Pyrethrum	4	„
Kid's fat	6	„
Euphorbium	1	*lîtrâ*

Vinegar as much as is necessary.

Work up thoroughly according to what is written above, and use.

Another [unguent], compounded with frogs, which is good for those who have their whole bodies twisted, and for pains in the excretory organs, and in the loins, and in hemiplegia and gout.

Sabînôn oil	12	*lîṭrê*
Root of wild cucumbers	12	„
Wax	1	*lîṭrâ*
Gum of cypress	1	„
Brain of the fat-tailed sheep	1	„
Galbanum	1	„
Frogs	12	*lîṭrê* Page 148

The oil we must divide into parts: in one part we boil the frogs, and in the other part the roots, and afterwards we clarify it, and rub smooth the other drugs which can be so treated, and we warm them, and then we smear the unguent on the parts which are diseased. Now others add to the root of the wild cucumber one and a half measures of sampsuchum marjorana (σάμψυχον), one *lîṭrâ* of crocus incense, one *lîṭrâ* of the oil of the laurel malabathrum, one ounce of balsam oil, and one small measure of tortoise blood. Mix all these together and work up thoroughly, and anoint the members therewith, | Fol. 72*a*. or use it on a plaster.

Another [unguent] which is used for pains in the excretory organs and in the joints, and in cases of gout and palsy, and for those who have the tremors, and for all the pains which take place in the nerves.

Sampsuchum marjorana	1	*lîṭrâ*
King's crown	1	„
Trigonella	1	„
Frankincense	1	„
Capsules of rosemary (κάχρυς)	1	„
Cypress (Berûthâ)	1	„
Cypress (Kûprôn)	1	„
Artemisia abrotonum (ἀβροτονον)	$^1/_2$	„
Costus speciosa	$^1/_2$	„
Cardamoms	$^1/_2$	„
Cassia	$^1/_2$	„
Pistacia lentiscus	$^1/_2$	„

Incense reed	$^1/_2$ *lîtrâ*
Chamaedaphne (?)	$^1/_2$ „
Wine with a bouquet	44 *ḳestê*
Arînôn incense	16 *lîtrê*
Oil of cypress	2 „
Oil of laurel	2 „
Glaucium	2 „
Wax	2 „
Cypress gum	1 *lîtrâ*
Galbanum	2 ounces
Opopanax	1 ounce
Ammoniac	4 ounces
Peppercorns	4 „
Frankincense	4 „
Red natron	4 „
Myrtle flowers	4 „
Inside of wild cucumbers	4 „
Centaureum	4 „
Tincture (or, juice) of mandragora	4 „
Myrrh	4 „
Stacte	4 „
Boiled honey	4 „
Frogs	6 in number (?)
. . . .	. . . . . .
Dried grass (?)	. . . . . .

Pound the dry drugs together, and rub down the liquid drugs with oil and wine; boil the frogs separately, and beat up together the ammoniac and the myrrh. Then mix all the above together and work them up thoroughly, and use the mixture either as a liniment, or as an unguent, on the parts of the body which are affected with rigidity or paralysis.

Page 149 **Another [unguent] attributed | to Demokrates, which is used in cases of debility, and palsy, and spasms, and for pains of long standing in the excretory organs, and for gout.**

White frankincense	1 *lîtrâ*
Aphronitrum	1 „
Copper oxide	1 „

Wet yeast	1 *lîṭrâ*
Ammoniac	2 *lîṭrê*
Myrrh	2 ,,
Galbanum	2 ,,
Opopanax	2 ,,
Euphorbium	4 ,,
Alkali of the fuller	4 ,,
Fine wax	$2^{1}/_{2}$,,
Cypress gum	1 *lîṭrâ*
Oil of laurel	1 ,,
Oil of cypress	1 ,,
Glaucium	1 ,,
Oil of roses	2 *lîṭrê*
Old oil	2 ,,
Oil of pistacia lentiscus	2 ,,

Vinegar as much as is necessary.

Crush the dry drugs, rub down the liquid drugs with oil, then add the sweet spices, mix and pound thoroughly well together, and apply in all cases of pains of the | nerves of long standing. Fol. 72*b*.

Now of the medicines which are used in cases of rigidity and paralysis, and which have been described [above], the constituents (or, matter) are well known, and also the manner in which they are to be used is well known, and the administering of them is not difficult, and is according to the lightness or obstinacy of the ailments. And we apply a medicine [sometimes] according to our own opinion, or a plaster, or an oil, which is beneficial for rigidity of the nerves of the whole body, or of one of the members thereof, according to the age and physical condition and special requirements of individual cases. It remaineth now for us to speak about the nerves which are smitten.

Of the wounds which take place on the nerves.

Now if the nerves receive a hard blow, or if they be swollen through an abscess, they cause very great pain, first because they possess exceedingly great sensitiveness, and secondly because their suffering is communicated swiftly to the brain.

And if a nerve be stabbed, or crushed, or cut, or become diseased through the bite of an animal of any kind, it hath need of medicines which are warm and delicate in character. We must take the greatest care that water, neither hot nor cold, doth not touch the wound, for water is antagonistic to the wounds which occur in the nerves, because it wetteth Page 150 them, and maketh them to run with moisture, | and because its natural composition is bound up with moisture and cold. Now for wounds of this kind warmings, by means of sweet oil wherein is no astringent property, are good, or one of the unguents in which delicate perfumes are boiled may be used with advantage. And great care must be taken that they do not apply to the wound any oil in a cold state, but it must be warmed, and wool must be dipped in it, and the parts which are diseased must be warmed by laying on them bandages which have been well dipped in hot oil. Afterwards dip the wool in the oil and tie the wound up with it. We must prevent the lips of the wound from sticking together, so that the matter may run freely therefrom and there may be no suppuration. And if the opening (or, mouth) of the wound be constricted, it is right for us to make it wider. If the body of the patient is seen to be overfull we must empty it by means of blood-letting, and by means of purging so as to reduce the belly, in order that material may be drawn away from it and the abscess be not increased. And the patient Fol. 73a. must follow a very | strict course of life, and he must guard himself from work and movement, and he must apply to the wounds the well-known, and well-tried medicaments.

Plaster (or, liniments) of euphorbium which are good for the wounds that take place in the nerves, and for the bites of evil beasts.

Euphorbium	1 part
Wax	2 parts
Old oil	6 „

Melt the wax with oil and let it cool, and then pound the euphorbium very carefully with a small quantity of oil, and mix with the wax and oil, and pour on the wounds in the nerves, and anoint with the mixture the whole member which is wounded. In the latter case thou must take some fresh euphorbium, for that which is old loseth its strength, and if a fresh supply is not available, put in a double quantity of the old.

Another unguent of euphorbium which is good for wounds of the nerves, and for abscesses of all kinds which are caused by colds and chills, and for wounds caused by evil beasts.

Fresh euphorbium	2 drachms
Wax	4 ,,
Gum of cedar	1 drachm
Gum of terebinth |	1 ,, Page 151
Chîà (stacte)	1 ,,
Old oil	7 drachms

Melt the wax with oil, and then throw on them the gums; pound the stacte and euphorbium and mix with them, and then use in the form of an unguent which shall be laid on the wounds.

Another unguent of opopanax and vinegar which is to be used for the wounds that come in the nerves, and for the bites of a mad dog.

Opopanax	5 drachms
Gum of cedar	20 ,,
Strong vinegar	30 ,,

Pound and dissolve in vinegar and use. This medicine doth not permit the sores from the bite of a mad dog to leave marks behind them, and for this reason it is used in cases of wounds which arise in the nerves; but inasmuch as it is very strong, it must [only] be used for those whose bodies are exceedingly robust. Now when thou wishest to use it on bodies which are tender, melt it down with some odoriferous unguent, especially with oil of chamaemelum (chamomile), or oil of *îsârê* (ἄσαρον), or oil of balsam, or some other oil of fine quality; but if these are not available use | old oil.   Fol. 73*b*.

Another, a fillet (σπλήνιον) of musk, which is used for the cutting of the nerves, and for wounds of the nerves although they be cut or crushed, and for the sores which occur through the breaking of bones, and for the collection of water (*i. e.*, dropsy), and for constriction (stricture?), and for abscesses in the anus; it relaxeth hardness, and it healeth the laxness of the joints from whatever cause it ariseth, and the shooting pains which come in the hands and feet. This medicine is beneficial and causeth no sore(?), and it is useful for very many ailments.

Spume of silver (protoxide of lead)	1 *mânâ*
Dust of frankincense	1/2 „
Gum of terebinth	1/2 „
Dry pitch (bitumen)	1 „
Calf's fat	1 „
Wax	10 *estîrê*
Verdigris	2 „
Oil of rice	1 *kaylâ*
Vinegar	1 „
Olive oil	1 „
	Moist pitch	1/2 „

Page 152

Boil the protoxide of lead with oil, and then pour in the fat and the frankincense powder, and shake up well. Then, little by little pour in the other drugs, boil them all together until the mass obtaineth consistency, and then allow it to cool. Then stir the oil with thy hands and work it up until it ceaseth to be sticky, and lay it on a strip of linen or a rag, and bind this on the wounds. Dip the rag in wine and squeeze it dry, and then lay it on the wounds and tie a bandage round it.

Another [unguent] which is called the "Persian"(?), and which is used for pains.

Ammoniac	4 drachms
Wax	4 „
Gum of cypress	4 „
Burnt copper	4 „
Protoxide of lead	4 „
Dry pitch	40 „
Aloes	4 „
Birthwort	4 „
Galbanum	4 „
Verdigris	1 drachm

Frankincense	2 drachms
Olive oil (in the summer)	$^1/_4$ *ḳiṭôlâ* (*kaylâ*)
Olive oil (in the winter)	8 drachms
Vinegar for dissolving	$^1/_2$ *ḳiṭôlâ*

Moisten and use.

Another [unguent] which is called the "Twelve", after the Twelve Apostles, and which is good for all wounds difficult to heal, which come in the nerves and in every member. It softeneth hard abscesses and dense secretions of viscous pus, and checketh scrofula, and dissipateth cancers, | and doeth good to old sores, and pain in the Fol. 74a. ears, and boils in the nostrils, and the severe pain which cometh in the womb.

Spume of silver (protoxide of lead)	24	*estîrê*
Ammoniac	7	,,
Bdellium	7	,,
Resin	16	drachms
Wax	16	,,
Cinnabar	9	,,
Galbanum	9	,,
Myrrh	8	,,
Opopanax	8	,,
Aloes	12	,,
Frankincense	12	,,
Birthwort (long)	12	,,
Olive oil (in the summer)	1	*lîṭrâ*
Olive oil (in the winter)	$1^1/_2$	,,

Page 153

Crush the protoxide of lead and beat it to a powder, then pour a little oil upon it, and crush it again until it becometh like a paste (σπλήνιον), and boil it over a fire until it dissolveth and becometh like honey. Then macerate the *hoshâḳ* and myrrh and frankincense and opopanax and bdellium in vinegar, and work them up together until they are dissolved. Then crush cinnabar, aloes, and birthwort and pour on the mixture, and work up and use for the pains which have been described. It will keep the wounds free from abscesses, and free from pain and disease, and will heal them. Now it is already well known that the material of the medicines which are used for wounds of the nerves is of a delicate nature, and that it must be liquefied and applied by means of the finest gums through hot and softening power, and by means of unguents of the most delicate nature, and by the fats of any animal or bird which is available, and by bitter and aromatic roots of all kinds, and by all kinds of vegetable products of a delicate nature, which are described in the Treatise on Medicines as being simple in their effects. And they must be worked up well together (*i.e.*, compounded) with proper skill, according as whether the wounds are recent or old, and according to the organs to which they are brought nigh, and according to the time of the year, and the condition of the atmosphere which surroundeth us. Now as concerning the nerves, and the symptoms of the injuries which happen to them, and concerning their healing, the things which are written in this Eighth Chapter are sufficient.

HERE ENDETH THE EIGHTH CHAPTER.

CHAPTER IX.

Of the disease [which is called] the strangles, and of all the ailments of the mouth.

Now the brain hath been already described, together with the five senses which are made effective by means of the nerves that proceed from it, and the injuries which happen to them, and the symptoms of them, and their healing, and it is meet that now, after these matters, we should speak about all the ailments which arise in the mouth and in the throat. And it is necessary that we should first of all describe the strangles and distinguish the various forms of this ailment; for that it is Page 154 possible for many kinds of the strangles to exist in this region of the body, is evident to us from the treatise which Hippokrates wrote on the "foreknowledge of the acute sicknesses", wherein he says:—Strangles are a very difficult and evil disease, and they bring a man to his end very quickly. One kind doeth nothing which can be recognized either in the throat or in the neck, but they bring on obstinate pains and defective breathing, and it is this kind which choketh a man on the day wherein it attacketh him, or the day after, or on the third day. Another kind likewise produceth pains, and it also produceth red abscesses in the throat, which are very destructive, but this kind lasteth longer, little or much, than the former. And the kind in which, during the attack, the throat is swollen and the neck red, lasteth longer still. Nevertheless, through one kind of the strangles more particularly patients can live, when the neck and the breast are red, and also when the hard red pustule itself doth not return within. From these things a man may know that all the diseases, of whatever kinds they may be, which occur in the region of the throat and which hurt the hearing, have one general name, and that the physician calleth them "the strangles". Now Hippokrates distinguished four kinds only, and he said that four diseases arose from them, and he described their symptoms; the following are the four diseases. The first kind existeth when there is a red abscess on the palate; now I call "palate" that part which is inside Fol. 75a. the mouth, to which reacheth the end of the stomach | and of the throat. The second kind existeth when there is nothing visible in the mouth, and nothing on the palate, and nothing outside which will shew that any abscess is formed, and the patient only knoweth that the strangles are there by feeling them in his throat. The third kind existeth when, in addition to these, there is an abscess in the region which is outside the Page 155 palate. | And the fourth kind existeth when there is an abscess both in the inside and on the outside of the throat. Now besides these there is another ailment in the neck when, as it sometimes happeneth, the vertebrae are ruptured towards the inside, and when the muscles that are associated with them come to be a kind of abscess, or ulcer, or sore, and when, as it sometimes happeneth, the stomach itself is deranged with them, and when the muscles that are fastened to the stomach in the throat are also diseased, and when the muscles of the throat by which the throat is set in motion are diseased.

Now all these diseases cause a certain amount of difficulty in breathing, but they do not bring in their train the danger caused by the strangles. Those, however, who suffer from them can only swallow what they do swallow with the greatest difficulty, and they suffer excessively when their drink is interrupted, and it cometh out through their nose. There are also cases in which the abscess creepeth (*i. e.,* moveth about), and the parts which are close to the palate, and the tongue, and the gums are swollen, even as Hippokrates wrote. When the tongue is diseased the destruction of the other operations taketh place. Now it seemeth from his treatise that Hippokrates also distinguished the incidents which are associated with the strangles, and which arise through the dragging of the neck, for he saith thus:—There are forms of the disease of the strangles which are like unto these, and in which the vertebrae of the neck incline inwards in some men to a very great degree, and in others to a lesser degree, and externally the necks of such men appear to have a depression in them, which cause them pain at that spot if a man presseth upon it with the hand. Now when this taketh place below the vertebra | which is called the "tooth" this doth not happen, Fol. 75b. but in some men a great *ḳĕnûdhrâ* (?) appeareth, and it is not on that vertebra which is called "the heavens of the palate", and it is not thick, and the parts which are below the | temples Page 156 are [not dissimilar] but are alike. Now these are not like unto the tumours of the neck which appear in man, but they exist naturally. And those that suffer from this disease cannot turn their tongue about easily, but they imagine that it is [too] large, and that it projecteth too much [from the mouth], and those veins which are under the tongue appear very prominent. They are unable to swallow anything at all, or at least only with the greatest difficulty; and if they force themselves greatly to attempt to swallow, the morsel escapeth to their nose and filleth their nostrils. The region about their temples becometh very hot, even though there be no fever. Many patients are not seized with an attack of the strangles except when they wish to swallow their spittle, or some other thing; and their eyes are not in a proper state. Some who suffer from strangles obtain relief speedily, but others have [to wait] for a day to obtain relief, and of these very many have no fever at all. Many are obliged to remain in this state for a long time, having in them some portion of an abscess, the existence of which their drinking and voice (or, speech) make known. Now their gums (or, the back of the throat) become perforated, and this is to them the sign of the evil, although it is not thought to be evil. Those whose neck is inclined out of the straight are paralysed in the region from the spot where the vertebrae in their neck are bent aside; and the opposite side of the neck is stretched tightly. All these things are made known clearly by the face and by the mouth and by the gums and by the region about them; and similarly the temples also are deflected (or, twisted). Now hemiplegia doth not take place in the whole body, but only in some of the members, and in some who are afflicted with the strangles it reacheth to the hand. Some of these could spit out only with the greatest difficulty anything which was cooked, and others could spit readily, and others | had fever. Of such patients the breathing was grievously interfered with, and whenever they wished to speak their mouth became filled with saliva, and the saliva of these was . . . . And the feet and legs of all of them were cold. Now to all the people whom I have seen who were attacked by this disease, there clung all these happenings which I have repeated, and they died straightway, some through the disease of the strangles itself, and some through the sympathetic action of the vertebrae of the neck and of the muscles that came from them to the parts which have been mentioned. Now there were others that came under my inspection and that died of suffocation caused by abscesses in those regions which have been described above, and it was very clear that it was the blood that caused the abscess, whether it were hot, or sluggish, or phlegmatic, as in the cases where there are inflated abscesses.

Page 157
Fol. 76a.

Of the healing of the strangles.

Now we have learned from [what hath been said] very often in the preceding pages, and from the matters which we have made plain, that abscesses are due to the flow of the blood. And we have also understood that at the beginning, when the vein floweth, it is right that the blood should be checked, and that we should dress it behind, and take care that it (*i. e.*, the blood) doth not run to another member which is in a healthier state, and cause danger. And we have also learned in this treatise, in the Chapter which treateth of headaches and in the Chapter which treateth of ailments in the eyes, how it is right to make ready the whole body for the healing of each of the members, and how we must first of all, if the body be overfull, make use of the letting of blood from those veins which are near to the member that is diseased, and must quickly draw out the superfluous blood, which is transformed into the

abscess. And after the letting of blood we must empty out
the chyme which is in excess in the body by means of some
cathartic that is suitable and by a reduction of the supply of
food. And when we have brought the | whole body low, and Page 158
have prepared it for the healing of the member [that is diseased],
we must then make use of the medicines that will do good to
the member. | Now it is possible that we may find those who Fol. 76b.
are about to heal the abscesses which arise in the mouth,
taking a proof (or, demonstration) from the nature of the
member that is to be healed by means of the simple and
compound medicines that they apply to it. The medicines,
then, which check the abscesses that spread (?) must be of the
kind which belong to the mass of those that cool and purify,
in order that they may check the blood which floweth and runneth down to the member [that is to be healed]. Therefore
the disease of this inner part which checketh the abscesses
that spread is found in a demonstrable manner from the pain;
but the other is found from the nature of the member which
is to be healed, even as ye are acquainted with the fact that
the skin of the tunics of the mouth is softer and more tender
than the skin which envelopeth the whole body, and it perceiveth immediately the medicines which are applied to it.
And it also possesseth two entrances, by which it is able to
make to flow and to make to enter into the belly and into a
certain part of the lungs a portion of the medicines that are
used for pains of the mouth.

For this reason it is meet to take the greatest care that we
do not apply for an ailment of the mouth the medicines that
nature abhorreth and that turn the stomach, but we must make
use of those that not only relieve the abscess, but that also
possess the power of food. Among all the principal medicines
which are used for the mouth the best is that which containeth
a preparation of the extract of mulberries when compounded
with honey or without, but there are several others that are
as efficacious as this. In the beginning we administer those
medicines that check the disease, and then, after the checking
of the rheum, we administer little by little those which check
somewhat and also alleviate the disease. And then, after [its]

severity [is passed] we administer those which alleviate it as
Page 159 well as cut it and | destroy it. And all these medicines are of a hot nature. And thus do we act in the case of the abscesses of the strangles, with which danger is closely associated, beginning with those medicines that cool and check them, and ending with those that are hot and which give relief, whether [the abscesses are filled] with pus, or contain gangrene, or pustules, or whether some part of the abscess, which hath
Fol. 77a. become hard, remaineth in some cavity of the ears. | We will [now] also write down the medicines which are to be used for these diseases.

Simple medicines which check the rheum.

Now the simple medicines that cool and check the disease are:—Dried roses, rind of pomegranates, gall nuts, rhubarb, lentils, alum, cisthus parasite, purple βαλαύστιον, plantain (ἀρνόγλωσσον), polygonum, oil of styrax, glaucium, aloes, lycium, mulberries, fresh walnut bark, *sanyâ* grapes, and all such like things. All these are used because they purify and check abscesses at the beginning of their growth; sometimes we press out their juices and administer gargles made of the juices mixed with water, and sometimes we make infusions of them, and use the water in which they have been steeped. There are some occasions also when we administer one of them only, and there are others when we administer two or three, and then we steep them all together, and make the patient use the water as a gargle. Now when the rheum which was flowing to the abscess hath been checked, and its violence hath begun to abate, we make use of the medicines which alleviate.

Simple medicines which give relief in cases of abscesses of the mouth.

These are they: water in which dried figs have been steeped, juice of myrtle berries, and the myrtle itself, fresh grape juice, sweet wine, water in which dates have been steeped, origanum (mint, or calamint), mallows, aniseed (or, dill), rue, mint and wild mint, licorice root, and all other such like things. When we wish to cleanse the mouth thoroughly and to relieve the pain, we sometimes sprinkle into infusions of these herbs | aphro- Page 160 nitron, or sulphur, gum from Cyrene, or myrrh, or crocus, or cinnamon, or peppercorns, and we give it as a gargle to the man who is sick. Or we give him as a gargle husked-barley water mixed with vinegar and honey, or we give him as a gargle sour wine, or vinegar in which the ears of barley have been macerated. The above form the mass of simple medicines which give great relief in cases where there are abscesses in the mouth and throat. Sometimes we mix the medicines that afford great relief | with a few of those that cleanse and Fol. 77b. purify; and sometimes we administer those that cleanse and purify greatly with a few of those that give relief, and this we do according to the time of the sickness and the manner of the chymes that produce it.

Compound medicines which are used for the strangles, and the medicine which is compounded of juice of mulberries, and which is called "Dîmrôn", and is used for the inflamed abscesses that arise in the throat and mouth.

Juice of mulberries, boiled and strained	5	*kestê*
Honey, with the scum removed,	1	*kestâ*
Fresh grape juice	1	„
Crocus	2	*kestê*
Myrrh	2	„
Juice of the cisthus parasite	2	„
Dross of the smelting furnace	$1^1/_2$	„

Boil the juice until it hath the consistency of honey, then mix with it the honey and the medicines, and use.

Here is the same prescription, only in another form.

Juice of mulberries	5	*kestê*
Honey	1	*kestâ*
Fresh grape juice	1	„
Myrrh	$1^1/_2$	*kaisâ*
Juice of the cisthus parasite	1	„
Crocus	1	„
Dross from the smelting furnace	1	„

First of all boil the juice, and then take skimmed honey and throw it on the juice with the fresh grape juice and the other [medicines], and use it as a gargle, for it checketh the abscesses and giveth the patient relief. Now other physicians do not pour in the fresh grape juice.

Another.

Juice of the fresh bark of the walnut tree which hath been boiled down to one half, and from which the scum hath been strained	5 *ķesṭè*
Honey and fresh grape juice	1 *ķesṭâ* each

Page 161 Boil | the juice with the honey until it becometh thick. Then take it off the fire and pour in one and a half measures of myrrh and the same quantity of alum and one measure of crocus, and mix all together and use. And in this manner may also be prepared the juice of pomegranates, and of *sanyâ* berries and tuber berries, and all the juices which are used for the mouth.

[Another] which is compounded of sanôkithâ, and is used for the strangles, and for all the diseases of the throat; and it checketh abscesses and affordeth relief.

Aniseed	1 ounce
Seed of rock parsley	1 "
Mint (ἄμμι), or *sîsôn ammî*	1 "
Sea spume	1 "
Cinnamon	1 "
Dross of the smelting furnace	1 "
Iris	1 "
Seed of wood rue	1 "
Persian (?) crocus	1 "
Dried roses	1 "
Costus (κόστος)	3 ounces
Fresh scoria of burnt	3 "
Crocus	1½ "
Nard	½ ounce
Amomum	½ "
Gall nuts, unslit	8 in number

Honey as much as is necessary.

Pound and mix these together, and use. When the abscesses are beginning use the medicines which cleanse and purify, such as the ordinary juice of mulberries, or of the senna plant, or of unripe pomegranates. And when the abscesses decline dissolve them in husked-barley water, or in honey water, or in fresh grape juice, or in Dîmrôn, and they will be found beneficial.

[Another] which is compounded of musk and is used for diseases of the throat, and for all the abscesses which arise in the membrane that extendeth upwards from "the member which secreteth".

Absinthium	8 drachms
Aloes	8 „
Myrrh	2 „
Musk	2 „
Malabathrum	2 „
Rhubarb	6 „
Crocus	4 „
Mint (ἄμμι)	4 „
Seed of rock parsley	4 „
Castoreum	$1^{1}/_{2}$ „

Honey as much as is sufficient.

Mix up well together, and use. When the abscesses go back administer as a potion or as a gargle one of these juices which appeareth to thee to be useful.

The Great Antîrâ[1] medicine, which is to be used for ailments of the throat, and which is to be blown into the mouth in the form of a dry powder. Crocus, root of the mountain rose, ammoniac (نوشادر), mîmrôn,[2] pyrethrum, peppercorns long and round, licorice root, | purple βαλαύστιον, leaves Page 162 of the green (or, yellow) rose, mîmîthâ,[3] leaves of the wood lettuce, berries of the incense plant, crocus root, green gall nuts, green myrobalanus chebula, lycium, glaucium, Persian *sathrê*, rind of the pomegranate, ferns, aloes, acacia, Indian salt, daucus gingidium (شيطرج), nard, amomum, ginger, aniseed,

---

[1] From ܐܢܬܝܪܐ. A medicine which causes abscesses to waste away and disappear.

[2] Mamîrân (?), or swallowwort.

[3] Glaucium phoeniceum (?).

seed of rock parsley, samtĕrîn, salsola fructicosa (cardamons), reed of the incense plant, lithargyrum, arsenic, ḳrôḳô maghmà, costus, myrrh, dog-excrement, îrîôn, (verdigris?), tamarix, caryophyllus aromaticus, vine mould (fungus?), seed of roses, balsam Fol. 78b. bark, and cassia. Pound all these together, | and reduce them to a powder, and apply sometimes in the form of a powder, and sometimes mixed with honey in the form of a gargle.

The Little Antîrâ, which is used for ailments [of the mouth and throat]. Myrrh, crocus, dross from the smelting furnace, purple βαλαύστιον, pyrethrum, dried roses, mîmrôn, ammoniac, licorice root, mîmîthâ, gall nuts, lycium, îrîôn (verdigris?), juice of the cisthus parasite, rhubarb, long peppercorns, stalks of the incense plant, acacia, aloes, tamarix, ferns, and the rinds of pomegranates. Pound all these together, and apply as before.

Another, which is good for ailments [of the mouth and throat]. Wolf's gall, elephant's gall, hyena's gall, bull's gall, glaucium phoeniceum, peppercorns, both long and round, crocus, flat lentils, and aloes. Take all in equal quantities, pound and reduce to a powder, and blow into the mouth of him that is nigh to suffocation. Or dissolve extract of husked barley, or mulberry juice, or pomegranate juice [in water], and let him use it as a gargle.

A medicine which is used for those who are nigh to suffocation.

Gall nuts	1	drachm
Rinds of pomegranates	1	„
Ashes of the burnt jawbone of a pig	1	„
Costus plant	1½	„
Peppercorns	1½	„
Myrrh	1½	„
Aloes	2	drachms

Pound and reduce to a powder, and blow into the nasal tube(?), or mix with honey water and administer in the form of a gargle.

For the throat which hath sores in it:

Myrrh	4 drachms	Page 163
Crocus	2 „	
Dross from the smelting fur-		
nace	10 „	
Iris	8 „	
Rose	8 „	
Green gall nuts, unslit	30 in number	

Pound and mix together, and use according to what is written.

[A medicine] which is used for the strangles and for the inner cavities of the ears which have become hard.

White excrement of a dog	
burnt in a potter's vessel	1 ounce
Peppercorns	2 drachms
Burnt gall nuts	1 ounce
Rinds of pomegranates	1 „
Ashes of a burnt jawbone	
of a pig	1 „
Costus plant	$^1/_2$ „
Myrrh	$^1/_2$ „

Pound and mix together, and apply by blowing, or work up in honey, and paint the inside of the throat with it by means Fol. 79a. of a feather. Or steep a river crab in dill water, and give to the patient in the form of a gargle.

Now if some portion of the abscess remaineth in the inner cavities of the ears, the following medicines will be useful:

Ammoniac (نوشادر)	5 drachms
Crocus	5 „
Gall of an elephant	5 „
Rhubarb root	5 „
Verdigris	2 „
Gall nuts	6 „
Peppercorns	6 „

Pound and reduce to a powder, and apply by blowing, or mix with honey and paint the inside of the throat with it by means of a feather, or dissolve in honey and let the patient use it as a gargle.

12

Another which is to be used for the strangles. Take
the excrement of a dog which is white and containeth many
bones, pound it and reduce it to a powder, and apply by
blowing, or work it up in honey and administer it in the form
of a gargle, or apply it with a feather.

Or, take

Dog's excrement	2 drachms
Aloes	1 drachm
Dried human excrement	1 „
Glaucium phoeniceum	1 „

Pound, and apply by blowing, or mix with honey, and apply
it to the inside of the throat by means of a feather.

And if the sore that is called a "pustule" be in the mouth,
now this cometh sometimes when there is fever and sometimes
when there is no fever, and there are sores(?) that break out
in the mouth and on the tongue, this is due to inflammatory
Page 164 chyme. Sometimes this pustule is red and sometimes it is
white, according to the [nature of] the chyme which causeth
it, and sometimes it putrefieth and gangrene cometh in it. In
its initial stage, when there is inflammation in it the following
medicines are to be used.

**A medicine to be used for an inflamed mouth with a
red pustule in it.** Take one drachm each of the seed of red
roses, amylum, sweet wine, ṭabshakîr,[1] seed of purslane and
crocus, pound them, and use. If there be sores in the throat
blow in the powder, and rub the whole of the mouth with thy
finger dipped in this medicine.

Another medicine for the same disease. Take rose leaves
and dross from a smelting furnace in equal quantities, pound
them and mix with honey, and [apply] as before.

**Another medicine which is to be used for a pustule in
the throat.**

Verdigris	3 drachms
Pyrethrum	3 „
Persian salt	3 „
Rhubarb	2 „

---

[1] *I. e.*, ܒ̈ܫ ܩܠܐ, the sugar of the bambusa reed.

Ginger	2 drachms
Podophyllum	2 ,,
Burnt peppercorns	2 ,,

Crush small and apply, sometimes by blowing, and sometimes by rubbing the whole of the throat with thy finger dipped in the medicine.

Another medicine which is used for the pustules and sores which come in the mouth, and for the abscesses in the throat which are hard.

Amomum	1 drachm
Crocus	1 ,,
Tragacanth	2 drachms
Poppies	1 drachm
Costus	1 ,,
Myrrh	1 ,,
Gum Arabic	1 ,,
Nard	1 ,,

Pound and apply by blowing, or mix with honey water and use in the form of a gargle, or lay it on the throat with thy finger and rub it in.

Another medicine for these diseases.

Shûshâ root	1 drachm
Iris	1 ,,
Tragacanth	3 drachms
Gum Arabic	3 ,,
Peppercorns	$^1/_2$ drachm
Dried roses	$^1/_2$,,
A little stacte.	

Pound and apply by blowing.

Another [medicine] which is good for pustules and running sores in the mouth.

Burnt vitriol	8 drachms
Iris	8 ,,
Crocus	8 ,,

Rub down and mix with honey, and use according to written instructions, or place it in a cotton cloth.

Another medicine for the mouth.

Dross of a smelting furnace	2 drachms

Gall nuts	2 drachms
Myrrh	1 drachm

Pound up, mix together, and use.

Another.

Amylum	1 drachm
Indian salt	1 ,,
Green myrobalanus chebula	1 ,,
Dross from the smelting fur-	
nace	1 ,,
Aloes	$^1/_2$,,

Page 165

Pound and use.

Another [medicine] which is used for an inflamed mouth.

Tabakshîr (طباشير, or bam-	
busa sugar)	1 *garmâ*
Amylum	2 drachms
Crocus	1 *garmâ*
Rose seed	1 ,,
Green myrobalanus chebula	5 drachms
Purple βαλαύστιον	1 drachm
Rock sugar[1]	2 drachms
Alum	2 *garmê*
Aloes	2 ,,
Glaucium phoeniceum	$^1/_2$ *garmâ*
Rhubarb	$^1/_2$,,

Crush and use according to the instructions written above.

Another [medicine] which hath been well-tried. Take
tûddhâ, that is to say, the tongue of a bull, burn it in a potter's
vessel, rub it down to a powder, and use by dipping thy finger
into it.

Another [medicine] for the red pustule. Take in equal
quantities crocus, gall of an elephant, glaucium phoeniceum,
Persian pot-herbs, cunila (satureia), myrrh, calamint, barley flour,
Fol. 80a. purple βαλαύστιον, dried roses, | and sugar (or, sweet wine),
pound them together, and use. Or dip thy finger in the mixture and rub it on the sides of the mouth and throat.

---

[1] "Sugar of the axe", because an axe is needed to break it.

[A medicine] for the pustule and the excessive inflammation which come in the mouth.

Burnt paper	3 drachms
Tincture of arsenic	1 drachm

Crush these and dip thy finger in the mixture [and rub it on the places], or lay it on a piece of linen dipped in oil of roses. It is good also for the abscesses which spread in the mouth, and also for the veins under the tongue when they are severed.

On the Uvula (κίων).

Now, if the uvula by itself hath become thick (*i. e.*, swollen), either in sympathy with the roof of the mouth or with the throat, and the root thereof be swollen and red, and the tip thereof be thin, at such a time it is difficult to cut it. If, however, after the emptying of the whole body there remaineth an abscess on the uvula, and the root thereof be thin, and the tip thereof be thick, and it holdeth back the spittle, this is a suitable time for cutting it, and there is no danger in doing so. Now the cutting of the uvula may be performed in | two Page 166 ways: by means of a knife and by means of caustic medicines.

Now as concerning the medicines which are good for abscesses in the throat, the following are useful only for the rheum which descendeth to the throat and to the oesophagus. Take two drachms of Cyrenean fat, and one drachm of dross from the smelting furnace, crush them and work them up with honey, and apply to the spot.

Or take green gall nuts which are unslit, and pound them in vinegar, and apply the mixture inside the mouth with thy finger, and smear some of the mixture on the growth of the head and flat part. Or burn the stalks of the fennel plant in the fire, and make the smoke of them to enter the mouth until the sick person feels that it is much purified. Or burn reeds and palm leaves and their stalks, and take the ashes and stir them up in cold water for a season; then let the mixture settle and strain off the water. Pound to a powder alum and gall nuts and rhubarb, and put into the water, and then let the patient take it in his mouth and wash it out

Fol. 80b. therewith. Or put salt in the water from sour milk | and the sap of the vine, and let him turn the mixture about in his mouth; and abscesses in the throat are relieved when the patient useth this mixture as a gargle. Or take sour berries and rhubarb and gall nuts, and pound and macerate them in sweet goat's milk, or ass's milk, from morning till evening, then strain the mixture, and squeeze out the sediment, and let the patient use it as a gargle. This will relieve all the ailments, and abscesses that come in the mouth and throat. Now Cyrenaean butter will give relief from the pain of hard abscesses in the uvula and throat when it is rubbed down with vinegar and used as a gargle.

Of the molars and the teeth and the gums round about them.

Page 167 Now the aches which arise in the molars and in the teeth are well known to every man. Sometimes they are split, or | eaten away (decayed), or black, or corroded with tartar, or they are unable to bear anything which is very cold, or acid, or very hot. Sometimes also one or two of them ache. Now some of them ache, although the causes of their doing so are hidden from our sight, and it is only the sensation of pain that indicateth which they are, thus we hold to be true the statements of the sufferers who say that they feel the pain throbbing inside their teeth, although no external injury in the teeth is to be seen. Now teeth appear to become diseased through over-feeding, and through insufficiency of nourishment. Nourishment in excess causeth pain in them when there is any sympathy with an abscess in the gums which is due to the flow of rheum, and they become diseased. And insufficiency of nourishment causeth pain in them, as, for example, the teeth of old men, which are denuded of their flesh and so lack nourishment, and sometimes the gums are eaten away by the acidity of the rheum which floweth on them, and the teeth are stripped bare. And sometimes the teeth totter (i. e., get loose) and fall out, and it happeneth also that they are sometimes not straight in their positions, or that those which are not necessary [for eating with] grow too long, and it happeneth sometimes that one or two of them grow larger than their companions. And again, teeth are very often broken by blows administered externally, | or they are rooted out, or they are Fol. 81 a. knocked loose, and then as a result there ariseth defective mastication, and drinking, and speaking. Now these, in brief, are the causes which produce injuries in the teeth and in the gums.

Of the healing of the teeth.

Now concerning that sympathetic aching of the gums because rheum is sent to them from the head, or from the stomach, it is meet that we should first of all exercise care over these members, lest, whilst applying remedies to the teeth or to their gums, | we cause excessive attraction of the rheum, and the Page 168 suffering become more severe. If it appeareth to be useful to let blood, we begin with this act, and we cut first of all that upper vein in the arm, and subsequently we cut the veins which are under the tongue, and then by means of a medicine which emptieth the belly (*i. e.,* an emetic), or by means of *pîkrâ*, we cleanse the stomach. If it be necessary to make the patient vomit, we effect this by means of a medicine which performeth the work sometimes after food hath been taken and sometimes before; then we cleanse the tongue (or, the tooth) by means of gargles, for we must cut the rheum which is sent into the mouth.

The medicines which are good for the lining of the mouth which hath become thick, and for the molars and the teeth. A medicine which is useful for a relaxed mouth, and for the blood which cometh from the gums of the teeth.

Take fresh myrtle leaves, pound them and sprinkle a little wine over them and squeeze out the juice; then let the patient take the juice in his mouth and move it about in it.

Or pound up fresh olive leaves and squeeze out the juice from them, and use it in the same way.

Or let him take the juice of fox-grapes in his mouth and gargle.

Or let him take sour milk in his mouth and gargle.

Or let him take the juice of the tongue of a lamb in his mouth and gargle.

Or let him steep rinds of pomegranates and gall nuts and olives and myrtle in water, or in wine, and let him work the liquid about in his mouth until he feeleth relief from the abscess.

Or let him make an infusion of rinds of pomegranates and roses and lentils and licorice root and husked barley, and hold **Fol. 81 b.** it in his mouth and work it about. |

Or let him steep in water mixed with wine valerian, senna, thorn root, and leaves of tamarix, and work the liquid about in his mouth; or macerate these medicines in wine, squeeze out the liquor, and work it about in his mouth.

Or let him chew myrtle or olive leaves, and collect in his mouth some of the saliva which they cause.

If, after the patient hath obtained relief from the abscess, it happen that boils appear in his mouth, or running sores (gangrene), the following medicines are to be used.

A medicine which is to be used for running sores (gan-
Page 169 **grene) and for all kinds of ailments | in the mouth.**

Verdigris	1 drachm
Pyrethrum	1 „
Persian salt	1 „
Ginger (Ζιγγέβερις)	2 drachms
Burnt peppercorns	2 „
Pîldalpôn (podophyllin?)	2 „

Crush, and apply by dipping thy finger in the powder and rubbing it on the teeth and gums. Then dip a strip of linen in vinegar, squeeze out the vinegar, and dip it in the medicine, and lay it on the places where the boils are.

Another [medicine] for gangrene.

Crocus root	4 drachms
Alum	4 „
Gall nuts, unslit,	3 „
Rhubarb	3 „
Purple balaustion	3 „

Pound, and use according to what is written above.

Another [medicine] for rubbing (*i. e.*, cleaning) the teeth.

Burnt horn of a fat-tailed sheep	2 ounces
Burnt pumice stone (κίσηρις, قيسور)	2　,,
Stacte	1 ounce
Sal ammoniac	1　,,
Sea sand	1　,,
Laurus malabathrum	6 *garmê* (*grammê*)

Pound and reduce to a powder, and rub with it the teeth that are decayed or black, first of all clearing away the impurity which clingeth about their roots by means of a suitable iron instrument.

Another. Break up verdigris with honey and put it in a strip of linen, or rub the places which cause pain with the mixture.

Another which is to be used for gangrene.

Dried olive leaves	1 *manyâ*
Acacia	40 drachms
Alum of the furnace	2　,,
Copperas water	2　,,
Vitriol	2　,,

Pound, and use.

Another [medicine] for teeth which are loose.

Round alum (rock alum?)	1 part
Salt	2 parts

Boil in wine and hold in thy mouth, and afterwards rub off the deposit from the gums of the teeth.

Another [medicine] for the debility of the teeth and for gangrene and for tooth-ache; it hath been well tried.

Flesh of sweet pomegranates	7 drachms	Fol. 82 *a.*
Thorn pods	7　,,	
Hellebore, green and black,	3　,, each	
Gall nut	one	
Rind (or, skin) of sepia	1 *denḳâ*	
[Peppercorns] long	1　,,	
Alum	2 drachms	
Palm (or, vine) ashes	2　,,	
Crocus root	3　,,	

Pyrethrum	3 drachms
Indian salt	$^1/_2$ drachm
Swallowwort	$^1/_2$,,

Pound, reduce to a powder, and use.

Page 170 **Another.** [Take equal quantities of] | pyrethrum, myrrh, Armenian nitre, round peppercorns, and a burnt linen rag, pound them all together, and apply.

Another [medicine] for gangrene. Take ashes of carob root and burnt sea-shells and crocus in equal quantities, and pound, and apply.

Another [medicine] for tooth-ache. Take mountain grapes, pyrethrum, the insides of gall nuts, the rind of pomegranates, laurel berries, myrrh, and lead, pound them together and rub the teeth with the mixture, and apply some of it to the place where there is a boil. Or rub down the root of dill in vinegar and rinse out the mouth with it. Or pour strong acid on the ground, and take the seething mud and smear on the spot.

Another [medicine], called "Adîtikôn", which is used for toothache.

Peppercorns, long,	5 ounces
Peppercorns, round,	5 ,,
Mountain grapes	1 drachm
Ginger	1 ,,
Armenian nitre	6 ounces

Pound, and apply.

Another [medicine], called "Ansûrentikôn", which is used for pains in the mouth, and for weakness, and as a tooth-powder.

Rinds of pomegranates	3 ounces
Crocus root	1 drachm
Purple βαλαύστιον	1 ,,
Rhubarb	$1^1/_2$ ounces
Persian salt	$^1/_2$ ounce

Pound with three (?) ounces of gall nuts, and rub the mouth therewith with thy finger, and also apply with a piece of linen.

Another [medicine], called "Pildalpôn", which is used for gangrene of the mouth. It hath been well tried.

Tincture of arsenic	6 drachms
Sandarkâ (sandarac)	6 ,,
Acacia	12 ounces
Unslaked lime	12 ,,

Pound and work up with vinegar, or with plantain juice, and make tablets (or, pills) and dry; when it is necessary break one up and use.

Another "Pildalpôn", which is used for gangrene of the mouth and for boils | of the nose; it hath been well tried Fol. 82*b*. and is a sure cure.

Arsenic (two kinds)	3 ounces
Acacia	3 ,,
Copperas water, fresh	3 ,,
Unslaked lime	4 ,,

Pound and mix, and do as in the preceding prescription, | Page 171 and use.

[Another] "Pildalpôn" which hath been well tried and is a sure cure.

Arsenic, both kinds	6 drachms each
Lime	12 ,,
Acacia	20 ,,
Myrrh	1 drachm
Samtĕrîn	2 drachms

Pound and work up with the juice of plantains (ἀρνόγλωσσον), make into tablets, dry, and use for gangrene of the mouth. And add it to the injection which is made for those who have dysentery, and it will be found very useful.

Another [medicine] which is good for gangrene of the mouth, and which is a well tried remedy.

> Take in equal quantities unslaked lime,
> Convolvulus[1] of the fullers,
> Leaves of wild bitter herbs (colocynth, or taraxicum?),
> Burnt shell of river shellfish,
> Red tile, or
> Scrapings from an old oven.

---

[1] *I. e.*, δορύκνιον. Three kinds of the plant are noted by PAYNE SMITH (col. 125).

Pound, reduce to a powder, work up with strong vinegar, and apply to every place in the mouth that is affected with gangrene, and it will be found to be a most excellent medicine.

Another [medicine] for gangrene of the mouth.

Pildalpôn	6 drachms
Myrrh	1 drachm
Ammoniac	1 „
Iris	1 „

Pound, and apply.

[A medicine] for cancer of the mouth and stinking secretions.

Fine myrrh	20 drachms
Red arsenic	6 „
Yellow arsenic	2 „
Ginger	2 „
Peppercorns	2 „
Costus	2 „
Acacia	4 „
Unslaked lime	6 „
Green gall nuts	3 „
Amomum	1 *grammâ*

Pound and work up with vinegar, make into tablets and dry; when necessary crush and infuse in vinegar. It is a well tried [remedy].

Another [medicine] which is good for gangrene and for the blood which cometh from the mouth, and it relieveth the debility (*i. e.*, it strengtheneth the mouth).

Pildalpôn	1 part
Surantîkôn [1] (sûrîtîkôn?)	2 parts

Mix together and put on a piece of linen [and apply].

Simple medicines which are used for the molars and [other] teeth. Take the root of *yathrâ* (nitre?), which is called "*tûthmalôn*" (τιθομάλον, a kind of euphorbium), and infuse it in wine, or in about half the quantity of vinegar, and let the patient hold it in his mouth three times a month, and he will Fol. 83a. never more have pain therein.

---

[1] It was a compound of arsenic.

Or let him make an infusion of the root of the camphor tree, and hold it in his mouth | and rinse it. Page 172

Or split a plant of bitter herbs and pour vinegar or wine on it, and keep it on the fire until it boils; and let the patient use the vinegar as a gargle. Now nigella sativa macerated in strong vinegar is good for the teeth which bite and knock together, and if it be laid on them, or in the cavities of the molars, it will relieve the pain.

Another [medicine] for the molars which knock together. Pound gall nuts which are not split and mix with vinegar, and lay the mixture on the molars and teeth that are diseased. And so also with all the medicines which are used for gangrene: mix them with pitch and apply, and thou wilt do well, for it will stick and will prevent the foetid matter from flowing.

Another [medicine] for teeth which have cavities in them, and knock together. Apply to them the finest assafoetida and they will cease to pain.

Or lay on them garlic in the same way and, if they do not [cease to pain], mix thyme with the garlic and apply to them, or the Philo medicine (φιλονεῖον φάρμακον, which containeth poppy juice and the juice of hyoscyamus), or put a little opium in the cavities of the molars, and they will cease to pain.

Or heat over a fire the roots of wild bitter herbs, or the seeds, and let the smoke go into the patient's mouth.

Or heat mustard in the same way and let the smoke enter into the patient's mouth. Or treat the hoof of an ass in a similar manner, or the teeth of a horse, or heat the seed of the hyoscyamus plant (سكّران, ὑοσκύαμος) over a fire [apply the smoke as before], and it will relieve the pain. Or inject into the nose of the patient extract of fresh ḳûnbârê (القنّابرى, or برغش, cannabis seed), or the extract of the plant "mouse's ears", or tincture of beetroot. Or pound the root of luzerne, press out the juice and inject it into the nose, and it will relieve pain in the molars. Or pour into the ear on the side of the face which is affected one of the tinctures that have been mentioned.

Or pour in the extract of the bark of the root of the camphor tree, but insert first of all a little oil so that it may not cause a sore.

Now, if the pains do not cease [after the application of the above medicines], and it is necessary to pull out the molar, and the patient cannot bear the pain of having it pulled out, release the gums on all sides and lay on them the drugs of **Fol. 83b.** which the names have been written above, | and then pull the **Page 173** tooth out with thy hand. |

Medicines which uproot the molars without [the help of] an instrument. Take the bark of the root of a mulberry tree and pyrethrum, pound them and macerate them in strong vinegar in the sun until the mixture hath the consistency of honey, and apply it to the molar once or twice a day, and it will become dislodged.

Or take pyrethrum and macerate it in strong vinegar for forty days, and crush it up, and put it in a vessel. When the time cometh for using it, put a layer of wax round the teeth which cause pain, and loosen them, and apply some of the medicine to them. Then wait one day, and lift the teeth with thy fingers and they can then be removed. Or take hold of them with an iron instrument, and they can be easily removed.

Or crush lead-ore (μολύβδαινα, الرصاص) in vinegar and apply to the teeth in a similar manner.

Or crush pyrethrum in strong vinegar for three or four days, and then lay it on the teeth which are causing pain, taking great care not to touch the other teeth with it, for it is meet that they should be loosened carefully and then smeared with the medicine. Do this for three or four days, for by these means they will become loosened and may be removed without pain.

Or take the roots of wild bitter herbs, and treat them like the pyrethrum, and use according to the instructions written above.

Medicines which clean the teeth and remove the tartar.

Indian salt	3 ounces
Dross of the smelting fur-nace	2 „

Pound them and work them up in honey, and wrap up in paper and roast; then crush them again and use.

Or   Burnt horn of a stag   1 *lîtrâ*

. . . .	. . . . . .
Tamarinds	3 *kestê*
Sal ammoniac	1¹/₂ ounce
White peppercorns	. . . . .
Costus	1 „

Rub down together and scrub the teeth and gums therewith.

Another [medicine] which scoureth the teeth, and whiteneth them, and sweeteneth the mouth, and preserveth the teeth from blackness.

Sepia	1 ounce
Ginger	1 „
Iris	1 „

Rub down and use, and then rinse the mouth with water.

Another [medicine] which scoureth the teeth and sweeteneth the mouth. Take in equal quantities cardamoms (قاقلّة), | caryophyllus aromaticus, bark, crocus, Persian thorns, sea-spume, scrapings of an oven, and ashes of palm leaves, and pound them, and scrub the teeth with the mixture.

Page 174
Fol. 84a.

Another similar [medicine] to be used in the same way. Take in equal quantities camphor, bark, cinnamon, amomum (خامامِ), cardamoms, caryophyllus aromaticus, nard, musk, wood of aloes, and tree fungus, pound and make into tablets with strong-smelling wine. When necessary to use crush and rub the mouth with the medicine, and at night let the patient hold some of it under the tongue.

Another [medicine] which is good for foulness of the mouth, and which was used by a queen.

Bark	2 *denkê*
Crocus	2 „
Cinnamon	2 „
Cardamoms	2 „
Musk	2 „
Camphor	1¹/₂ „
Gall nut, unslit	1 *denkâ*

Pound and put into strong-smelling wine, and work up and make into tablets, and use in the same way as the preceding.

Another [medicine] which sweeteneth the mouth.

Salsola fruticosa	1 drachm
Nuts of incense	1 ,,
Betel nut (قوفل)	1 ,,
Caryophyllus aromaticus	1 ,,
Camphor	1 ,,
Cinnamon	1 ,,
Galanga (خلنجان)	1 ,,
Musk	2 drachms

Pound and work up with wine, and make into pills, and use according to the written instructions.

Or crush aniseed in wine and honey, and rub and sweeten the mouth therewith.

Or	Pistacia flowers	1 drachm
	Myrrh	$^{1}/_{2}$,,
	Musk	1 *grammâ*
	Lead	1 ,,

Pound and work up in wine of lilies, and let the patient swallow [some] and use the rest as a gargle.

Or	Nard	1 drachm
	Crocus	1 ,,
	Pistacia flowers	1 ,,
	Salsola fruticosa	1 ,,
	Caryophyllus aromaticus	1 ,,
	Myrrh	$^{1}/_{2}$ *denḳâ*
	Lead	$^{1}/_{2}$,,
	Musk	1 ,,

Crush and work up in wine of lilies, and make into pills of the size of chickpeas. And let the patient swallow one in the morning and one in the evening, and let him place one under his tongue [during the night].

Another [medicine].

Alum	4 drachms
Licorice	4 ,,

Peppercorns   1 drachm

Origanum majorana   1   „

Pound and mix together, and rub the mouth therewith.

Another which scrubbeth the teeth.

Nitre

Sea-spume   } in equal quantities

Myrrh

Pound, and scrub the teeth therewith.

Another [medicine] for scrubbing teeth which are very black.

Burnt, soft *kisrâ*	2 ounces	Page 175
Peppercorns	2 „	
Amomum	2 drachms	
Laurus malabathrum	2 „	
Lime, burnt,	8 „	

Pound, and rub on the teeth.

Another [medicine] for the toothache.

Mountain grapes   1 drachm

Fennel root   1   „

Costus   1   „

Pound, and use both morning and evening. Of this medicine it hath been said that it must have been given by God.

Another [medicine] which acteth as an astringent on loose gums.

Take in equal quantities yellow hellebore, raisins, the rinds of sweet pomegranates, and pound, and use.

Or infuse the root of sorrel (الحماض) in vinegar, and let the patient hold it in his mouth.

Or infuse raisins in wine, and let the patient use the mixture as a gargle.

Or let the patient hold the salt water of old olives in his mouth three times a day.

Or let him hold the salt water of old fish in his mouth and gargle, and it will purify foul breath of long standing.

Or let him use brine in a similar manner.

Or let him infuse willow bark in vinegar, and hold the mixture in his mouth.

Or infuse the leaves of the *agîs* plant in vinegar, and use.

13

Or mix alum of the furnace with the palm fronds(?), and put them on the gums, and they will strengthen both teeth and gums.

Or infuse calamint in vinegar, and use as a gargle.

Or crush nitre, and rub it on the teeth and gums.

Or pound Cyrenean vinegar and peppercorns and calamint and pyrethrum, and work up with galbanum (χαλβάνη), and lay on the teeth, and the mixture will cure cold pains in the teeth.

[Medicines] for stupor (numbness) of the teeth.

Purslane is very good [for this ailment] when chewed, or gum of the terebinth, or wax, or the pitch which is found in a wine vessel; each by itself will cure stupor of the teeth. Now if after using the above medicines there still remaineth pain in the mouth and in the teeth, burn them in the following manner. Boil some sheep-oil or cow-oil in a saucepan. Take some wool and wrap it round the needle of the *kohl* pot, or round some suitable piece of wood, and then dip it in that oil whilst it is boiling, and apply it to that part of the gums where the teeth Page 176 are affected until | it becometh white, or apply it in the cavities of the teeth. When the burning hath gone to the roots, apply *surantîcôn* on a piece of linen rag. If the cauterizing iron be Fol. 85 a. necessary, | cauterize the cavity of the molar with a finely pointed iron instrument, and cauterize the gums round about the teeth which ache with an instrument having a broad, flat end.

Teeth that are superfluous and are growing irregularly it is meet to remove.

Now as concerning the teeth which are longer than the rest of the teeth, the physician must file them with a suitable file· If they be denuded of flesh (*i.e.*, gums), and are loose, we take hold of the tooth with fingers of the left hand from within and from above, and we file the tooth with the right hand until it is reduced in height to the level of its fellows.

The above remarks on the symptoms of all the diseases which occur in the throat, and on the healing of the same, must suffice for the Ninth Chapter.

HERE ENDETH THE NINTH CHAPTER.

AND TO GOD BE THE GLORY!

CHAPTER X.

On the injuries which happen to the organs of speech.

Now, inasmuch as we have already written about the strangles, and about all the diseases which take place in the mouth, it is time for us to give also some instruction, as indeed the plan of the present treatise demandeth, about the injuries that take place in the voice. We do not refer here to the actual voice which speaketh when we speak of giving instruction, but to the organs of speech that produce the voice. Of all the organs that produce speech, the most important is the tongue, and I remind you of what is already well known to you, that the nose and lips and teeth help the tongue. And similarly the throat, and the muscles that work it, and all the nerves that bring down to it power from the brain, are also organs of the voice. And if the muscles which open and close the throat are deprived of motion, then the man who suffereth from this disease is deprived of his voice, [and his throat] becometh closed entirely. And as also when their motions are impeded, or when they only act at intervals, as in the case of paralysis, | **Page 177** injuries to the voice appear similar to those that take place through disease in them, so also do the same injuries appear when they act only tremblingly or palpitatingly. Now because | **Fol. 85b.** their power may be weakened sometimes through some disease in them, or in one of the nerves which set them in motion, even so they produce indistinctness and feebleness in the voice. If one of those other muscles which set in motion the throat be diseased, or be injured in any way whatsoever, the voice is in this case also small, but he who suffereth in this way doth not become entirely voiceless, and moreover, the voice is not

13*

extremely feeble. As therefore the muscles which are the most important among those that set the throat in motion are those that shut it and open it, so also among the nerves of the voice those that turn upwards are the most important, and they [receive] a portion from the sixth exit from the brain, just as do those others that grow out of it. Now they are different from them because they are not separated in the neck from the sixth exit, until they reach the region of the breast. Therefore as concerning the sixth exit there is no physician who is so stupid as to sever it ignorantly by an operation of cutting; but as concerning the nerves that turn upwards, there are occasions when physicians extract them inadvertently, and there are also occasions when one or two of them are severed, together with the oesophagus, when it is struck violently. Moreover, if they meet with any severance, they impede the voice until they are repaired and until their natural constitution is restored to them. Now it may happen that the voice becometh injured obviously through the severity and seriousness of the severance, both when the muscles of the palate are unable to stretch [the throat] and also when that tunic which is inside the palate and the throat is over-filled with moisture, Page 178 and at this time also | it injureth the voice very seriously. And the setting free of the rheum which they call "catarrh" likewise deadeneth the voice, a fact which is well known to every man, even as also doth violent shouting. And this disease also Fol. 86 a. produceth a kind of disease in the tunic | of the throat, which hath been mentioned, and in the muscles thereof which resembleth that which is caused by an abscess.

Now it is well known also that, when there is an abscess in the muscles which are in the throat, the disease of the strangles ariseth, and that this disease injureth both the voice and the breathing. And, to speak briefly, all the abscesses that grow in the channels of the respiratory organs, and those that press upon them externally, injure both the voice and the breathing. And thus also is it in the case of the stomach, which is a tube within the oesophagus by which food goeth down into the belly; when an abscess groweth in this, it presseth upon the tunic of the oesophagus which uniteth from behind the cartilages

of the oesophagus, which are fashioned in the form of half a
hole (semicircular?), and then both the voice and the breathing
are injured, just as is the case when the vertebrae that [control]
the voice are doubled over inwards, as we have already said
above. Therefore all kinds of diseases happen to the voice
when any one of its members is injured, either actually or
sympathetically. Now there are other diseases which affect
the voice when there is any defect in the voice itself. Now
he (Hippokrates?) sheweth us in the Discourse on the Voice,
that this matter is, speaking generally, the action of the
respiratory organs in drawing in the breath, but speaking more
particularly, and making the distinction that befitteth it, it is
the action of the breath itself in going out, for it is the great
force of the spirit (or, breath) which ariseth from the muscles
that are between the ribs. Thus then do the great impulses
of the breast, or the relaxing of one of its sides, cause ˌpri-
marily | in a man the diminution of the voice to one half. Page 179
Now when the voice is injured in cases of apoplexy, and
stupor, and the falling sickness, and cold, this generally happeneth to all the operations connected with the will, because
then all parts of the spinal column are affected, and with them
the brain also, because the following five operations are closely
connected with each other, namely, inhalation, | and exhalation Fol. 86b.
without sound, and exhalation with sound, and voice, and
speech. It is right to know that, together with the first mentioned, all are injured, but with that mentioned last none of
them is injured.

With the second the three following are injured, with
the third, the two others [which follow], with the first, the
last; that is to say, the fifth is injured by itself. If a man
doth not breathe at all with the various organs of his
chest, or with one of them only, or if he doth not breathe
with two of them, he becometh suffocated immediately; if
he breatheth with one only he arriveth at the state of half
breathing and of half voice, and at the same time the
two operations which follow after, that is to say, a portion
equal to one half of them (*i. e.*, the four), perish also. To
one of these two I am accustomed to give the name of

"exhalation with (?)[1] sound", and to the other "exhalation without sound." If now respiration standeth [still] exhalation is destroyed, and the three other operations perish, that is to say, exhalation with sound, and voice, and speech. If exhalation with sound (or, voice) perish, the voice also perisheth with it, and speech also, just as when the voice perisheth, speech also perisheth with it.

Since therefore ye know what are the muscles which perPage 180 form these operations, | it is quite easy for you to calculate which of them is suffering (or, diseased), and in which of them the operative faculty is injured. And if ye make a mistake in any particular in respect of those that have been demonstrated to you in dissections, the facts about the causes of respiration, and the facts about the voice should remind you. All the muscles collectively produce inhalation (or, inspiration), but it is the muscles between the ribs that produce exhalation (or, expiration), which is the great breathing. The muscles of the palate produce exhalation with sound, and the muscles of the Fol. 87 a. throat are those which produce the voice. |

Now it is the tongue that regulateth the voice and produceth speech, and the teeth and the lips help it to perform this work; and the nostrils and the roof of the mouth and the uvula likewise help, and also the regulating ligament of the tongue. Those who stutter and those who stammer, and those who, like them, speak defectively, have in their organs of speech some element that containeth a defect which is due either to natural formation, or to something resembling a natural defect which hath happened to them subsequently. And to those also who have their nostrils blocked up by the sore called "many feet" (*i. e.*, polypus), which [sometimes] groweth therein, or by something else, and to those whose front teeth have fallen out, or whose lips are sore, there cometh a defect in their speech. These are the distinguishing features between the organs of voice and the organs of respiration and the organs of speech, and these are the causes, which are briefly described above, of defects in the organs of the voice.

---

[1] Read ܡܦܘܚ (?)

Of the healing of the injury to the voice.

Now it is good for those who have some injury in their voice, either through the thickening of the muscles of the palate, or through one of the causes that have been mentioned, before all things to abstain from the drinking of strong, | rough Page 181 wine, and from every kind of acid or salt substance, and from excessive speaking, and from uttering loud cries and shouts. For them also are good soft (*i. e.,* well-cooked) food and fatty broths made with sheep-oil or sesame-oil, and the drinking of sweet soft wine, or must, and hot soups made with almonds and butter and honey and the extract of wheat, that is to say, starch (amylum), and if starch be not available use the finest flour (*glûskâ*). Now it is well known that by means of such foods as these the pain caused by an abscess in the muscles and in the tunics of the trachea of the throat is relieved. If, however, there be found[1] moisture in a quantity so excessive that the use of the voice is impossible, | it is right to administer Fol. 87 b. foods compounded with honey and cummin, or with bean-flour, or with mint water, or with the extract of husked barley mixed with honey, or hot, boiled eggs. If there be hoarseness of the voice, medicines which are placed under the tongue must be administered, for even though they cause no pain [they will cure?] the roughness [in the throat]. Some of these are of an exactly opposite character, for they are pungent, and cause pain, and even when they are applied to a perfectly healthy member they cause it irritation. Of an intermediate character to these are the medicines which possess the power of cleansing, and they do not cause pain. Now these, which are painless remedies and are of a glutinous nature, are compounded of new wine (must), gum Arabic, gum tragacanth, and extract of licorice root. And the medicines which are pungent are compounded of peppercorns, cassia, cinnamon, galbanum, myrrh, frankincense, gum of the terebinth tree, and fir-cones (στροβιλος, אָצְטְרוּבְּלִין). And the following, which are of an intermediate character, are compounded of extract of husked barley, almonds,

---

[1] Reading ܐܬܬܫܟܚ.

bean-flour, and others of a similar nature. The following palliate this disease: trigonella, oak-galls freed from their husks

Page 182 (or, pods), an infusion of dried fat figs, | and chopped flax seed. The following clear the throat better than the above: gum of the terebinth tree, frankincense berries, boiled honey, bitter almonds, and, still more than these, flour of the bitter vetch, elecampane, husked barley, resin, and ferula opopanax (الۇربى). Of all the things for softening the throat already named the two most important in all the compounds are must (new wine) and sweet wine, and even more important than these is an infusion of dried figs, which oileth the throat by its power and cleareth it painlessly. Must which hath been boiled clingeth to the throat and causeth no pain, but an infusion of figs giveth it relief and cleanseth it. Now it is meet that he who placeth any of these medicines under his tongue should watch carefully when he taketh into his throat that which is dissolving gradually, and swalloweth it, and should

Fol. 88a. use his utmost power | to prevent himself from coughing. If, however, he doeth this, he must put a reed down [his throat], and it will cure him straightway. And in addition to all these things, it is meet for him to be well acquainted with the powers of simple medicines. Every one who wisheth to use medicines which are compounded, must be able to know from his own experience which medicine is most needed, and to know which medicines moisten the throat and which clear it, and how to abate or how to increase [the quantity], according to the nature of the ailment.

The medicines which are used for the voice.

A medicine which is good for the voice that is husky, or hoarse, and which is to be placed under the tongue. It relieveth rheum of the throat, and cleanseth it and softeneth it.

Seed of Orsîmôn, roasted	20	drachms
Flour of chick-peas	20	,,
Fir-cones	8	,,
Flax seed, roasted	20	,,

Grapes, with the skins removed,	8 drachms
Crocus	1 drachm
Gum tragacanth	8 drachms
Honey	$^1/_4$ of a measure

Pound and work up with honey, and make into pills the size of chick-peas, and give them to patients to put under their tongues.

Another, which is used for excessive moisture of the throat, and by those who bring up pus from the chest, or lungs.

Nard of Kulaṭa (?)	3 drachms	
Pistacia lentiscus	2 ,,	
Cassia	8 ,,	Page 163
Indian nard	3 ,,	
Cinnamon	10 ,,	
Frankincense	8 ,,	
Myrrh	4 ,,	
Costus	1 drachm	
Compound malabathrum	2 drachms	
Root of mandragora	3 ,,	
Stacte	3 ,,	
Crocus	5 ,,	
Fir-cones	30 in number	
Dates	30 ,,	

Steep the dates in must, strain off the liquor and boil it until it hath the consistency of honey; pound all the other medicines, mix them with the liquor, and make them into pills the size of a chick-pea. Each morning give [one to the patient] in an infusion of husked barley, or in an infusion of hyssop. Let him drink three [drachms] in the morning, and in the evening let him hold one pill under his tongue.

Another [medicine] which is good for the throat, and for those who bring up with difficulty pus from their lungs, and for excessive rheum | of the chest, and for hoarseness. Fol. 88*b*.

Gum of terebinth	4 drachms
Frankincense	4 ,,
Myrrh	4 ,,
Crocus	4 ,,

Cinnamon	4	drachms
Amomum	4	,,
Tragacanth	4	,,
Fir-cones	4	,,
Tincture of licorice root, or the root itself when pounded	4	,,
Nard	2	,,
Acacia	2	,,
The fleshy part of dates	2	,,
Galbanum	1	drachm
Costus	1	,,
Fine honey	4	measures

Boil the gum with the honey in a double vessel, and when it hath been [sufficiently] boiled put in the galbanum, and when they have all become dissolved and well mixed together, take [the vessel] from the fire that [its contents] may cool. Then mix with them the other drugs, and make into pills the size of chick-peas, and let the patient hold them under his tongue. Leave some of the mixture not made into pills, and administer to the patient one drachm in an infusion of hyssop, or the extract of husked wheat.

Another [medicine] which is used for pain in the throat, and for hoarseness of the vocal chords, and for loss of voice, and for the pus which collecteth in the lungs or chest. And I myself use this medicine.

Page 184

Gum Arabic	2	drachms
Gum tragacanth	2	,,
Frankincense	2	,,
Myrrh	2	,,
Crocus	2	,,
Tincture of licorice [root]	2	,,
White peppercorns	45	in number
Fat dates	5	drachms

Rub down the dates with must, pound and sift the other drugs, and mix them together and work them up, and make into pills the size of beans. Let the patient put them under his tongue, and let him drink as a draught of this medicine, mixed with an infusion of licorice root, one drachm by measure.

Another [medicine] which is used for loss of the voice caused by study. Take dried, fat figs and origanum, steep them carefully in water, mix the liquor with gum Arabic, and let the mixture be as thick as honey. Administer in the morning and evening, and let those who are affected lick some of it.

Another [medicine] which is used by those who take care of their voice, and for pain in the throat, and for coughing, and for boils in the ears, and in the privy parts, and for assisting the digestion, and for relieving the hoarseness that is caused by cold.

Frankincense	4 drachms
Myrrh	4 ,,
Must (new wine)	3 measures

Boil the must until it is as thick as honey, | pound the myrrh Fol. 89a. and frankincense, mix them with the must, and use in the manner written above. This medicine is used also by those who have dysentery, and colic, and pain in the belly.

Another [medicine] which is used for a harsh-sounding voice, and for loss of voice caused by excessive shouting, and for hoarseness of the throat, and for the uvula from which rheum floweth. Take two drachms of crocus, and one drachm of gum from Cyrene, and six measures of fine honey; boil the honey until it becometh thick, and, having pounded very small the other drugs, mix them with the honey and make into pills, and give to the patient to hold under his tongue.

Or crush a carob nut, squeeze out the juice, and add thereto honey equal in quantity to the carob nut; then boil them over the fire until the mixture becometh thick, and administer some of it as a linctus to those who have lost their voice.

[A medicine] which is used for pain in the throat, and for loss of voice, and for hoarseness caused by cold.

Crocus	3 drachms	
Myrrh	$1^1/_2$,,	
	Juice of licorice root	1 drachm	Page 185
Frankincense	1 ,,	

Infuse and work up in as much must as is necessary and

make into pills, and use according to the instructions written above.

Another [medicine] of Dioscorides, which is good for pain in the throat and for loss of voice.

Licorice juice	8	drachms
Myrrh	42	„
Gum of the terebinth tree	36	„
Gum tragacanth	38	„

Crush these singly and keep separate and reduce to a powder. Then throw the gum of the terebinth tree on a little honey and rub down into a paste, add to it the other drugs and work up, and make into pills the size of beans. Dry in the shade and use.

Another [medicine] which is good for pains [in the throat].

Myrrh	4	drachms
Crocus	4	„
Gum tragacanth	4	„
Juice of licorice root	4	„
Seed of stinging nettles	4	„
Fine fir-cones, cleaned,	60	in number
Almonds, cleaned	60	„

Work up with boiled honey and make into pills, and use.

Another [medicine] which is good for the throat and for Fol. 89b. **loss of voice | caused by excessive heat, and it quencheth thirst.**

Gum Arabic	1	drachm
Gum tragacanth	1	„
Seeds of the garden cucumber	1	„
Licorice juice	2	drachms
Starch	2	„

Pound with must, and make into pills; administer in the evening, and let the patients hold them under their tongues.

Another [medicine] which is good for fluxes from the head, and for the uvula and the throat, and it quieteth the voice.

Frankincense	1	drachm
Gum of the terebinth tree	1	„
Purple βαλαύστιον	1	„

Gum Arabic	3 drachms
Seeds of poppies	3 „
Gum tragacanth	3 „
Crocus	1 gramma
Myrrh	$^1/_2$ drachm

Pound and make into pills with must, and use.

Another refreshing [medicine], which is good for a cough caused by over-heating, and for loss of voice.

Seeds of the garden cucumber	3 drachms
Licorice juice, or juice of the root	8 „
Purslane	1 drachm

Page 186

Pound and work up with some fine white of an egg, and make into pills the size of chickpeas, and use.

Another [medicine] which is good for huskiness of voice caused by excessive moisture [getting wet?]. Pound and mix together in honey three drachms of burnt mustard and one drachm of peppercorns, and make into pills the size of chickpeas, and use [by placing] under the tongue.

Another linctus, which is good for hoarseness of the voice.

Trigonella	4 drachms	
Almonds, bitter and sweet	4 „	each
Starch	2 „	
Gum Arabic	2 „	
Gum tragacanth	2 „	
Licorice root	2 „	
Fir-cones	2 „	

Pound and work up in must, and give it to the patient; let him use it as a linctus, taking a spoonful evening and morning.

Another linctus which is good for hoarseness of the voice.

Bitter almonds	8 drachms
Myrrh	6 „
Galbanum	4 „
Styrax	4 „
Crocus	2 „
Butter of ferula opopanax	1 drachm
Honey, as much as is necessary.	

Mix well together, and use.

This is the mass of medicines that are good for the voice, and they must be administered with the accurate knowledge of the skilled physician. Now therefore, in accordance with our systematic plan [of this book], it is right to set down in writing following the description of the organs of the voice Fol. 90a. an account of the organs of breathing. | As concerning the organs of the voice and the medicines to be applied in healing the diseases of them, what is written in this Tenth Chapter is sufficient.

HERE ENDETH THE TENTH CHAPTER.

CHAPTER XI.

Of difficulty of breathing, and shortness of breath, and asthma.

Now we have sufficiently described shortness of breath in the Chapter on the injuries of the spinal column, but we must recall in the present Chapter what we have said, and we will then add what there is left to say, that is, we will describe the theory of the matter. And in a general way, inasmuch as ye have a good memory, ye are well acquainted with all the muscles of the chest, and also with all the nerves which come thereto. When therefore ye go | into the presence of one who Page 187 is suffering from shortness of breath, look ye first of all and see if the muscles of the breast are in motion and those between the ribs, as well as the upper muscles, and the organs of digestion, and the muscles between the ribs. If ye see that they are all in motion, then ye must know that one of the upper muscles of which I have spoken, is primarily the cause of motion of this kind, and afterwards ye must try to determine which is the cause in the place. If, however, they are not all in motion, then it is right for you to arrive at quite other decisions. If, for example, we take the case of a sick man, all of whose muscles are in motion in such a way that his chest expandeth obviously and his shoulders move, this man, I say, must be suffering from one of the three following causes: enfeebled power, or constriction of the passages of the organs of respiration, or excess of heat in his heart and lungs. It is possible that two of these causes may be in operation at the same time, but there may perhaps be one only, or there may be all three operating at the same time; but when all three

are present at the same time, the man that is thus afflicted
Fol. 90b. dieth.   When two | of them are in operation, he may live, but
only with difficulty.   And by the characteristics and symptoms
which are closely associated with [the cause] it may be de-
termined whether the sick man is decreed to die or to live, I
mean in the case of weakness of the chest when it existeth
by itself.   For this reason he who is suffering [from disease of
the chest] cometh to the three kinds [of disease], and it setteth
them all in motion, because it is impossible for one kind only
to move continuously.   For if the digestive membrane were
moving sufficiently, as it were in a natural manner, he would
have no need either of the muscles which are between the
ribs, or of those upper muscles; but, because of enfeebled
Page 188 strength, he moveth them all | little by little, and doth neither
work them to the full extent of which they are capable, nor
to the least possible extent of which they are capable.   Therefore look attentively on these symptoms, and understand clearly
what they indicate, for by means of some of them ye shall be
able easily to determine the cause of the shortness of breath
(or, difficulty of breathing).   For when excessive heat (inflammation) is collected in the organs of respiration, the sick man
maketh use of all the muscles of the chest, to a small extent
and with frequency, and he setteth them in motion continuously,
that is to say, when in addition to his weakness of the chest
there ariseth excess of heat.   Now feebleness of power, neither
very little nor very abundant, doth the movement possess when
it taketh place, with the exception of the heat of inflammation;
moreover, it doth not greatly inflate (or, expand) all the organs
of the chest.   Therefore one [movement] only doth it possess
at once with that shortness of breath, which ariseth through excessive inflammation, that is to say, that which it performeth with
all the muscles of the chest.   Now closely associated with this
is the shortness of breath that taketh place through excessive
heat vehemently, frequently, and swiftly, with certain expiration,
for with a hot and inflamed breath inspiration is associated.
And with that which taketh place through feebleness of power
Fol. 91a. is closely associated inspiration, as well as expiration, | for it
taketh place by means of the mouth or through the nose only.

Now the nostrils make this very clear, for they contract inwards at the drawing in of the breath, which is itself a great proof of feebleness of power. In attacks of difficulty of breathing which take place in the organs of respiration the breast is inflated abundantly, quickly, and frequently. And besides inflation, those who are afflicted also perform the function of breathing out.

Now when heat and oppression of the chest both happen to take place together | in the organs of respiration, as, for Page 189 example, in those sicknesses caused by inflamed abscesses of the lungs, not even that great, frequent, and rapid breathing is sufficient for them, and the sufferers therefore hold themselves upright (or, straighten their bodies), and sit up, because they feel that in this position the whole chest can be more easily inflated. When they lie down the chest falleth on itself, because the upper portions of the chest come down upon the other portions of it which are close to the spinal column. When the spinal column itself is stretched out (or, extended) in an upright direction, the breast also standeth in an upright position with it, and doth not rest heavily upon it.

In a similar manner also taketh place the breathing of those who have the oesophagus of their lungs filled by the copious discharge of the rheum that cometh down from the head, or from one of the places that are round about the lungs, and they are different in one matter only, namely, [their chests] are not inflated, and they do not respire in a heated manner. In like manner also is it in the case of those in whom copious pus is collected between the chest and the lungs, now such sufferers the physicians call "purulent", and they inflate the whole of their chests; but they do not breathe in a heated manner, moreover, they do not gasp. On the other hand, if it happen that they have burning fever, they are straightway choked by the pain, and inasmuch as their power is enfeebled, they, in any case, would choke through the excess of pus. Now when Fol. 91 b. the rheum cometh to the lungs, or when there is in the lungs the inflamed abscess which is called "inflammation of the throat (lungs?)" (περιπλευμονία), or when there is asthma, the strength of the patient doth not become necessarily enfeebled,

14

for the exact opposite is the case and his power is firm and strong; moreover, the pains of asthma, which arise from the thick and viscous chymes, become joined to the lungs. And when also, because of something which it is difficult to dissolve (or, set free), there is pain therein, it happeneth on such an Page 190 occasion that the chest is greatly inflated | in an outward direction, although it doth not draw in a large quantity of air, and therefore patients feel difficulty in breathing regularly, except as regards drawing in the breath, and this is the sure symptom of the presence of excessive heat (or, inflammation).

Now we come to [the consideration of] another form of difficulty of breathing, and also of those who have the "strangles", when there is difficulty in breathing, not in those chests that take in breath, but in those that expel it. Now to those who have an abscess, and pus, and excessive fluid, either in the empty places of the chest or in the lungs, neither their throat nor their oesophagus being obstructed, for it is well known that difficulty of breathing ariseth in those breasts that receive the air which is drawn in, there must necessarily cling the great expansion of the chest in an outward direction, although very little air is drawn in, and for this reason the afflicted man respireth quickly and abundantly. Now, I wish you to remember before everything the distinguishing characteristic of the great breathing, which is of two kinds; for at one time it taketh place with great movement outwardly, and at another it consisteth of a great quantity of air drawn into the chest. And ye must understand that the things of which I here remind you it is absolutely necessary for you to know [in connection with] this treatise. All those who, apart from an Fol. 92a. abscess, or some unnatural swelling, or some difficulty | in the organs of respiration, [suffer from] shortness of breath in fevers accompanied by inflammation, in these [I say,] the greatness of the air-essence that is drawn into [the chest] is much [or little] in proportion to the greatness of the inflation of the chest. In those in whom the air-essence that is drawn into the chest is great, and in those who have some kind of abscess, or difficulty in the organs of respiration, besides a burning Page 191 fever, the motion of the chest from outside is great. | Now the

air that is drawn in doth not become less through the measure
of the inflation of the chest, but is reduced in no small degree
by the natural drawing of the breath.

Now besides the ordinary definitions, the inflation of the
chest maketh known very clearly one indication (or, mark),
namely, when it possesseth facility in drawing breath. And
even in such a case as this it is meet for you to observe
accurately, lest ye make a mistake on some occasion and
imagine that the kind of breathing that taketh place when all
the muscles are helping, because of feebleness of power [in
the chest], is in very truth the "great breathing". This kind
of breathing, on one occasion when I was wishing to explain
it, I called the "upper breathing", because this appellation
came into my mind at that moment. And Hippokrates [also]
called inspiration by means of the upper portions of the chest
"upper breathing", and on the occasion when he did so he
spake thus:—"When the great breathing taketh place." Sometimes "excessive" indicateth a change of opinion, and it is well
known that he calleth "great" that breathing which is "excessive",
and which also is possible to take place in two ways, that is
to say, in addition to the operation of the upper muscles, which
is also frequently accompanied by the operation of the muscles
that are between the ribs, and that also of the travelling membrane of the chest, which inflate it to no small degree, they
having no need of the upper muscles in the great drawing in
of the breath.

That a drawing in of the breath of this kind indicateth a
change of mind Hippokrates sheweth us in [his] treatises on
shortness of breath (or, difficulty of breathing), as hath also
been said by us in the section treating of the disease of difficulty of breathing, of whatever kind it may be. Now, as a Fol. 92*b.*
certain collection [of facts] concerning the other diseases [of
the chest] hath been compiled in that work, and they resemble
those that have been described by us in other treatises (or,
sections), and we have even described briefly the different
kinds of shortness of breath, what we have already said in an
incomplete manner we will here say in a complete form. For,
as it appeareth that the "great breathing", | changeth frequently, Page 192

14*

when the drawing of the breath is frequent it indicateth the presence of other diseases, and when it is infrequent it indicateth the presence of one disease only. Thus we say concerning the "little breathing" that, whenever it is infrequent, that is to say, the breath is only drawn at long intervals, it indicateth cold, not only in the organs of respiration, but also in those other members that move with them, that is to say, the liver, the spleen, the belly, and the stomach. Now, since there are many diseases to which pain clingeth when a man moveth frequently, it is meet for us to examine and determine accurately by means of these signs, so that we may know if he hath an abscess, and whether it is red and hard, or whether it be full of pus, or whether it be an open sore, when it causeth him pain. Now ye have also learned that pains are often associated with constitutions that are unequal, and with the inflation of the chest with a superabundance of breath, sometimes when it distendeth the region that is shut up, and sometimes when the breath escapeth forcibly. Similarly also when a superabundance of hot, acid chyme, or cold and viscous chyme, is imprisoned in that region, or when it distendeth it, or when it escapeth forcibly, there cling to them pains more abundantly than when they are imprisoned in the members that are diseased, even though they be not in motion; when they are in motion pains are present in a small degree.

Fol. 93a.   It hath been said by us | in those sections on the shortness of breath (or, difficulty of breathing) that the cause through which the lesser respiration, namely that which is abundant, taketh place, indicateth pain in those members that are moved

Page 193 by the breathing; and that the breathing that taketh place | at long intervals maketh manifest great cold in the organs of respiration only, and more particularly in the lungs and heart.

There is also another kind of shortness of breath which taketh place when the movement of the chest is impeded and killed by a short cough, sometimes when drawing in the breath, and sometimes when expelling it. This kind happeneth either through the disease of rigidity of the muscles of the chest or through a superfluity of heat, which compelleth the sick man to draw in his breath, or to expel it, in this manner.

Now there is another disease which taketh place in the function of breathing and which they call "breathlessness", because, so far as one can see, at the time of the attack there is no breath at all in the patient; but according to nature, it is not to be believed that there is no breath at all, for nature testifieth to the fact that it is impossible for a man to be wholly deprived of breath and yet live. Notwithstanding this, the reptile lieth up in his hole in the winter season, and doth not move his breast at all. We are forced, then, by this fact to think one of two things: either that the reptile breatheth so slightly that our perception is deceived and we think there is no breath in it, or that the reptile hath no need of any breath at all at that season, and that the evaporation which taketh place in his whole body is sufficient for him. Now this is effected by the heart through the arteries, and breathing is effected by the brain through the breast. Thus also is it with the disease called "breathlessness", for it appeareth to be common to all the members, and to resemble that which happeneth in apoplexy, and in attacks of stupor and of the falling sickness and of chill. For in all these is no special disease in the organs of respiration more than there is in the voice, or in the speech, or in walking; but, because disease of the breathing is the most important matter to them, all the members | that are ministered to by it must necessarily suffer Fol. 93b. with it, since they derive from it the power that governeth each one of them.

Now inasmuch as we have | already made mention of asthma, Page 194 together with shortness of breath, we will describe what asthma is. We call "asthmatic" those who, though having no fever, draw their breath in gasps as do those who run, and such people, because they draw their breath in this manner, physicians are in the habit of calling "asthmatic". These, however, so far as facts are concerned, would be more clearly described if they were called "those who draw their breath in an upright position", because they are compelled at all times to keep their chests in an upright position owing to their fear of being choked, and they always cause the parts of their beds against which they support their backs to be in an upright position, so that they may not be choked whilst they sleep. Now the air which they draw in when they breathe inwardly causeth less pain than when they breathe it out, and from this fact it is known that in the interior of their chests there is some unnatural oppression, even as do feel those who suffer from the disease of asthma.

Let us now collect the causes which we have described above through which shortness of breath, and difficulty of breathing, and asthma, take place, and, as we may say shortly, those by which the various kinds of cure are to be distinguished. Now asthma and shortness of breath are caused by moisture which is confined between the lungs and the chest, when patients cough and bring up pus after the manner of those who are called "[spitters up of] pus"; or there is an inflamed abscess in the lungs or in some region of the chest, as is the case with those who suffer from hard, red sores, or pustules, or weakness of the nerves, or of the muscles that work the chest and throat and that have their source either in the marrow of the spinal column or in the marrow of the head (*i.e.*, brain); or when these nerves and muscles are attacked by some kind of abscess; or when irregular changes in the conditions of the body take place in the chest and lungs. Or asthma is caused by some inflation of wind which Fol. 94a. cannot escape, or | when the mouth of the lungs or of the Page 195 oesophagus is filled with | thick and sticky fluid. These things are, stated briefly, the causes that produce shortness of breath and asthma.

Of the cure of asthma and shortness of breath.

Now that fluid, which is confined between the lungs and the chest and which hath been said to produce shortness of breath, is removed by coughing; and the medicines that have been described in connection with the diseases of the lungs and chest are good for those who cough and bring up pus, and for those who are called "[spitters up of] pus." For cases of the difficulty of breathing that ariseth through an inflamed abscess, or through a pustule, or through ulceration of the lungs, medicines, and [certain kinds of] foods, and plasters which cool and relieve the pain of the abscesses, are good. For the difficulty of breathing that taketh place through the weakness of the nerves or of the muscles those medicines, and plasters, and unguents that are prescribed in the Chapter on the diseases of the nerves are good. Similarly, even though an abscess be present, the plasters (or, medicaments) that are prescribed therein are beneficial to them, and bathings, and fomentations, and wet bandages applied to the spinal column, and to the nerves and muscles, when they are laid on the places in which the injury is situated. The irregular mixtures which take place in the lungs and chest, and through which asthma ariseth, are cured by mixtures of an opposite kind. And moreover, if there be any wind which causeth inflation, we cure it by means of agencies which relieve [the pain], and open the pores, and which are applied externally to the chest, and have an effect inwardly. Now the asthma that ariseth through a thick and sticky fluid which filleth the mouth of the lungs or of the oesophagus, and is caused by a discharge of rheum from the head, or which descendeth into the lungs from another place, we heal by means of medicines which cleanse and cut the fluid (phlegm?) that appeareth through coughing; these medicines are those that are set down in this Chapter. | Fol. 94 b. Page 196

Medicines which are good for shortness of breath and asthma.

Sulphur, unburnt	1 drachm
Seed of wild rue	1 ,,
Birthwort	1 ,,
Absinthium	1 ,,
Incense	1 ,,
Ammoniac	1 ,,

Pound and work up with vinegar, and make into pills the size of beans, and give three morning and evening; and let him take and swallow after them some digestive [medicine].

Another [medicine] which is good for difficulty of breathing and asthma.

Aniseed	6 drachms
Sulphur	6 ,,

Ammoniac	4 drachms
Castoreum	4 „
Garden nigella	4 „

Pound and work up with water, and make [into pills] the size of beans, and give five in the evening and five in the morning, and [let the patient take] one measure of oxymel after them.

Another [medicine] which is good for the aforementioned pains.

Artemisia abrotonum	4 drachms
The tops of branches of ab-	
sinthium	4 „
Aristolochia (birthwort?	
rhubarb?)	2 „

Crush and work up in vinegar, and use in the manner of the preceding, or in honey water, or in warm oxymel.

Another [medicine] for the pain.

Artemisia abrotonum	2 parts
Fleshy, dried grapes	4 „
Branches of rue	1 part
Rain water	3 *ķesţè*

Boil until two-thirds have been evaporated, and pour into what remaineth three ounces of honey; then boil again until the mixture becometh thick, and administer as a draught when warm three measures from time to time.

Another [medicine] which is good for asthma, and for difficulty of breathing; it giveth relief immediately.

Dried grapes cleaned from	
stones	1 handful
Trigonella	1 „
Rain water	1 *ķesţâ*

Boil until the ingredients are cooked, then strain off the juice and pour into a potter's vessel, and give generous doses of it in the form of a warm draught containing four measures to the patient from time to time.

Another [medicine] for the pains.

White peppercorns	1 *lîţrâ*
Ligusticum	4 ounces
Calamint	3 „

Dried mint	3 ounces
Rock parsley	1 ounce
Smilax (bindweed? wind-weed?)	1 ,,

Pound and work up in honey, and give to the patient a piece Fol. 95*a*.
the size of a | chestnut in honey water or oxymel.   Page 197

Another [medicine] for the pains.

Origanum (calamint?)	8 drachms
Smilax	6 ,,
Wood mint	6 ,,
White peppercorns	6 ,,
Roasted aniseed	4 ,,

Pound and work up in honey, and use.

Another. Take fresh extract of squills, pour on it a like quantity of honey, put on the fire and boil until it becometh thick. Take one spoonful (*tarwâdâ*) of it, and administer to the patient before and after meals.

Another. Throw two measures of the insects that collect under the water jars and are called "millepedes" into a potter's vessel, heat them over the fire, and when they are white rub them down and mix the powder with boiled honey, and administer in the form of a linctus one spoonful (*tarwâdâ*) [to the patient when] fasting.

Another [medicine] which is good for asthma and shortness of breath.

Inside of a bitter cucumber	$1^1/_2$ drachms
Iris	2 ,,
Roots of ferula opopanax	2 ,,
Artemisia abrotonon	1 drachm

Pound and work up in vinegar, make into pills the size of sheep droppings, and administer four in a draught; after them let the patient drink one measure of warm honey water.

Another [medicine] which is good for loss of voice and for difficulty of breathing.

Cardamoms	3 drachms
Iris	2 ,,
Cassia	4 ,,
Quince	$1^1/_2$,,

Styrax	2 drachms
Root of ferula opopanax	1 ¹/₂ „
Smilax	4 „
Seed of wild rue	4 „
Aresṭachyâ (Aristolochia?)	4 „
Sulphur, unburnt	4 „
Ammoniacal incense	4 „

Crush and work up in strong vinegar, and administer a portion as large as a chestnut in hot water.

Another [medicine] which is good for asthma of long standing.

Sulphur, unburnt	1 drachm
Ammoniacal incense	2 drachms
Aniseed	1 drachm
Garden nigella	2 drachms
Castoreum	2 „

Crush and work up with water, and make into pills each containing one drachm. Then dry and crush, and administer to the patient | at the time of [going to] sleep in one | measure of hot water.

Page 198
Fol. 95 *b*.

Another [medicine] which is good for breathing in an upright position.

| Opopanax | 2 drachms |
| Insides of colocynth berries | 1 drachm |

Crush and administer one drachm in hot water.

Another [medicine] for the above diseases. Take castoreum and ammoniac in equal quantities, and administer half a drachm in a draught of honey water.

Another [medicine] which is good for difficulty of breathing, and for bringing up vomit.

Mustard	1 drachm
Common salt	1 „
Elaterium (a purgative)	1 *denḳâ*

Pound and work up in water, make into pills, each containing one drachm, and administer one in a draught every two days; after taking it let the patient drink of honey water one measure.

Another [medicine] which is good for shortness of breath, and for asthma, and for hardness of the liver, and for obstruction in the bowels.

Rhubarb	3 ounces
Aristolochia	3 ,,
Flour of trigonella	8 ,,
Myrrh	2 ,,
Crocus	1 drachm
Madder	1 ounce
Agrimony eupatorium, or the juice thereof	2 ounces

Pound and mix together, and administer in the form of a draught, if there be no fever, in wine mixed with water, and, if there be fever, in hot water.

Another. Take the lungs of a fox, dry and pound them, and administer in the form of a draught with wine. It is beneficial.

Another. Catch a tortoise from the river and kill it, and pour some of its blood into a clay vessel: then cut open the body, take out its inside, roast it in the fire, mix a few peppercorns with it, and work up with honey, and use as a linctus morning and evening.

Another, [ascribed to] Dioscorides, which is good for coughs and asthma. Take mint, origanum, warrior's hair (Venus's hair?), horehound, iris, and licorice root, in equal quantities, soak in water until one half hath evaporated, mix with honey, and administer in the form of a draught.

Another [medicine] for asthma of long standing. Take fat figs, origanum, and rue in equal quantities, steep them in water, add honey to the liquor from them, and administer as a draught.

All the above medicines are good for difficulty of breathing, and for the kind of asthma that ariseth from the thick and sticky fluid that is confined | in the mouth | of the lungs, and that I have described above. And since this kind of asthma demandeth that the fluid that causeth pain be reduced and cut, medicines and [suitable] foods are necessary. Now the medicines that are beneficial in cases of abscess are finer and

Page 199
Fol. 96a.

drier than those that possess aromatic odours, and the medicines that are good for breathing in an upright position and
for asthma are those that cut [the phlegm] and cause no pain.
For this reason those that give relief especially are the vinegar
that is called "*askil*" (squills), that is to say, the essence of
squills, and oxymel that is compounded of vinegar of this kind.
The medicines that cause great pain are not good for this
ailment.

Now therefore, as sufficient hath been written about all kinds
of difficulty in breathing, and the means which are to be
employed in healing them, it is right for us to pass on to
describe the other diseases that are closely connected with
them, and these things must suffice for this the Eleventh
Chapter.

HERE ENDETH THE ELEVENTH CHAPTER.

CHAPTER XII.

Of the bursting forth of blood from the internal organs.

Now in the preceding Chapter a description hath been written of all the various kinds of difficulty of breathing that take place in the organs of respiration, and in the present Chapter we will speak about all the various kinds of bleeding that take place in the organs of respiration themselves, and about all the causes that produce them. Now bleeding from the internal organs ariseth from certain predisposing causes, and the primary causes, for so I am accustomed to call them, are, as ye well know, falling from a height, or the falling of a man upon his neighbour in some athletic contest, or the falling upon the chest of some heavy substance, which is itself inanimate, such as a stone or a log of wood, or excessive anger, or violent shouting, or overstraining in the performance of some work. | These are the causes to which I am accustomed Fol. 96*b*. to give | the name of "primary". Bleeding internally also Page 200 ariseth from causes that are called "predisposing", that is to say, either from superabundance of blood, or from some such like thing. In the case of superabundance of blood, this ariseth when the veins and the arteries are unable to hold the blood, and [the pressure] of it causeth effusion. Now sometimes [the pressure] causeth an opening to appear in the members that have been mentioned, or it happeneth through the natural structure of the members becoming cold and hard, and then bleeding from it ensueth. And bleeding also ariseth through gangrene, which attacketh internal parts. These, briefly, are the causes of bleeding internally. And all physicians have already admitted that the blood goeth down from the stomach and belly to the excrement, and that in vomiting it goeth upwards. From the organs of respiration it goeth up by means of coughing, and similarly it goeth up from the palate and from the uvula, and by means of emissions it is expelled from the mouth in the spittle. Now we often see a sudden bleeding and blood descending from the head, and especially from the uvula, and from the palate, and it is spit out by coughing, for immediately it goeth forth on the palate it setteth the patient coughing. And it is meet to watch this kind of bleeding with special care, lest we make a mistake and think that this blood cometh up from one of the organs of respiration. Now I myself know certain physicians who honestly thought this; they, however, made a mistake, thinking in their error that certain of the skilled and famous physicians had not decided correctly, and that bleeding of this kind was exceedingly serious, inasmuch as it indicated that the lungs were in an advanced state of disease, whilst actually it was caused by ruptures in the small veins. Now it is possible also that the difficulty of a sudden excess of blood is sometimes caused by some kind of Fol. 97a. food, | and this is accompanied by coughing.

Page 201　　Now when, in the early stages of the disease, | a man spitteth up a very little blood after certain intervals, and when it happeneth later that he bringeth up large quantities of blood during fits of coughing, although he hath not fallen from a height, and hath not suddenly fallen in a struggle or contest, and no heavy thing hath fallen upon his chest, it remaineth to us to think that he cougheth and bringeth up much blood by reason of some large gangrene. Very many who have suffered in this way have brought up some portions of the lungs themselves in fits of coughing. For this reason it is right to observe most carefully when any viscous matter cometh up with the blood, for this is a sure sign that the blood cometh from the lungs, or whether there cometh up any part of the base of the spinal column, or of the oesophagus, or of the coating of the arteries or veins, or of the substance of the lungs themselves. None of these things appeareth in the matter expectorated by those who bring up blood from the chest, just as those who bring up blood from the lungs do not suffer pain, because two small nerves from the sixth exit from the brain come to the membrane that envelopeth it on the outside, but to the substance [of the lungs] down below they do not reach. In the chest, however, as ye well know, there are many nerves that come from within, and there are also many that ascend to it outside, and it is because of these latter that a man perceiveth very quickly a painful disease. Moreover, the chest is muscular and bony, but the lungs are soft and plastic, and because of this disease of the chest is extended to them, and what is in them becometh paralysed.

When therefore any portion of the chest causeth a man pain, and he spitteth up blood, not in a large quantity, and not in a pure state, but which is | dark-coloured and is only Fol. 97b. brought up by the exercise of great force, it is chiefly his chest that causeth him pain, | and he spitteth blood from his Page 202 lungs. And thus also is it in the case of the man who bringeth up pus, in the diseases that produce pus, and in which the pus that is between the breast and the lungs becometh so solid that it can be perceived by the touch, and likewise, what is spit up hath the colour which is seen in matter expectorated during pleurisy. And in the lungs there appeareth quickly a wound (ulcer?), which, according to some physicians, is incapable of healing, but it hath been thought by others that it can be healed with very great difficulty. Now to the chest are attached many veins which may become ruptured, and of these some contain much blood; and however long the rent which taketh place in them may last, it can never be entirely healed. On the other hand, when the rent which taketh place in the lungs subsisteth for a very long time, although it is sometimes healed, yet to the very end there will remain in it something which is obstinate and fistular in character, and which, even after the lapse of much time, will be easily denuded of its skin by the smallest cause. And there cometh up sometimes, with the things that are spit up from this region, that which is called by the physicians "slough". Now the breakings forth of blood which take place in the lungs are recognized by the fact that a copious flow of blood cometh from the ruptures, and by the fact that the primary cause of the rupture is well known and is easily perceptible, such as a fall, or shouting, or some one of the causes which have been mentioned above, besides some manifest cause which is external. Moreover, there sometimes taketh place a rupture in the veins and arteries of the lungs through excess of blood, just as in any other member, whatever it may be, when the vein or artery becometh hardened and unpliant, either because of some chill which hath attacked it, or through some evil condition of its natural structure.

Now one cause of rupture of the lungs is frequent movement, and excess of the chymes by itself will cause it, or some cold and inflating wind, which cannot be disposed of; this

Fol. 98a.
Page 203

distendeth the | veins and, with the chymes, causeth rupture. Now of the rupture which taketh place through these things the sudden discharge of a large quantity of blood is a symptom of no small [importance]; but the cause of the opening of a vein is the opposite of cold. And the changes caused by heat precede it, and they take place when a man batheth frequently in hot water, and when he is in a hot country, and when the season of the year is hot, and when he partaketh of hot foods and hot drinks. The sudden effusion of blood which is the result of gangrene (or, cancer) taketh place after a slight spitting of blood, and after the acid rheum which moveth about and cometh down from the head to the lungs. There are times when there is brought up with it a portion of the lung, or some slough, and therefore it is impossible for disease of this kind to be mistaken. For if there be an ulcer in the oesophagus, and that which is expectorated from it be of the nature of pus, and even if the wound be in some diseased spot, the sensation of pain that ariseth in the patient, and the smallness of the quantity of that which is ejected [indicate the place where it is]. The physician must distinguish between an ulcer of this class and the ulcers that exist in the lungs.

Now as when a part of the lungs cometh up it serveth as an indication [of the existence] of ulcers of the lungs, so similarly do we often see the substance of the "daughter of the tongue" (ὑπογλωσσίς) coming up and being expectorated, because of the existence of some ulcer. And the substance

that cometh up is not the only indication of the existence of
a diseased spot, but there is also the sensation of pain that
ariseth in the lungs, and together these shew, as we have said,
that there is also an ulcer in the oesophagus. Therefore in
the lungs, besides the sensation of pain, all [kinds of] diseases
arise; now in other places the diseases create a sensation of
pain, but as this is very small, so far as those whose sensitiveness to pain is not acute [are concerned], it is not very clearly
known.

Now in all these diseases from which wounds arise, and
which I have enumerated | here, there is bleeding continually,
and in them there come up, accompanied by coughing, small
clots of blood, which come from within, and which may be
due to ulcers in another part of the body. Those which are
from the belly and stomach come up by vomiting, those which
come from the kidneys and bladder go forth in the urine, those
which come from the bowels descend through the anus, those
which come from the womb emerge through the vagina, those
which come from the head descend by the palate and nostrils;
but those which come from the organs of respiration cannot
be ejected except by coughing. Now there is another kind
of spitting of blood, [which may be described] thus:—If a man
spitteth and breatheth in the ordinary way, and blood cometh
up several days in succession, and he hath neither felt beforehand nor feeleth then any heaviness or pain in his head, and
he hath not received a blow of any kind, it is meet that we
should examine with the greatest care the orifice of his nose,
and that part of his mouth where it is hollowed out towards
his nose, for in a case of this kind it sometimes happeneth
that a diseased growth closeth up this place. In the early
days of this growth, by reason of its small size it escapeth notice,
but after three or four days it is easily discernible. From a
similar cause blood is evacuated from the belly, when, as it
sometimes happeneth, a man is attacked by [the disease called]
"'alaḵĕthâ"; now blood of this kind is thin and inert. Whether
the blood is seen to come from the belly, or from the nose,
or from the mouth, when the physician seeth it he must enquire
into the constitution of the man, and ask questions as to what

Page 204
Fol. 98 b.

15

symptoms have previously shewn themselves, and from all these he will be able to decide concerning the truth of the matter, except in cases of this one kind of bleeding only, namely the bleeding caused by the "ʿalakĕthâ". All other cases of bleeding that occur, from whatsoever spot the blood Page 205 cometh, | are due to the three following causes: the rupture of some blood-vessel, the opening [of a vein], and gangrene (or, Fol. 99a. cancer). The above | are the symptoms of the bringing up of blood that taketh place from the organs of respiration to the man who doth not disregard them contemptuously.

Of the healing of the bringing up of blood (haemorrhage).

Now the first danger in connection with the bringing up of blood, which ariseth through the immoderate vomiting (or, evacuation) that taketh place through the opening or bursting of the veins or arteries, is that which ariseth through the eating away of the region [which is affected]. The cure for arteries that have become opened is to stop up the patient's mouth, and the cure for the burst artery (or, vein) is to bind together [the sides of the wound]; in the case of the artery that hath been eaten away the wound must be cleansed and the substance of the artery made to grow again. The vein that is open must be treated with medicines that are of an astringent nature and bind together; the vein that is rent open must be closed by medicines of an astringent nature mixed with those that make the sides of the wound to unite and cling together, as, for example, alcyonium (i. e., bastard sponge), and gâsestîr.[1] The part of a vein that is eaten away is made to grow again by means of foods containing chymes that are beneficial, and by the medicines called "flesh increasers", such as frankincense, caryophyllus bark, and starch. These are the means which are to be employed in curing any kind of haemorrhage in the lower parts of the members.

Now to cure haemorrhage in the organs of respiration, especially of the lungs, is very difficult, so also are cases of

---

[1] I. e., talc ܐܪܥܐ ܟܘܟܒܐ = γῆς ἀστήρ.

haemorrhage in the breast, and in the membranes of the digestive organs, and in the travelling membrane of the ribs, and in the oesophagus. When there is bleeding from these, or when there is a wound (*i.e.,* perforation) in any one of them, it is possible sometimes to heal it if care be exercised in the beginning. To heal wounds in the lungs is very difficult, and frequently it is not only exceedingly difficult to heal them, but when they have existed even for a short time, it is absolutely impossible to cure them at all. For this reason when haemorrhage of the organs of respiration taketh place, in the earliest stage of it, | and provided that the patient is healthy Page 206 and strong, and his physical condition and age be favourable, a reduction of the blood, by letting blood from the middle or | Fol. 99*b.* lower vein in the arm, will be useful; in addition to this there should be applied tight bandages to the hands and feet, and powerful rubbings (massage?), and plasters that adhere should be placed on the whole of the chest, or on the spot that is over the part that is affected, and over the whole head, and food of an astringent and binding character and drink of vinegar mixed with water should be administered, and if there remaineth a drop of blood it will be set free and will come up. And medicines that are far stronger than those used for other parts of the body are required, because neither the foods which are eaten nor the medicines which are drunk are able to penetrate these parts. Therefore, with the medicines that are good for the bringing up of blood from the organs of respiration are to be mixed those medicines that are hot and subtle by nature, and those that are antidotes for the three kinds of ailments that have been mentioned, namely, the opening of the arteries, the tearing of the arteries, and the eating away of the arteries.

Now the medicines which brace up and bind stop the arteries, and these need to possess the power of making a passage so that their strength may be transmitted to the organs of respiration. Now when haemorrhage occureth in the belly, or in the stomach, or in the bowels, it doth not require medicines such as these, on the contrary, medicines possessing exactly opposite qualities are required. There is another kind of

15*

medicine which cleareth and reviveth the faculty of sensation, and this also is mixed with those medicines that are good for haemorrhage, and they produce deep sleep, and are beneficial for those who are tortured by coughing, and by means of cold they prevent the blood from flowing to the artery that causeth pain. The following is a list of the compound medicines that Page 207 are good for haemorrhage of the | organs of respiration.

A medicine which is good for those who bring up blood when coughing, and I use it myself.

Acacia	4 drachms
Dried | roses	4 „
Purple βαλαύστιον	4 „
Gum Arabic	2 „
Gum tragacanth	1 drachm

Fol. 100*a*.

Pound and work up in water, make into three-drachm pills, dry in the shade, and give the patient one to swallow in a draught of rain water.

Another [medicine] which is good for haemorrhage of any kind whatsoever, and for bleeding from any place whatsoever.

Seed of hyoscyamus	4 drachms
Frankincense	4 „
Opium	4 „
Talc, الطلق	4 „
Amylum (starch)	2 „
Sea gems	2 „
Coralium (κοράλλιον)	4 „
Purple βαλαύστιον	2 „

Pound and work up in juice of polygonum or of plantains, make into pills each containing half a drachm, and let the patient swallow [one] in rain water, or in juice of "lamb's tongue", or in juice of wall-wort (comfrey, or black bryony).

Another.

The fleshy parts of grapes	16 *lîṭrê*
Red mulberries	16 „
Fleshy figs	16 „
Rain water	16 measures

Place these in water and boil until two-thirds of it have

evaporated, pour the remainder into a potter's vessel, and administer it in doses of one measure, fasting.

Another, for the [same] disease.

Frankincense bark	2 drachms
Talc	2 ,,
Starch	2 ,,

Crush and reduce to a powder, and administer it in doses of one spoonful, each in one measure of water.

Another.

Purple βαλαύστιον
Talc
Gum Arabic
Bark of frankincense
} in equal quantities

Crush and reduce to a powder, and administer according to the instructions written above.

Another, which is to be taken on the tongue, and which is good for those who bring up blood and for violent vomiting. Take one *lìtrà* of clarified honey and one measure (ξέστης) of best wine, pour them into a potter's vessel, and work up together. Put the mixture on the coals of a fire and boil it until it becometh thick, then take it off the fire and let it cool. Pour on it crushed gum Arabic, | four spoonfuls, and Page 208 two drachms each of purple βαλαύστιον and talc, then heat it again over the fire, and administer it to the patient in two-drachm doses.

Another, which is good for those who bring up blood, and which is called "sublime" and "splendid".

Starch	2	drachms
Purple βαλαύστιον	2	,,
Talc	2	,,
| Juice of hapḳaṭedis (goat's beard)	2	,,
Gum Arabic	2	,,
Crocus	2	,,
Opium	2	,,
Juice of "lamb's tongue" (plantain)	3	,,

Fol. 100*b*.

Mix well together, and administer one drachm in rain water.

Another, which is compounded of coralium, and which is good for disease and pain of the belly, and for diarrhoea.

Plantain (psyllium), pure,	6 drachms
Iris	30 „
Stacte	30 „
Coralium	30 „
Amber	30 „
Crocus	30 „
Opium	16 „

Macerate the plantain in hot water, and let it steep, and when the liquid hath become glutinous and sticky, strain it off, mix in it well the other medicines, and work up the mixture and make it into pills each containing half a drachm. Administer at bed time in some convenient vehicle as a draught.

Another, which is good for those who bring up blood, or pus, or rheum.

Seed of white hyoscyamus	10 drachms
Bark of mandragora root	10 „
Sweet wine	10 „
Frankincense	10 „
Opium	10 „
Styrax	10 „
Cypress fruit	10 „
Stacte	30 „
Amber	30 „
Germander (teucrium polium)	30 „
Water for macerating the germander	—

Mix well together, and use according to the instructions written above.

Another, which is compounded with coralium, and is good for the same disease. A most excellent medicine.

Coralium	8 drachms
Talc	8 „
Starch	4 „
Purple βαλαύστιον	4 „
Lycium	4 „
Sea gems	4 „

White hyoscyamus	4 drachms
Opium	4 ,,
Juice of goat's beard (trago-pogon pratense)	4 ,,

Pound and work up in plantain juice, and make into pills each containing half a drachm, and use in some convenient form of draught.

Pills for the disease.

Coralium	8 drachms	
	Costus	8 ,,	Page 209
Juice of goat's beard	2 ,,	
Purple βαλαύστιον	2 ,,	
Acacia	2 ,,	
Rhubarb	2 ,,	
Talc	2 ,,	
Opium	2 ,,	
Frankincense	2 ,,	
Sumach	2 ,,	

Pound and work up in plantain juice, and use.

Pills (or, tablets) which are good for those who bring up blood, and for pain in the belly, and diarrhoea, and rheum of the stomach.

Pounded pomegranates	6 drachms
Egyptian cotton	6 ,,
Purple βαλαύστιον	6 ,,
Lycium	4 ,,
Rhubarb	4 ,,
Opium	4 ,,
Myrrh	2 ,,

Pound and work up | in the juice of myrtle berries, make into Fol. 101 *a.* tablets, and administer them in rose water according to the strength of the patient.

Pills (or, tablets), [which are good] for these diseases.
Take in equal parts:

Purple βαλαύστιον
Goat's beard
Opium
Acacia

> Frankincense
> Myrrh
> Crocus
> Dried grapes
> Aloes
> Rhubarb
> Lycium
> Myrtle berries
> Rind of pomegranates

Pound up together and make into tablets with the best wine, each containing one drachm. Administer in wine to those who, have no fever, and in honey water to those who have fever.

Tablets which I myself use for those who bring up blood or pus, and for diarrhoea.

Myrrh	6 drachms
Frankincense	6 „
Crocus	6 „
Bark of plantain root	4 „
Opium	4 „
Seed of hyoscyamus	4 „

Make into pills, each containing half a drachm, and administer in wine, or in juice of "lamb's tongue", or in juice of myrtle berries, or in juice of black bryony, before meals, and they are beneficial at bed time.

Tablets, which are compounded of madder, and are good for those who bring up blood and pus.

Best wine	$^1/_2$ measure
Infusion of madder	I „

Pour the madder into a glass vessel, with the wine, and let it soak for a whole night and day, and the following day. Work the madder well with thy hand until it emitteth its juice; then
Page 210 squeeze out the juice and pour into it half a *litrâ* of honey, | put it on the fire and let it simmer, and when the mixture is as thick as honey take it off, and administer one spoonful as a linctus.

Tablets which are good for those who bring up blood and pus, and who have protracted fits of coughing, and

who bring up something viscous and sticky, and for difficulty in breathing, and for women whose courses do not return after childbirth. And it bringeth down the fat, and it is good also for young children who are vexed with pains, and it may be administered to those who suckle them.

Juice of madder	6 measures

It is meet for this to be squeezed out | when the plant is Fol. 101*b*. moist, and it is then to be dried; or take it when fresh.

Fine wine	1 measure
Fine honey	1 ,,
White peppercorns	4 drachms
Frankincense	4 ,,
Myrrh	4 ,,

Boil down the madder juice with the wine to one-third, then pour in the honey, and when it is thick sprinkle the dry drugs over the mixture, mix them together carefully, take it off the fire, and pour it into a glass vessel. Administer as a dose one spoonful when [the patient] is fasting.

Pills which are good for those who spit blood. Take in equal quantities purple βαλαύστιον, gum Arabic, and sagapenum, pound them and make into pills (?), and administer a dose of six grammes in a draught of hot water. These have been well tried.

Pills which are good for those who bring up blood. Take in equal quantities styrax, sea-sulphur, licorice root, castoreum, and gum of the terebinth, pound them together and make into pills the size of vetches, and administer a dose of two grammes in a draught of hot water, and let the patient also keep them under his tongue.

Pills for the disease. Take two measures of purslane juice, and add to them a little sugar (or, sweet wine), and administer as a draught.

Or mix the juice of lamb's tongue with a little vinegar, and administer as a draught.

Or mix together one measure of the juice of the Greek leek with half a measure of vinegar, and administer as a draught.

Or burn a sponge in the fire, and crush it in wine mixed with water, and administer as a draught.

Or mix one measure of plantain juice with half a measure of vinegar, and administer as a draught.

Page 211 **Pills for those who bring up | blood.**

Hyoscyamus seed	5 drachms
Poppy seed	5 ,,
Purple βαλαύστιον	2 ,,

Pound together, and administer one drachm in a draught of vinegar mixed with wine.

Another, which is good for those who bring up living blood.

Acacia	4 drachms
Dried roses	8 ,,
Purple βαλαύστιον	4 ,,
Gum Arabic	2 ,,
Opopanax	1 drachm

Pound and work up in honey, and administer one drachm in a draught of cold water.

Another. Take in equal quantities amber, *samtĕrîn*, and Fol. 102a. gum Arabic, | pound them and give to the patient to drink in vinegar mixed with water.

Or take one drachm of sea gems, and crush it well in vinegar mixed with lukewarm water.

Or crush myrrh, and frankincense, and *samtĕrîn*, and administer as a draught with plantain juice, or with the plantain [itself], or with the juice of calamint, mixed with a little vinegar.

The above are the medicinal materials which are good for every kind of bringing up of blood.

Now, the foods which are good for haemorrhage are those that purify (or, are of an astringent character) and check it, and of which the chymes are beneficial (nutritious?), as, for example, all strong-smelling herbs, and pomegranate water, and the infusion of husked barley, mixed with juice of asparagus or the juice of myrtle berries, and mallows, both wild and cultivated, and sumach water, and vegetables, lettuces, chicory, and especially purslane, and every herb which is slightly astringent. And milky substances are also good for patients: if it be necessary to thicken the body administer cows' milk, and if to reduce it, ass's milk; and, if milk of an intermediate nature be required, take goat's milk. These must be administered in a warm state, and if it be possible the patient should drink the milk of sheep or goats. Now, for those whose stomachs are weak and cold, it is meet that a small quantity of honey should be mixed therewith. And the animal from which the milk is taken must be supplied with food that possesseth strengthening properties, and is slightly astringent, for then the milk will not curdle quickly in the stomach. And patients must abstain from every kind of exciting work of the body, and they must keep | continually under their tongues Page 212 some one of the medicines which are good for the bringing up of blood, and must struggle to their utmost power not to cough; and they must suck the medicine little by little, and by doing this, the strength of the medicines will descend to the place that is diseased, and will cure it.

Now the plasters that are laid on their chests must be composed of medicaments that are binding and astringent, as, for example, those that are compounded of aloes, and frankincense, and acacia, and purple βαλαύστιον, and these are to be laid on the chest, exactly over the places from which we think | the Fol. 102 b. blood cometh. And in addition to these things we must exercise the greatest care in respect of the whole body, that is to say, we must pay attention to the reduction of the belly by means of blood-letting and drenches, and we must bandage the hands and the feet, and rub them.

Of plasters which are good for those who bring up blood.

A plaster which is good for haemorrhage of the organs of respiration; it is to be placed on the chest, and it is good also for diarrhoea.

Smelters' dross from the furnace	5	drachms
Dried grapes	5	„
Aloes	10	„

Vine flowers (οἰνανθίς)	3 drachms
Stacte	3 „
Juice of fleshy grapes	3 „
Dried roses	3 „
Wax	4 ounces

Oil of asparagus as much as is sufficient.

Pound the dry drugs, rub down the wax with the oil, mix them together, and smear on a piece of linen, or on a bandage, and bind on the chest.

Another [which is good] for the diseases of the chest.

Dog's bane (ἀπόκυνον, aconitum lycoctonum)	3 drachms
Vine flowers	3 „
Dried roses	3 + 6 „
Aloes	3 „
Scoria of smelters	3 „
Dried grapes	3 „
Acacia	3 „
Purple βαλαύστιον	3 + 6 „
Stacte	3 „
Absinthium	3 „
Flesh of dates	18 „
Wax prepared with myrtle oil	36 „
Crocus	3 „
Glaucium	3 „

Page 213 Pound them all and dissolve in vinegar. Then dip cotton | wool in this mixture, or smear it on a strip of linen, and lay it over the place whence cometh the blood.

Another.

Dried myrtle	1 drachm
Dried roses	1 „
Rind of pomegranates	1 „
Senna	1 „
A hand of sumach	1 hand
Purple βαλαύστιον	¹/₂ „
Dried grapes	100 in number
Smelters' dross	6 drachms

Vinegar, or wine with a "bite" in it	2 *dulḳî*

Make an infusion of these, then dip a sponge in the infusion, place it on the breast, and fasten it with a bandage. Others, however, dip a sponge, or cotton wool, in vinegar, and place it on the chest, and keep it in position by means of bandages. Others smear the whole chest with the juice of unripe grapes or olives, and then sprinkle over it a thick layer of the powder of burnt pomegranates. Others smear the chest of those who suffer from haemorrhage | with well known tablets, having first Fol. 103 *a.* dissolved them in ordinary vinegar, or in very pungent vinegar.

Pills of various forms (πολυειδής) which are good for many diseases.

Smelters' dross	3	drachms
Myrrh	8	„
Flowers of copper	2	„
The contents of a pomegranate	12	„
Gall of an ox	4	„
Aloes	4	„

Pound and work up in old wine, make into pills, and, when the occasion demandeth, dissolve in vinegar or in sour wine, and then smear on the breast and, if necessary, on the head also. These relieve the gangrene (or, cancer).

Pills of Andronius, which are good for the disease.

Finely chopped pomegranates	10	drachms
Dried grapes	4	„
Smelters' dross	4	„
Myrrh	4	„
Aristolochia	4	„
Vitriol	2	„
Flowers of copper	2	„
Gall of an ox	6	„
Aloes	8	„

Pound and work up with new wine, make into pills, and use according to the written instructions. These are also good for the gangrene which cometh in the external | members. Page 214

Now others make these drugs into small pellets with must

or wine, and administer them to patients who suffer from haemorrhage of the chest, ordering them to keep them under their tongues.

A plaster which is good for haemorrhage from any part of the body, and for diarrhoea.

Myrrh	4	drachms
Aloes	4	„
Frankincense	4	„
Crocus	4	„
Purple βαλαύστιον	8	„
Samterîn	8	„
Acacia	8	„
Glaucium	8	„
Lycium	8	„
Opium	3	„
Hyoscyamus seed	3	„
Root of plantain	3	„
Smelters' dross	6	„
Sea gems	6	„
Copper shavings (or, rust)	6	„
Vitriol	6	„
Rock parsley	2	„
Mint	2	„
Mustard	2	„
Lepidium	2	„
Aniseed	2	„
Sumach	6	„
Dried grapes	6	„
Insides of pomegranates	6	„

Crush small and work up with sour wine, make into pills and dry in the shade. When required for use dissolve them in vinegar or wine, and smear on the breast and head when there is haemorrhage from the organs of respiration, and on other Fol. 103*b*. parts of the body from which there is | haemorrhage, or in which there is any other ailment, and they will afford relief; they have been well tried.

The above form the mass of medicines which are good for those who suffer from haemorrhage in the organs of respiration.

Of the healing of [the disease called] "'alaḳĕthả". Now
if the patient be suffering from that kind of spitting of blood
which hath already been mentioned, and which is caused by
a leech that hath been swallowed, set the sufferer in a position
opposite the sun, and let the physician press down his tongue
with a specillum (σπαθομήλη), and look into him very carefully,
and if thou seest a leech clinging to the oesophagus, lay hold
of it with the instrument with which the uvula is held. Do
not lift it off suddenly, lest the creature break, but take hold
of it gently and lift it up until its mouth letteth go its hold
upon that to which it is fastened, and then take it out. If the
creature doth not fall within the range of vision [the following]
medicines are good for it:

Vinegar	2 measures
Nitre	3 drachms
| Garlic	3 „

Page 215

Work up the garlic and the nitre well together, shake them
up well in the vinegar, and let the patient use this as a gargle,
and the creature will come out.

Or crush into a powder two drachms of mustard, and four
drachms of nitre, and blow it into the mouth in a dry form.

Or crush to a powder absinthium and garden nigella, and
blow [into the throat].

Or take one part each of garden nigella, absinthium, artemisia abrotonon, lupins (or, chick-peas), costus, and the inside
of orichalchum (بريق), and *sarkash* (?), and two drachms of
garlic. Make a strong infusion of these with vinegar, and let
the patient use it as a gargle. If the leech be in the stomach,
let the patient drink glaucium pheniceum, and it will fall out.

Or let a man eat garlic in large quantities, and let him drink
no water. And let him set before himself a tank of cold
water, and bend over it, and then let him stir up the water,
and open his mouth above it, and the leech will go forth. Let
him do this in the sun.

Or let him pound up vitriol, and blow it into the patient's
mouth, and the leech will go forth.

Or pound mountain hyssop and madder and blow them into
the patient's mouth, and it will go forth.

Or pound nitre and salt, and mix them with urine, and let the patient take the mixture in his mouth and gargle, and it will go forth.

If after the blood-sucker hath gone forth from its place Fol. 104a. there be haemorrhage, | make use of the medicines that are of an astringent nature, such as *surantikôn* [1] (collyrium refrigerans) and many others that are equally good. And let the patient eat foods which are of themselves strengthening.

Now, concerning all kinds of haemorrhage, and their symptoms, and the means to be employed in healing them, sufficient hath been written, and these things must suffice for the Twelfth Chapter.

---

[1] It contained arsenic.

HERE ENDETH THE TWELFTH CHAPTER.

CHAPTER XIII.

Of the symptoms of the injuries that take place in the lungs, and in all the organs of the breast.

Now the path of those who are skilled in [the knowledge of] the symptoms of the places which are diseased is of two kinds: the one is the path that ariseth from happenings which are visible, and the other ariseth from the places in which these things happen. If now a man be trained | in both paths, he Page 216 hath two wherein to walk, and I say that these are not the same; but two entirely distinct paths. Behold, we have already spoken a little about the diseases of the lungs in the Chapter on shortness of breath, and also in the Chapter on haemorrhage, but we must also speak again about them here, since the subject of the present Chapter is all the members of the chest. Now severe pain never taketh place in the chest without there being a feeling of weight in it. Sometimes there is also a kind of tension of the chest that reacheth to the spinal column, because it is there that the membranes which envelop it are made fast. Moreover, those who suffer often feel oppression, and for this reason they gasp frequently and rapidly, and they move their chests greatly, but do not draw into them much air. If these symptoms take place, except when there is fever, they indicate that there is in the chest an ulcer (or, abscess), or an excess of thick and viscous chymes, or something else of this kind, or that the parts round about it are clogged with pus | Fol. 104 *b*. or with thick or viscous chymes, or they indicate that there is an excess of some kind.

Now all these may be defined by means of the happenings which precede them. If a man, who is in good health and

16

performing his ordinary duties, beginneth to feel a shortness of breath and a pain increasing in him, in addition to the appearance of a certain restraint in his breathing, it is meet to think that there is some hard ulcer (or, abscess) in his lungs; and if, besides these things, he hath a wheezy sound in his breathing, and the matter which he expectorateth is copious and thick, or viscous, these shew that the ulcer clingeth to the oesophagus of the lungs. If, however, shortness of breath cometh upon him suddenly, together with a feeling of oppression, then these things must be to thee a proof that rheum descendeth to the lungs, either from the head, or from some of the other members that are near to it. When these things

Page 217 take place through the excessive effusion of some viscous | or thick fluid, which descendeth to the lungs—I mean shortness of breath in a man—it is to be understood that the fluid is confined in the oesophagus of the lungs. Now the oesophagus, which is situated in the throat and is also called the "rough artery", is distributed throughout the lungs in the same way as the artery that cometh to it from the heart, for it is well known that the nature (i. e., substance) of the little oesophagi is in all the lungs. Now those who dissect the body call the little tubes (or, canals), which are composed of cartilages, "little oesophagi", and they have a kind of circular [form].

And if perforation (i. e., pleurisy) seemeth to progress, and if the strength of fever maketh itself manifest therewith, and if there remaineth a certain amount of sensitiveness in the ribs at the bottom of the chest, in addition to that which accompanieth a sudden change in the position of him that is lying down, and as some liquid floweth from this side to that, especially at the time when patients are turned from one side to the other, and as there is a certain superfluity, and they appear to have [inside them] water pots possessing the faculty of feeling, and as moreover a certain sound is heard perceptibly in their chests, this, and the things which have been

Fol. 105 a. already mentioned, show that nothing of that which | cometh up from the great perforation, which was originally in the ribs, is ejected in expectoration. Now, if some other inert or phlegmatic chyme, in addition to the fever, transferreth itself

suddenly to the empty places of the chest and causeth shortness of breath, then patients spit up something which is not
thick; what they spit up is at first little, and the spitting is
accompanied by violent coughing, but when it is dissolved to
a moderate degree, they spit copiously, and the matter expectorated is copious, and is accompanied by violent coughing.
Now patients spit little for two reasons when there is violent
coughing; some because of the thickness and viscosity of the
chymes, and some because of their extreme thinness. When
the chyme which cometh up during coughing is extremely
thin, it is emptied out by the breath of the cough which
presseth upon it, | and it floweth downwards and doth not go Page 218
up. When the chyme is thick and viscous, it goeth up with
difficulty, because it cannot separate itself easily from those
bodies to which it adhereth, and it is impossible for it to be
expelled by a little (*i. e.,* weak) breath; for, if the breath which
is coughed up be abundant and frequent, it is unable to drive
up anything with it. When then the chyme is not very moist
or watery, and also when it is not very thick or viscous, it
goeth up easily, and especially when the strength of the
patient is in a healthy condition; for unless the chest be contracted to a great degree it is impossible for there to be much
coughing, and, unless there be frequent coughing, it is impossible for the thick and viscous chymes to go up.

When there is shortness of breath accompanied by oppression
and weight, and fever, and inflammation, there is an inflamed
abscess in the lungs; but if there be very much inflammation
and the sensation of weight and oppression be little, there is
a hard, red abscess in the lungs. And because these two
states can be clearly defined, it is not difficult to recognize
whether that which is in the lungs be a hard, red abscess,
[or not].

Now there are other diseases that attack the lungs besides
the effusion of chymes during the regular or | irregular change Fol. 105*b.*
in the composition of the body; that change which is irregular
produceth coughing, whilst that which is regular and even,
when it is moderate, only changeth the regularity of respiration,
even if it becometh strong and increaseth. If there be great

16*

heat (or, inflammation) in it, it produceth an eager desire for cool air and for cool drinks, and sometimes this also clingeth to fever. And the change in the composition of the body which is cool, and which is the opposite of the things that accompany it, that is to say, the desire for warm air [and warm drink, is good] so long as it is moderate; but if even this increaseth, air filleth the lungs.

All the diseases, then, which I have enumerated hitherto, and which take place in the lungs and in the empty places of Page 219 the chest, and also in the oesophagus | and in the throat, as we have seen them often, even so have we described them; but the following diseases, about which I am now going to speak, have been only seen by me on very rare occasions. Now, a man [beginneth] to cough suddenly, and he bringeth up chyme which is like fresh bile, and the colour of which is between red and yellow (*i. e.,* reddish-yellow), and which containeth no acidity whatsoever. From this time onward the man spitteth more and more each day, and at length attacks of subtle fevers seize him, his body beginneth to diminish, and he beginneth to spit pus, and after a period of about four months he bringeth up a little blood with the pus, and he is attacked by fever and inflammation. And again, after a little, he spitteth copiously, and he also wasteth away rapidly, and after this the fever increaseth, and in this way his strength is turned into weakness, and he dieth, in the same way as those who are attacked by the sickness of phthisis die.

I saw another man also who was ill in exactly the same way for six months, and I saw another man also who was ill in exactly the same way, only more seriously. He whom I had seen first of all did not seem to me in the beginning of his illness to have anything bad in his condition, but subsequently it became well known that he had, and I knew only too well that he was in evil case. When, however, I saw the second man I knew at once, from the very beginning, that I Fol. 106 a. must treat him with the greatest possible care, | and henceforth I treated both of them with the greatest care. Neither of them lived, however, nor any other man after them who was attacked by the same disease, for they all spat [blood] before their end, and they brought up portions of the lungs that had become rotten. Therefore I knew well that a certain disease had attacked them which was like unto that which taketh place in the external members when they become saturated and decay through the fluid of corruption; only, in the case of the external members it is possible for the physician to cut them off, and, before this, to cauterize them, but with the lungs it is not possible for him to do | either of these Page 220 things, and therefore they all perish.

Now when I saw the second man, it seemed to me that I might be able to dry up the lungs forcibly by means of the aromatic drugs and draughts that were good for them, and I therefore ordered him to draw in the odour of that drug which is called "hedrôkhnôn" (ὑδράργυρος, hydrargyrum) the whole day long, and I held it to his nose, and he inhaled the odour of it continually. And I also ordered him to keep his nostrils anointed at all times with one of the unguents that are compounded of drugs of very sweet smell, namely, that which they call "Pôrâṭôsîpîḳà". And I gave him to drink draughts of the potion which is called "Metdôrîṭôs", and "Ambrosia", and "Immortality", and "Tûrḳê", but even with these things and drinking these drinks he only lingered for a year, and at length, in the same manner as those who have phthisis, he too died. Now he lingered a long time because of the course of treatment which we have described.

And I saw also another case of disease in the lungs which resembled this. This man had been coughing for a very long time, and he used to spit up a small quantity of thick matter which was like unto small hailstones, and he coughed some of them up and brought [them] and shewed them to me. And again, after a few days, he coughed up some more, and these also he brought and shewed to me. Now, I thought that what he had coughed up the first time was some viscous chyme, and when it was dry it resembled this. However, through drinking a certain medicine which I gave him | to Fol. 106b. drink, and which contained the things that were good for shortness of breath, he coughed much less, and brought up much fewer of the things like hailstones, in fact, he only coughed them up after intervals of many days; but for many years he never ceased to suffer in this manner, and he did so until the end of his life. Now, the size of the greater number of the hailstones which were coughed up was about that of Page 221 chick-peas, but sometimes | they were larger, and sometimes they were smaller. And I have also seen many other men who spat up matter as did this man, and they lived for many years, and I have also seen many who died. And I have also seen others who had some kind of disease in their organs of respiration, but not one of them spat blood. Concerning the bleeding that taketh place in the lungs and the wounds that appear in them, we have already spoken in the preceding Chapter, wherein I have spoken about all kinds of effusion of blood which take place in the organs of respiration. It is now meet for us to speak about the natural mixings of the lungs.

Of the natural mixings of the lungs.

Now the lungs do not make the belly only to thirst or not to thirst, and to crave for drink, hot or cold, but they act on the chest in a similar manner, and they change each of the mixings when it is in excess. Those who drink are not immediately satisfied, and a cold drink quencheth their thirst more than a hot one, and also cold air refresheth them through being drawn into them, but this in no way relieveth the thirst which ariseth from the stomach. Similarly also those which are felt by the senses to be the opposites to these cause pain through the indrawing of cold air, and this is a great sign (or, symptom), namely, that they spit up and bring up by means of coughing and speaking the cold matter which is in the lungs and the excess of phlegm. Now dryness of the lungs doth not produce superfluity of and freedom from phlegm, just as also this superfluity increaseth moisture, and produceth unclearness and dryness of the voice, and superfluity preventeth Fol. 107a. men from speaking in a clear, high voice, | when they wish to do so. These are the symptoms of the mixings of the lungs.

Of the diseases of the chest.

In this member, as also in all the other members, some diseases exist independently, and some by virtue of sympathy with other diseases which exist in it, but they all | most cer- Page 222 tainly tend to the injury of the faculty of breathing, because the breast maketh manifest that it is the true organ of respiration. The painful diseases which appear in it cause respiration to be feeble and gasping (or, frequent), even as we have already shewn in the sections on shortness of breath, just as they do in each of the other members that are closely connected with it and move with it. And we have described in the preceding sections how respiration is injured also by the diseases of the spinal column, although the chest itself is in no wise diseased, but only one of the nerves which come forth from the spinal column, or the spinal column itself. Now as concerning the diseases of the chest, some of them arise through the muscles, and some of them subsist in the membrane that is attached firmly to the ribs. Concerning those other diseases that arise in the skin, or in the bones of the ribs, it is not our intention here to speak, because the places in which they exist are beyond our faculty of perception. As concerning the muscles thereof, some of them ascend to places that are between the ribs, and some of them are outside these, and are situated outside. Therefore when they are attacked by painful diseases, or when they are crushed, or when they have in them wounds, or pustules, or a hard, red ulcer, or abscesses, these are clearly visible to the physicians, and they can lay their hands upon them. On the other hand, the inflamed abscesses which exist in the muscles that are between the ribs, and especially in the lower part of the back, are, as ye are well aware, full of inflammation, yet they do not fall under our power of inspection. Now, these are very much more painful than those that are external, and they drain inwards, as doth also some portion of that rheum which formeth the abscess, through that membrane which is firmly fastened | to Fol. 107 b. the chest, and which must, of sheer necessity, become thick with them, just as with this membrane, when, as sometimes

happeneth, it is diseased beforehand, there doth also thicken
Page 223 the inner | portion of the muscles which is between the ribs.
And true perforation (*i. e.,* pleurisy), when it becometh an abscess, taketh place in this membrane, and therefore extendeth
to the clavicle or to the epigastric region. Now, fever accompanieth all abscesses of this kind, inasmuch as the region that
is diseased is near the heart, and the heart itself is the machine
(*i. e.,* motive power) of that membrane, as it is also of that
other membrane (diaphragm?) which is called the "digestive
apparatus". And the throbbings (*i. e.,* the pulse) of the arteries
especially shew thee if the membrane that is firmly attached
to the ribs is diseased, or if the muscles that cleave to it are
diseased, for the muscles stretch to a very small degree, but
the membrane which moveth (diaphragm) stretcheth to a very
large degree, and it hardeneth the arteries, just as in the
diseases that take place in the lungs hardening of this kind
never appeareth at all.

Now that putridity, which is poured out into the empty
places of the chest, being taken into the oesophagus of the
lungs, and spit up as by the oesophagus, maketh known what
kind of chyme it is that hath produced the abscess of the
perforation (or, pleurisy), that is to say, whether it is red bile,
or black bile, or phlegm, or blood. When the abscess is
caused by red bile, patients spit up matter which is reddish
yellow; they spit up what is black when the abscess is caused
by black bile; and they spit up yellowish-white rheum when
the origin of the abscess is phlegm in excess; and when the
cause of the disease is blood, they spit up matter which is
whitish-red. In coughing there cometh up something which is,
manifestly, from the empty spaces of the chest. And it is
possible to know by those diseases that [extend] from the
outside to the inside, when a wound hath been hollowed out
in them, or when the chest itself is torn and the wound will
not close, or concerning some great abscess which hath broken
inside and outside, or hath been cut so that it may discharge
externally when it is able to do so. But when the membrane
Fol. 108 a. that moveth hath been eaten into | by gangrene (or, cancer),
Page 224 or, when a rib hath been cut off | by gangrene, the physicians

are unable either to keep the membrane free from disease, or to find out when it hath gangrene and is rotten. For in all diseases [of the chest] when honey water is poured into the chest from outside, it is immediately spit up by the process of coughing, and it is known to the patient by means of the taste that what hath been poured into him is honey water. Now how this happeneth, and by what paths the liquid cometh to the oesophagus, is easy to learn for him that wisheth [to do so]. Therefore it is manifest that the operation of breathing hath its origin in the chest, and that the lungs possess no independent motion; but at the time in which they are inflated, the breast also by the attraction towards the emptying out of its breath is inflated outwardly. Now by the contractions of the lungs the breast also is contracted, and it is squeezed together just like sponges that we hold in our hands and squeeze. In this way, then, there appeareth the throbbing through that which goeth forth from the membrane of the diaphragm which [is felt] in the large members. Nevertheless, even when it throbbeth not, and the chest is contracted greatly, it presseth upon the lungs, and driveth it inside the oesophagi which are in the lungs; but not as it cometh shalt thou hearken to that which we have spoken at great length in our discourse. For, if the chest doth not on all sides press with very great force on the lungs, the pus is not taken into the mouths of the oesophagi, and for this reason there hath been given to the animal creation a power of pressure that is called "coughing", which is a natural happening and is like unto sneezing, and hiccoughing, and vomiting. Now as regardeth the cause of this natural happening, because those who are sufferers have need of strength, rightly [do we say from our experience] of many who have coughed up pus, that this taketh place before their strength hath declined; because it is the healthy and frequent contraction of the chest that produceth coughing, for when the strength hath become weakness, a man cannot cough immediately, | or healthily, or frequently. | Now as for Page 225 Fol. 108b. the spittings that come from the empty spaces of the chest what hath been said here must be sufficient.

Of the disease which is called "perforation" (pleurisy).

Now the abscesses that appear in the moving membrane of the chest, and in those muscles that are joined thereunto, produce the ailment called "perforation" (pleurisy), and there accompany it happenings which are not remote from it, that is to say, acute fever, and stabbing pain, as if the parts in which it taketh place were being stretched out and stabbed through. And the breath is little, and is drawn in gasps, and the pulse maketh the artery hard, and distendeth it, and there is coughing which often produceth expectoration that is coloured. Sometimes, besides the expectoration, there are perforations from which nothing cometh up, and these physicians call "indissolvable"; these either kill the sufferer quickly, or are only disposed of after a very long time. Now the pain that patients suffer, in the majority of cases, either ascendeth to the clavicle, or descendeth to the epigastric region. There are also pains of the ribs which arise with the fever, and in such cases the breath is, necessarily, very little and is drawn in gasps; but nothing is expectorated. For this reason they resemble those perforations that are indissolvable, but they are easily distinguished from them, since during their existence patients do not cough at all, for when indissolvable perforations exist, although the sufferers from them do not spit at all, yet they have a dry cough, and moreover, in pains of this kind, there is no extended, or hard, pulse at all. And in addition to these things also, the fever in them is not acute, and the shortness of breath vexeth them in a very slight degree. Now some patients of this class suffer pain immediately the part of the body which is outside the diseased part is touched. Such patients, moreover, do not expel their pus by means of spit-

Page 226 ting, because there doth not come | to the empty places of the chest at all any of the chyme that produceth the abscess;

Fol. 109a. but if this abscess become ripe, and the pus | that is formed therein do not abate beforehand, it goeth forth to the skin outside, and is cut off and emptied out.

Of disease of the membrane which discerneth.

All the early physicians have called the lower boundary of the chest "parnôs", because when an abscess existeth in it the understanding (or, knowledge) of those who suffer becometh injured. Now [the word] "parnôs" (πρόνοος), being translated into Syriac, means "that by which we carry on the process "of thought and the process of making calculations (or, ar- "rangements), and that by which we understand." Others have called it the "diaphragm", and others the "understanding", for they thought that this filled the need for boundaries in animals, because it distinguisheth and defineth the ferocious (or, wrath- ful) part of the soul, which is situated in the heart, from the lustful portion, which dwelleth in the liver. The word "dia- phragm" (διάφραγμα), being translated into Syriac, meaneth "barrier" (or, "boundary"), and for this reason physicians have acquired the habit of naming the "discerning" membrane "dia- phragm", [a name] which [hath] also [been applied to the membrane of] the spinal column. Now, the correct name of this membrane is "marrow", and at length physicians have distinguished it by the name of "marrow"; and this membrane of the spinal column is the most important member of the organs of respiration. Rightly then is the breathing impeded when [the marrow] is diseased, even as we have said above in the section that treateth of shortness of breath. Now, we have also spoken in the Chapter on the nerves which are diseased about the diseases that take place in the marrow when it becometh diseased in sympathy with the spinal column, and also about the diseases that take place in it independently; therefore we will here speak concerning the happenings that take place in it independently, and concerning those that take place in it when it becometh diseased with the marrow. Before, then, the understanding is changed[1] by inflammation of the brain, it produceth gasping and a short supply of breath to the organs of respiration; and even when there hath been | Page 227 inflammation of the brain before, it doth not produce in them

---

[1] *I. e.*, before a man goes out of his mind.

uniformity in various ways, even as [Galen] hath shewn in the Discourse on shortness of breath.

There taketh place also a change in the understanding through the mouth of the belly, when it is diseased, and in cases of fiery fevers, and during pleurisy, and when there are

Fol. 109*b.* abscesses in the lungs; but the attacks of inflammation | of the brain that take place through this membrane (*i. e.*, the diaphragm), are so severe that they may be compared to those that come upon the insane, and also to those that appear in the diseases of the other members. When in fiery fevers the violence of the diseases passeth away, immediately the attacks of inflammation also become quiet. It is a characteristic of the attacks of insanity, which supervene only during fevers, that they do not subside; neither doth the violence of the fevers persist, because the brain doth not suffer sympathetically in that disease, but independently. Now, in the early stages of this affliction, the disease is established little by little (*i. e.*, gradually), and men do not get attacks of inflammation of the brain quite suddenly, as they do when there existeth disease in the other members which I have mentioned above. And happenings not a few precede the coming into being of this disease, and they are all called "signs of inflammation of the brain" (phrenitis?), which are, light sleep, or a little sleep which is disturbed by visions so vivid that the sufferer sometimes crieth out and leapeth up [in bed]; first of all, there is seen in such patients forgetfulness which is unusual, as for example, when they ask for a vessel into which they may void water, and they forget to do so; or they void water and forget to do it in the vessel. And again, we have the case of the man who, having been hitherto of a peaceful disposition, will utter threats of violence in a fierce and truculent manner. Moreover, all patients of this class drink very little, and their respiration is short and feeble, and their pulse is weak and hard, and sometimes they feel pain in the place which is between their head and their neck.

And when the time draweth nigh that their understanding

Page 228 is to be changed, | their eyes become exceedingly dry, or from each of them is squeezed out a tear of a very acid nature,

and afterwards they are covered with a film, and the veins in
them become filled with blood. Now, when this condition of
things happeneth, such patients are unable even to answer a
question properly; moreover, they imagine that they are gathering grass and woolly plants from their beds. And very acute
fever overtaketh them, but there are no great | variations in it Fol. 110a.
from one phase to the other, as there are in other fevers which
have in them frequent accesses of violence, and after the violent
attacks are over, the patients have relief. What now is it
meet to say? And moreover, as concerning the tongue, it is
parched (or, rough) exceedingly; at one time they begin to
babble gladly, and at another they are cast down and miserable,
and will scarcely answer a question. Sometimes when they
are suffering from a disease in one of their members they are
unconscious of the pain, even if one press upon the member
very hard.

Of the various kinds of inflammation of the brain.

Now the attacks of phrenitis which take place when the
membrane of the diaphragm is diseased are different from those
that arise through the brain or through the other members.
When the brain is diseased the attacks are slight, but when
another member is diseased, the attack doth not last continuously, the only exception being when the membrane of the
diaphragm is diseased. In diseases of the diaphragm the
attack lasteth for no short time, although it doth not last so
long as when the brain is diseased. For this reason it appeared to the early physicians that the great attacks which
take place were caused by abscesses in the diaphragm, and
they gave the diaphragm the name of "parnôs" (πρόνοος)
because of this opinion, inasmuch as it helped in no small
degree that member through which we have understanding
(or, knowledge). Now the attacks which take place through
this disease are defined (or, declared) by the other severe
attacks which occur during the happenings that take place in
the eyes, and by the droppings that flow from the nose, and
by the character of the respiration. Respiration [becometh]

Page 229 full and | easy to those who, because of disease of the brain, suffer from attacks at all times; in the case of those who suffer from disease of this membrane (*i. e.,* the diaphragm), the respiration becometh so irregular, that at one time it is feeble and gasping, and at another it is very full and resembleth the heaving of sighs. Now, in the early stages of the abscess that appeareth in this membrane (*i. e.,* diaphragm) and before the attack cometh, patients draw their breath feebly and in gasps, which is the exact opposite of that which taketh place when there is disease in the brain, for before such abscesses burst, Fol. 110*b.* patients draw their breath easily, | and take deep breaths. And, to speak briefly from the things which have been said here, either small matters precede the attack of delirium which cometh through disease of the brain, or none at all. Now they happen when abscesses begin to appear in this membrane (*i. e.,* diaphragm), just as it happeneth also that the epigastric region is distended, and this is a sure symptom of the beginning of disease in the diaphragm. When the disease is in the brain, these things appear at the end and not at the beginning of the disease; and heat increaseth in the head and face when the cause of disease is in the brain. Now the other diseases of the diaphragm, both those that are sympathetic, and those that are of independent origin, have been described by us in the preceding Chapter, and it is therefore now meet for us to append some account of the means to be employed in healing the diseases that have been mentioned.

Of the healing of ailments of the lung, and of all the members of the chest.

Now the healing of these is very difficult, as it appeareth, both in theory and in practice. This is manifest in theory, because the lung is always in a state of motion through respiration and coughing. If there be in the lung a wound, or an abscess, or any other kind of disease whatsoever, when it is about to be healed it hath need of coughing. And it is also manifest in practice, because no man, who hath actually been through serious diseases of this kind, hath ever been cured. Now in treating | all the various kinds of abscesses Page 230 which come in the lungs, and in all the organs of the chest which have been described, the most helpful thing to do first of all is to let blood and to clear out the belly. The second part of the beneficial treatment is to administer the foods that are good and suitable for the ailment; the third is to administer, in the form of draughts, the medicines which are suitable for the diseases; and the fourth is to apply to the chest externally plasters, and cataplasms, and fomentations, and unguents (or, salves). Now if there be an abscess in the lung, we let blood from the left forearm, and from the middle vein or from the third. | If there be an abscess of the muscles that are between Fol. 111a. the ribs, or in the membrane that surroundeth the ribs on the inside, and there be the disease of perforation (i. e., pleurisy), we let blood from one of the arms of him that is diseased, from the lower vein of the curve of the forearm, or from the self-same vein in the palm of the hand. One findeth continually [that the place chosen] is above the little finger, close to the third finger. If the perforation is on the right side we let blood from the right arm, and if it is on the left side, we let blood from the left arm. Great care, however, must be taken to enquire into the strength of the sick man, and his constitution, and his age, and the season of the year, and the country, and the condition of the air (i. e., climate), and the kind of life which the man hath led hitherto, and the nature of the sickness, and time [of its duration]. The most important of all these is the strength of the sick man, and the time of the sickness; there are occasions (or, cases) in which it is unnecessary to consider the other matters, because of the serious nature of the illness. Similarly we must make enquiries as to the clearing out of the belly; there are cases in which this must be effected by the foods which are of a purgative nature, and there are others when it can only be done by means of strong drugs or clysters. We must use the means which, in accordance with our observation of the sick man, appear to be suitable.

The second aid in these diseases are those foods which ‘are beneficial, according to the time and character of the sickness.

Page 231 There are occasions during the early stages of the illness | when there are fever and thirst, and we administer foods which are cooling and light and easily digested, such as extract of barley, or pieces of bread soaked in water, or broth of mallows, or broth of spinach (ﺍﻧﺎﺧﻰ), or broth of gourds, or broth of orach, or broth of mallows (?), or broth of sorrel, [made] with chick-peas and sesame oil, and we administer draughts of hot water if the season of the year be that of winter. There are cases in which it is good to give patients honey water or new wine (must), in order to reduce the sickness, provided there be no fever (or, inflammation).

The third aid in such diseases consisteth of medicines which Fol. 111b. are good for producing coughing and for bringing up the pus, | and which are prepared with two objects in view. The medicines which effect the first object are those that relieve roughness, and soften, and give warmth in the first place, and that check the abscesses; and the second is effected by those that incite the patient to cough, and that cleanse and purify the oesophagi of the lungs and the empty places of the chest, and that open and cleanse pustules. The first that have been mentioned are given in the early stages of the illness, when as yet the abscesses are not ripe, and when they are still un-opened, and the substances of these medicines are such things as licorice juice and licorice root. Now if one administereth to the patient any of the following it will procure for him relief: "hair of the warrior" (or, Venus's hair), and "teats of the bitch" (myxa, or plum tree), seed of asparagus, gum Arabic, gum tragacanth, seed of poppies, an infusion of figs, grapes, dates, and pepper, roasted flax seed, beans and chickpeas and everything of this kind. The following, however, soften and warm [the chest] more than the preceding medicines: extract of fircones, extract of sweet and bitter almonds, galbanum, myrrh, styrax, and gum of the terebinth tree. The following warm and cleanse [the chest] more than these, and are more efficacious in bringing up pus; and they open [the passages] and cleanse [them], and stimulate coughing: flour of the bitter Page 232 vetch, asarum (asarabacca), | aristolochia (rhubarb) of both kinds, and peppercorns, the three kinds (white, round, and long), mustard, pepperwort, cardamoms, root of ferula opopanax, costus, cassia, cinnamon, and spikenard, and every other drug of this kind.

Now there are other kinds of drugs that are mixed with the above, and they deaden feeling (narcotics), and are very helpful when there is much coughing, and when there is sleeplessness, because they produce sleep, and they relieve pains when they exist in excess. Strictly speaking, the physician should not use them, for they cause injuries through their excessive coldness in the persons to whom they are administered, for they cool and quench the natural warmth of the members. Now these are: | opium, hyoscyamus, the bark of mandragora Fol. 112a. root, and hemlock, and all the other drugs of this kind.

In the fourth kind of treatment which is helpful there are used plasters, cataplasms, and ointments, which mollify and warm, and which are applied to the chest externally. Sometimes we make infusions of herbs which soften and warm, such as chamomile, and dill (anise), and the seed of flax, and trigonella, and leaves of marsh mallows (hibiscum); and we sponge the chest with such infusions. And sometimes when these have been pounded, and pressed, and mixed together, we boil them in water and add the oil that is good for the particular kind of sickness, and we apply them on the part of the breast that is affected, after the manner of cataplasms. And sometimes we make use of those famous plasters which are compounded with wax, and which we are about to describe in this Chapter, according as the order demandeth. Thus, although only one in eight of those irregular conditions be in the members [of the chest], we heal it by means of that [drug] which is opposite in character. Moreover, if we suspect that there is some inflating wind confined therein, we heal it by means of those [drugs] which check it, and ease it, and open the pores [of the body], that is to say, by means of plasters, | and Page 233 washings, and unguents, and fomentations. This, then, is the art of healing diseases of the lungs and chest; but because the medicines that are good for coughing and the bringing up of pus must be set down in this Chapter [we give them further on].

17

Now coughing is a natural action which giveth information concerning the diseases that arise in the organs of respiration, and the pus also, according to the rule of the things that are to be ejected [from the body], indicateth the places from which it cometh up. Let us therefore collect in a brief form all the causes thereof, and also say what the places are from which pus is brought up by means of the action of coughing. Now the bringing up of pus taketh place in those in whom fluid floweth down from the head to the lungs and throat, whereby the oesophagi of their lungs are filled, according to what hath Fol. 112b. been said in that Chapter on shortness of breath. | And the bringing up of pus also taketh place in those who have on some former occasion suffered from haemorrhage of the organs of respiration, which hath not been healed completely, even as it hath already been said in that Chapter on the bringing up of blood. The bringing up of pus therefore taketh place in those who have an abscess, which produceth a pustule in the organs of respiration, and it becometh perforated, and [the pus] falleth round about the lung. And the bringing up of pus also taketh place in those in whom fluid in excess is collected in the empty spaces of the chest or in the lungs; and it also happeneth through some irregular condition (or, mixture), or from some inflating wind when it is confined in the organs of respiration. These things cause the action of coughing, but they do not cause any [pus] to come up. And again, caused by the atmosphere by which we are surrounded, how often doth coughing arise through the cold condensation of dampness and the excessive fieriness of the heat therein! Now, we will indicate the places, and the causes, which have been described, of the actions (or, happenings) that accompany Page 234 them, together with | the sensations in them.

If there be an inflamed abscess in the viator membrane of the ribs (*i. e.*, diaphragm), or in the membrane which effecteth digestion, or in the muscles between the ribs, burning fever, and stabbing pain, and difficulty of breathing, and coughing, will be closely associated with it. Again, if there be in the lung the abscess which is called "περιπλευμονία" (*i. e.*, tumour), it is accompanied by a fiery fever, and difficulty of breathing, and redness of the cheeks, and the ticklings of coughing, and loss of feeling in the region of the chest, for the whole of the chest of patients of this class is inflated. And the fluid, which descendeth from the head to the throat and to the oesophagi of the lungs, is accompanied by constant coughing, and change of voice, and difficulty of breathing, and asthma, and some-times also by running of the nostrils, and sneezings, and ulcers in the lobes of the ears. These are the causes that produce coughing, and these are also | the places from which pus is Fol. 113a. brought up, and these are the actions (or, happenings) that accompany them, summarized briefly. Now, the cures of all the diseases that arise in the organs of respiration have one mark in common, even as hath already been said in the Chapter in this treatise on the diseases of the head; for if there be diseases in the brain, or in the empty spaces which are in the head, or in the membrane, or in the veins, or in the arteries, the various means which are employed in curing them possess one mark in common, for it is necessary to apply all kinds of relief to the head. And thus also is it in the case of the organs of respiration, for since these are all placed in-side the chest, it followeth of necessity that every possible relief (or, assistance) must be applied to the chest, either in-ternally or externally. Only in the case of the bringing up of blood, or pus, which taketh place through that fluid that des-cendeth from the head to the lungs and throat, doth the head participate in the helps [given], *i.e.*, in the form of plasters which give warmth and which are applied to the head, | and Page 235 the bathing with aromatic compounds, and the fomentations. Sometimes we use the medicines that impart warmth slightly, and produce dryness, and strengthen the head, and sometimes those that are pungent (or, acid), such as nitre, and mustard, and euphorbium, and Thapsia Asclepium, and aromatic herbs, and spices which produce exceeding dryness. Now, in the other diseases, besides taking due care for the head, it is meet that all the remedies (or, helps) should be applied to the chest, and if there be disease in the lungs or in the diaphragm, [we must effect healing] by means of draughts, and pills that are to be placed under the tongue, and plasters, and by the foods

17*

that are beneficial. And the compound medicines that are good for coughing and for bringing up pus are these:

A medicine which is called "Dîyâ̄ḳôdâ" (*i. e.,* "Syrup of Poppies"), and which is good for coughs and for the runnings that flow down from the head to the chest; and it is beneficial in cases of sleeplessness and diarrhoea.

Two hundred Egyptian poppies of moderate size

[Two] measures of rain water.

Fol. 113*b.* Crush the poppies and macerate them in the water for one day and one night, then boil the mixture until one half of it hath evaporated. Strain the liquid and squeeze out the juice of the poppies and throw them away. Pour into the liquid two measures of honey and two measures of new wine (must), and boil the mixture until it becometh thick. Then take one drachm each of acacia, sumach, crocus, myrrh, extract of goat's beard (tragopogon pratense), purple βαλαύστιον, and gum tragacanth, and pound them to a powder, throw it into the mixture, and stir it up, and boil it until it becometh like honey. Then pour it into a glass vessel and administer as a linctus, one spoonful in a draught. If there be fever and thirst dissolve it in extract of barley, and administer it in the morning and in the evening at bed time.

Another common Syrup of Poppies which is good for the diseases. Take fifty Egyptian poppies and one hundred and fifty measures of rain water, or water from a spring. Break the *rûmânê* (poppies), which are also called *maykûnê,*

Page 236 and steep them in water for one whole night | and day. Boil the water and the poppies in a metal pot until one half hath evaporated, then take the pot off the fire. Pound up the poppies, and having squeezed out their juice throw them away. Pour the liquor into the metal pot, and add to it thirty-eight measures of honey and new wine (must), then boil the mixture until it acquireth a good consistency, and use it in the same way as the preceding medicine.

Another [medicine] which is called "Akûmlî", and which is good for coughs, and it reduceth fluid in the chest when it is in excess. Pour fifteen measures of rain water, or spring water, into a metal pot, and boil until only five measures remain; then pour two measures of honey into it, boil the mixture down to one half, and then take it and pour it into a glass vessel, and administer one spoonful in a draught.

Another linctus, which is good for phthisis, and for those who bring up much pus.

Crocus	3	drachms
Asaron (ἄσαρον)	3	,,
Enḳath aylâ (tragacanth)	3	,,
Seed of hyoscyamus	3	,,
Myrrh	2	,,
Fir cones	2	,,
Starch	10	,,
Fat dates, cleaned,	4	ounces
Dry dates, | cleaned,	4	,,
Styrax	1	ounce
Honey	35	measures
New wine	35	,,

Fol. 114*a*.

Pound and mix together, and administer a portion the size of a chestnut as a linctus.

Another [medicine] which is good for the diseases.

Crocus	4	drachms
Frankincense	4	,,
Myrrh	4	,,
Cinnamon	4	,,
Amomum	4	,,
Licorice juice	4	,,
Gum of terebinth	4	,,
Fir cones	4	,,
Dates of Hîrôn	4	,,
Poppy seed	4	,,
Bark of cinnamon	2	,,
Spikenard	2	,,
Costus	2	,,
Tragacanth	3	,,
Galbanum	$^{1}/_{2}$	drachm

Steep these when moist in new wine for a whole night and day, then pound them and work them up in honey, and give the patient a portion | about the size of a dried Page 237 grape, and let him take the dose both morning and evening.

Another linctus which is good for coughs, and for those who bring up pus.

Fir cones,	2 ounces
Flax seed, roasted,	1 ounce
Sweet almonds	1 „
Round peppercorns	$^1/_2$ „
Honey as much as is necessary.	

Pound and mix with honey, and use.

Another, which is good for phthisis.

Dried roses	2 drachms
Frankincense	2 „
Laurus malabathrum	2 „
Resin	2 „
Myrrh	$1^1/_2$ „
Oil of balsam	$1^1/_2$ „
Bdellium *biyôrâ*	1 drachm
Amomum	1 „
Spikenard	1 „
Esṭûmkâ (stomachic?)	1 „

Dissolve the bdellium in wine, add as much honey as is sufficient, mix together, and use.

Another linctus, which is made of bamboo sugar, and which is good for those who bring up pus, and for those who suffer from dry rheum on the chest.

Incense bark	3 drachms
Shûshmîr (amomum, هَيْل)	3 „
Long peppercorns	6 „
Bamboo sugar	200 „
Sugar	30 „
Sheep oil ⎫ Honey ⎭	as much as sufficeth.

Pound together and administer for a cough, old or new, and for hardness of the breast and lungs, and administer to him that bringeth up dry rheum, and for weakness of the liver and stomach, and for phthisis, a draught containing a portion the size of a gall nut in an infusion of licorice juice, or of Venus's hair, or in extract of barley, or in ass's | milk, or in hot water. Fol. 114*b*.

Another, which is made of fir cones, and which is good for those who cough with difficulty and bring up pus.

Flax seed, roasted,	4 ounces
Sweet almonds	4 ,,
Gum Arabic	4 ,,
Fir cones	4 ,,
Enḳath aylà (tragacanth)	4 ,,
Fat dates, cleaned,	20 in number
Honey Sheep oil }	as much as sufficeth.

Pound and mix together, and administer.

Or pound and work up together in honey roasted flax seed, and administer as a linctus a portion as large as a gall nut, morning and evening, for a dry cough.

Another, which is made with | licorice juice, and which Page 238 **is good for him that bringeth up thick pus and blood, and which cleareth the chest.** Take in equal quantities juice of licorice root, *enḳath aylâ*, galbanum, and bitter almonds peeled, pound and mix together in as much honey as is sufficient, and administer.

Another, which is made of vetches, and which is good for those who bring up pus, and for ulceration of the bowels.

Flour of vetches	4 ounces
Asaron (ἄσαρον)	4 ,,
Mountain mint, or River mint }	4 ,,
Mint	4 ,,
Peppercorns	4 ,,
Dried hyssop	3 ,,
Spikenard	5 ,,
Gum of terebinth	5 ,,
Oil of roses, or Cow oil, fresh }	3 *lîṭrê*

Pound and mix together, and administer a portion as large as a gall nut in a suitable draught, or let the patient use it as a linctus and gargle. Or crush the frankincense, and boil it in honey, and let the man, from whose head rheum descendeth to his chest, use it as a linctus.

Another, which is good for pains in the chest and for phthisis, and which produceth sleep.

Flour of vetches	5 *liṭrê*
Opium	5 drachms
Myrrh	5 „
Peppercorns	20 „
Dried hyssop	40 „
Frankincense	1 drachm
Madder	40 drachms
Licorice root	40 „

Honey as much as sufficeth.
Administer in the morning and evening a portion as large as a chestnut.

Another, which is good for a cough and for hoarseness (or, sore throat) of long standing, and for a sudden attack Fol. 115a. of the ailment. It hath been well tried: do not despise | it.

Cassia	1 ounce
Frankincense	1 „
Licorice root	2 drachms
Myrrh	4 „
Spikenard	4 „
Enḳath aylâ	4 „

Pound and work up with honey, and administer.

Another, which is made of fir cones, and which is good for coughs, and asthma, and for the rheum that descendeth from the head to the chest.

Dried hyssop	4 drachms
Peppercorns	10 in number
Fir cones	2 ounces
Oil of nard	10 drops

Honey as much as sufficeth.
Pound well together, and administer.

Page 239   Another, | which is good for a cough, and sore throat (*i. e.,* roughness of voice, or hoarseness), and for [bringing up] pus.

Myrrh	6 drachms
Bitter almonds	6 „

Galbanum	4 drachms
Styrax	4 ,,
Fir cones	4 ,,
Crocus	2 ,,
Opopanax	3 ,,

Pound and mix together with honey, and administer a portion about the size of a gall nut.

Another, which is made of lupins, and which removeth obstructions of the liver and stomach, and relieveth a cough of long standing.

Lupins	1 ounce
Cyperus	1 ,,
Trigonella	1 ,,
Seed of fennel	1 ,,
Flax seed	1 ,,
Rhubarb	1 ,,
Laurel berries	1 ,,
Stacte	1 ,,
Aloes	1 ,,
Petroselinum (rock parsley)	1 ,,
Pulp of colocynth	1 ,,
Extract, or juice, of agrimony	1 ,,
Spikenard	1 ,,
Costus	1 ,,
Long peppercorns	7 drachms
Honey as much as sufficeth.	

Mix well together, and administer a portion as large as a chestnut in some convenient draught.

Another, which is used by the Indians, and which is good for fluid in excess in the lungs and chest.

Root of the thorny caper	100 drachms
Dûr	100 ,,
Sesame seed	100 ,,
Cyperus berries (?)	100 ,,
Orichalchum	100 ,,
Sugar	100 ,,
Myrobalanus chebula (black)	3 ,,
Fennel (?)	3 ,,

Amlag	3 drachms
Round [peppercorns]	3 ,,
Ginger	3 ,,
Long [peppercorns]	3 ,,
Honey ⎫ Oil ⎭	as much as sufficeth.

Mix well together and administer a portion as large as a gall nut, and it will relieve wonderfully.

Another, which is made of mint (or, calamint), and which is good for the thick viscous chymes that ascend from the chest, and it will bring on the menstrual flow in women.

Mint of the river	200	drachms
Mint of the mountain	200	,,
Petroselinum (rock parsley)	200	,,
Parsnip	200	,,
Parsley (water?)	4	,,
Epithymum (ἐπίθυμον, thyme)	4	,,
Peppercorns	20	,,
Ligusticum	6	,,

Fol. 115*b*. | Honey as much as sufficeth.

Mix well together and use, [administering] a portion the size of a gall nut in [some convenient] draught.

Another, which is good for those who cough and who bring up watery and frothy pus.

Sweet almonds	4 drachms
Flour of beans	2 ,,
Trigonella	1 drachm
Sheep oil	1 measure
Honey	$1/_2$ ounce

Page 240 Mix well together and administer | one spoonful in a draught.

Another which is made of licorice, and is good for a cough and sore throat.

Licorice juice	10 drachms
Bitter almonds	4 ,,
Pulse	4 ,,
Cotton wool	4 ,,
Herbs	4 ,,

Cooked eggs ⎫
Cream   ⎬ as much as sufficeth.
Honey   ⎭

Mix well together, and use.

Another which is made of gum of terebinth, and which is good for sore-throat, and for the throat, and for loss of voice, and for the bringing up of pus, and for tightness of the chest, and for the removal of obstructions.

Flax seed, roasted,	3	*liṭrê*
Raisins, cleaned (stoned?)	3	,,
Fir cones	6	ounces
Almonds, sweet,	6	,,
Almonds, bitter,	6	,,
Roasted chestnuts	2	,,
Gum of terebinth	2	,,
Licorice juice	2	,,
Gum Arabic	2	,,
Flour of trigonella	3	,,
Enḳath aylâ (tragacanth)	3	,,
White peppercorns	1	ounce
Flour of beans	1	,,
Flour of chick-peas	1	,,
Rhubarb	1	,,
Starch	1	,,
Mint	1	,,
Iris	1	,,
Pepperwort	1	,,
Styrax	1	,,
Myrrh	$1/2$	,,
Crocus	$1/2$	,,
Frankincense	$1/2$	,,

Pound and work them up in ass's milk, make into pills and dry them in the shade, then crush them and work up with honey, and administer as a linctus. And in the evening let the patient put some of the medicine under his tongue, and it will give relief.

Another, which is made of starch, and which is good for a cough, and it purifieth the chest.

Sweet almonds	20	drachms
Bitter almonds	20	,,
Starch	20	,,
Sugar	20	,.
Amomum	3	,,
Cinnamon bark	3	,,

Honey as much as sufficeth.

Pound well together, and use.

Another, which is made of the juice of fennel (μάραθρον), and which is good for a dry cough, and for hoarseness.

Almonds of both kinds	2	drachms
Styrax	2	,,
Flax seed	2	,,
Aniseed	2	,,
Tragacanth	2	,,
Gum Arabic	2	,,
Licorice juice, or root,	1	drachm
Sugar	4	drachms
Panîd	4	,,

Fol. 116a.

Pound and work up in juice of fresh fennel, and administer.

Another, which is good for coughs, and for difficulty of breathing, and asthma. |

Page 241

Fir cones
Chestnuts
Sweet almonds
Stacte
Gum of terebinth } in equal quantities.
Peppercorns
Thyme
Persian pot-herbs
Calamint

Honey as much as sufficeth.

Pound together and administer.

Another medicine which is called "Bàkîḵôn" (βήχιον, calf's foot), and which is good for frequent and violent coughing (*i.e.,* a racking cough), and for sleeplessness, and it produceth sleep.

Galbanum (χαλβάνη)	1 ounce
Honey	1 ,,
Seed of fennel	2 ounces
Opium	$^1/_2$ ounce

Pound and mix together, and administer a portion about the size of a chick-pea in hot water.

Another, which is good for a frequent cough, and which will quiet it immediately.

Aniseed	8 drachms
Styrax	6 ,,
Opium	6 ,,

Boiled honey as much as sufficeth.

Mix well together, and administer for a dose a portion the size of a bean at bed time.

Another [medicine] which [is to be placed] under the tongue, and which is good for a pain in the chest, and for a cough which is caused by inflammation (heat?).

Juice of licorice root	3 drachms
Fat dates, cleaned, (stoned?)	3 ,,
Starch	1 drachm
Crocus	1 ,,
Tragacanth	$2^1/_2$ drachms

Pound and work up in ass's milk, and make into pills the size of chick-peas, and administer as a dose one, which the patient is to place under his tongue at bed time.

Another, which is good for pain in the throat (sore throat?), and roughness of the vocal cords, and for bringing up blood and pus.

Fir cones	6 drachms
Pulp of pulse (or, carobs)	4 ,,
Cotton wool	4 ,,
Myrtle berries	4 ,,
Sea gems or, Viscous red mud	4 ,,
Frankincense	4 ,,
Tragacanth	4 ,,
Goat's beard (tragopogon pratense)	4 ,,

Gum Arabic	1 drachm
Crocus	1 „
Poppy seed	8 drachms
Scoria of the furnace	1½ „
Myrrh	1½ „

Pound and work up in licorice juice, and make into pills the size of beans, and give one to the patient to put under his tongue.

Another, which is good for frequent coughing, and for excessive dryness of the chest (or, tightness of the chest).

Fol. 116*b*.

Almonds, sweet ⎫ Almonds, bitter ⎬	4 drachms each
Roasted flax seed	4 „
Fir cones	4 „
Aniseed	1 drachm
Tragacanth	1 „
Gum Arabic	1 „

Page 242

Licorice juice, or root	1 „
Panîd	4 drachms
Sugar	4 „

Pound and work up in juice of fresh fennel, and make into pills the size of beans, and administer.

Another, which is called "Pónîḳâ", and which is good for phthisis of long standing, and for those who cough and bring up blood and also pus, and for pains in the ribs, and it is a very fine medicine indeed.

Opium	4 ounces
Crocus	4 „
Acacia, black	4 „
Iris	4 „
Purple βαλαύστιον	4 „
Costus	2 „
Gum of terebinth	6 „
Styrax, dissolved,	4 „
Galbanum (χαλβάνη)	6 „
Starch	6 „
Choice dates, cleaned (stoned?)	6 „

Dried fruits (?), cleaned
(stoned?)   6 ounces

Break up the dry constituents and reduce to a powder, then
pound them in a mortar and pour them out on honey from
which the scum hath been removed, and work up, and use.
Administer as a dose a portion as large as a bean in a draught
made of an infusion of licorice root and hyssop.

**Another, of Menestius, which is good for pleurisy, and
difficulty of breathing, and coughing, and pain in the belly,
and for the liver and spleen.**

Gentian	2 drachms
White peppercorns	2 ,,
Crocus	3 ,,
Opium	3 ,,
Aniseed	3 ,,

Honey as much as sufficeth.

For coughs take a dose equal to one bean in a suitable draught
made of an infusion of licorice root; for pain in the liver take
a dose in an infusion of rue, and for pain in the belly take a
dose in an infusion of myrtle berries.

**Another, which is made of madder, and which is good
for coughs, and for pain in the throat, and for pain in the
lung, and in the breast, and for the voice, and for bringing
up of pus, and for removing obstructions of the liver, and
it relieveth a wet or a dry cough.**

Madder	20 drachms
Plantain	20 ,,
Mountain mint	20 ,,
Dried hyssop	20 ,,
Eskil (squills)	20 ,,
Iris	20 ,,
Fir cones	20 ,,
Aristolochia, both kinds	20 ,,
Licorice root	20 ,,
Poppy seed	20 ,,
Rain water	16 (?) measures

Crush the | dry drugs to a fine powder, and put it into a metal | Fol. 117 a.
pot. Pour water on it until the mixture hath evaporated to Page 243

five measures. Then strain the liquor, squeeze the drugs dry and throw them away, and pour two measures of honey water into it until it becometh as thick as honey. Add ten drachms of starch to it, and mix [the whole] together. Take the mixture off the fire, and when it hath ceased boiling a little time, pour into it five *estîrê* of gum of terebinth; work up and take and use. Administer one spoonful to the dose as a linctus, or let the patient swallow it in a draught.

Another, which is good for every kind of cough, and for excessive rheum of the chest, and for pain in the interior of the body.

Sagapenum (σαγάπηνον, saco-penium)	4 drachms
Gentian	4 „
Myrrh	4 „
Jackal fat (or, fat of ferula opopanax)	4 „
White peppercorns	4 „
Laurel berries, clean	4 „

Crush them, and work up with new wine (must), and administer as a dose one drachm, in a suitable draught, to the patient when fasting.

Another, which is made of mandragora, and which is good for phthisis, and for those who bring up much pus.

Rind of mandragora root	4 drachms
Seed of wild parsley	4 „
Myrrh	4 „
Crocus	4 „
Aniseed	4 „
Carrot (or, parsnip)	4 „
White peppercorns	4 „
Castoreum	4 „
Sagapenum	4 „
Jackal fat (or, fat of ferula opopanax)	4 „
Honey as much as sufficeth.	

Give a dose in amount equal to a chestnut, in some convenient draught.

Another, which is good for those who bring up pus or blood, and for coughs, and for pain in the lungs, and for pleurisy, and for turning of the stomach (nausea), and for pain in the belly, and for diarrhoea, and for pain in the bladder, and for pain of the womb, and for intermittent fever, if drunk one hour before an attack, and it relieveth those whose condition is destroyed (*i. e.*, those who need a tonic), and it is an antidote to poison, and for the bites of wild animals.

Cinnamon	1	ounce
Costus	1	,,
Galbanum (χαλβάνη)	1	,,
Castoreum	1	,,
Poppy juice	1	,,
Round peppercorns	1	,,
Long peppercorns	1	,,
Styrax	1	,,
Honey as much as sufficeth.		

Dose, ‖ one portion as large as a bean, in some convenient draught. Page 244
Fol. 117*b*.

Another, which is made of sulphur, and which is good for intermittent fever with shivering, and coughs of long standing, and for those who bring up blood and pus, and for difficulty of breathing and ague, and for swollen sores, and for the liver and spleen, and for dropsy, and for the colon, and it bringeth down the stones that come in the kidneys, and it serveth as an antidote to poison, and to everything that produceth weak health.

White peppercorns	5	drachms
Seed of hyoscyamus	12	,,
Cardamoms	12	,,
Myrrh	12	,,
Frankincense	12	,,
Opium	10	,,
Crocus	10	,,
Sulphur, unroasted,	6	,,
Aristolochia, the long	3	,,
Fat of ferula opopanax	3	,,

18

Euphorbium	3 drachms

Run honey as much as sufficeth.

Mix well and administer in hot water to those who suffer from intermittent fever one hour before the attack seizeth them. Dose, as much as a chestnut, dropped on thy finger which hath been dipped in a little vinegar. Administer in some suitable draught to those who bring up pus.

Another, which is made of Cyrenean gum, and which is good for the diseases [of the chest].

Peppercorns	3	ounces
Gum from Cyrene	3	,,
Gum from Syria	3	,,
Gum of poppies	3	,,
Crocus	4	,,
Sulphur, unroasted	4	,,
Myrrh	4	,,
White hyoscyamus seed	4	,,
Mandragora apples	2	,,
Cardamoms	1	ounce

Honey as much as sufficeth.

Mix well together, and administer in doses similar to those of the preceding medicine.

Another, which is called "Metdôrîṭôs", and which is good for those who suffer from shortness of breath, and for coughs of long standing, and for excessive rheum of the chest, and for obstructions of the liver, and for the putrefying of the chymes, and for diarrhoea, and it stirreth up the desire for food (*i. e.*, createth an appetite), and beauti-fieth the colour of the face (*i.e.*, complexion), and purifieth Page 245 **the body, | and dissolveth stones of the kidneys, and cureth retention of the urine, and abateth the affliction of the soul** Fol. 118 a. **(*i. e.*, depression), and preventeth the | black bile from smiting the body, and protecteth the child in the womb, and cureth the greater part of the diseases incidental to women, and it produceth keenness of vision, and doth not permit a man to be injured by drugs which kill, and it delivereth from poison. Every one who hath tried it calleth it "immortality".**

Myrrh	10	drachms
Crocus	10	,,
Agrîḵôn (ἀγαρικόν, fungus?)	10	,,
Ginger	10	,,
Cinnamon	10	,,
Spikenard	8	,,
Frankincense	8	,,
Tragacanth	8	,,
Babylonian pepperwort	8	,,
Pistacia lentiscus	8	,,
Balsam	8	,,
Lavender (στοιχάδος, اسطوحودوس)	8	,,
Meadow saxifrage (seseli officinale)	8	,,
Costus	8	,,
Galbanum (χαλβάνη)	8	,,
Gum of the terebinth	8	,,
Long peppercorns	8	,,
Castoreum	8	,,
Juice of Tragopogon (goat's beard)	8	,,
Styrax	8	,,
Opopanax	8	,,
Laurus malabathrum	8	,,
Black cassia	7	,,
White peppercorns	7	,,
Sûrîngân (colchicum autumnale)	7	,,
Poley-germander (teucrium polium, spleenwort)	7	,,
Water-germander (scordium)	7	,,
Carrot (or, parsnip)	7	,,
Incense berries	7	,,
Berries of ḵûpîôn	7	,,
Mû (μῆον, meum athamanticum, or, bear-wort?)	5	,,
Stacte	5	,,

Gum Arabic	5 drachms
Opium	5 ,,
Rock parsley	5 ,,
Cardamoms	5 ,,
Seed of fennel (μάραθρον)	5 ,,
Roses	5 ,,
Gentian	5 ,,
Dittany (δίκταμνος)	5 ,,
Aniseed (or, dill)	4¹/₂ ,,
Valerian (phu)	4¹/₂ ,,
Eupatorium	4¹/₂ ,,
Acacia	4¹/₂ ,,
Belly of the Egyptian croco- dile (σκίγκος)	4¹/₂ ,,
Asarum (wood nard)	3 ,,
Sagapenum	3 ,,
Old, good wine, wherein the drugs which are soluble may be macerated	. . . . .
Honey as much as sufficeth.	

Dose, a portion as large as a chestnut in a suitable draught.

Another, which openeth the pustules that appear in the internal organs, whether they be in the organs of the chest, or in any other place.

Ammoniac	1 drachm
Sarcocolla (Persian gum)	1 ,,
Galbanum (χαλβάνη)	2 drachms
Paníd	3 ,,
Honey	1 spoonful
New wine	2 measures

Crush and mix up with honey and pour into a shallow metal pan, and then pour in the wine and boil until they are all melted and mixed together; then strain the liquid, and let the patient drink it all in two days. And let him give some soothing, fermented drink, and it will help [him].

Another, which is good for coughs of children.

Boil some mint in goat's milk, and use it as a linctus.

Page 246

Or melt tragacanth and sugar in milk, and | give it to the Fol. 118b. patient to drink.

Or mix fennel juice with goat's milk, and warm it, and give it to the patient to drink.

Another, which is good for a cough of long standing, and asthma.

Steep mint, rue, and fleshy figs in water until the liquid becometh thick, then strain, and administer in the form of a draught.

Or crush and work up irises with honey, and administer in doses of one spoonful, to be taken as a linctus.

Or pound and work up wood of the balsam tree in honey.

Or pound and work up eķrôn in honey, and let the patient take it morning and evening, in doses of one spoonful in hot water.

Or take three fresh eggs and double their bulk in honey, and the bulk of one egg in the cream of sheep's milk, and forty peppercorns; pound the peppercorns and mix with the other ingredients, and heat up, and administer in the form of a linctus.

Or crush sweet cyperus berries, and work up in honey, and let the patient take as a linctus every day four drachms, and it will cleanse and purify the chest.

Or make an infusion of cardamoms, and administer in draughts.

Or crush cardamoms, and work them up in honey, and use as a linctus, and [it will be found to be] beneficial for a cough which ariseth from cold, and for runnings [from the head].

For a cough which is due to stoppages in the organs of the chest.

Pound asarum and spikenard, and work them up with honey, and use as a linctus.

For pleurisy and pain in the side.

Pound costus, and administer one drachm in an infusion of dill and one spoonful of sweet oil, and it will be helpful.

For a cough caused by chill.

Administer oil of balsam in a draught of hyssop water, and anoint the breast therewith externally.

For a cough of long standing.

Let the patient place under his tongue each evening a piece of myrrh as large as a chestnut, and it will be [found] beneficial.

For a cough, and a running from the head caused by cold, and for huskiness of voice (or, hoarseness).

Melt together honey-like styrax and honey, and use as a Page 247 linctus; | let the patient inhale the same for runnings at the nose, and it will be helpful.

For ulcers on the lungs of long standing.

Let the patient be rubbed with oil of tar every morning, and it will heal pains of long standing.

Or melt down honey-like gum of the terebinth with honey, and administer one spoonful as a linctus in the morning and evening.

Or melt down honey-like galbanum and honey, and ad-Fol. 119a. minister as a linctus in the morning | and evening.

An infusion which is good for a pain in the chest and for coughing.

Venus's hair	
Dried hyssop	
Seed of fennel	each one hand
Root of pistacia lentiscus	
Root of licorice	
Dried irises	

Infuse well in water, and administer two measures fasting. This medicine is also good for the reduction of acute sicknesses.

Another, which is good for fever and cough, and it purifieth the stomach and chest.

Dried hyssop	
Irises	
Venus's hair	
Teats of a bitch (myxae)	each one hand
Licorice root	
Stacte	
Aniseed	

Infuse in water, and administer in draughts.

Another, which sootheth the chest and refresheth.

Licorice root
Figs
Dried hyssop
Black thorns ⎬ in equal quantities
Myxae
Husked barley, cleaned,
Venus's hair

Infuse in water and administer, either with syrup of poppies or by itself.

Another, which is good for coughs, and pleurisy of long standing, and which purifieth the breast.

Licorice root	1	hand
Asarum	1	„
Resin	1	„
Root of pistacia lentiscus	1	„
Thyme	1	„
Mint	1	„
Teats of mouse	1	„
Fat figs	1	„
Dried herbs, cleaned,	1	„
Dates	1	„
Hîrôn, cleaned,	1	„
Husked barley	a little.	

Boil well in water, and administer with one of the lozenges that have been described above.

Powders for burning which are good for those who have phthisis and a cough of long standing.

Green arsenic	1 spoonful
Peppercorns	10 in number

Crush well and pour on them a little white of an egg and | Page 248 sesame oil, then mix them all well together, and spread out the mixture on a piece of new cotton wool and dry in the shade. And, when it is required, bring a vessel like a thurible with a cover, and make a hole in the cover, and fix a reed therein, and put hot coals in the thurible, and cut off a piece of the cotton wool on which the medicine hath been spread, and throw it on the fire. Then let the patient take the reed

in his mouth and inhale as much of the smoke as he is able
Fol. 119b. to bear for seven days. Let him eat bread [soaked] in fatty |
broth, and nothing else, and he will become well.

Another, which is good for the diseases [of the chest] although they bring up nothing foetid.

Leaves of the sweet service tree (sorbus)	
Pig's excrement	
Fat of the kidneys of the antelope	All in equal quantities
Arsenic	
Hare's excrement	

Pound and work up into a paste with sweet wine, and make
into pills the size of a bean, and when the need demandeth,
put in a fire and do according to what hath been prescribed
above. And afterwards let the patient partake of fine flour
boiled in water with sheep oil and honey.

Another, which is good for diseases [of the chest].

Take bark of the service tree (sorbus), and arsenic, and
kidney fat, and mix them together, and use in a similar way.

Or reduce red arsenic to a powder with sheep oil, and
spread it on the leaves of the service tree and dry [them].
When a case of need ariseth make a fumigation with them
according to what is prescribed above.

Another, which is good for a cough of long standing, and for foetid pus which is mixed with blood.

Take green arsenic, and long aristolochia (rhubarb), and the
bark of camphor root in equal quantities, pound them and
work into a paste with cow oil, and make into pieces the size
of a chestnut. For use, take one of these and wrap it up in
cotton wool, and let the patient burn two or three under his
mouth every day. And let the patient take after it *ardalagh*
boiled in sheep oil.

Another, which is good for the diseases [of the chest].

Take in equal quantities fir cones, galbanum, and gum of
Page 249 the cedar, | and pound them and mix them together, and make
into tablets, and burn as before.

Another. Take a very little arsenic and thyme, and

pound and work into a paste with sesame oil, and burn as before.

Plasters which are good for pain of the chest, and for pleurisy and phthisis of long standing.

The plaster which is called "Môtôn", and which is good for pleurisy, with or without fever.

Dill, or anise,	1 hand
Chamomile	1 „
Flax seed	1 „
Trigonella	1 „
Beans	1 „
Vine flowers	1 „
Barley flour	2 portions

Crush, reduce to a powder, mix together, and boil in water with some convenient oil, and spread on a piece of linen, and bind it on the part that is affected. When inflammation, and fever, and excessive thirst | are not present, we add to it honey or Fol. 120 a. one of the medicines that are especially good for allaying the disease, and we also add oil, which giveth warmth.

Another, [which is good] for pleurisy and coughs.

Wax	10 drachms
Resin	20 „
Pitch	4 „
Oil of sheep, or Sesame oil	as much as sufficeth.

Melt the wax and oil over a fire, then take them off and add the other ingredients when crushed and reduced to powder; work up and spread on a strip of linen and use as a plaster.

Another, which is good for acute pleurisy accompanied by fever. Take wax, and resin, and pitch, all in equal quantities. Melt the wax with oil of jasmine or oil of narcissus, mix with them the resin and pitch, and stir up well, and use as a plaster.

Another, which is called "Dekhîr", and which is good for pain in the chest; and it softeneth the nerves.

Wax	20 drachms
Cow oil	16 „

Sheep oil	12 drachms
Fresh hyssop	18 ,,
Gum of the terebinth	4 ,,

Oil of dill, or
Oil of narcissus } as much as sufficeth.

Melt and stir up well, and use.

Another, [which is good] for the diseases [of the chest].

Wax	8 drachms
Gum of terebinth (?)	3 ,,
Galbanum		6 ,,
Flowers of dill	2 ,,
Irises	2 ,,
Fresh hyssop	2 ,,
Laurel berries	2 ,,
Bean flour	4 ,,
Trigonella	4 ,,

Page 250

Oil of dill
Honey } as much as sufficeth.

Melt the ingredients that can be melted, and mix the dry ones
with them, and stir up well, and use as a plaster.

**Another, which is made of figs, and which is good for
pain in the ribs, and for pleurisy, and which softeneth the
hardness of blains in every part of the body.**

Bdellium	8 drachms
Ammoniac	4 ,,
Flour of beans	2 ,,
Flour of chick-peas	2 ,,
Flour of vetches	2 ,,
Flour of trigonella	2 ,,
Flax seed	2 ,,
King's crown (melilotus offi-cinalis)	2 ,,
Dill	2 ,,
Anise	2 ,,
Flour of lupins	2 ,,
Figs	24 *estîrê*

Macerate the figs in new wine one whole day and night, and
then boil them until they become a pulp. Crush the *hûshâk* (?)

and the bdellium, and macerate in new wine, | and crush them Fol. 120*b*.
until they are dissolved, and pound the dry ingredients and
reduce them to a powder.   Then stir up the figs, and mix
the ammoniac and the bdellium with the powder of the dry
ingredients, and work them up in the new wine and oil of dill
until they become moistened and are one mass, and apply as
plasters to the parts that are affected.

**Another, which is good for consumptive folk and for a
cough of long standing.**

Flax seed	12	drachms
Trigonella	12	,,
Flour of barley	12	,,
Resin of shells (?)	4	,,
Pitch	4	,,
Wax	5	,,
Oil of jasmine, or } Oil of nard }	8	,,

Melt and mix together, and use as a plaster.

**Another, which is made of laurel, and which is good
for diseases of long standing.**

Wax	
Gum of cedar	
Pitch	
Nitre	
Sumach	in equal quantities
Laurel berries	
Ammoniac	
Bdellium	
Goat's fat	

Oil of narcissus as much as sufficeth.

Melt and stir up well together, and use as a plaster.

**Another, which is made of flax seed, and [which is good]
for pain in the chest, and for pain in the foot, and in the
excretory organs, and for all pains of the nerves.**

Flax seed	4 drachms		**Page 251**
Flour of barley	3	,,	
Resin of shells (?)	30	,,	
Fresh pig oil	30	,,	

Wax	4 drachms
Gum of cypress	4 „
Flowers of dill	40 „
Fruit of cypress	40 „

Crush the dry ingredients and reduce them to a powder, melt the wax, and gum, and resin, with oil, over a fire; take them from the fire and let them cool. Then mix the other ingredients with them, and stir up well, and use as a plaster.

Another, which is called "Kenâr", and which is good for excessive dryness of the chest, and for coughs of long standing, and for torturing pains and paralysis of the nerves.

Wax	40 drachms
Fresh hyssop	4 ounces
Gum of terebinth	4 „
Resin of shells (?)	4 „
Stacte	1 ounce
Goat's oil	
Pig's fat (lard)	
Oil of nard	a very little of each.
Oil of dill	

Melt down, mix together, stir up well, and use as a plaster.

Another, which is made of myrrh, [and which is good] for the diseases.

Myrrh	16 drachms
Gum of terebinth	16 „
Ammoniac	16 „
Bdellium	16 „
Dry pitch	16 „
Styrax	16 „
Irises	16 „
Wax	18 „
Oil of irises, or	
Oil of nard	as much as sufficeth.

Fol. 121 a.

Dissolve the soluble ingredients in wine with a strong odour, and melt down the wax, and pitch, and styrax with oil, and mix together and stir up well, and use as a plaster.

Another, which is made of terebinth gum, and which is good for pains of the chest of long standing, and for pleurisy,

and for the viscera, and it softeneth the hardness of abscesses, and healeth pain in the excretory organs of long standing, and it is good for all protracted diseases due to colds.

Gum of terebinth	4	drachms
Wax	4	,,
Galbanum	4	,,
Myrrh	4	,,
Bdellium	4	,,
Ammoniac	4	,,
Frankincense	4	,,
Stacte	4	,,
Styrax	4	,,
Dry pitch	4	,,
Resin of shells (?)	4	,,
Opopanax	4	,,
Sagapenum	4	,,
Castoreum	4	,,
Euphorbium	4	,,
Flour of irises	5	,,
Flowers of bdellium	5	,,
Mallows	5	,,
Mint (ammi)	5	,,
Seed of parsley	5	,,
Aniseed	5	,,
Mustard	5	,,
Cardamoms	5	,,
Doves' droppings	5	,,
| Nitre	2	,,
Pig's fat, unsalted	3	,,
Goose fat	3	,,
Cow fat	3	,,
Chicken fat	3	,,

Page 252

Oil of narcissus, or } as much as sufficeth
Oil of irises

And wine for the soluble ingredients.

Macerate the soluble ingredients in wine and dissolve [them], and melt with oil those which can be melted, pound the dry

ingredients and reduce them to a powder, mix them all together and stir them up well, and use them as a plaster on
every part that is affected, ... and it will check the pain.
And the following will also help [patients]: Baslîḳôn which is
made of four medicines, and the other one which is made of
seven medicines, and those plasters which have been written
down in the Chapter on protracted diseases. And fomentations
also help them, and anointings with unguents which warm and
soften when applied to the chest and head, and rubbings of
the hands and feet, and foods that are beneficial. These are
the means to be employed in healing all the diseases that
arise in the chest, and the descriptions thereof must suffice
for this, the Thirteenth Chapter.

Fol. 121 *b.*   HERE ENDETH THE THIRTEENTH CHAPTER. |

CHAPTER XIV.

Of the injuries which happen to the heart, and of their symptoms.

IT hath already been shewn that the heart is the dominant member of the body, and that from it, as from a fountain, by means of the arteries life is transmitted throughout the whole body, and that, as it were by a kind of hands, the throbbing of the arteries (*i. e.,* the pulse) which proceed from it maketh manifest to us all the diseases that happen in the body. The heart is a kind of nervous substance, and it hath in it two hollows, which are filled with fine blood and living spirit. It is suspended in the lungs in the middle of the chest, and a nervous tunic surroundeth it. It hath two "ears" which are opposite to its two hollows, and it is inclined rather to the left side than to the right, and the fine point (or, head) thereof is opposite to the spleen; for there cometh to it from the liver that great vein which is filled with blood, and then | it entereth Page 253 the lungs, and [both] are nourished by it. And as it is with all the [other] members of the body, so is it with the heart, and it is meet that we should define what happenings 'take place therein, either through independent suffering, or through suffering to which it is predisposed, whatever may be the name by which a man is pleased to call it, and those that take place therein when it suffereth because other members are suffering.

Now there are three "heads" by which the whole body is governed: the heart, which is also a kind of head, and the brain, which hath been shewn to give sensation and motion to all the members of the body, and also the liver which is the "head" of the faculty of nourishment; but with any immoderate change of condition (or, composition) of the heart, death is always associated; for all the [other] members are so constituted that they must suffer with it, because, as we can show, they all work through the regular conditions of the heart. And consequently, it is necessary that, when the measure (*i. e.,* balance) of its condition is destroyed, the operations (or, faculties) of [all the other] members are destroyed likewise, and with them also [those of] the brain and the liver; with the Fol. 122 *a.* operations of these | two, however, the destruction of the operation of the heart is not closely associated. And although a man may not be in possession of these faculties of action and sensation, and his nutritive powers may be inactive, as is the case with creatures that hibernate, it is possible for a man to think that he is alive so long as the heart is in no way diseased. Moreover, it is often seen that for several consecutive days a man knoweth nothing, and suffereth no pain, and remaineth without sensation and motion, yet he is alive notwithstanding. And it also happeneth sometimes that a man liveth for a long time, even when, owing to the failure of the power of the liver, he remaineth without nutrition; but if one depriveth the heart of breath, the man is destroyed immediately.

Page 254   All those, then, | who think that the chest is wholly useless in the operation of breathing, are incapable of discovering the reason why those men who fall quite suddenly into violent apoplectic fits, through some injury to the upper "head", die immediately. To you, however, this is not a matter of difficulty, for you are correctly acquainted with the fact that the chest is set in motion by the muscles that derive the source of motions from the nerves which come to them from the [part of the] spinal column that is close to the neck, because the chest is then opened by that power which descendeth from the brain, and setteth in motion the muscles. And even if these two "heads" were held fast together, it would still be possible for a man to live, although they were deprived of the upper "head".

Now perhaps through this thing wherefrom we find the cause whereby the heart is injured, when the brain is suffering from great and violent fits of apoplexy, perhaps, I say, from this same thing, the enquiry falleth upon us in another manner, so that we have to discover what is the reason why the man who suffereth from slight attacks of apoplexy, and "fallings", and faintings (?), and long periods of unconsciousness, doth not perish. And, when considering this matter, a man must also find out how it is | that those who are attacked by these Fol. 122 *b*. diseases succeed in drawing their breath; for the breast, which is as it were bound tightly with some cord, and can only ex-hale its breath with the greatest difficulty, ejecteth mucous matter. (Now we have already spoken sufficiently about this question in the Chapter on difficulty of breathing). Inasmuch as men draw their breath in times of fever, although all the other muscles of the body are in a state of rest, and we say that their breathing power is in operation, even so during those attacks of exceedingly violent apoplectic fits, and also during those other attacks of the diseases which are akin to them, do we declare that their breathing power is in operation. But, if the fit of apoplexy, or the attack of some other disease, be so severe that | the operation of the muscles that are in the Page 255 chest is destroyed, the power of breathing is necessarily brought to an end, and with it also the natural condition (or, state) of the heart, for with this state also is necessarily closely associated the destruction of the whole man. Therefore through these causes doth a man perish, namely, through injury to that upper "head".

And he perisheth in another way, that is to say, when that lower "head", which is the liver, is diseased, for he then dieth because the liver faileth through lack of nutrition; now animals suffering from diseases of this kind last a very long time [before they die]. And there is yet another way in which a man can die, that is to say, when he suffereth from serious diseases of the stomach, and from oft-repeated sicknesses; and he also dieth through great fear, and others die of over-great joy. In all those whose life-giving power is weak, if they suffer to a very serious degree through want of knowledge by reason of some mental disease, the substance of their soul (or, mind) is dissolved very quickly. Now some men of this class have died of grief, though not immediately after the manner of those

19

who have been mentioned here; for the man whose mind is intelligent and great doth neither die of grief nor of any other mental ailment, for he who is mightier than grief never falleth [by reason of it] into death, for the power of the mind of those who are of this class is exceedingly healthy, and their diseases are slight.

Of the special diseases of the heart.

Fol. 123 a. | Now, the special disease which taketh place in the heart is that which subsisteth often through a simple change of condition, whether it be regular or irregular. If an abscess or a hard, red sore beginneth to grow in the heart, straightway, before the disease increaseth, the man is destroyed. And there also accompany diseases of this kind very strong involutions (cramp?), which resemble those that are closely associated with pains (or, ailments) of the stomach; but happenings of this kind in the heart are due to diseases of both the mouth of Page 256 the belly and | the heart. The diseases themselves are caused either solely through a severe change in the condition, or through the moisture produced by foul chymes. Sometimes, however, they are produced by a hard, red sore, or by some other abscess, whatever its nature may be. Now the small changes of condition which take place in the heart change the throb of the arteries (i.e., pulse) according to their nature at every time (or, season). In severe diseases when the change in the condition of the members becometh one that causeth heat (or, inflammation), and it reacheth the heart, death doth not accompany it sharply (i.e., immediately); when it taketh place in the organic members, death rusheth in suddenly, but it is accompanied by premonitory signs. One of these signs hath been mentioned by Hippokrates thus:—"To those who "are doubled up violently, very often doth death come suddenly, "and without any clearly understood cause." Another symptom is palpitation of the heart, which taketh place either by itself, or, using a simile, like that which is in troubled waters. And it is not a thing to wonder at that the quantity of liquid, which is collected in the tunic that holdeth the heart, is sometimes so considerable that it abateth (or, destroyeth) the motion of the heart externally. For behold, we often see in the animals which we dissect that there is a very considerable quantity of liquid in the membrane that surroundeth the heart. There was once an ape the body of which began to dwindle away by degrees, and we did not order him to be dissected because of our complete incompetence to do so. When he was dead, all the other members | of his body were without disease, but Fol. 123 *b.* there was seen on the tunic of his heart something that was like an abscess of some kind, which was filled with a bloody liquid, similar to that which is confined within the cysts that come forth under the eyes. And again in a cock that we once dissected we found a thing like this abscess, and it also | was placed on the tunic of the heart, and it was hard, and Page 257 resembled many layers of fat placed one above the other. Therefore it seemeth likely that something of the same kind may take place also in men, and that an inflamed abscess may exist in the heart. And we have sometimes seen in cases of those persons who contended each against the other in a contest (or, war) that death hath supervened immediately, and that it resembled the deaths of those into whom stones entered, that is to say, through pains of the heart. If, then, that which cometh to (or, attacketh) the heart reacheth one of the [two] hollows (sinuses) that are in it, we agree that blood must be let immediately, especially when the left hand hollow happeneth to be attacked. If, however, that which attacketh doth not reach the hollow, but is retained in the body (or, substance) of the hollow, men will not suffer only on the day whereon they were attacked, but also in the night that followeth, and if they have an abscess death will supervene straightway. Now all such men are able to use their faculties of understanding and to think so long as they are alive, and therefore it is evident, a fact which testifieth to [the correctness of] the confession of the ancients, that the reasoning and thinking powers of the soul are not situated in the heart.

Now palpitation of the heart appeareth both in the young and in those of mature years, and it cometh to them when they are in a healthy condition, quite suddenly, and without

19*

any [apparent] cause whatsoever. All such are greatly benefited by the letting of blood, and many of them have been
completely freed from this occurrence thereby. [Others again,]
who after the letting of blood have lived very carefully and
employed helps of this kind, have been cured, and some of
them, even though the disease hath returned to them sometimes, have been [ultimately] cured. I once knew a man, to
whom this disease of palpitation of the heart used to come

Fol. 124a. every year, in the spring, | who had learned by the experience
of many times that the letting of blood helped him. Early in
the fourth year he had himself bled before the disease attacked

Page 258 him, and this he did with the best results for many years. | But
though in addition to the other remedies which were good for
his disease he employed the letting of blood, his health failed
and he too died before he became an old man, even as died
all the others. Some of these fell into bad health through
violent (or, sharp) fevers, and died suddenly, and some of them
[died] through [this] method of healing, but one or two of them
died from a cause other than that of weakness (heart-failure?).
Now, many of the men who suffered in this way were over
forty years of age, and less than fifty.

The covering of the heart is [one] of the important members, whatever may be the kind of disease by which it is
affected; but if there be in it an inflamed abscess, and it
reacheth the heart itself, the heart will suffer sympathetically.
When it (i. e., the covering of the heart) suffereth by itself it
doth so in the same way as all the others which are made to
suffer; when there is disease in it through the protective tissues
and covering of the principal members, it causeth no danger
[to life]. Now the heart doth not produce any other shortness
of breath, or any other ailment besides those that are mentioned
in the preceding Chapter, and in that which is on shortness of
breath. When the heart is hot (or, inflamed) it causeth the
breath to be drawn in very deeply and rapidly, and it also
causeth the expirations to be hot; and it causeth the exact
opposite of these things to take place when there is cold in
the heart, for by reason of it the breathing becometh short
and light. Now, it hath been said above that if there be in

the heart an inflamed abscess, or a hard, red ulcer, or if there
be in it anything that causeth a division in its substance, or
cancer, a speedy and sudden death overtaketh a man. Therefore it followeth that the heart itself alone receiveth a change
of condition, when in addition to a certain [amount of] liquid
some deadly chyme either is collected therein, or in the space
between it and the tunic that surroundeth it, and that through
this there ariseth either palpitation of the heart, or the torture
of the whole | body | that hath been mentioned a little way
back; but, inasmuch as the whole body suffereth together with
the heart, when the condition of the heart is bad, and we are
about to bring forward the means for healing, it is necessary
for us to diagnose what are the natural conditions of the heart
from the unnatural conditions which arise in it.

Fol. 124*b.*
Page 259

Of the symptoms of the conditions of the heart.

The simple conditions of the heart. Now, the symptoms
of the hot condition of the heart are made known by the
greatness (*i.e.*, depth) of the breathing, and by the lightness
and rapidity of the feel of the arteries, and by truculence, and
wrath, and boldness, and by the hair that is on the breasts of
[men of] this kind, and especially by a whistling sound over
the heart, and by [the state of] the members which are by the
side of it; and in the majority of cases the whole body is hot;
but, if the liver standeth much against it, and there is breadth
of chest, it is a sign of the heated condition of the heart, and
if in this case the brain standeth against them to a great extent [it is also a sign of the heated condition of the heart].

Now an immoderately cold condition of the heart diminisheth
the touch (or, feel) of the arteries, and the men who possess
this temperament are naturally timid, and languid, and careless, and their breasts are smooth.

A dry condition of the heart maketh the feel of the arteries
hard, and causeth a man to be slow to anger, but when his
anger is roused up it becometh a savage passion, and in the
majority of cases such anger subsideth with difficulty. And
the whole of the bodies of men possessing this temperament

is dry, provided that the liver be not found to be standing against it.

A wet condition of the heart causeth the arteries to feel soft to the touch. Those who possess this temperament are stirred up quickly to wrath, and this wrath subsideth quickly; and the whole of their bodies is moist, provided that in this case also the liver doth not stand against it.

A hot and dry condition of the heart causeth the arteries to feel hard, and large, and rapid, and light, and the breathing Page 260 of men who possess this temperament becometh great, | and light, and frequent. Their breasts and the upper parts of their bellies are hairy, and their actions are fierce and savage, and Fol. 125 a. they are disposed to wrath, and are truculent, | and tyrannical, and they are swift to wrath, and their anger doth not quickly subside. If heat and moistness be present in the condition of the heart, they are less hairy than the class of men that hath just been mentioned. Men possessing this temperament are prompt in action, and their wrath is not followed by cruel acts, for they are only prone to be angry, and their arteries are large, and full, and light, and rapid to the touch.

If cold and moistness be present in excess in the condition of the heart, they cause the arteries to feel soft to the touch. Men who possess this temperament are timid, and languid, and careless in their works. And their whole body shaketh with their chest, and they do not nurse their wrath, because they are not prone to anger.

If cold and dryness be present in excess in the condition of the heart, they cause the arteries to feel hard and small. Men who possess this temperament do not become angry on small provocation, but when once they are moved to anger, their wrath endureth for a long time, and they nurse their ire. And their bodies and breasts are wholly without hair. These are the distinguishing marks of the natural conditions of the heart, and if there be found characteristics which are contrary to these they must be described as unnatural (*i. e.*, abnormal).

The compound conditions of the heart.

The rectification of the hot, evil condition of the heart and of the whole body. Now the hot and bad condition of the heart is caused through the red bile being in excess. When this goeth downwards it is not right to fear, but when it descendeth into the stomach it is necessary to bring it up by means of vomiting after exercise, before food, and after food, and after bathing. Now it is good for those who suffer in this manner to undergo a course of exercises (or, training) which is not violent and severe, but gentle and | easy. Those who Page 261 possess a very ardent temperament do not need any course of treatment, but easy walking and bathing are sufficient for them, and they will therefore rejoice in bathing after meals. Those who possess | an exceedingly dry temperament need Fol. 125 b. abundant nourishment, and they must abstain from violent and excessive exercise, such as it is well known that men abstain from in the season of summer. And frequent washing (or, bathing) is beneficial for them, and especially draughts of cold water after a meal, for these are the opposite of dry constitutions. And frequent copulation injureth them, and consequently it is meet for them to guard themselves against anger, and fatigue, and anxiety, and sleeplessness.

Now a constitution that is moist naturally during the period of youth casteth its possessor into the pains of rheum, and of superfluity, and into the sicknesses of fluxes, and therefore young folks require much exercise, and complete digestion by the stomach; and they should make water once, twice, and thrice when they wash before taking a meal, and they will find themselves much benefited if they use water that is warm naturally. Now it is meet that we should be most careful about their evacuations which are to be brought about by means of frequent bathing and a purgative, or about the emptying of their bodies by means of the urine and faeces. And there is nothing to prevent them from making use of the helps that bring down phlegm, and purgative medicines, and foods that contain good chymes, and they may drink wine that will cause them to make water.

The rectification of the cold, evil condition of the heart.

There are three changes of the cold, evil condition of the heart, but the most evil of them all is the dry change, when it hath existed for a long time in those who possess a cold, dry constitution; therefore it is necessary to moisten and warm such men. This may be brought about by moderate exercises, and by the foods that are moistening and warming in character, Page 262 and by the drink of | hot, vinous drinks, and by abundant sleep, but we must take care that the body is emptied each day of the superfluity that is produced in their bodies. And copulation is injurious to all men with dry constitutions, and especially to Fol. 126 a. those who possess coldness as well as dryness; | and it is only good for those who possess ardent and moist constitutions.

Now, the cold constitution is evil, even when moistness predominateth in it, and it casteth the possessor thereof into the pains of rheum. Abundant and light exercise, and the abstaining from washing, and the use of warming unguents, help men of this kind; but as to those who are constitutionally of a cold nature, and who possess a constitution which is moderately dry and moderately moist, it is meet to stimulate their warmth and strength by means of the moistness and dryness of nutritive agents, and it is meet to select those that are intermediate in nature.

Of the healing of the diseases peculiar to the heart.

For palpitation of the heart, which hath been mentioned above, the writer declareth that the letting of blood from the left arm, and from the middle vein, is good, and in addition to this, the patient must take careful and exceedingly easy exercise. Now the means to be employed in healing the changes of condition that take place in him have been mentioned a little way back. The evil moisture, which is collected either in the heart, or in the covering thereof, or any dense and flatulent wind, or any obstructions, if there be such in him, may be healed by means of bathings, and aromatic unguents and cataplasms, and by foods that possess health-giving chymes.

And to the patient will be useful also the medicines which are gentle in their nature and are transmitted to the heart; such are the following:—

A medicine which is made of musk, and which is good for palpitation of the heart, and for shortness of breath, and for all the diseases that arise through black bile.

Zôrbadh (amomum zedoaria, or amomum zerumbeth) Dornâgh	1	
Unpierced pearls	1 drachm	
Electrum	1 ,,	
Coralium	1 ,,	
Living silkworms (?)	1 $^1/_2$,,	
White Behmân (centaurea behen, or salvia haematodes)	$^1/_2$,,	
Red Behmân	$^1/_2$,,	
Laurus malabathrum	$^1/_2$,,	Page 263
Spikenard	$^1/_2$,,	
Cardamoms (cardamomum vulgare)	$^1/_2$,,	
Caryophyllus aromaticus	$^1/_2$,,	
Tree moss (?)	2 *denkè*	
Castoreum (?)	1 *denkâ*	
Long [peppercorns]	2 *denkè*	
Ginger	2 ,,	
Musk	$^1/_2$ *gramma*	

Crush and work up into a paste with unboiled honey, | and administer in doses [the size] of one chick-pea, in strong-smelling wine. Fol. 126*b*.

The Caesar Antidote, which is good for palpitation of the heart, and for evil winds, and for fevers of long standing, and for the stomach which doth not digest food, and for those who bring up blood, and for shortness of breath, and for violent hiccoughs, and for pain in the liver and spleen, and for griping in the belly, and for delayed menses, and for headache, and for the medicine of death (*i.e.*, poison), and for the bites (or, stings) of noxious reptiles and insects.

Castoreum	3 drachms
Licorice juice	3 ,,
Cassia	3 ,,
Costus	3 ,,
Peppercorns, round,	3 ,,
Peppercorns, long,	3 ,,
Styrax	3 ,,
Opium	3 ,,
Crocus	3 ,,
Spikenard	3 ,,
Opopanax	1 drachm
Musk	$^1/_2$,,
Zôrbadh	$^1/_2$,,
Dornagh (doronicum scor-pioides)	$^1/_2$,,
Unpierced pearls	$^1/_2$,,
Myrrh	8 ,,

Honey as much as sufficeth.

Dose: one [the size of a] chick-pea in a draught as convenient for the disease.

The Gold Antidote, which is called "Question"(?), and which is good for all diseases of the heart and head, and for shortness of breath, and for continued fear (or, apprehension, or nervousness), and for grief (depression?) which is without cause.

Euphorbium	20 drachms
Opium	20 ,,
Peppercorns (round?)	$4^1/_2$,,
Cypress seed	$4^1/_2$,,
Black cummin	$4^1/_2$,,
Coralium	$4^1/_2$,,
Pearls	$4^1/_2$,,
Peppercorns, white,	4 ,,
Peppercorns, long,	4 ,,
Pyrethrum	4 ,,
Caryophyllus aromaticus	4 ,,
Mandragora seed	2 ,,
Gold and silver	$4^1/_2$ *denkê*, each

Jackal fat	2	drachms
Sagapenum	2	,,
Laurus malabathrum	4	,,
Zrôbadh	2	,,
Drônâgh	2	,, and 4¹/₂ *denḳê*
Crocus	3	,,
Camphor	3	,,
Castoreum	3	,,
Spikenard	4	,,
Seed of wood rue	4	,,
Costus	2	,,
| Alpûrâ (Alpaohrâ? "bryony")	4	,, Page 264
Pasharshtîn (فاشرستين, *i. e.*, black bryony, or dioeca)	4	,,
Aristolochia, round	7	,,
Amomum (grain of Paradise)	4	,,
Spikenard	4	,,
Asarum	4	,,
Stacte	2	,,
Paeonia (peony)	2	,,
Seed of dill	2	,,
Cabbage stalk	2	,,
Seed of the Egyptian myrtle	2	,,
"King's crown"	2	,,
Incense berries	7	,,
Cinnamon	7	,,
"Stag's horn"	1	drachm
Greek sulphur	6	drachms and 4 *denḳê*

\| Crush the dry ingredients and reduce them to a powder, and Fol. 127 a. dissolve those that can be dissolved in strong-smelling wine. Rub down the gold and the silver in a mortar, and mix the powder with the dry ingredients, then mix them all together in honey from which the scum hath been removed, stir up well together, and pour the mixture into a vessel made of lixivium, or glass. After a year hath passed since the compounding of this medicine, administer as a dose a quantity equal in bulk to that of a bean, in wine mixed with water. And place some of it in the nostrils of those who suffer from headaches of long standing, and to those who have attacks of the falling sickness administer about half the size of a vetch, mixed with the milk of a woman and oil of jasmine.

The **Wild-rue Antidote**, which is good for diseases of the heart.

Wild rue	I	drachm
Nigella sativa	I	,,
Camphor	I	,,
Castoreum	I	,,
Aristolochia	I	,,
Cypress	I	,,
Seed of hyoscyamus	I	,,
Bryony	I	,,
Black bryony	I	,,
Pyrethrum	I	,,
Peppercorns	I	,,
Persian thyme	I	,,
Serpent medicine, which is, in my opinion, gentian,	I	,,
Spikenard	I	,,
Wild parsley	I	,,
Seed of rue	I	,,
"Killer of its father"	I	,,
Carraway seed	$^1/_2$	,,
Opium	$^1/_2$	,,
Crocus	$^1/_2$	,,
Incense berries	$^1/_2$	,,
Cassia	$^1/_2$	,,
Costus	$^1/_2$	,,
Sagapenum	4	drachms
Opopanax	4	,,
Sugar (?)	2	,,

Honey as much as sufficeth.

Dose: for delicate folk, one drachm; for healthy folk one drachm and two *denḳê*.

The **Musk Antidote**, which is good for fear (nervousness?), and for those who suffer from attacks of the falling sickness at the beginning of each month, and for palpitation of the heart.

	Cassia	2 drachms	Page 265
Spikenard	2 „	
Laurus malabathrum	2 „	
Amomum	$1^1/_2$ „	
Costus	$1^1/_2$ „	
Resin	$1^1/_2$ „	
Gold	2 carats	
Silver	2 „	
Musk	3 drachms	
Camphor	3 „	
Leeks	1 drachm	
Coralium	1 „	
Electrum	1 „	

Honey as much as sufficeth.

Dose: a portion as large as a bean in wine and water, at the beginning of the month, at the middle of the month, and on each of the last three days of the month. And place some, mixed with beet juice, or with the juice of conyza odora (ele-campane, or fleabane), in the nostrils of those who suffer from attacks of the falling sickness, and they will be restored to health.

The Pearl Antidote, which is good for diseases of the heart.

Zrôbadh	3 drachms	
Drônâgh	3 „	
	"Sea seals"	$1^1/_2$ „	Fol. 127*b*.
Unpierced pearls	1 drachm	
Stony plant	1 „	
Caryophyllus aromaticus	1 „	
Silk (?)	1 „	
Bamboo	1 „	
Castoreum	$1^1/_2$ *denḳâ*	
Sugar (?)	3 drachms	

Honey as much as sufficeth.

Dose: about a drachm and a half.

Another Pearl Antidote, [which is good] for every pain of the heart, and for all the diseases which are produced by black bile.

Unpierced pearls	1 *gramınâ*
Musk	1 „
Camphor	1 drachm
Marmahôz (origanum)	1 „
Electrum	3 drachms
Coralium	3 „
Gold	1 carat
Silver	2 „
Zûrembâr (*sic!*)	1 drachm
Drônâgh	1 „
Laurus malabathrum	1 „
White hellebore	1 „
Cassia (or, cinnamon bark)	1 „
Costus	1 „
Castoreum	1 „
Incense seed	2 drachms
Mountain pepperwort	3 „
Rock sugar (*i: e.,* loaf sugar)	4 „

Crush and mix together, and administer dry in doses of one drachm, or half a drachm, in wine mixed with water, when the patient is fasting, for three days.

Another [antidote] for fear and palpitation of the heart.

Dried ox-tongue, pounded	1 drachm
Amomum zerumbeth	. . . . . .
Drônâgh	. . . . . .

Crush and mix together, and administer at the beginning, and middle, and end of each month, one drachm, in wine and water when the patient is fasting. And if it be found that the ox-tongue is stale, put it in wine without water, and let the patient drink the wine very frequently for seven days.

Page 266 ### Another [antidote] for | palpitation of the heart, and for fear, and for attacks of the falling sickness.

Spikenard	2 drachms
Cardamoms	2 „
Amomum zerumbeth	2 „
Drônâgh	2 „
Dried ethrôg	1 drachm
Seed of dill	1$^1/_2$ drachms

Pound, and administer in doses of one drachm in one and a half measures of wine wherein the tongue of an ox hath been macerated, and let the patient drink it at the beginning of the month for three days.

Or crush samsîkôn (sampsuchum) and administer one drachm in warm water.

Or crush three gramma of caryophyllus aromaticus, and administer one measure in goat's fat.

Or crush the root which is called "gahzîr", and administer the dose in wine mixed with water.

The Coralium Antidote which is good for palpitation of the heart, and for fear, and for the winds | which hide in Fol. 128 *a*. **the members, and for meagreness of body, and for fevers of long standing, and for all obstinate pains.**

Laurus malabathrum	2 drachms
Polium	2 ,,
Abrotonum of the field	2 ,,
Gahzîr	2 ,,
Mountain rue	1¹/₂ ,,
Cassia	1¹/₂ ,,
Pistacia lentiscus	1¹/₂ ,,
Dill	1¹/₂ ,,
Stacte	1 drachm
Electrum	1 ,,
Caryophyllus aromaticus	1 ,,
Silk (?)	1 ,,
Fennel	1 ,,
Unpierced pearls	3 drachms
Castoreum	1 drachm and 4 *denḳè*
Seed of hyoscyamus	1 ,, and 4 *denḳè*
Rue, wild,	1 ,,
Rue, garden	1 ,,
Bamboo	8 drachms
Loaf sugar	15 ,,
Behmân, white,	1 drachm
Behmân, red,	1 ,,
Honey as much as sufficeth.	

Dose: as much as a bean in an infusion of rock parsley, or in wine mixed with water, three times in the month. And for those who suffer in this way are also beneficial the medicines that expel from the body the chymes of black bile and of the living phlegm that is not digested. And great care must be observed about the whole body. It is now meet that we should pass on from this point to the diseases of the stomach, for so the ordering of our work demandeth. What, then, we have said about the diseases which attack the heart, and about their distinguishing features, and about the means to be employed in healing them, must suffice for this the Fourteenth Chapter.

HERE ENDETH THE FOURTEENTH CHAPTER.

CHAPTER XV.

Of the diseases which attack the stomach, and of their distinguishing features, and of the means to be employed in healing them.

Now this member of the body was divided by the ancients into three parts, each of which had its own name. The first part, which beginneth with the palate and goeth down to the stomach itself, they called "stomach". The [second] portion, which is annexed thereto, they called "belly", or "heart". And they call "belly" the whole of that part of the body that extendeth to the end of the lower mouth, which is attached to that upper fullness which is named "sâumâ". Now, here are attached the veins that come from the liver, and that are called "mâsriķê" (*i. e.*, those that are connected with the mesentery, or, the membrane to which the intestines are attached), and by the operation | of the natural power that is Fol. 128*b*. called "attraction" they draw and bring up that juice, which the stomach maketh from the food, to the liver, and there it is changed into blood. It is a natural product, and by means of the nerves it is associated with the head, and it hath a function that concerneth all the members, for if it be supplied with nourishment, and if it be in a sound condition, the whole body is developed thereby through the food [taken]. The food entereth first of all the stomach, wherein it is cooked (*i. e.*, digested) until it becometh fit for absorption by the liver. Now, it is well known that this member performeth a very important function in us, and as concerning the statement which hath been made that its operation is general, this is so truly the case that when it is injured all the members suffer.

20

Let us, however, speak first of all about the injuries that attack that first portion of the body which is called "stomach". Now the stomach hath two kinds of use: in the one it is the path taken by everything that is swallowed, and in the other it operateth in the two following ways, that is to say, it taketh down into the belly the things that descend from the mouth, and it bringeth up to the mouth the things that ascend from the belly. The stomach hath two parts, which are called Page 258 "tunics" (or, "coats") | by the dissectors of the body. With one of these it lifteth up and bringeth down food and drink into the belly, and with the other it bringeth up that which is ejected when men vomit. Therefore rightly are the diseases of the stomach of two classes; one class ariseth in connection with the bringing up of things that have been swallowed, and the other ariseth in connection with it as the organ of those things that are evacuated. Now this also is well known to every man that of the two needs which it supplieth [that is to say, of vomiting and of evacuation, that it performeth the latter frequently], and the former only occasionally. The class of diseases which I have referred to as arising in connection with the bringing up of things swallowed, is the result of the stoppage of the passage of the stomach through an abscess of some kind, and the other ariseth because the body itself is in an enfeebled state, and is unable to perform the operations whereby it is maintained in a sound condition. Now, it hath Fol. 129 a. been said by us | in the Chapter on the Strangles, that with the change which taketh place in the vertebrae of the neck is associated another affliction (or, disease), which taketh place from without when they are afflicted by these things; let us keep this fact in mind in the present case. And let us remember also the abscesses which arise in those muscles that resemble the lower part thereof, in whatever form they may exist therein. Now, all the abscesses which arise therein are known when there cometh upon them, from the inside, that condition, wherein if patients swallow anything, they often experience such a feeling of compression that their drink is cut in twain, and it riseth up through their nostrils. When, then, there is an abscess in it, and the affliction ariseth from

this, and not from any other place, severe pain ariseth therein
when they swallow anything, and there is difficulty in making
the food to go down into the belly, more especially when the
patient wisheth to swallow anything whilst he is lying on his
side, and they therefore desire to change the position in which
they are lying, and to sit up. Then, having learned from that
which happeneth to them that they are able to swallow easily
in their [new] position, | it seemeth that their act of sitting up Page 269
helpeth in no small degree their power to pass down the food
into the belly, and [in this position] it is often possible for this
to happen solely by pressure from the mouth to the stomach.
And no help cometh from the proper descent [of the food],
for the operation of swallowing is due to the power that is
inherent in the stomach.

Now, ye know that when all the members that have abscesses in them are in operation they feel severe pain, and
that when they are in a state of rest they feel refreshed.
And, because there is in the stomach the faculty of sensation,
we also make more use of this, which overcometh in the
[diseased] member, as a symptom of the diseases which exist
therein, than of the judgement of the sufferer when he doth not
understand [his case] sufficiently well to be able to explain
(or, describe) it. | When the patient suffereth severely, the Fol. 129 b.
faculty of sensation that is in the stomach is of very great
value, and it helpeth us powerfully in obtaining symptoms of
the diseases from which he is suffering. Now, very often in
cases wherein men suffer from a certain debility they say that
they possess sensation and, after a long time and with difficulty, they make to pass over to the stomach foods which
descend into it; but there are times when at the first attempt
they go down easily, and they straightway remain there so
firmly that they might be thought to be held there by pressure;
and in like manner they descend without trouble to the lower
region. Now, the first kind that hath been mentioned maketh
known weakness in working [of the stomach], and the second
maketh manifest some trouble in one of its parts; but we are
able to define each of them by means of the other happenings. We recognize weakness, then, when there is in it a

20*

change of condition only, besides an abscess, from the superfluity of the descent of the things that they swallow, when
this existeth simply without any pain whatsoever; and this
descent is delayed very much by lying on the side, but it is
lessened when the neck is in upright positions, for then, with-
Page 270 out | any feeling of oppression, the descent of the food is ready
to take place. Therefore the weakness, which existeth con-
jointly with an abscess of some kind, causeth more oppression
in one of the parts of this member (*i. e.*, the stomach) than
its companion in the stomach; moreover, the descent of the
food is delayed more by it than by the abscess. And if the
abscess is hot, or if it is hard and red, there are associated
with it pain, and thirst, and a feeling of great heat, and fever
which doth not burn overmuch, for its intensity is not in proportion to the thirst. And if the abscess be not hot, in addition to the fever, and inflammation, and thirst, very great
delay in swallowing taketh place in one of the members [of
the stomach], and foods are held back, especially when the
patient swalloweth something that is hard and that hath not
been thoroughly well cooked, but there is little pain. And
Fol. 130a. when | attacks of this kind have taken place for a long time,
and these have been accompanied sometimes by daily fever,
and sometimes by shiverings, then we conclude that there is
a pustule which is hard to abate. Now, after a time, there
cometh in the patient a feeling of rupture, and immediately
he vomiteth pus, and he doeth this the next day and on the
day after; moreover, there follow closely other indications, all
proceeding from the stomach, which shew that there is a
wound in it. And he feeleth a biting sensation, or one of
compression, whenever he swalloweth anything of an acid, or
salt, or pungent nature, and he feeleth pain in that region in
a moderate degree, even though he swalloweth nothing at all.
Now, the foods that bite (*i. e.*, are acid), and are pungent are
especially liable to produce this sensation and pain. A patient
of this kind can only be cured after a very long time (*i. e.*,
course of treatment), and with the greatest difficulty, even
though his time of life be in his favour; all such patients who
are old die.

Now all those who have had | in the stomach a painful Page 271
disease, have also had immediately in the region that is between
the shoulders a pain which is the cause thereof; and it is
known to us, for we have seen it in dissections, that the
stomach is extended by the side of that part of the spinal
column which is in the neck, and that it goeth downwards.
And this also is evident from the same considerations, that
when patients vomit pure blood, the blood cometh from the
veins that are in the stomach; but when a coughing up of
blood taketh place because of some rupture of the veins, the
pain which ariseth from this indicateth the place in which the
rupture hath occurred. Similarly, when blood cometh up
through cancer of some kind, or some tear [in the veins], the
place where either of these things hath happened is well
known. Now, when some vein hath opened of its own accord,
and blood cometh up, the stomach feeleth no pain whatsoever.
And in like manner | two other kinds of vomiting of blood Fol. 130b.
may be distinguished: that which is not the result of a severe
blow, and that which is not the result of a great fall, according to what usually happeneth in ruptures [of the veins]. Sometimes ruptures of the veins take place through some external
cause and through excess of blood. The bringing up of blood
which taketh place through a cancer indicateth that there was
originally a wound on the spot wherefrom the blood came up.
And ruptures of the veins usually produce such a wound, as
doth also the transference of the evil chymes which come to
the spot, even as we see this frequently happen externally in
the skin.

Now, that second portion of the stomach, which cometh
after this and which was called by the ancients "mouth of the
belly", or "heart", was thus named by them because of the
effect that it produced on the diseases of the stomach; for men
are not only attacked by the fatigues that extend to the heart
because of the diseases thereof, but also by fits of apoplexy,
and fits of deadly unconsciousness, and insanity, and by the
disease that is called "the falling sickness", and also | by fits Page 272
of dimness of vision, even as we have already said in the
Chapter on the Eyes, "when all these suffer, the other members

"suffer with it." Now, when the "mouth of the belly" is itself diseased, no appetite existeth, and it turneth to corruption the food that is in the stomach, and this becometh exceedingly like unto the matter that is ready to settle down in the fundament of the belly; when the foods are difficult to reduce to this condition, nothing of this kind happeneth to them.

It is very right, then, that we should examine most carefully the diseases that take place sympathetically, and that we should distinguish them from those arising from independent disease in those members of which the operative power is defective through the inaction of the "mouth of the belly." As in old times the physicians were accustomed to call the heart the "mouth of the belly", so now all the physicians call the stomach the "mouth of the belly"; but the boundary line between the diseases of the heart and those of the stomach is extremely

Fol. 131 a. well known to him that is practised | and trained in the art of making diagnoses (or, the art of symptoms). For certain kinds [of symptoms], which cannot possibly be described, are intermixed with others which it is quite possible to explain, and these confirm the symptoms of the members that are diseased. Now of these every man can discover the symptoms for himself; but those symptoms, which form, as it were, the general foundation of all symptoms, and of which there are some of the same kind as the effects that cannot possibly be described, the lovers of learning can learn. Now, I will myself describe the case of a certain sophist who, before he had tasted any food at all, used to be seized with a fit of the falling sickness, more particularly when he was engaged in teaching strenuously, or when he was studying, or during a prolonged fast, or when he was angry. Thereupon I thought out the matter carefully, and concluded that the "mouth of the belly" was diseased to such an extent that it had become exceedingly sensitive, and that in addition to this the brain was diseased, and that this caused his whole body to be affected by the

Page 273 falling sickness. | Therefore I ordered him to be very careful about the proper digestion of his food, and to eat meat, which had been prepared with the greatest care, every two or three hours, and if he was not thirsty to eat it without drinking, but that if he felt thirsty he was to drink some wine mixed with water, but that the wine must be one of those that are wholesome, for these strengthen the belly and do not make the head hot, as do very strong wines. And having obeyed these orders, it fell out that the sophist did not suffer any more, and in this way I acquired certain knowledge about those things which I had only imagined before. And I used to give him doses of the medicine that is made of aloes and is called "pîḳrâ", in draughts, twice or thrice every year, because it cleanseth the whole belly of the superfluity that is therein, and strengtheneth it to perform its functions. By these means this man lived for a period of about twenty years, and enjoyed good health the whole time; but if it happened, as it did now and again, that, owing to pecuniary | needs he remained for no ^{Fol. 131b.} short time without eating anything at all, he was seized immediately with slight shiverings. And I have also seen other men who fell down and had attacks of the shiverings caused by the "mouth of the belly", when they did not digest their food satisfactorily, or when they drank too much wine without water in it, or when they indulged in copulation immoderately.

And I have seen other men who were attacked by the disease of the shiverings during fevers, without any symptom of apoplexy or of the shiverings appearing before the attacks, and immediately they had vomited gall, at that same moment they were delivered from all the things that were vexing them. Now, some among those who suffered in this manner vomited something which was of a pale colour, and others | something ^{Page 274} which was like unto the colour of leeks. And others again, through an excess of bad (or, unsuitable) foods, whose stomachs were chilled (i.e., were inactive) fell into a stupor, which continued until they had vomited the things that were causing trouble to the "mouth of the belly." And of all these things no man could think that any one of them was caused by the "mouth of the belly", or by serious failures of strength, and we ourselves should not have done so unless we had seen them happen many times.

There is a very large number of nerves in the stomach which come down from the brain, but they are not so many that a

man should think that it is because of them that the brain
arriveth at such a state of sympathy with the stomach that
fits of apoplexy and shiverings take place. Neither as concerning the heart must a man think that it suffereth so specially
with the "mouth of the belly" that a severe failure of strength
taketh place thereby. Now, there take place in very many
patients not only deep sleep and visions that startle (or, dis-
turb) them, but also change of mind (*i. e.,* they go out of their
minds), because of the noxious chyme that is collected in the
"mouth of the belly." And as concerning the disease called
"epigastric", and "flatulent", there is no man [living] who could
not tell how great is the number of those who suffer from it,
and whom it hath turned into miserable, afflicted, and ob-
Fol. 132*a.* stinate-minded folk, | who are only a very little short of being
insane; and these are also seen to be attacked violently after
digestion [hath begun] by the fits which have been mentioned.
Now, very many of them are so splenetic that a man must
think, having regard to this fact, that some noxious, foetid
matter floweth from the spleen and goeth down into the belly,
for all the attacks that are caused in the brain or in the eyes
by the belly are followed closely by the smoky vapours of the
chymes. Now those failures of power, which take place because
of the stomach, produce quaking through some kind of pains
which arise in it; but perhaps when this change of condition
Page 275 which hath arisen therein passeth over, | and arriveth at the
heart, in order to operate on it also by means of a great
change of condition, it is at that movement perhaps, I say,
that a sudden failure of the strength followeth. The absence
of craving, and asthma, and also lusts for abominable things
doth the stomach itself create. Now I call "absence of craving" the state wherein it doth not desire foods at all; and I
call "asthma" the state wherein it lusteth for them darkly
(*i. e.,* secretly); and I name "abominable lusts" the state wherein
it lusteth for meat and drink to an inordinate degree, and
wherein it longeth with a greedy longing for abominable and
varied kinds [of food]. We will now describe all these, one
after the other, beginning with the happenings that take place
in the stomach itself.

Now all these things take place in the stomach itself, and they follow very closely its natural operations, because of which it hath need of the great nerves, which come to it from the brain, and through which it possesseth this faculty of sensation to a degree that is far greater than that possessed by any of the other members. For when any body, no matter of whatsoever kind it may be, is deflated by means of the skin into the air whereby we are surrounded, those members that are under the skin are first of all emptied, for the power of these members, even as that book on "Natural Powers" sheweth us, draweth the foods from those members | that are close to Fol. 132*b.* them, and filleth therewith the place that hath been made empty. And they also derive [their] foods from the members that are near them, and in this way this process taketh place at all times, as it were according to a certain plan, and thus this food is made to reach quickly the veins that come to the belly, in order that they also may be emptied out. Now these veins are at the same time constructed to draw the foods from the belly, and they resemble the roots of plants that descend into the earth. All this work doth not appertain | to the soul Page 276 (or, mind), but to the nature of the body, and it is performed therein as it is in plants and animals. For the earth ministereth at all times to plants, and it giveth to them, as from a belly, food in abundance; and it is always ready to give generously, according to the natural revolution of the seasons, which is ordered by God. If sometimes through an excess of drought moisture is lacking in it, then through [want of] nourishment the plants dry up. Now, in animals, because they are not directly akin to the earth, Nature fashioneth, though certainly in a very few of them, moisture of the belly which is the storehouse of food, even as the earth is for plants, and giveth unto them a feeling when food is lacking which maketh them to feel hunger, and to run in quest of the meat and drink, whereby they may be filled at one and the same time. And this craving after fullness is called "appetite". Now [this] taketh place in the faculty for feeling want when the veins require food to be given to them from the belly, and when they wish to suck in nourishment and to have power over the

same. The destruction of this sensation taketh place either
when the power of suction (or, absorption) is destroyed, or
when the members are not empty. And in the very same way
injuries to their operative faculties arise, not when they perish
wholly, but when they become near (or, akin) to nature.

Now abominable lusts arise when the lust [for food] increaseth in the intensity of the greedy craving, and this is
Fol. 133a. called the "lust of the dog", | which taketh place when the
bitter, noxious chymes gnaw the belly, and when the whole
body is deflated beyond measure. For the cold, noxious chyme
produceth lust by means of its gnawing after the manner of
the suction of the veins, and it stirreth it up in the similitude
of natural disease, which, because of coldness, craveth greedily
Page 277 for food. Now, when a saltish and bitter kind of | noxious
chyme gnaweth at the belly, then they lust for drink more
than food, and at the same time the belly becometh hot and
dry, and these two things are the cause of thirst; and, when
these chymes are discharged in the belly and in the veins, the
same thing happeneth. And the discharge of the chymes
filleth the veins that are round about it, just as an attack of
cold, according to my own opinion, emptieth them. Therefore
cold in the belly and in the adjacent parts of the body in-
duceth hunger which is not little, because the cold emptieth
the bodies (i. e., organs), and contracteth their coatings, and
condenseth them, and inciteth them to sensitiveness. And as
concerning the fact that no hunger to an excessive degree
accompanieth the other [organs], of this result heat is the
cause; for it softeneth the bodies that are hard, and it dissolveth them, and it maketh them to cleave to them weakly.
Heat disperseth substances that are moist, because it is near
akin to them. Now, one cause that produceth this "lust of
the dog" is a kind of bitter and noxious chyme; but a second
cause that produceth it, even as we have said, is excessive
emptiness, which taketh place through the exhaustion of the
whole body, and which ariseth either through the intensity of
heat, or through the healthiness of the attractive power. Now,
in the first disease it happeneth that evacuations from below
of the foods that have been received into the body take place.

In the second disease, according to what happeneth through the exhaustion of the whole body, | the food itself goeth up Fol. 133b. from the belly into the body, but this happeneth because of the lust for meats and the hunger which ariseth and cannot be measured. Now the foul things which are craved for are caused by those substances that are in the coats of the belly, whereto the noxious chyme adhereth, or those that are between the | Page 278 coats thereof, which have in them some foul kind of residuum. For this reason this longing is common, at least in the majority of cases, among women, whose chyme is noxious, because they are bearing children, and this disease, which attacketh them, is called "ḳiṭâ", (i. e., a large, voracious fish, or, whale), and whilst it is on them they long for everything that is sour and acid, and sometimes also for anything that is pungent, and at others they long for St. John's bread, or dust, or potter's loam, or cinders, or something else of this kind that is nasty. In the case of the majority of women this longing attacketh them right on to the end of the period of gestation, but some of them feel it during the second or third month [only]. In the fourth month they [usually] cease to feel it, because the noxious chyme is ejected in vomitings, and because she passeth it in her water, the woman herself being emptied thereof because of her emptying, and because she is emptied of the noxious-ness of the chyme. For during the first two months of her gestation the child draweth to himself (i. e., absorbeth) but little [nourishment] from the blood; inasmuch as he is yet small, for at that time he cannot, strictly speaking, be called a "child", only a "foetus"; but when he increaseth in size he useth up a large quantity of nourishment, and not only doth he absorb the blood (?) in the veins that is exceedingly good, and draw it to him, as I have already said above, but because he feedeth abundantly he absorbeth and sucketh in some of the foul matter that is in the veins. Therefore the body of the woman hath rest from the longing caused by the excess of the foul matter of the chyme that is in her, which itself diminisheth therein.

Now, as concerning the child, the residuum of the foods (i. e., nourishment) that he draweth to himself he collecteth in Fol. 134a. two membranes, | and there existeth in him, that is to say, in Page 279 the majority of cases, the noxious chyme and a foul state, just as if he were being nourished on foul blood, when the woman who is pregnant useth foods that are not beneficial during the whole time of her pregnancy. Now this disease sometimes attacketh men, whereupon there becometh collected in them the noxious chyme in the "mouth of their belly", and all such attacks make the lust for food to subsist. In a similar manner other men are attacked by the lust (or, craving) for drink, and the number of those who are attacked by this lust is as large as the number of those who are attacked by the lust for food. And in this case also the removal of the lust taketh place when the body hath no need whatsoever of drink because of moisture, or because of an excess of cold, inasmuch as the belly doth not feel the disease that is therein; now this feeling taketh place in a small degree when these things attack the body in a lessening degree. And the abominable lust for drink existeth in the same degree as the abominable lust for food, which also taketh place sometimes when we long for drink in immeasurable quantities, because the noxious salt or bilious chyme is collected in the coats of the belly, and sometimes because the natural moisture which is therein becometh hot and burning. Now [some] men crave for foul drinks, just as they do for foul meats, in proportion to the [amount of the] noxious chyme that is in them, and as such men are badly constituted, the attacks [of such cravings endure] for a long time; and others fall into a condition of unquenchable (or, never-ceasing) thirst, and some of these perish.

Now I know of men who have eaten vipers of the kind called "danpasdês", that is to say, "producers of thirst", and also reapers who have drunk wine wherein a viper of this kind had been drowned, and of others who drank old wine and became drunk, and of others who wished to endure thirst
Fol. 134 *b*. without drinking at all, | and of those who were in a ship and Page 280 who were so short of water that they were driven to drink sea water. And I have known of other men who did not drink at all, and of whom some were attacked by violent thirst to such a degree that their bellies dropped, and they were severely bitten by their thirst, and they died deaths that

were far more agonizing than those of the above-mentioned men. And I have also known of others, that is to say, of some of those who were attacked by a burning fever, and who, even though their malady had not reached the gravest stage, craved greedily for cold water to drink, and who up to the time of their death [the cold water having been given to them], were never filled (*i.e.,* satisfied). These, and other things like unto them, happen in connection with the appetites (or, lusts) of the belly, and besides these also [must be mentioned] the attack that is called "pain of the heart", which is also an attack that subsisteth in the mouth of the belly, and that causeth a painful sensation because of the chymes that are therein and that cause pain.

Besides these diseases there existeth also that disease called "Bôlîmôs" (βούλιμος, bulimy, that is to say, faintness of the stomach), which ariseth through cold, and through weakness, and through emptiness of the stomach. And it is not a matter to wonder at if debility and failure of strength are associated with pains of the mouth of the stomach, for men are sometimes seen to lose their strength even through a stumbling of the toes, and therefore it is not wonderful if this happeneth also through the stomach, because of the sharpness of the sensation, and because of the close proximity of the stomach, for because of these the two "heads" of the body are easily made to be in sympathy therewith.

Now, we find that great attacks take place in the stomachs of those in whom the sense of feeling is great, and that this causeth them to suffer more acutely, and to feel more keenly the gnawings of the stomach caused by painful diseases, than other folk in whom the sense of feeling is not so great. Therefore we see that injuries are transmitted to the two "heads" of the body, that is to say, to the brain and to the heart, when a particular class of nerves | in man is exceedingly _{Page 281} sensitive, or when it feeleth pain with great readiness, and that when it is in this condition, it bringeth, with the greatest ease, the two "heads" [of the body] into | sympathy with the stomach,_{Fol. 135 a.} especially when by reason of some disease, or through some natural quality, they are found to be in a weak state. Now, when these four things are gathered together serious disease must necessarily be the result; and the four things to which I refer are the following: [1.] The disease which afflicteth the stomach when it is severe. [2.] The sensation in the stomach when it is clearly defined. [3.] A disease of special nerves and arteries. [4.] A weak condition of the brain, or heart. For the disease of the falling sickness also attacketh many because of the stomach, and, because of the sympathy of the brain, the source of the nerves, with the stomach, a man is attacked by silence (*i. e.*, insensibility), stupor, delirium, and madness. These diseases are called " weaknesses (or, faintings) of the stomach" when that "head", that is to say, the heart, and the arteries suffer sympathetically with the stomach; and when the two "heads " (*i. e.*, the brain and the heart) are in great sympathy with it, there take place neither throbbings of the veins, nor slight throbbings, nor severe throbbings.

Difficulty of breathing (or shortness of breath) taketh place for various reasons, that is to say, sometimes it is due to compression of the viator membrane of the chest, and sometimes through sympathy with these two diseases. Now, the causes that produce these diseases in the body are not a few in number. Thus severe cold will cause these diseases, sometimes by itself, and at other times in conjunction with the phlegm that is very cold, and of this class is the chyme that resembleth glass; but they are more especially caused either by winds that are flatulent and cold, or by cold foods, or by some medicine that is exceedingly cold. When the mouth of the belly is cold, in no small degree the brain also, through the nerves, becometh cold along with it, and the heart becometh

Page 282 cold through the great artery; | for then, through the participation of the family of nerves, the stomach maketh cold the brain, and, because the heart is placed in close proximity to the stomach, it maketh that cold also.

Fol. 135*b*. For when the great artery goeth forth from | the heart, it cometh first to the spinal column, and thus is immediately joined to and united with the oesophagus by means of the membrane [thereof], and after this it is extended lengthwise in a downward direction over the stomach and belly. And by means of this great artery, which is extended in this wise downwards, and which goeth forth from the heart, this "head" of the body, together with the mouth of the belly, feeleth pain, and the brain, through the nerves, suffereth in sympathy. Therefore it is not a thing to be wondered at when in the course of its ailments great and violent attacks overtake the whole body. In like manner those attacks, which are called "vomitings" and "hiccoughs", are diseases of this member. Now all the diseases that attack it and are called "organic" possess symptoms which are clear and well-defined. And it is quite impossible for you to make a mistake about any one of them if ye keep in mind the general indications that have been described above in this treatise, and that we have also mentioned in other places. For neither pustules, if they happen to exist in this member, nor inflamed abscesses, nor hard red blains, are able to lead us astray, any more than can wounds if they happen to exist in it, because their symptoms are general, and also they exist in addition to those that have been described above with that other [member] the stomach. Now, the symptoms which appear in this are clearer, inasmuch as this member is more sensitive, and it falleth more under the sense of feeling [than the other]. In like manner is it also with the blood that goeth up therefrom, for this is an indication of a general symptom, and, in addition to the various kinds of symptoms which have been mentioned above in connection with the stomach, is in itself a clear and well known symptom. And it seemeth that in the sinus also | of the whole Page 283 belly diseases arise which are like unto these.

Now, the fact that when diseases arise in the mouth [of the belly] they are exceedingly severe, is one which hath been scornfully neglected by the physicians, who, as if such a thing never happened in the lower regions of the belly, have one and all confessed that it is due to the inactivity of the digestion which happeneth in the parts (or, regions) that are below the mouth [of the belly]. Therefore the cause of this | indigestion Fol. 136 a. is the lower sinus of the belly, when it is not well constructed (or, placed), that is to say, when through irregular foods, or through excess of food, or through various causes, it happeneth that a man cannot digest them. Now, all the other diseases that are common to all the organic members, and those also that occur through a change of condition, and that possess indications of the diseased places, are exceedingly well known to every man.

Now, about the blood that is vomited from the belly, many physicians disagree among themselves. Certain of them have thought it to be impossible that through it should take place the emptying of the blood of the lung from the liver and from the spleen, similar to that which taketh place, namely, the emptying of the pus which floweth from these [two members]. And because I have tested the statement (or, confession) of both sides on the subject, I am of the opinion that the empty-ing of the liver and of the spleen taketh place sometimes through it. Since, then, all the symptoms of the unnatural attacks that take place in the stomach have been described, it is necessary to speak now about the symptoms of the natural conditions of the stomach, in order that they may be distin-guished from those that are unnatural.

Of the natural conditions of the belly.

Now if the belly be of a dry nature, it quickly thirsteth, but a small quantity of drink sufficeth it, and an excess of drink weigheth heavily upon it, and it is made putrid thereby; and it rejoiceth in dry foods. A moist (or, wet) belly is known by the fact that it doth not thirst quickly, and it receiveth much Page 284 fluid | without injury [to itself] and it rejoiceth in wet foods. The belly that is hot by nature digesteth more than a wet one, and especially the things that are hard and difficult in their transformations (or, changes). It rejoiceth in foods and drinks that are hot, but it remaineth uninjured by those that are cool provided that it maketh use of them in moderation. The Fol. 136b. belly which is cold by nature digesteth more | than a wet one. It digesteth but little, especially the things that are difficult in their transformations and are cold, for they turn sour in it easily, and therefore eructations with a sour taste take place in this kind of belly. It rejoiceth in cold foods, but it is easily injured if it maketh use of them immoderately. The evil conditions of the belly that occur through sicknesses are different from its natural conditions, because they lust for the opposites of them and not for those like unto them. If the belly digesteth satisfactorily it is well known that its condition is satisfactory, but if it belch forth smoky or hot eructations, it is well known that the heat thereof is great, and that it exceedeth the normal. If the eructations thereof have a sour taste, it indicateth that cold is in excess therein, and also that the things are difficult to digest. The belly, the hot condition of which exceedeth the normal, digesteth much, but the belly that is cold will not digest [much]. Now, it is meet to examine and see lest, peradventure, this condition attack it through the chyme which moveth about and cometh to it from the other side; for in the belly wherein phlegm is in excess eructations with a sour taste take place. The eructations of the belly wherein is bile are smoky and stinking, and with these there cometh also depression of spirit (or, soul); but if the evil chyme be opened and it escape into the belly, it will immediately be vomited. The patients who suffer in this manner are moved in vain to the retchings of ordinary vomiting, | because of the chyme that is Page 285 confined in the coats of the mouth of the belly, and they bring up nothing. Now therefore, inasmuch as the differences which exist between the natural conditions of the belly, and those which occur therein unnaturally, are well known, let us take the belly as a proof thereof, and let us heal all the changes of condition that take place therein after the manner of skilled physicians. | Fol. 137 a.

The correction of the dry, evil condition of the belly.

Now, if the dry, evil condition of the belly existeth when the members that are of a firm nature and that resemble each other are dry, it cannot be healed. When the wet nature perisheth with these entirely, together with the moisture (or, liquid) whereby the members are nourished, which is confined in all the members of the body, and is dispersed in them like dew, it is impossible to restore to a man [these] members without food,

21

and the healing of ailments of this kind is difficult, and the means of healing can only be applied through the small veins and arteries. It is, then, right to heal dryness of this kind by means of moist food, and by filling each one of the members that resemble one another with its essential moisture, and by contriving for him baths of a moderate temperature, wherein the patient must remain for a considerable time. And after a bath ass's milk mixed with honey must be given to him immediately, and he shall be made whole forthwith. And after this he must be allowed to repose in peace until the time for the second bath, and then he must be rubbed gently and softly (?) And if he retaineth the milk that he hath drunk, the digestion is complete, of which fact he receiveth an indication from his eructations and from the size of his belly. It is meet that between his first and second bath a period of four or five Page 286 hours shall elapse, | provided that thou art prepared to give him three baths, but if not, then the period between the baths may be a little longer. And the patient must be anointed with oil after each bath before he putteth on his clothes. And if he retaineth the milk [after his first bath], let milk be given Fol. 137 b. to him to drink after the second bath also, but if not, | let the physician give him sweet barley water properly boiled, or meal boiled in water like barley. Then let him partake of a meal composed of pure bread from the oven and of rock-fish stewed in white sauce, and, to speak briefly, let the food on which he feedeth be easy to digest and nourishing, and let it not be either thick (or, viscous) or over-rich; let the wine which he drinketh be white, and pure, and insipid (dry?), and sharp. The above is a great cure for over-dryness of constitution. Now, the constitution that lacketh dryness hath no need of all this careful dieting because the patient may eat much or little [as he pleaseth].

Now we may note another kind of dryness, which is due to some cold substance, and which resembleth the preceding. And we must add the following to the things that we have said about the use of milk and honey and wine: Let not the physician give the patient wine that is too insipid (dry?), and let him not give to him foods that over-heat in their natural

state, but let newness of flavour be imparted to them, and let
the physician cause the patient to be anointed continually with
oil of nard or with pure oil.

Now when cold in excess existeth in conjunction with dryness, we consider this state to constitute a serious ailment
which cannot be corrected, and we employ the very same
things, and also pure, unboiled honey, and very old wine, for
this is the principal medicine for one who is in this state. And
we anoint the belly with oil, and we moisten it, | and we lay on Page 287
it a plaster made of hot pitch, which we remove frequently
[and replace by a new one]. And it helpeth the patient greatly
if a well-grown, plump child sleepeth with him and lieth continually close to his epigastric region; but if a little warmth be
mingled with the dryness it is meet in this case to be exceedingly careful in respect of the treatment mentioned first of all,
for honey is quite | useless. Wine that is not old is beneficial Fol. 138a.
for such patients, and foods that are milky and digestible. And
we must anoint the belly of the sick man with oil of roses or
oil of incense, for this cooleth this pain, especially when there
is heat in excess therein.

Let us assume that the evil effect of a hot condition is in
excess in the belly, and that moisture is mingled therewith;
now, an evil condition of this kind can, without fear, be healed
by means of draughts of cold water, and by the use of acid
foods that are not hot. When the moist, evil condition is in
excess, foods that are dry and that neither over-heat nor over-
cool are helpful to the patient, who must at this time abstain
from his ordinary drinks. The moist evil condition wherein
cold is in excess is healed by means of pungent [foods] wherein
are mixed constituents that are acid but not cold. Patients
of this class also may use sparingly wines that are warming
and strengthening. Now the means to be employed in healing
these are different, for very often an evil condition of the
stomach ariseth through some fluid, which is confined in the
stomach, and which hath been sucked out from the coats
thereof. Therefore it is meet to speak concerning these things.
If it happeneth once, it is meet to purify the stomach by
means of vomiting; but if the fluid be from the other members,

it (*i. e.*, the stomach) hath need of the symptoms with which Page 288 healing is associated. Now, it is meet | for us to apply the means of healing to the member that giveth pain, and to take the greatest care that the members that receive the fluid when it dischargeth itself suddenly are not injured. Injury of this kind taketh place through the things that are sharp (*i. e.*, acid), and through those things that are administered to the body in connection with the condition that existeth in the coats of the stomach. For the stomach hath need of the medicines that purge gently, as for example, something that hath aloes in it, Fol. 138*b*. and of the medicine called "pîkrâ" (ἱερα πίκρα, or bitters (?), | which is made of aloes. When the stomach is filled with viscous phlegm, it hath need of the medicines that cut this, and then it becometh cleansed; but, if it be convenient, let the patient eat radishes and honey, and vomit. If the chyme that is confined in the stomach be neither viscous nor thick, let the patient imbibe sweet barley water, and vomit, or honey water, or a decoction of absinthe with honey. And, similarly, we must find [other] exits by means of which the liquid that is imprisoned behind them may be disposed of. If we cannot find the mouth or some [other] exit that is suitable, it is meet to expel that excessively evil chyme in the same way as windy flatulence, when it is confined [in the body], is expelled, until the whole body is healed.

Let us once more briefly summarize the things that have been said, in order that we may know how to cure each of the evil conditions that have been described. It is meet to cool the hot condition, and to warm the cold condition; similarly it is right to dry that which is wet, and to wet that which is dry. If the condition be compound in nature, it is meet to mingle the two diseases; that which is both wet and cold we must warm and dry, and in the same manner we must moisten and cool that which is hot and dry. That which is dry and cold we must moisten and warm, and that which is hot and wet we must cool and dry. It is right for us to know that Page 289 the condition which is by far the most serious of all | the simple conditions is the dry one, and of the compound conditions that which is dry and at the same time hot. These,

then, are the eight evil conditions, and this is the art of healing them briefly described. If they arise because of diseases
of the stomach, or of one of the other members, which induce
an evil condition in the whole body, then it is meet to come
to the healing of the other diseases of the stomach that are
followed by | other attacks, as, for example, bestial (or, dog- Fol. 139*a*.
like) lust for food, and hiccoughs, and nausea (or, vomiting),
and eructations, and the other diseases, the symptoms of which
have been described above.

The healing (or, cure) of the bestial lust for food.

Now to those who have that lust for food which is called
"bestial" (*i. e.*, dog-like), provided that they hunger at all times
unceasingly, it is meet to give to drink as the chief medicine
wine that hath been heated sufficiently; a wine suitable is that
which is deep red or purple in colour, and which is not acid,
for it is right that wine of this kind should cure the bestial
lust for food which, as hath been said above, ariseth through
the excessive coldness of the whole stomach. And it is meet
when patients sit down to eat that first of all should be given
them to eat a larger portion of the food which is fat and
greasy, than of the other foods which form the meal. And
to speak briefly, all their food must contain a very large
proportion of oil, and it must have in it nothing sour and acid,
and after this they must drink wine, even though they are not
as yet thirsty; for their hunger (or, craving) will be diminished
by the use of such a diet as this, and after the lapse of much
time it will be entirely abated. And they must eat flesh of
every kind that is fat and nutritious, and every kind of vegetable that is pungent and moist, and all kinds of seeds that
possess the same characteristics. In the case of that lust for
food which is due to evaporation of the whole body, | it is Page 290
meet to fine down the body from without, by means of the
plasters (or, medicaments) that restrain it and cool it, and by
bathings in cold water, provided there be nothing to prevent
this, and by foods of a delicate character, which are its opposites. Let them eat everything that is fatty and difficult to

digest internally, and [let them apply] everything that restraineth the evaporation externally.

The cure of hiccoughs.

Fol. 139b.   Now, to those who have hiccoughs, | which are due to full-ness (*i. e.*, excess) of cold chymes, it is meet to give seed of rue [mixed] with wine, or natron [mixed] with honey, or that medicine which is called "Dîspôlîṭôs" (*i. e.*, a preparation of cummin), or a medicine containing three peppercorns, or the seed of petroselinum (rock parsley), and cummin, and aniseed, or vinegar of squills, or calamint, or nard, or *esrôn* (asarum). These may be administered separately, or mixed together, after they have been crushed, and pounded, and worked up well in honey. And castoreum given to the patient in vinegar and water is good for the hiccoughs, or a dose of *ṭlônàyà* (betel, or areca nut?) about the size of a chick-pea given in wine, and this medicine is good also for the hiccoughs that arise through windy flatulence. And strong purges, which clear out the chyme that vexeth them, are also very good for those who suffer from hiccoughs. In many cases sneezings will put an end to the hiccoughs that are due to the over-fullness of the stomach, and these may be brought about by setting the nostrils in motion by means of some medicine that provoketh sneezing.

The cure of the turning of the stomach (*i. e.*, nausea).

Now, the turning of the stomach is due sometimes to the weight of the chyme, which causeth discomfort, and sometimes to the bitings of its acidity. Some people are moved to vomit, but they vomit nothing, for the chyme, which causeth the dis-comfort, is confined in the coats of the stomach, but in others it is fixed [in the stomach] like water in a sponge, and in others it clingeth tightly, and is very hard to clear out from the stomach. The cure of nausea may be effected in three ways: [1], by fasting, and toil, and sleep, whereby digestion is pro-duced; [2], by the emptying of the fatty membranes of the stomach by means of vomiting caused by honey water only; Page 291 [3], by means of radishes and honey, administered either after or before food, or by means of some medicine that exciteth vomiting (*i.e.*, by an emetic). If the chyme clingeth tightly to the coats of the stomach, and it possesseth a bilious character, the medicine that is made of aloes and is called "pîkrâ" (ἱερα πίκρα) will help patients who are suffering in this manner, but [medicines] that are acid are injurious. And if a cold | liquid saturateth the coats of the stomach, in such Fol. 140*a*. a case the hot medicines that have been prescribed for hiccoughs will help the patient. Now, it hath been stated above many times that a sensation of heat and smokiness is associated with bilious stomachs, and acidity with the cold stomach, and for this reason it hath been said by Hippokrates that thou shalt not give honey to bilious subjects, or sweet barley water to those whose food containeth things of an acid (or, sour) nature.

The cure of abscesses (or, ulcers) which attack the stomach.

If there be an abscess (or, ulcer) in the stomach, and it be hot (or, inflamed), and fever, and thirst, and if a sensation of great pain be associated therewith, these, in addition to the sensation in the abscess itself, give a sure indication [that the abscess existeth]. Now, for an abscess in the stomach which doth not make its presence known either by nausea or by the "coming of the belly", it is useful to administer, during the early stages of its growth and in the later stages of its development, cooling medicines and foods that cool (*i.e.*, refresh) and that soften the belly, and ointments, and plasters made of medicines that are slightly acid, and that purge, and those that soften slightly, in such a way that whilst we employ a complete course of softening treatment, we may not destroy the natural power of the stomach. Now, the liver and the stomach, because they are the organs of digestion and of nourishment, require to have applied to them the medicines that are purgatives, and are bitter and aromatic, when it happeneth that they become thick. | Such are, for example, Page 292

absinthe and oil of nard, or asparagus soaked in wine and
mixed with chalk; and among plasters (or, ointments) may be
mentioned the plaster made of *ânantîs*, and aloes, and wax,
and oil of nard.   Sometimes, according as the occasion
demandeth, we mix therewith drugs that purge and are extremely acid in nature, and sometimes those that soften,
Fol. 140*b*. according to the weakness, or strength, of the stomach. | Now,
if the abscess lasteth for a long time, and becometh hard, it
is more proper to apply to it those medicines that are aromatic
and that will soften [it], as, for example, some preparation that
containeth "king's crown" (μελίλωτον, *melilotus officinalis*).

He whose stomach burneth, or hath nausea, or loatheth [its]
food through the hot and bad condition thereof, or from any
cause whatsoever except fever, is helped by cold water mixed
with the juice of meat (extract of meat), or by extract of
asparagus, or by [the extract of] the seeds of cucumbers mixed
with cold water, or by fresh branches of wood mint, which
are to be crushed and laid upon the stomach, or by a vessel
(or, bag) filled with cold water [laid upon the stomach], or by
snow, or by the shell of a fresh gourd being filled with water
and placed upon him.   And the following also help those
whose stomachs burn and who thirst greatly; tablets (or, pills)
made of fresh roses, and those made of cucumber seed, and
those made of gum tragacanth, and wine of roses, and oxymel,
when it is mixed with aloes.   As concerning the stomach that
produceth black bile and is flatulent (or, inflated), place on its
outside, especially when the pain is most intense, sponges
dipped in strong vinegar.   And afterwards, if the pain continueth, take wet dross from the furnace and burnt copper
(*i. e.*, oxide of copper), and crush them, and work them into
a paste with honey, and lay the mixture upon it.   Or give the
patient a draught made with an infusion of calamint, mixed
with a little honey, and one drachm of peppercorns.   For the
Page 293 stomach that is loose | and weak, and that vomiteth from the
mouth, or that letteth down the food below because of its
weakness, are good the ordinary tincture or preparation of
asparagus, and the juice of apples, and tight bandages applied
externally, and foods that are acid and purgative.   As concerning

the stomach that produceth much wind and flatulence, thou must know that the [cause] thereof is lack of warmth, for much warmth, or much cold, doth not produce wind. Cold, because it is not able to rub down small | and to dissolve [the food], Fol. 141a. is incapable of producing wind, and great heat, because it rubbeth down small, and relieveth [the stomach] setteth free wind. But the heat that is feeble is capable of producing wind because it is able to produce partial digestion, but it is unable to rub [the food] down small and to relieve [the stomach]. Wherefore it is right to rub the food down small by means of warming medicines infused in oil, as, for example, some preparation of cummin, and the seed of rock parsley, and mint (or, aniseed, *ammi copticum*), and parsnip, and rue, and laurel berries, and seed of dill, and garden nigella. Infuse these in oil and anoint the patient therewith, or mix some of it with the plasters (or, ointments), which are laid on the body externally. Or pour some of it into a purge, with a little castoreum, and give the patients draughts of an infusion of *teucrium polium* (*i. e.*, the germander plant, or spleenwort), or calamint, mixed with a little honey. Now, for diseases [of this kind] anointings with oil are good, and warming unguents prepared from millet and the seed of flax, and large cupping glasses which draw up the navel and draw suddenly by means of fire (?). The following are helpful in cases of flatulence of the stomach and gripings of the belly; aristolochia, both kinds, the round and the long, and castoreum, when macerated in vinegar. Now, these are to be applied externally, together with the other medicines that are good for these diseases. This is the [simple] system of cure | of the diseases of the stomach, wherefore it Page 294 is [now] necessary for us to write down the compound medicines that are good for the diseases thereof.

A medicine prepared from wild mint, which is good for nausea, and for looseness of the bowels, and for burning of the stomach.

Juice of sweet and bitter pomegranates, pounded with the inside of their rinds	2 *ķestê*
Juice of wild mint	1 *ķestâ*
Honey, with the scum removed	1 „

Boil together until the mixture acquireth a tolerably thick state of consistency, and administer a spoonful with cold water or with extract of barley.

A medicine made of fruits, which is good for turning of Fol. 141b. **the stomach, | and for diarrhoea and burning (heartburn?).**

Bitter pomegranates	10 in number
Asparagus	10 ,,
Gûîkhsaḳ	50 ,,
Shîkê	50 ,,
Sumach	1 ḳesṭâ
Water	1 ,,
Kûmathrê	50 in number
Myrtle berries	5 ḳesṭê

Remove the outer rind of the pomegranates, and pound the fruits, and then pour them all into a stone pipkin with water, and boil them up together thoroughly, and then squeeze out the solid matter and throw the refuse away. Strain the liquor, then add to it one lîṭrâ of honey, and boil it until it acquireth the consistency of honey. Administer in doses of one spoonful, with cold water.

And in like manner the juice of asparagus, or of apples, or of pomegranates, or of myrtle berries, will help patients, when it hath been boiled in the manner that hath been described.

A medicine made from a preparation of asparagus, which is good for weakness of the stomach, and for looseness of the belly, and for thirst.

Extract of asparagus	2 dûlḳê
Honey	1 dûlḳâ

Boil these together until they are reduced to one half, take the mixture off the fire, and pour into it two drachms of amomum (shushmîr), and twenty drachms of round peppercorns, and twenty drachms of long peppercorns; pound, and mix together, and use. Add also to this ten drachms of ginger, and administer as a dose one spoonful.

A medicine made of a preparation of asparagus, which is good for weakness of the stomach, and for cold (or, chill).

Extract of asparagus, boiled and well strained	4 ḳesṭê
Gum (?)	1 ḳesṭâ

Mix together, and boil until the mixture becometh thick; then pound

Caryophyllus aromaticus	2 ounces
Peppercorns	3 „
Cinnamon	1 ounce
Ginger	1 „

and pour them into the liquid, and let it simmer for a little time. Then set aside and use [when necessary].

Another medicine, made of wild mint, which is good for Page 295 **excessive looseness of the stomach (bowels), and for nausea, and for cramp (?).**

Juice of sweet pomegranates	20 ḳesṭê
Juice of bitter pomegranates	20 „
Juice of wild mint	2¹/₂ „

Mix these together and boil them until they are reduced to one half, add one ḳesṭâ of honey, and boil the mixture until it becometh thick. Then pound one liṭrâ of aniseed and one liṭrâ of stacte, | and pour into the mixture, stir all up together, Fol. 142 a. and set aside, and use [when necessary].

Another medicine which is made of pomegranates, and which is good for cramp (?) and for violent nausea. Take ten dûlḳê of sweet pomegranates, boil them, and strain the juice. Pour into this three dûlḳê of honey, and boil the mixture until it is thick. Crush together three ounces each of peppercorns, and ginger, and bark, and spikenard (?), and two drachms each of long peppercorns, and cardamoms, and cinnamon, and cassia, and mix with the honey and pomegranate juice, and having taken the mixture off the fire, stir it up well and pour into a vessel, and administer in doses of one spoonful.

Tablets made of a preparation of cucumber seed, which are good for burning of the stomach (heartburn?), and for thirst, and for a burning fever.

Seed of garden cucumbers	8 drachms
Seed of purslane	8 „
Licorice root	8 „
Tragacanth	8 „

Crush and work up into a paste with the whites of fresh eggs, and stir up well and make into tablets, each containing [one] drachm. Administer in cases of acute sickness with extract of barley, and make this paste also into pills, and let the patient keep them under the tongue.

Pills, which are made of roses and spikenard, and which are good for the diseases.

Fresh roses	20 drachms
Licorice root	4 "
Spikenard	4 "

Crush and work up into a paste with new wine, and make Page 296 into | tablets and pills, and use in the same way as the preceding medicine.

Tablets made of a preparation of the bamboo, which are good for a burning fever, and for heat in the stomach, and for the heat that appeareth in babes when they suck.

Bamboo	8 drachms
Fresh roses	8 "
Seed of garden cucumbers	20 "
Agrimony	3 "

Pound and work into a paste with the juice of fresh coriander seed, and make into tablets, each containing half a drachm, and administer in some convenient draught.

Others, however, take three drachms of fresh roses, and one drachm of bamboo, and mix them together and make into tablets, which they administer to patients suffering from these diseases with beneficial effects.

Tablets made of roses, which are good for fever and burning.

Fol. 142 b.

Fresh roses	6 drachms
Spikenard	2 "
Crocus	2 "
Licorice root	3 "
Cucumber seed	3 "
Liquid honey (manna, *i. e.*, طراجبين, or ترنجبين, Pers. ترنگبين)	3 "
Tragacanth	1 drachm
Gum Arabic	1 "

Pound, and work up into a paste with new wine, and make into tablets, each containing one drachm, and also into pills. Administer in cases of acute sickness with extract of barley, and in cases of fever with oxymel, and in cases of burning heat with tincture of chicory, or tincture of fennel.

Amazon tablets, which are good for the natural turning of the stomach that ariseth from stoppage (?) of the bowels; and they are beneficial in cases of disease of the sun (sun-stroke), and for every general sickness.

Seed of petroselinum (rock parsley)	6 drachms
Aniseed	6 „
Cinnamon	6 „
Absinthe	4 „
Myrrh	2 „
Peppercorns	2 „
Opium	2 „
Castoreum	2 „

Crush and work up into a paste with water, and make into tablets, each containing a drachm, and administer as a dose one in some draught that is suitable for the disease.

Tablets made of a preparation of costus, which are good for pain in the stomach caused by burning, and for nausea caused by stoppage (or, twisting of the bowels), and for violent hiccoughs, and for the thin watery saliva that is produced in the stomach.

Costus	4 drachms
Spikenard	4 „
Cassia	4 „
Fresh roses		4 „
Stacte	4 „
Crocus	4 „
Asarum	2 „
Aloes	2 „
Opium	1 drachm

Page 297

Work up into a paste with extract of psyllium (plantain), and make into tablets, each containing one drachm, and administer [one] in some draught that is convenient and is good for the disease.

Tablets of the Invincible Star; they have been well-tried, and they cause no pain, and produce sleep. They relieve sufferings, and they cure in a most marvellous manner pain of the stomach, and eructations that leave a bitter (or, sour) taste, and difficulty, and colic, and gripings, and flatulence; and pains in the head of long standing, provided that they are macerated [in water] and rubbed on the forehead,—now when they are macerated in vinegar they are very beneficial—; and pain of the ears (earache), when dissolved in tincture of origanum majorana, or tincture of origanum pulegium; and [tooth-ache] when they are placed on decayed teeth, having been worked up into a paste with galbanum (χαλβάνη). They are also good for a flow of blood, and for colds and coughs of long Fol. 143a. standing, and for dysentery, and for intermittent | fevers, and for attacks of deadly diseases.

Castoreum	4	drachms
Myrrh	4	„
Spikenard	4	„
Cassia	4	„
"Sea-seals"	4	„
Rind of the bark of the mandragora	4	„
Costus	6	„
Crocus	6	„
Opium	6	„
Seed of poppies	6	„
Aniseed	8	„
Seed of rock parsley	8	„
Dôḳôn	8	„
Meadow saxifrage (tordylium officinale)	8	„
Seed of white hyoscyamus	8	„
Esṭumkâ (Stomachic?)	8	„

Crush and rub down the dry substances to a powder, and dissolve those that are dissolvable in wine with a pungent smell; add the powder to the liquid and stir up well together, and make into tablets, each containing half a drachm, and administer in some convenient draught.

Tablets compounded of aromatic herbs, which are good for violent turning of the stomach, and for the bile that clingeth to the coats of the stomach, and for all the diseases for which the drug called "Pîḳrâ" (ἱερα πίκρα) is prescribed, and which are described [in the Chapter on] headaches.

Cinnamon	3 ounces
Reeds of the incense plant	3 ,,
Cinnamon bark (?)	3 ,,
Balsam wood	3 ,,
Flowers of pistacia lentiscus	3 ,,
Incense bark |	3 ,,

Page 298

Pound and crush exceedingly fine, and pour them into a new earthenware vessel, and pour over them six measures (*ḳesṭê*) of water, and boil until only three measures remain. Then take one *lîṭrâ* of the finest aloes, and pour as much rain water as is necessary over it, and crush it up at noon, when it is very hot, until the mixture is dried. Wash the powder until it is clean, and then leave it so that the liquor may become clarified, and pour it off. Dry the aloes, and when they are dried pour the infusion of aromatic herbs over them, and crush them in the sun until they are dry. Then crush and mix with them three ounces each of myrrh, crocus, and stacte (now some add only one ounce of each), then crush them up all together and put into a vessel, and use when dry. The dose is one or two drachms in some draught that is convenient for the kind of the disease. Others, however, make the mixture into tablets, and use \| according to what is written. For this Fol. 143 *b*. medicine cutteth through, and cleareth away, and removeth stoppages, and cleanseth the coats of the belly from the evil chyme. Especially good for such diseases is the drug *pîḳrâ*, which hath been described in the Chapter on the head. Now for a burning pain in the stomach, which is accompanied by fever, oxymel prepared in the following manner is extremely beneficial.

Oxymel, which is good for a burning pain in the stomach, and for a burning fever.

Root of rock parsley	1 *lîṭrâ*
Bark of the root of anethum foeniculum (dill)	1 ,,

Seed of rock parsley	5	*estîrê*
Seed of dill	5	„
Root of pistacia lentiscus	5	„
Vinegar	8	*dûlķê*
Honey, with the scum removed	2	„

Steep the roots and the seeds in vinegar for one day and one night, then boil the vinegar down to one half; squeeze out the roots and the seeds and throw them away, mix the honey with the vinegar, and boil the mixture until it acquireth a stiff consistency. Administer in doses of one spoonful with cold water, Page 299 or with extract of barley. Now, some | mix with it a little crocus, in order to improve the colour and to relieve the patient.

Oxymel prepared with squills, which is good for the stomach, and for shortness of breath, and for pain in the sides, and for those who do not digest their food, and for eructations that leave behind a bitter (or, sour) taste, and for pain in the liver and spleen.

Inside of squills	2	*lîṭrê*
Ginger	1	ounce
Seed of dill	1	„
Aniseed	1	„
Rock parsley	1	„
Silphium (ferula asa foetida)	1	„
Pyrethrum	1	„
Costus	1	„
Dried hyssop	1	„
Origanum pulegium	1	„
Peppercorns	2	ounces
Seed of rock parsley	2	„
Cummin	2	„
Dâķôn	1/2	ounce
Ammi copticum	1/2	„
Ligusticum	1/2	„
Laurus malabathrum	1/2	„
Amomum (grain of Paradise)	1/2	„
Spikenard	1/2	„
Cardamoms	4	drachms
Fresh rue	4	„

Wild mint	1 bundle
Strong white vinegar	6 *ķesţê*

Pound and throw them all into vinegar, and leave them there for seven days; then boil for a short time, and strain and squeeze out [the roots], and let the patient drink it | fasting. Fol. 144 a. After food give a dose of one spoonful in cold water.

Oxymel catharticum, which is good for fever, and for thirst, and for pain in the stomach.

Licorice root	4	*lîţrê*
Warrior's hair	4	„
Root of dill	4	„
Dried hyssop	4	„
Absinthium	4	„
Root of rock parsley	4	„
Calamint (?) of the river	9	ounces
Agaricum (ἀγαρικόν) fungus (?)	1	*lîţrâ*
Squills	1	„
Vinegar	5	*ķesţê*
Honey	5	„

Crush to a fine powder, and boil in vinegar, and clarify, and strain; and mix the vinegar with the honey, and boil until the mixture becometh thick, then throw in four ounces of scamonia (convolvulus) and nine drachms of euphorbium, and mix with them the kind of agaricum (fungus) that is written down among the medicines. And administer as a dose one spoonful in some convenient draught.

Rose water, which is good for fever, and for excessive thirst, and for a burning pain in the stomach. Take two *lîţrê* of fresh rose leaves, with the thick stems removed, | and Page 300 throw them into a well-made metal vessel, and pour on them rain water, or water from some pure fountain, [and set the vessel on the fire to boil]. After the water hath become greatly reduced over the fire, [add] ten measures (*ķesţê*) of water whilst the leaves are still boiling. Cover the vessel carefully, and let it stand for one day and one night, then strain out the rose leaves, and squeeze them dry, and throw them away. Then pour into the water two measures of honey from which the scum hath been removed, and boil it over the coals until

22

it is reduced to one half, and then pour into a glass vessel, and set it in the sun for forty days. Administer as a dose one spoonful in cold water, or with extract of barley.

Rose wine, which is good for fever and thirst, and for pain in the stomach.

Extract of roses	5 measures
Old wine, with a strong odour,	3 „
Honey, with the gum removed	4 „
Seed of rock parsley	2 drachms
Cummin	2 „
Crocus	6 „
Spikenard	6 „
Incense bark	6 „
King's crown (melilotus)	6 „
Laurus malabathrum	5 „
Stacte	5 „
Dried roses	17 (?) „
Cummin	4 „

Fol. 144 *b.*

Pound, wash, and mix well together when wet, then put in a glass vessel, and hang it up in the sun for forty days. And wash it every other day, and after six months administer as a dose two drachms to those who suffer from ulceration of the bowels, and from fever, and from thirst that ariseth from a burning pain in the stomach.

Another kind of rose wine, which is good for the disease.

Fresh red roses	5 *lîtrê*
Honey, with the scum removed	10 „

Pound the roses and mix them with the honey, put the mixture in a glass or metal vessel, and set it in the sun for forty days, and shake the contents thereof once each day. Administer as a dose a portion as large as a nut, in extract of barley or in cold water.

Page 301 **Apḵûmlî, that is extract of gûrḵâ,** | which is good for fever, and for thirst, and for nausea, and for cramp (?) in the stomach. Boil ten measures of the juice of the flesh of grapes until they are reduced to six measures, then add to it one measure of honey and let it boil on the coals gently until the mixture is reduced to four measures. Take it off the fire and pound two drachms of cardamoms and two drachms of crocus, and add to it, then mix all these well together, and administer as a dose one spoonful.

Another common [medicine]. Mix five parts of meat juice, which hath been boiled and clarified three times, with one part of honey from which the scum hath been removed, and boil them together until they acquire consistency, then put the mixture in a vessel and set it in the sun for forty days, and administer in doses as above.

Common extract of asparagus which is good for fever, and turning of the stomach, and for thirst and for looseness (or, cramp?). Mix ten measures (*dûlķê*) of asparagus juice, which hath been boiled and clarified, with one measure of vinegar, and one measure of honey, and boil until they acquire consistency, and administer as a dose one spoonful in cold water.

| **Extract of myrtle berries, which is good for debility of** Fol. 145 a. **the stomach, and for looseness of the bowels, and for nausea, and for acute sicknesses.** Mix ten drachms of extract of myrtle berries, which hath been boiled, and from which the scum hath been skimmed, with one measure (*dûlķâ*) of honey, then boil until the mixture becometh thick, and use.

Wine of lilies, which is good for irregular throbbing of the arteries, and for great debility of the stomach, and for chill on the liver, or heart, and for those who faint from emptying (*i. e.,* the loss of) **blood, or for the turning of the stomach and sudden pains therein. When drunk or when used in anointing the body, or when used as a liniment, it produceth great benefit for the patient. When mixed with extract of . . . . . it is used in fomentations, which are good for the liver, and the spleen, and the stomach.**

Costus	2	ounces
Laurus malabathrum	2	,,
Caryophyllus aromaticus	2	,,
Reed of incense	2	,,
	Cassia	3	,,
Bark	3	,,
Salt of Cappadocia	3	,,
Balsam wood	4	,,

Page 302

22*

Amomum (grain of Paradise)	1 ounce
Fennel	1 „
Crocus	4 ounces
Musk	2 drachms
Oil of balsam	1 ounce
Styrax, liquid,	3 ounces
Lilies	400 [in number]
Mountain wine, which hath not fermented in the sun }	4 measures.

Remove the thick stalks of the lilies, wipe their heads with a piece of thin, clean linen, lay them out on a clean cloth in the shade, spread a clean cloth over them, and let them remain there for one day and one night. Pound the dry drugs very small indeed, pass this powder through a sieve with wide meshes, and work up all the [other] medicine with styrax and oil of balsam. Bring a glass vessel with a mouth sufficiently large to admit the hand, and arrange in it the drugs and the lilies alternately, and cover the mouth of the vessel, and leave it for one night and one day. On the second day dissolve
Fol. 145b. the musk and crocus | in wine, and pour it over the lilies and the drugs and mix them well together, and smear the mouth of the vessel, and the whole rim of it, with mud which hath been 'made into a paste with strong-smelling wine, and then set the vessel in the shade, in a place where there is a moderately good supply of air, and which faceth the north. After six months use, and administer in doses of one spoonful in some convenient draught, in cases of sicknesses of every kind which have been described above.

Myrtle-berry wine, which is good for great debility of the stomach, and for looseness of the bowels, and for the turning of the stomach.

Wine, sour and black	4 *kestê*
Crushed myrtle berries	2 „

Put the berries into a glass vessel, pour the wine over them, and let them remain in it seven days. Then strain out the berries, and squeeze them dry, and throw them away, and pour the wine into a vessel and keep. Administer in doses of one spoonful in some useful draught.

Wine of Demaḳrâṭis, | which is good for weakness of Page 303
the stomach, and for an uncomfortable feeling therein.

Fine irises	9 drachms
Seed of rock parsley	2 ,,
Cinnamon	2 ,,
Bark of cinnamon	4 ,,
Laurus malabathrum	3 ,,
Myrrh	3 ,,
Twigs of *apsentîn*	3 ,,

Pound all these to a very fine powder, pour six measures of
strong-smelling white wine over it, and put the mixture into a
glass vessel and seal it with lime plaster (gypsum). After a month
of days administer one cup of it before and after food. Now,
it is said that Demaḳrâṭis (Democrates) kept this wine fit for use
all the days of his life.

Wine of absinthe, which is good for sluggishness and
weakness of the stomach, and for the evil condition thereof, and for the liver and spleen that have become hard,
and it softeneth the belly.

Strong-smelling sweet wine	8 *ḳesṭê*
Honey, with the scum removed	2 ,,
Laurus malabathrum	2 drachms
Fennel	2 ,,
Dried roses	2 ,,
Aloes	2 ,,
Agaricum	2 ,,
Pistacia lentiscus	2 ,,
Costus	4 ,,
Stacte	4 ,,
Twigs of Greek absinthe	7 ,,
| Crocus	1 drachm

Fol. 146a.

Pound the drugs to a very fine powder, tie up in a clean linen
rag, steep in wine, and let it stand for seven days, but each
day squeeze the bundle in the wine once, the wine being placed
in the sun. Then take, and administer as a dose one measure
before and after food.

Wine of squills, which is good for an evil condition of

the liver and stomach, and for those who collect water
(*i. e.*, the dropsical).

Squills	2 ounces
Bark of pyrethrum	4 „
Seed of nettles	2 „
Rock parsley	1 ounce
Laurus malabathrum	4 ounces
Fennel	4 „
White peppercorns	4 „
Old, strong-smelling wine	3 *dûlḳê*
Honey	¹/₂ *ḳestâ*

Mix well together, and boil, and administer as a dose one
measure in hot water.

Page 304   **Wine of roses,** | **which is good for fever, and for burning
in the stomach, and is exceedingly beneficial.** Pour as much
honey, from which the scum hath been removed, as thou
wishest, into a vessel, add to it as many red roses, partly
crushed and with their roots removed, as the honey will cover,
mix them together well, and place in the sun for forty days.
The wine that thou makest this year shall be for use next year.
When necessary bring ten flasks of black wine, and put into
it eight *liṭrê* of the roses and honey which thou hast prepared,
and set it aside in the shade for seven days, and shake the
vessel with thine hand each day twice.   On the seventh day,
open the vessel carefully, and squeeze the roses dry and throw
them away.   Then strain the liquid through a piece of clean
linen, and pour it into a vessel, and administer one measure
as a dose.

   **Wine of aṭrûghê fruit (quinces?), which is good for cold
in the stomach and debility.**   Take the leaves of fifty aṭrûghê,
and wipe the dust (bloom?) off them carefully with linen rags,
Fol. 146 b. and pour six flasks of strong-smelling | wine over them, and
let them steep therein for seven days.   Then strain out the
fruit, and squeeze it dry, and throw it away.   Then bring one
measure of honey, of which the scum hath been removed, stir
it up well with thy hands until it becometh white (frothy?), pour
on it good wine, stir them up together, and boil them very carefully.   Pour it into a vessel, and administer one measure as a dose.

Now the above are all medicines that are prescribed as suitable for hot diseases of the stomach, and for those that arise both with and without fever. It is meet that we should now compile a list of the medicines that are good for cold diseases of the stomach, and for those that turn the foods sour in the stomach.

The medicine called "Dispôliṭis" (*i. e.*, "Heliopolitan"), which is good for excessive cold of the stomach, and for eructations which leave a bitter taste, | and for the lust of Page 305 the dog, and for hiccoughs due to over-fullness, and for protracted fevers. And it hath been handed down by tradition from the priests and old men (or, sages).

Cummin steeped in vinegar and roasted	16	*estîrê*
Round peppercorns	20	drachms
Ginger	20	„
Rue	20	„
Natron	20	„

Pound, and reduce to a powder, and work up into a paste with honey from which the scum hath been removed, as much as sufficeth. Administer as a dose a portion as large as a chestnut with hot water, or with wine mixed with water. Now, some add to this two drachms each of cinnamon bark, cinnamon, incense bark, myrtle berries, fennel, and stacte, and it then becometh a most excellent medicine.

A medicine which is made of the three kinds of peppercorns, and which is good for the diseases.

Peppercorns, round	3	ounces
Peppercorns, long	3	„
Peppercorns, white	3	„
Wood of balsam	1	ounce
Spikenard	4	drachms
Amomum (grain of Paradise)	4	„
Ginger	1	drachm
Seed of rock parsley	1	„
Seseli officinale	1	„
Bark	1	„
Asarum	1	„

Resin   1 drachm

Honey as much as sufficeth.

The dose and manner of use as before.

A medicine composed of seeds, which is good for the diseases.

Bark	
Amomum (grain of Paradise)	
Spikenard	
Ammi copticum	
Seed of dill	
Seed of rock parsley	
Aniseed	all of them in equal
Lacerpitium (silphium?)	quantities
Castoreum	
Aristolochia (rhubarb?), long	
Stacte	
Asarum	
Seed of fennel	
Cummin		

Fol. 147 a.

Honey as much as sufficeth.

Dose: two drachms in wine mixed with water, or with honey water. This medicine is also good for the cold pains that take place in the liver.

A medicine made of asparagus itself, which is good for cold in the stomach, and debility thereof, and for sudden diarrhoea (?) and for the rapid turning of the stomach which is due to indigestion.

Fine asparagus, peeled inside and outside	2	manahs
Honey, with the scum removed	1	manah
Peppercorns, long	5	drachms
Peppercorns, round	5	,,
Amomum (grain of Paradise)	4	,,
Ginger	12	,,
Cardamoms	4	,,
Caryophyllus aromaticus	4	,,
Cinnamon	2	,,
Crocus	1	drachm

Boil the asparagus in new wine (living wine), and crush it and Page 306 pound it well, and then throw it into a frying-pan in order that all the moisture therein may be dried out of it. Then pour on it honey and boil it, shaking the pan frequently, let it boil until it becometh a round mass and doth not stick to the pan, then take the pan off the fire. Pound the medicines, and reduce them to a powder, and stir up the asparagus with the medicines, and leave it for five days, or for seven, until it is dried. Then cut it up into cakes, each weighing a *mithkâl.* Dose, to be eaten four drachms, and let him drink after it wine mixed with water.

A medicine made of myrtle berries, which is good for sudden looseness of the bowels, and for nausea due to cold in the stomach, and indigestion.

Fine myrtle berries, dried	1	manah
Myrobalanus chebula, black	4	manahs
Bĕlilkê	4	,,
Amlagh (depilatory)	4	,,
Castoreum	4	,,
Costus	12	,,
Root of the thorny caper	12	,,
Copper (oxide of?)	12	,,
Loaf sugar	12	,,
Laurel berries	12	,,
Cypress berries	8	drachms

Pound well the caper root by itself, and pound and reduce to a powder the dry medicines. Boil oil and honey in equal quantities, and pour upon the medicines as much as sufficeth, and stir them up well together. After six months administer in doses of two drachms, in an infusion of rock parsley and dill. Let the man who taketh this medicine keep himself from much wine, and from | fatigue, and from anger, and from Fol. 147*b.* copulation, and let him live on thin soup.

A medicine made of Bêldôr (root of the thorny caper?), which hath been handed down from Solomon. It is good for protracted pains in the stomach, and it maketh the face to shine and brighteneth the intellect. Behold, there is a way of preparing it [mentioned] above, at the side, in addition to the preparation with myrtle berries. The preparation with myrtle berries is as follows:

Fine myrtle berries, dried	1	manah
Myrobalanus chebula, black	20	drachms
Bĕlîlḳê	20	,,
Amlagh (depilatory)	20	,,
Ṭalîspar[1]	20	,,
Caryophyllus aromaticus		12	,,
Bark	12	,,
Reed of incense	12	,,
Amomum (grain of Paradise)	12	,,
Incense berries	12	,,
Peppercorns, round	10	,,
Peppercorns, long	10	,,
Ginger	10	,,
Stacte	6	,,
Cardamoms	6	,,
Cummin (karwâyâ)	6	,,
Aniseed	6	,,
Cummin	6	,,
Spikenard	6	,,
Bark	6	,,
Cardamoms (ḳâḳôlâgh)	6	,,
Costus	6	,,
Nuts of incense	5	,,
Seed of rock parsley	5	,,
Ammi Copticum	5	,,
Laurus malabathrum	4	,,
Amomum	4	,,

Page 307

Pound all these and reduce to a powder, and steep the myrtle berries in mountain wine for one night and one day, then pound them well, and mix with them all the medicines, and pour on them honey, from which the scum hath been removed, as much as sufficeth.

A medicine made of Ṭalîspar, which is good for great cold of the stomach, and for those whose food turneth

---

[1] Meaning unknown.

sour [therein] and doth not digest, and for flatulence in
the stomach and liver.

Ṭalîspar	3	drachms
Bark	3	„
Amomum (grain of Paradise)	3	„
Peppercorns, long	8	„
Ginger	12	„
Cummin	1	drachm
Seed of rock parsley	1	„
Seed of dill	1	„
Ammi Copticum	1	„
Cinnamon	2	drachms
Sugar	11	„ ¹

Pound, mix together, and administer in doses of two drachms
in wine mixed with water.

A medicine made of amomum (grain of Paradise), which
is good for the diseases.

Amomum (grain of Paradise)	4	drachms
Bark (of cinnamon?)	2	„
Nagbûsht (نابوشك)	3	„
Peppercorns, long	4	„
Peppercorns, round	4	„
Ginger	6	„
Caryophyllus aromaticus	6	„
Cinnamon	6	„ ²

Dose, one drachm.

Gasrashân, which is called "Hûzâyâ", which is good
for indigestion of the stomach, and is very helpful.

Costus	10	drachms
Incense berries	10	„
Spikenard		10	„
Bark of incense	10	„
Bark (of cinnamon?)	10	„
Nuts of incense	4	„
Cardamoms	4	„

Fol. 148 a•

---

¹ Here follow the figures 50, 700.
² Here follow the figures 30, 70.

Page 308

Caryophyllus aromaticus	4	drachms
Aniseed	4	„
King's crown (melilotus)	4	„
Ṣiṭragh (lepidium latifolium)	4	„
Oil of cypress |	10	*estîrê*
Peppercorns, round	5	„
Peppercorns, long	5	„
Ginger	10	„
Bĕlîlḵê	6	drachms
Myrobalanus chebula, black	10	*estîrê*
Myrtle berries	1	*lîṭrâ*

Pound and reduce to a powder, and make up into a paste with honey; dose, two drachms in hot water.

Paddîḵôn[1] (بنداديقون), which is good for the liver and for weak stomachs that produce wind and indigestion.

Ginger	6	drachms
Seed of rock parsley	6	„
Spikenard	6	„
Aniseed	6	„
Peppercorns	6	„
Stacte	6	„
Wild mint, or calamint	5	„
Cummin	2	„
Berries of incense	2	„
Bark (of cinnamon?)	2	„
Pyrethrum	2	„
Honey as much as sufficeth.		

Dose, one drachm in some convenient draught.

Esṭamkîḵôn (στομαχικός), that is, a stomachic, which is good for hardness of the stomach and the liver, and the spleen, and for protracted fever, and for internal wind.

Ambrosia, which is good for the stomach that will not digest, and for a hard liver and spleen, and for excessive inflation of the whole body.

Dôkôn	1	drachm
Cummin, Indian,	1	„

---

[1] A medicine called after the name of its first maker.

Wood of balsam	1 drachm
Myrrh	2 drachms
Peppercorns, white	$^1/_2$ drachm
Costus	$^1/_2$,,
Peppercorns, long	$^1/_2$,,
Bark	1 ,,
Cardamoms	1 ,,
Flowers of the pistacia lentiscus	1 ,,
Seed of rock parsley	1 ,,
Aḳrôn (ἄκορος, acorus calamus)	2 drachms
Crocus	2 ,,
Laurel berries	10 in number
Honey as much as sufficeth.	

Dose, a portion as large as a chestnut, in hot water or in wine mixed with water.

A medicine made of ḳôrnîthâ (origanum pulegium?), which is good for cold diseases of the stomach and liver, and for protracted fits of shivering, and for fever.

Ḳôrnîthâ of the river	12 drachms
Ḳôrnîthâ of the mountain	12 ,,
Rock parsley	12 ,,
Meadow saxifrage	12 ,,
Seed of rock parsley	4 ,,
Thyme	4 ,,
Lybasticum		16 ,,
Peppercorns	48 in number
	Honey as much as sufficeth.	

Page 309

Fol. 148 b.

Dose, twenty drachms.

A medicine made of Anḳardyâ, that is to say, Bêldôr, which is good for every ailment of the stomach, and for those who are delirious, and who are attacked by vertigo by reason of the stomach, and for delirium, and for the spleen and kidneys, and for an evil condition of the stomach, and for the malady of the gout, and for pain of the womb, and for elephantiasis, and for all the ailments that are produced by the black bile.

Spikenard	1	ounce
Bear-wort	1	,,
Cassia	1	,,
Laurus malabathrum	1	,,
Crocus	1	,,
Epithymum (ἐπίθυμον)	1	,,
Pistacia lentiscus	1	,,
Rhubarb	1	,,
Myrobalanus (palma un- guentariorum)	1	,,
Caryophyllus aromaticus	1	,,
Stacte	8	*germê*
Honey of bêldôr, that is, ankardyâ	8	,,
Incense berries	1	ounce
Ginger	1	,,
Aloes	1	,,
Bdellium	1	,,
Myrrh	1	,,
Oil of balsam	1	,,
Irises	2	ounces
Bark of the root of shamrâ (mint?)	2	*lîṭrê*
Vinegar	. . . .[1]	

. . . . three days, and then put in a saucepan and boil a little; strain out the root, squeeze it and throw it away. Then pour one and a half *lîṭrê* of honey on the vinegar, and boil gently over the coals until the mixture becometh thick, then mix with it all the other medicines. Administer in doses of one drachm in some convenient draught.

A medicine in which pearls are compounded, which is good for those who suffer from phthisis, and it strengtheneth the stomach, and fatteneth the body, and it is good for fistula, or haemorrhoids in the anus.

Unslit pearls	5	drachms
Caryophyllus aromaticus	5	,,

---

[1] Some words have been left out here.

Peppercorns, white	5 drachms
Cummin	6 „
Armenian bôlos (βῶλος Ἀρ-μενιακός)	7 „
Cinnamon	7 „
Amomum (grain of Paradise)	8 „
Kôlîghân	8 „
Lekâ	8 „
Hellebore, black	9 „
Hellebore, white	9 „
Stacte	10 „
Bamboo	10 „
Cummin (karwâyâ)	11 „
Coriander	12 „
Bark	12 „
Chinese ginger	16 „
Gum Arabic	2 „
ǀ Loaf sugar	30 *estîrê*

Page 310

Pound, and mix well together, and administer in doses of one drachm in some draught which is convenient.

A medicine which containeth oxide of iron, and which is good for cold of the stomach, and for fistular wind, ǀ **and** Fol. 149a. **for the kidneys, and it increaseth the seed which is insufficient, and it beautifieth the complexion.**

Myrobalanus chebula, black	2 drachms
Bĕlîlḳê	2 „
Amlâgh (depilatory)	2 „
Peppercorns, round	2 „
Peppercorns, long	2 „
Ginger	2 „
Seed of rock parsley	2 „
Seed of dill	2 „
Coriander seed	2 „
Sesame	2 „
Sea spume	2 „
Cummin	2 „
Sal ammoniac (نوشادر)	2 „
Convolvulus turpethum	6 „

Sandaracha (سندروس, σανδα-ράκη) white	6	drachms
White clary (sage?)	6	,,
Seed of leeks	6	,,
Seed of gourds	6	,,
Seed of malva officinalis	6	,,
Seed of parsnips	6	,,
Daucus (a kind of parsnip)	7	,,
Inside of copper (*i. e.,* pure copper)	10	,,
Leaves of the caper plant	14	,,
Oxide of iron	10	,,
Panîdh	10	,,

Pound and reduce to a powder, and work into a paste with honey, and administer in doses of two drachms with extract of leeks. This medicine hath been well tried.

Another medicine which was [prescribed by] Galen, and which is good for protracted fevers accompanied by attacks of shiverings, and for cold of the stomach and of the liver and kidneys.

Peppercorns, long	12	drachms
Seed of rock parsley	12	,,
Ferula asa foetida	8	,,
Rock parsley	6	,,
Origanum of the mountain	4	,,
Origanum of the river	4	,,
Thyme	4	,,
Seseli	4	,,
Honey as much as sufficeth.		

Dose, a portion as large as a chestnut, in some convenient draught.

Parânôsh, a medicine which containeth iron oxide, and is good for fistular winds, and for cold of the stomach.

Myrobalanus chebula, black	10	drachms
Bĕlîlķê	10	,,
Amlâgh (depilatory)	10	,,
Ginger	10	,,
Daucus (a kind of parsnip)	10	,,

Peppercorns, round	10 drachms
Peppercorns, long	10 ,,
Spikenard	10 ,,
Cypress berries	10 ,,
Seed of fennel	4 ,,
Iron oxide	11 ,,
Pepperwort, white	12 ,,

Pound the iron oxide and reduce it to a powder, steep it in vinegar for one night and one day, wash it in vinegar twice | and in water ten times, and dry. Pound and reduce to a Page 311 powder all [the other medicines], and mix them with as much cow oil and honey as suffice [to make them into a paste]. Administer in doses of two drachms in wine mixed with water. Let the patient partake of fatty broth, and it will do him good, and let him keep himself from everything that is acid | or salt. Fol. 149 b.

A medicine which is made from roasted berries (?), and which is good for looseness of the bowels that ariseth from debility of the stomach.

Roasted berries	8 drachms
Sumach	8 ,,
Flour of sorbus (service tree)	10 ,,
Ginger	4 ,,
Coriander seed	4 ,,
Ammi Copticum	4 ,,
Seed of dill	4 ,,
Loaf sugar	4 ,,
Peppercorns, long	1 drachm

Pound and mix together, and administer in the form of a dry powder with extract of asparagus, or in cold water.

Ṭarsûm, a medicine which is good for indigestion, and for want of power in the stomach, and for colic, and for straining at stool.

Myrobalanus chebula, black	12 drachms
Sugar	12 ,,
Ginger	6 ,,

Dose, one spoonful in wine mixed with water, or in hot water.

23

The Little Ṭripal, which is good for want of power, and for liquid in the stomach, and for fistular wind, and it beautifieth the complexion.

Myrobalanus chebula, black	
Bĕlilḳê	in equal quantities
Amlâgh (depilatory?)	

Pound and mix together in as much cow oil and honey as are sufficient to make them into a paste, and administer as 'a dose a portion as large as a nut in wine mixed with water.

Barnagsar, a medicine which containeth Panîd, and which is good for cold and liquid in the stomach, and it restoreth the belly to a normal condition, and cleanseth the stomach. It is to be taken before and after food.

Myrobalanus chebula, black	3 drachms
Bĕlilḳê	3 ,,
Amlâgh	3 ,,
Peppercorns, long	·1 drachm
Copper	1 ,,
Scamonia	1 ,,
Convolvulus turpethum, white	8 drachms
Panîd	12 ,,

Pound and mix well with the panîd, and make into portions the size of small ṇuts. Administer one as a dose to be swallowed. And after it let the patient drink wine mixed with water. If he be in sore pain let him take a spoonful.

Page 312   A medicine made of | myrtle berries and ferula asa foetida, which is good for protracted inflation of the stomach, and for prolonged looseness of the bowels.

Ferula asa foetida, black	4 drachms
Ammi Copticum	4 ,,
Black cummin	4 ,,
Crocus	3 ,,
Myrtle berries	5 *estîrê*
Stacte	2 drachms
Honey as much as sufficeth.	

Fol. 150a. | Dose, a portion the size of a raisin, in some convenient draught.

A medicine made of irises, which is good for an evil condition of the stomach, and for dropsy, and for dropping of the belly.

Irises	44 drachms
Peppercorns	20 ,,
Ginger	12 ,,
Aniseed	4 ,,
Stacte	4 ,,
Seed of dill	4 ,,
Ferula asa foetida, black	12 ,,
Ammi Copticum	8 ,,
Seed of rock parsley	8 ,,
Honey as much as sufficeth.	

Dose, a portion as large as a raisin, in wine mixed with water.

The medicine Angabhdîgh, which is good for cold of the stomach and liver, and for pleurisy, and for pains in the back, and for fistular wind.

Costus	8 drachms
Acorus calamus	2 ,,
Ginger	5 ,,
Lepidium latifolium	7 ,,
Myrobalanus chebula, black	7 ,,
Indian salt	5 ,,
Ammi Copticum	3 ,,
Honey as much as sufficeth.	

Dose, two drachms in wine mixed with water.

Pelpelmôr (Pepperroot), a medicine which is good for pain of the stomach, and it restoreth the tone of the belly, and increaseth the seed.

Myrobalanus chebula, black	3 drachms
Bĕlîlḳê	3 ,,
Scamonia	4 ,,
Ammi Copticum	1 drachm
Seed of rock parsley	1 ,,
Cummin	1 ,,
Incense bark	1 ,,
Pepperroot	2 drachms

Sugar 2 ounces

Panîd 2 „

Dose, one spoonful, in wine mixed with water.

Another medicine which is good for pain of the stomach, and for protracted fever.

Green myrobalanus chebula, peeled	20 drachms
Greek absinthe	20 „
Eupatorium	20 „
Epithymum	20 „
Parsnip	20 „
Dâûradh	20 „
Root of rock parsley	or, one hand of each
Root of pistacia lentiscus	1 hand each
Thyme	1 „
Persian thorns	1 „
Dried herbs	3 estirê

Page 313

Infuse all these in water, and administer two drachms for a dose.

[Another medicine which is good for] **fits of shivering, and for flatulence of the stomach and belly.**

Make an infusion of thyme and ginger, and give to the patient to drink.

Or make an infusion of the berries and root of rock parsley, and coriander seed, and ammi Copticum, and administer as a Fol. 150b. dose two measures to the patient when fasting. |

Or dry a gourd, and pound it fine, mix it with water and administer it in draughts, in doses of one spoonful. It is very helpful in inflations of the belly and colon, and for . . . . This is a well tried medicine.

Now, for stomachs in which bile is frequent, or phlegm, and it is necessary to clear them out by vomiting, the medicines which are written below are useful.

Medicines which produce vomiting.

Common salt	1 part
Elaterium (a purgative)	1 „
Natron	1 „
Mustard	1/2 „

Pound, and administer it in a draught made of an infusion of fennel and honey, and the patient will vomit.

Or take saponaria officinalis (or, ptarmica), which is kîndôsh (کندس, gypsophilla struthium), and cucumber seed in equal quantities, pound them, and mix with an infusion of fennel, and administer a draught, and the patient will vomit.

Or pound up black hellebore, and roll a radish in it, and let the patient eat it, and he will be violently sick.

Or take two drachms of "kangrîs", which is the gum of artichokes, pound them, and administer them in an infusion of fennel, and it will clear out the stomach by means of vomiting.

Or take three or four roots of narcissus, cut them up and steep them in water with fennel, and let the patient drink two measures of the infusion, and they will clear him out by vomiting.

Or take two drachms of the nuts used for producing vomiting, pound them, and administer them in an infusion of fennel, and the patient will vomit without trouble, and this medicine will also clear away jaundice.

Or let the patient eat fish which is tolerably salt, or radishes and honey, and let him drink wine with a very little water in it, and let him thrust a feather down his throat, and he will get rid of the thick chyme by vomiting.

Or	Mustard	2 drachms	
	Borax	1 drachm	Page 314
	Saponaria officinalis	2 *denķê*	

Mix and administer in honey water, and it will clear away the chymes that cling to the coats of the stomach.

Or cut up radishes with their leaves, and let the patient eat the mixture with oxymel, abundantly; this medicine will clear away jaundice also.

Or take two drachms of the root of the garden cucumber plant, dry it, and pound it, and administer it in honey water after supper, and it will cleanse the stomach by painless vomiting.

Or crush sixteen dried herbs of the mountain, and administer them mixed with honey water, and let the patient take every Fol. 151 a. hour an infusion of fennel and sweet oil, and it will clear away the thick chymes by vomiting.

Or crush some wild hellebore, and place one drachm of it in a radish; let the patient swallow it, and drink honey water after it. This is to be done before and after meals.

Or let him make from the root thereof pieces like unto oblong pills, and insert them in the anus, and they will produce a stool.

Or cut up in pieces bulbs (ܒܘܠܒܐ = βολβός) [of lilies], and make an infusion of them, and administer two measures, and they will cleanse the stomach.

Now if through these medicines, which produce vomiting from the stomach, excessive vomiting taketh place, administer one or other of those drugs that have been written down above, and that are good for violent vomiting from the stomach, [mixed with] wood mint (or, wild mint), and chopped up pomegranates.

And as concerning the bringing up of blood that cometh from the stomach, thou must cure this by means of the treatment and medicines that are written down in the "Chapter on the Bringing up of Blood". Similarly, use the same means in cases where there is the 'alûkâ disease in the stomach; but the means to be employed in these cases are described further on.

Now, as concerning the medicines which are to be drunk for all the diseases of the stomach, it is my opinion that too many of them have been described in this Chapter; but the physician, who is well acquainted with the power (*i.e.*, effect) of simple medicines and with the treatment to be used when certain symptoms of the diseases appear, will find it quite easy to prepare a medicine for himself that will be suitable for use in the case before him. He can use simple medicines, and he Page 315 can add to or take away from those that are written down, | for the means to be used in healing diseases depend upon the symptoms thereof. For it is said, If a man is unable to make a diagnosis of a disease he is unable to effect the cure of it, wherefore it is meet for us to write down also a description of the helps that are to be applied externally in cases of disease of the stomach.

The medicines, and fomentations, and plasters, and cataplasms which are good for pain in the stomach.

If there be an inflamed ulcer in the stomach, the presence of which will be made known by the fever that accompanieth it, and by thirst, and by the sensation of excessive heat, make an infusion | of cabbage (κράμβη), and fox-grapes, and solanum Fol. 151*b.* halicacabum (or, *alkâkengî* الكاكنج), and sponge the patient therewith, and let the physician anoint him with oil of roses.

Or pound these drugs up when fresh, and mix with them flour of barley, and oil of roses, and lay upon the patient bandages dipped in hot water wherein these have been mixed.

Or crush the rinds of pomegranates, and lentils, and dried roses, and husked barley, and make an infusion of them, and sponge the patient therewith.

Or press these thoroughly, and pound them well, and add to them oil of roses, and lay bandages spread with the mixture on the stomach.

Now, if there be a sensation of heat in the stomach without the inflamed ulcer of fever and of bile, pound chicory, and mix therewith barley flour and oil of roses, and bind a cold plaster made of these things on the stomach. Or use purslane in the same way. And if there be much pain present with the inflammation, pound leaves of hyoscyamus, or leaves of mandragora, and mix with them flour of barley and oil of roses, and use as a plaster.

Or pound the shell of a fresh gourd, and mix with it oil of roses and flour of barley, and a little malva officinalis (hibiscus), and use as a plaster.

Or steep psyllium (plantain) in cold water, and when the liquor becometh thick, mix oil of roses therewith, and use it as a plaster.

Or steep flax seed in vinegar, and mix them up, and use as a plaster in the same way.

Or mix together polygonum, and flour of barley, and oil of roses, and use as a plaster.

And if, without the presence of fever and inflammation, there be debility only | in the stomach, wherefrom arise looseness of Page 316 the bowels and nausea, make an infusion in water of absinthe, chamomile, spikenard, barley, and "king's crown" (melilotus), and sponge the stomach therewith

Or make an infusion in water of chamomile, absinthe, purple βαλαύστιον, the rinds of pomegranates, spikenard, cassia, and origanum majorana, and sponge the patient therewith, and anoint him with oil of nard.

Fol. 152a. Or use extract of myrtle, or extract of asparagus, | or stacte, or any of the finest unguents which is available, or pour over him hot water in which the following drugs have been dissolved: Take stacte, absinthe, spikenard, frankincense, barley (?), myrtle berries, laurus malabathrum, rinds of pomegranates, dried grapes, dried myrtle berries, clematis, aromatic reeds,[1] wood of aloes, sumach, purple βαλαύστιον, dried roses, ammi Copticum, cummin, seed of rock parsley, all of them in equal quantities, mix them together, and pour into a vessel one measure of strong-smelling wine, and a little wine of lilies, and oil of jasmine, or oil of nard, let them simmer over the coals for a short time, and then throw into them the dried medicines which thou hast pounded. Then dip into the mixture a bundle of clean wool, and foment the stomach, and also all the belly, if the patient be weak, until the wine is exhausted, and then spread some of the above-mentioned drugs when dried over the stomach until it regaineth its healthy condition. And thou must take care that the patient eateth and drinketh only such things as are healthful, according to the nature of the disease.

And if there be a hard ulcer of long standing in the stomach, make an infusion of chamomile, fennel, trigonella, "king's crown" (melilotus), flax seed, and malva officinalis (hibiscus), and sponge the patient therewith, and anoint him with oils that produce warmth, and cause cessation of the pain. And thou must also remember well this fact: if the disease be in that upper stomach which descendeth from the mouth to the belly, the patient Page 317 hath need of fomentations, or of plasters, | on the back, and between the shoulders, and on the vertebrae of the neck. The

---

[1] Read ڢڛٮٮٮ.

writer ordereth these things to be done because the spinal column is situated near [these members].

Now therefore it is meet for us to describe also those well known plasters that are good for the stomach, and also the means to be employed externally in healing diseases of the liver, which are the same as those that are adopted when there are debility and ulcers (or, abscesses) therein. | Fol. 152*b*.

[The medicine] "Pôlkarîôn of the Kingdom", which is good for pain in the stomach and liver, and for hardness of the same, and for the ulcers and pustules which come therein, or in any of the members.

Wax	3	*lîṭrê*
Incense	2	„
Ammoniacal extract	2	„
Cedar gum	1	*lîṭrâ*
King's crown (melilotus)	10	*lîṭrê*
Dry pitch (bitumen)	12	„
Esṭûmka (stomachic?)	12	„
Myrrh	12	*esṭîrê*
Spikenard	12	„
Cypress berries	12	„
Irises	12	„
Cardamoms	12	„
Opopanax	12	„
Crocus	8	drachms
Cinnamon bark	8	„
Flowers of pistacia lentiscus	8	„
Stacte	8	„
Wood of balsam	8	„
Amomum	8	„
Oil of nard	$^1/_2$	*lîṭrâ*

Wine as much as sufficeth.

Dissolve the dissolvable drugs in strong-smelling wine, and rub down in oil those which can be rubbed down, and pound the dry drugs, and mix them, and use for plasters on the stomach and liver, and on every part where there is a hard ulcer.

A medicine made of thorn pods, which stimulateth the stomach, and checketh looseness of the bowels.

Thorn pods	1 ounce
Cummin steeped in vinegar and roasted	1 ,,
Dry myrtle berries	1 ,,
Flour of lotuses	1 ,,
Sumach	1 ,,
Coriander seed	1 ,,
Oak apples	1 ,,
Roasted grain	1 ,,
Roasted berries	1 ,,
Stacte	4 drachms

Pound, and administer [as a dose] one spoonful in extract of myrtle berries, or in extract of asparagus.

The Great Ṭrîpal, which is good for cold and dampness of the stomach, and for fistulae, internal and external, and which warmeth the whole body, and beautifieth the complexion.

Page 318

Myrobalanus chebula, | black	1 ounce
Bĕlilk̬ê	1 ,,
Amlâgh	1 ,,
Shahṭragh (parsnip)	1 ,,
Ammi Copticum	1 ,,
Seed of rock parsley	1 ,,
Persian thorns	1 ,,
Spikenard	3 drachms
Amomum (shôshmîr)	3 ,,
Amomum	3 ,,
Cinnamon	4 ,,
Flower of bûsht	4 ,,
Peppercorns, round	4 ,,
Peppercorns, long	4 ,,
Indian salt	4 ,,
Mustard	1 drachm
Pepperwort	1 ,,
Sal ammoniac	1/2 ,,
Acorus	3 drachms
Iron oxide, washed	3 ounces
Cow oil } Honey } | as much as sufficeth.	

Fol. 153a.

Dose, a portion about the size of a walnut, in wine mixed with water.

Another [medicine] for the consumptive, which is good for looseness of the bowels, and for debility of the stomach, and for the wasting away of the whole body.

Ammi Copticum, roasted	2 drachms
Seed of rock parsley	2 ,,
Peppercorns, long	2 ,,
Ginger	2 ,,
Cummin	1 drachm
Carman	1 ,,
Bark of incense plant	1 ,,
Acacia	1 ,,
Dried green grapes	1 ,,
Seeds of dried herbs, roasted	1 ,,
Roasted furnace dross	$^1/_2$,,
Roasted rice	3 drachms
Thorn pods	3 ,,
Caryophyllus aromaticus	$^1/_2$,,
Dried herbs (?), stoned	11 *esţîrê*

Pound the dry drugs, and mix them with the dried herbs, and make them into Avella balls, about the size of raisins, and administer one as a draught. Let the patient drink after it tincture of asparagus, or tincture of myrtle berries.

[Another medicine] made of sorrel, which is good for torpidity of the stomach, and for looseness of the bowels, especially in children.

Sorrel seed	4 drachms
Seed of dried grapes (?), roasted	4 ,,
Oak apples	2 ,,
Poppy seed	2 ,,
Sugar	2 *esţîrê*

Dose, one spoonful, in some useful medium.

A medicine which is made of laserpitium (juice of silphium?), and which is good for cold in the stomach, and for protracted fits of shivering, and for coughs which are due to cold, and for intestinal worms.

Juice of silphium ⎫
Garden nigella ⎪
Peppercorns ⎬ in equal quantities
Mustard ⎪
Pepperwort ⎭

Pound, and work into a paste with honey, and administer as
Page 319 a dose a portion as large as a chestnut. | If this medicine is
required for a cough, make it into tablets, and let the patient
hold them under his tongue; if for cold, give it in wine and
water, and if for worms, give it in an infusion of nigella, lupins,
and costus, in rain water. This is a well tried remedy, and is
a sure [cure].

**A medicine made of Kôlinghân, which is good for severe
cold in the stomach and liver, and it promoteth the digestion
of the food, and breaketh up wind.**

Fol. 153*b*.

Kôlînghan (galanga)	4 drachms
Incense bark |	4 ,,
Peppercorns, white	4 ,,
Amomum	3 ,,
Peppercorns, long	5 ,,
Cinnamon	3 ,,
Flowers of bûsht	3 ,,
Ginger	8 ,,
Seed of rock parsley	1 drachm
Aniseed	1 ,,
Black cummin	1 ,,
Cummin	1 ,,
Ṭalîspar	1 ,,
Panîd, or sugar	4 (?) drachms

Dose, two drachms in wine mixed with water.

**A medicine made of cummin (kârwâyâ), which is good
for stoppages of the stomach, and for the liver and indigestion.**

Cummin	3 drachms
Ammi Copticum	3 ,,
Seed of rock parsley	3 ,,
Dried herbs (or, grapes)	3 ,,
Fennel (seseli)	3 ,,

Seed of zûprâ (Pers. زِرُّقا)   3 drachms
Bitter almonds, cleaned   6   „
Honey as much as sufficeth.
Dose, a portion as large as a bean.

Medicines which are good for pain in the stomach, and which are made into infusions.

Root of rock parsley	1 hand
Bark of shâmrâ root (anethum foeniculum)	1 „
Root of pistacia lentiscus	1 „
Stacte	4 drachms
Aniseed	4 „
Spikenard	4 „
Dried herbs (or, grapes)	12 „
Water	1 *dûlkâ*

Infuse these until they are reduced to one half, and administer two measures as a draught, and the medicine will warm and purify.

Another [medicine] which is good for the stomach that hath become hard, and for pleurisy and coughs.

Root of rock parsley	1 hand
Bark of shâmrâ root	1 „
Root of pistacia lentiscus	1 „
"Warrior's hair"	1 „
Licorice root	1 „
Fleshy dates	10 in number
Fleshy figs	10 „
Dried herbs (or, grapes)	1 hand
Aniseed	4 drachms
Stacte	4 „
Dog's nipples	4 „

Make an infusion, | and administer in doses of one measure. Page 320

Another [medicine] which warmeth, and doeth away fits of shivering of long standing.

Stacte	3 drachms
Aniseed	3 „
Seed of rock parsley	3 „
Seed of shâmrâ	3 „

Thyme	3 drachms
Thorns	3 ,,
Spikenard	3 ,,
Bark of cinnamon (?)	3 ,,
Barley (?)	3 ,,
Berries	3 ,,
Ginger	4 ,,
Kôlînghân (galanga)	4 ,,
Spikenard	8 ,,
Root of pistacia lentiscus	1 hand
Root of rock parsley	1 ,,
Origanum majorana	1 ,,
Dried wood mint	1 ,,
Rue	1 ,,

Make an infusion of these in water, and administer [a dose] with honey for a fit of shivering, and it will give relief.

Fol. 154a. | Another Pôlkariôn, which is good for pain of the stomach, and for the liver, and for a hardened spleen, and for pain in the lungs, and for the back, kidneys, bladder, and womb when diseased, and for nervous pains.

Wax	1 *lîṭrâ*
Terebinth gum	1 ,,
Bdellium	1 ,,
Incense	1 ,,
Ammoniacal preparation	1 ,,
Cardamoms	1 ,,
Barley	1 ,,
Amomum	8 *lîṭrê*
Nard	8 ,,
Crocus	8 ,,
Myrrh	8 ,,
Frankincense	8 ,,
Balsam wood	8 ,,
Oil	20 ,,
Kûprînon	20 ,,

Strong-smelling wine as much as sufficeth for dissolving the ammoniac, bdellium, frankincense, and myrrh.

Pound these and steep in wine, and crush them until they are dissolved. First melt the wax with oil, then add the gum of terebinth, and, when this is melted, take the mixture off the fire, pound the dry drugs and reduce to a powder, pour them all into a mortar, mix them well together, and use as a plaster.

The Piâgaryás Plaster, which is good for the diseases.

Crocus	6 drachms
Wax	45 ,,
Aloes	4 ,,
Stacte	4 ,,
Bdellium	4 ,,
Ammoniac	4 ,,
Liquid styrax	4 ,,
Goose fat	16 ,,

Oil of roses as much as sufficeth.

Wine for the drugs which are to be dissolved.

Mix well together, and use.

[The medicine] Milîṭis, which is good for the hard ulcers that come in the stomach and liver. Now the word "Milîtis" | may be translated "King's Crown" (*i. e.,* melilotus).　Page 321

King's crown	50 drachms
Ammoniacal incense	11 ,,
Gum of terebinth	11 ,,
Wax	12 ,,
Spikenard	20 ,,
Barley	16 ,,
Cardamoms	16 ,,
Irises	16 ,,
Myrrh	16 ,,
Crocus	8 ,,

Oil of nard ⎫
Wine ⎭ as much as sufficeth.

Pound well together, and use.

Adisparmaṭón, which is, being translated, "a medicine made of seeds", and which is good for the diseases which have been enumerated, and for dropsy.

Flour of trigonella |	6	*estîrê*
Seed of shâmrà	6	„
Seed of rock parsley	6	„
Ammi Copticum	6	„
Cummin	6	„
Irises	6	„
Root of ferula opopanax	6	„
Aniseed	6	„
King's crown	6	„
Wax	30	drachms
Ox fat, unsalted	30	„
Honey	36	„

Oil of nard as much as sufficeth.

Work up well according to the directions written above, and use as a plaster.

Another Diâsparmaṭôn, which is good for the diseases.

Ammoniac	3	*lîṭrê*
King's crown	3	„
Spikenard	3	„
Amomum	3	„
Laurel berries	3	„
Origanum majorana	3	„
Seed of rock parsley	3	„
Aniseed	3	„
Natron	3	„
Irises	3	„
Barley	3	„
Crocus	3	„
Cassia (?)	3	„
Frankincense	3	„
Myrrh	3	„
Styrax	3	„
Opopanax	4	„
Wax	12	„
Honey	6	„
Ox fat	6	„

Oil of cedar as much as sufficeth.

Pound well together, and use as a plaster.

Another Pôlkaryôn, which is good for the obstinate (or, severe) diseases that take place in every part [of the body], and for nervous spasms.

Castoreum	I	ounce
Amomum	I	,,
Bdellium	I	,,
Frankincense	I^1/$_2$	ounces
Peppercorns, round	I^1/$_2$	,,
Peppercorns, long	I^1/$_2$	,,
Hydrargyrum (mercury)	3	,,
Fruit of balsam	3	,,
Costus	3	,,
Laurus malabathrum	3	,,
Cassia (?)	3	,,
Spikenard	3	,,
Opopanax	3	,,
Wood of balsam	3	,,
Flowers of pistacia lentiscus	3	,,
Cinnamon	3	,,
Pyrethrum	3	,,
Crocus	3	,,
Cardamoms	6	,,
Aristolochia, round	6	,,
Aristolochia, long	6	,
Irises	6	,,
Barley	3	,,
Wax	4	*lîtrê*
Gum of terebinth	2	,,
Aphronitrum (?)	3	,,
Stacte	3	,,
Aloes	3	,,
Styrax	3	,,
Oil of balsam	3	,,

Page 322

Oil of nard
Strong-smelling wine } as much as sufficeth.

Mix carefully, and use.

A Stomachic, which is good for indigestion in the stomach and in the liver, and for wind.

24

Crocus	2 drachms
Stacte	—
Absinthe plant, Greek	12 ,,
Spikenard	4 ,,
Styrax	4 ,,
Aloes	2 ,,
Wax	18 ,,
Oil of nard	6 *estîrê*

Mix well together, and use.

A medicine made of Ônantîs (οἰνανθίτης), which is good
Fol. 155 a. for | ulcers of the stomach, and of the liver, and of every
member, and for debility and languor, and severe nausea
of the stomach, and for the rheum that cometh down from
the head to the belly.

Ônantîs (*i. e.*, wine made from the flowers of the wild vine)	6 drachms
Crocus	4 ,,
Stacte	4 ,,
King's crown	4 ,,
Bark of mandragora root	4 ,,
Wine of myrtle berries Oil of roses	as much as suffice
Flesh of dates	3 ounces
Stacte	4 drachms
Aloes	4 ,,

Dissolve the dates in strong-smelling wine, pound the dry drugs
to a powder, mix them with the wine, and pour on them as
much oil as sufficeth, and lay as a plaster on the stomach.

Plaster of Esclapiades, which is good for severe nausea
of the stomach, and for acute diarrhoea, and it produceth
sleep.

Seed of hyoscyamus	4 drachms
Seed of rock parsley	4 ,,
Aniseed	4 ,,
Flowers of roses	4 ,,
Juice of Hypocistis (tragopogon pratense)	4 ,,
Myrrh	4 ,,

Crocus	$1^1/_2$ drachms
Opium	$1^1/_2$ „

Pound the dry drugs to a powder, and dissolve the juice of | Page 323
tragopogon, myrrh, and opium in wine of myrtle berries, and
mix the dry drugs with them, and work up into a paste, and
make into tablets (or, pills) and dry in the shade. When
required for use, pound and dissolve in some suitable medium,
and use as a liniment; or dissolve in honey, and smear on a
piece of linen and use as a plaster. If the attack of the rheum
be severe, add juice of tragopogon and roses to the ônantìs,
and sumach, each one drachm.

**Another medicine which is made of Ônantìs, and which
is very powerful indeed. It is good for the stomach which
is relaxed through the rheum that descendeth into it from
the head, and for severe nausea, and for looseness of the
bowels, and for debility of the liver.**

Ônantìs	1 ounce	
Anpîḳnôn (cantharides?)	1 „	
Roses	1 „	
Aloes	1 „	
Green grapes	1 „	
Furnace dross	1 „	
Acacia	1 „	
Flesh of dried pomegranates	1 „	
Purple βαλαύστιον	1 „	
Flesh of dates	6 ounces	
Opium	2 drachms	
Myrrh |	2 „	Fol. 155 *b*.
Frankincense	2 „	
Hyoscyamus	2 „	
Mandragora root	2 „	
Wax prepared with oil of pistacia lentiscus	1 *lîṭrâ*	

Add as much wine of myrtle berries as will suffice to work
the dry drugs into a paste, and to dissolve those that are
capable of being dissolved, and prepare and use. This is a
very excellent medicine.

The Solon plaster, which is good for hardness of the

24*

stomach and liver, and spleen, and for ulcers in the epigastric region, and for every nervous pain.

Wax	100	drachms
Ammoniacal incense	100	„
King's crown	200	„
Myrrh	8	„
Bdellium	8	„
Cypress oil	1	*lîṭrâ*
Vinegar } Wine }	for dissolving purposes	

Prepare and use.

A plaster made of asparagus and Greek bread, which is
used for excessive debility of the stomach, and for emission
of blood, or bile, and for diarrhoea.

Page 324

Asparagus, peeled within
and without
Greek bread
Leaves and fruit of the
myrtle
Flowers of the vine
Cummin
Santalum |
Wood of aloes
Pragnagh (Indian cucumber)
Crocus
Rinds of pomegranates
Acacia
Stacte
Frankincense
Wine of lilies
Oil of roses

all [in equal
quantities]

Boil the asparagus in strong-smelling wine, steep the Greek
bread in the wine, pound all [the other ingredients], and work
up, and mix together, and use as a plaster.

A plaster made of dates, which is good for weak stomachs
and livers, and for those who vomit their food.

Unripe dates	3	drachms
Asparagus, cleaned	3	„

Dried roses	3 drachms
Absinthe plant	1 ounce
Stacte	1 „
Ladanum (λάδανον) gum	3 ounces
Aloes	2 „

Boil the dates, asparagus, and wine, and pound them, and pound to powder the dry drugs and mix with them, and add to them oil of jasmine, or oil of nard, and work up well together and use.

Bread plaster, which is good for ulcers that come in the stomach. Pound bread | made of the very finest flour, and Fol. 156a. steep it in water, and crush it, and mix with it flour of trigonella, and pig's fat, and fresh leaves of hyoscyamus, and boil over the fire. First anoint the patient with hot oil, and then put on a plaster.

A plaster made of king's crown (melilotus), which is good for hard ulcers of the stomach, and liver, and spleen, and it checketh hard pustules and ulcers.

King's crown	4 drachms
Trigonella	4 „
Flowers of chamomile	4 „
Myrtle berries	4 „
Liquid unguent (or, honey)	4 „
Root of *adam*	4 „
Absinthe	3 „
Ammoniac	4 „
Bdellium	4 „
Fleshy figs	—

Boil the figs in new wine, dissolve the ammoniac and bdellium in the wine, mix together and work up, pour on them a little cow oil, and use as a plaster. Ligatures and bandages are also very helpful, for they all assist in checking the ulcers.

Now, for the physician who knoweth the medicinal substances that are to be used for external application, and | who keepeth Page 325 in his mind the remembrance of their effects, it is not difficult to make use of them in a proper manner and with understanding. In addition to this, he must be especially careful, if he be unable to help the sick, not to do them harm. Therefore it is necessary for us to write down a description also of those oils from which the plasters (or, ointments), wherewith patients are anointed, are compounded, and which benefit them, and then we will bring to an end this Chapter also.

Fine oil of nard, which is good for every sickness, and especially for cold in the stomach, and liver, and womb, and nerves, and it openeth (*i. e.*, removeth) **obstructions, and it helpeth the colon also.** In the case of cold in the stomach or liver, it may be drunk with wine, or used as an unguent. For disease of the womb it must be used as an unguent, and injected, and applied to the vulva by means of a pad (of soft rag or wool. For wind, and for the colon it must be used in the bath, that is, a measure by weight of two drachms mixed with an infusion of dill. For protracted pains Fol. 156b. in the head | it must be poured into the nostrils, and it will make the countenance (or, face) beautiful. For cold in the bladder it must be injected into the member that maketh progeny to flourish, and it will diminish the pain therein.

Costus	30	drachms
Cassia	30	,,
Incense berries	30	,,
Reed of incense	20	,,
Peppercorns, long	8	,,
Ginger	6	,,
Cardamoms	6	,,
Mountain grapes, dried	6	,,
Barley	10	,,
Root of pistacia lentiscus	200	,,
Irises	200	,,
Leaves of incense	10	,,
Leaves of myrtle	10	,,
Origanum majorana	10	,,

[The first boiling.] Pound them all very finely, and throw them into strong-smelling wine for a whole night and a day. Then pour the mixture into a saucepan, and pour upon them a quantity of water equal to that of the wine in which they have been steeped. Pour in ten measures (*kestê*) of sweet oil, and boil over a slow fire that produceth no smoke for six hours, and then take it off the fire and set it aside to cool. Skim off the oil from the water. This is the first boiling. **The second boiling.** Take

Asarum |	5	drachms
Spikenard	5	,,
Amomum	5	,,
Laurus malabathrum, and	3	,,
Myrrh	10	,,

Page 326

Pound these and steep in wine for one whole night and day, and boil as at first, and let the mixture cool, and then skim off the oil. **The third boiling.** Take

Caryophyllus aromaticus	3	drachms
Leaves of incense	3	,,
Bark of incense	3	,,
Styrax, liquid, and	30	,,
Oil of balsam	$^1/_2$	ounce

Pound the dry drugs, throw them into wine, and boil as at first. Before thou removest the mixture from the fire, dissolve the chamomile in wine, and pour it into the vessel, and, finally, dissolve the styrax in oil of balsam, and pour it in also, and let it boil for one time (hour?). Take it off the fire, let the mixture cool, separate the oil from the water and the drugs, and put it into a vessel, and use according to the instructions written above.

Oil of nard, which is good for the diseases, prepared in another manner.

Incense berries	5	*estîrê*
Costus	5	,,
Irises	5	,,
Myrrh	4	,,
Musk	2	,,
Wood of balsam	6$^1/_2$	,,
Peppercorns, long	5	,,
Oil of balsam	5	,,
Bdellium	2	,,
Spikenard	3	,,
| Mountain grapes, dried	2	,,
Asarum	6	,,

Fol. 157a.

Styrax	5 *estîrê*
Leaves of camphor	5 „

Pound the dry drugs, and steep in wine for a whole night and a day, and dissolve the drugs which can be dissolved and pour them in. Add two measures (*dûlḳê*) of sesame oil, boil over a gentle fire until the liquid of the wine is driven off, clarify, keep, and use.

Shḳîlâ oil, which is good for the diseases.

Caryophyllus aromaticus	2 ounces
Spikenard	2 „
Malabathrum	2 „
Styrax	2 „
Irises	2 „
Incense bark	2 „
Cassia	1 ounce
Bark of cinnamon (?)	¹/₂ „
Myrrh	¹/₂ „

Pound to a fine powder, steep in strong-smelling wine for a Page 327 whole night and a day, put in a boiling pot, | pour on them sweet olive oil or sesame oil, three measures (*hêmînê*), boil on the fire for a whole day; and when water is lacking in the pot add to it another boiling thereof. In the evening strain the mixture, and pour on the drugs three more measures of oil, and boil as at first. Before thou takest the pot from the fire, pour the second boiling into the first, and let the mixture boil once or twice. Before thou takest the pot from the fire, dissolve the myrrh in wine and pour into it, and then the styrax, and set it aside to cool. Separate (*i. e.,* skim off) the oil, strain it, put into a vessel, and use.

Oil made of lilies, which is good for pain in the stomach, and for pain of the womb, and it softeneth the nerves and warmeth [them].

Cassia	1 ounce
Costus	1 „
Styrax	1 „
Incense berries	1 „
Stacte	1 „
Caryophyllus aromaticus	¹/₂ „

Incense bark $^1/_2$ ounce
Crocus $1^1/_2$ drachms

Pound these to a powder, and pour into a glass vessel. Then take one measure of oil, and thirty white lilies, the roots of which have been cut off, with their flowers (?) cleaned | and Fol. 157b. dried, and throw them all into a vessel, and set it in the shade in a place which faceth the north, and where the air is neither too hot nor too cold. After six months dissolve and use.

Asparagus oil, which is good for excessive heat in the stomach and liver, and which may be used for purposes of anointing, and it may be drunk, and it may be mixed with plasters. Take three measures (*ḳesṭê*) of the finest oil, and eight measures (*ḳesṭê*) of cleaned asparagus, place them in a glass vessel and set them in the sun for thirty days, and use.

Pure oil, which is good for debility of the stomach, and for hardness thereof, and for nervous spasms.

Pounded stacte   6 *ḳaisê*

Boil in a double vessel with two measures (*ḳesṭê*) of oil, | and Page 328 take and use.

Myrtle oil, which is good for sluggishness and debility of the stomach and liver.

Oil   3 *ḳesṭê*
Fresh myrtle leaves   4 *lîṭrê*

Pound the leaves and macerate them in wine until the wine barely covereth them. Put them in a boiling pot, and then lay oil over them, and boil over a fire until the wine hath evaporated; take off the fire, strain, and use. Now, some throw into the oil fresh extract of myrtle to [improve] the colour.

Oil of costus, which when used as an unguent, or drunk, is good for cold of the stomach and liver. When used for anointing purposes it is good for the fits of shivering that accompany fevers. And it beautifieth and preserveth the hair.

Costus   20 drachms

Pound to a fine powder, steep in wine for a whole night and day, pour on it one measure (*dûlḳâ*) of oil, boil over the fire [in a pot] or in a double vessel until the liquid of the wine is evaporated, then take and use. And in the same way we may also treat irises, or any other root thou mayest be pleased to boil in oil, when oil properly prepared is not available, either for purposes of anointing or for mixing with plasters. These remarks on the symptoms of diseases of the stomach, and Fol. 158a. on | the means to be employed in healing them, must suffice for this, the Fifteenth Chapter.

HERE ENDETH THE FIFTEENTH CHAPTER.

CHAPTER XVI.

On the injuries (or, diseases) which attack the liver.

Now the liver is a substance that is charged with blood, and it is placed on the right side [of the body]. A small portion of it is attached firmly to the membrane that effecteth the digestion, but the greater part of it is spread over the stomach, and the whole of it is intersected (or, perforated) by large veins. It receiveth the juice from the stomach | by means Page 329 of the veins which are called "masrîḳê" (*i. e.*, appertaining to the mesentery), and which are firmly fixed to the part that joineth the lower mouth of the stomach to the higher intestine, which is called "ṣâwmâ". And it transformeth that juice into blood by means of the natural powers which it possesseth, and it purifieth the substance thereof, and it despatcheth it by that great vein in it which is called "tarʿaâ" (*i. e.*, "door"). Some of this juice it separateth above, about the heart, and some of it below, over the spinal column, about the kidneys, and these portions of the juice are distributed and propagated throughout the body, even to the most remote members. Now the function of the liver is a dominant one. As the power that giveth to the body motion and sensation descendeth from the head through the nerves and muscles, and as the blood is sent through the whole body, by means of the arteries, through the living power which cometh from the heart, so doth that nutritive power, which watereth the whole body with blood by means of the veins that go forth from it, come from the liver.

Now, it hath been said that the food is digested three times before it becometh transformed into a substance suitable for the nourishment of the members. The first digestion taketh place in the stomach, and the residuum thereof is the faecal matter which is ejected by means of the bowel and anus. The second digestion taketh place in the liver, and five different

Fol. 158*b*. kinds of products are the result. The first, which resembleth | the scum on broth, is the phlegm; this is transmitted to the lung, the chest, and the brain. The second, which resembleth the fat on broth, is the red bile; this the bladder (gall-bag?), which is in the liver, draweth away, and it becometh the bile. The third is blood, and it is taken by the veins, which have already been mentioned, and transmitted by them to the whole body. Now, the impurity thereof, which resembleth the impurity (sediment?) in wine, doth the spleen draw away by

Page 330 means of its veins. This is black bile. | The fourth is the residuum of the digestion. The fifth is the watery product of the whole digestion, and it resembleth the water which riseth up under the milk that cometh from sour cheese. This is the urine, and the kidneys draw it away and transmit it to the bladder, and it is emptied out by the urinary canal. This is the way in which the liver worketh, and these are the products that result from the digestion that taketh place therein. The third digestion taketh place in all the members that transform the food (or, nourishment) into substances suitable to their natures.

In the stomach food is transformed into a white juice, according to its nature, and in the liver food is transformed into blood, according to its nature, and in each member food is transformed into a substance of a nature similar to its own. The natural powers [first] transform the food, and then they nourish themselves thereon. The residuum of the third digestion, that is to say, of the digestion which taketh place in all the members, is found in the urinary organs, and it is the white substance which is found in the urine. And there is the residuum which is excreted by the nose and palate from the lungs and chest, which cometh up through coughing and spitting. And there is the residuum from the flesh of the whole body, and from the skin, which is excreted by the pores of the skin, that is to say, the sweat, and there are certain fumes

| that arise through respiration, which are imperceptible. Fol. 159 a. These, then, are the operations of the natural forces which are in every member. Now, inasmuch as the liver itself doth not keep itself ¦continually in the same regular condition, but subsisteth in one [or other] of the eight irregular states, and as it is sometimes in the condition in which it was originally formed, and at other times in a condition that is quite different, it is necessary for us to speak also about the symptoms of these things, which can be known from those that take place therein in an unnatural manner, and to describe how far remote they are from its original nature.

Of the symptoms (or, indications) of the natural conditions of the liver.

The existence of a hot condition in the liver is known by Page 331 the flatness of the veins, and by the excess of red bile, which is sometimes strong, and in a smaller degree by the black bile also. And the blood of those who are in this condition is exceedingly hot, unless it happeneth that the heart formeth an obstacle thereto, and their epigastric region, and belly, and their whole body is hairy.

A cold condition of the liver is known by the narrowness of the veins, and by the excess of phlegm, and by the coldness of the blood of the whole body, unless it be made hot by very great heat of the heart.

A dry condition of the liver thickeneth and reduceth the blood, and drieth up the veins and all the body.

A moist condition of the liver increaseth the blood, and dilateth and softeneth the veins, and moisteneth the whole body, always provided that the heart be not in a state which is the contrary to this.

A hot and dry condition of the liver maketh hairy the epigastric region, and diminisheth and thickeneth the blood, and with violence it increaseth the red bile, and, according to the strength, the black bile [also]. It wideneth and hardeneth the veins, | and it heateth and drieth up the whole body. For the Fol. 159 b. great heat of the heart overcometh the coldness of the liver, and thus also is it in the case of the other conditions [of the liver]. Now it is well known that when the conditions of [these] two members are united they make all the rest of the body to be like unto them.

The hot and moist condition of the liver maketh the epigastric region to be less hairy than the hot and dry, and, apparently, it maketh the veins wider, and maketh the whole body hotter and moister, always provided that the condition of the heart doth not stand as an obstacle in the way. Now, if all be turned aside in an unnatural manner in these two [kinds of] stomachs, | those who possess these characteristics are easily attacked by the sickness that is due to decay of the evil chyme, and their chymes will be far more evil, especially if moisture be in excess of the heat.

Page 332

The wet and cold condition of the liver maketh smooth (*i. e.*, hairless) the epigastric region, and maketh phlegm to overcome the blood, and causeth narrowness (or, shrinkage) of the veins and also of the whole body, always provided that it be not drawn from the heart towards a contrary condition.

Now the cold and dry condition of the liver maketh the blood to be little, and the veins narrow, and the body cold, and the epigastric region hairless, always provided that in this case also the heart doth not overcome it.

Of the symptoms (or, indications) of the diseases of the liver.

In this member also, even as in all the other members [of the body], two kinds of diseases arise: one kind is due to a change of condition only, unaccompanied by an ulcer (or, abscess), and the other is due to a change of condition accompanied by some kind of hard, or hot, ulcer or pustule, and the distentions which take place in the liver are due either to this, or to the wind, as also are the obstructions that are due to the thick and viscous chymes which block up the ends of the veins therein, and which grow out of and proceed from the vein that is called the | "door". Now these are accompanied by a sensation of heaviness (or, weight) which weigheth down

Fol. 160a.

the right side; and when a great quantity of vapoury wind is collected therein, and it hath no means of escape, there cometh not only a feeling of heaviness, but also a feeling of distention. Moreover, of all the hot ulcers, in addition to those that arise in the liver through weakness, which arise in the upper regions of the body, and especially those which are large, thou canst easily obtain indications by the mere touch; whilst those that arise in the lower parts and are deep down, are made known chiefly by the attacks that follow their appearance, and not by the touch.

| I will therefore begin to speak about those which arise Page 333 in the upper parts, and which we call "rounded", but I must first call to your memory the muscles that ye have seen in dissections, and that extend inwards from the skin; these are eight in number, and [are arranged] in four pairs. One of these pairs is situated in a visible position. These two muscles are fleshy, and the pair extendeth from the breast to the bone of the lumbar regions. The three other pairs consist of a fleshy substance until they approach the first, simple pair, but when they come near this pair they turn towards certain nervous membranes. One pair of these muscles, that which is above all the rest, descendeth obliquely, and cometh down in front. The second pair is situated opposite the first pair, for it beginneth below, and goeth upwards in a slanting direction towards the front. Below this pair there is also placed another pair of muscles, the situation of which also is from side to side, and it is attached to the inner membrane itself, and is called "appertaining to the peritoneum".

Now the ulcers (or, abscesses) that arise in these straight muscles are long, | and they extend into the middle of all the Fol. 160b. belly, embracing the region of the navel, even as the pair of muscles embraceth it. Therefore the symptoms of these are well known, both by their position and by the form of the ulcers, and besides these also by the fact that outside these there is no other muscle. Now these are inside, as are also the muscles that descend from the breast in an oblique direction. Of the muscles that underlie these it is difficult to obtain indications (or, symptoms), but it is the most difficult thing of

all to obtain information about the other interior muscles that
Page 334 lie | from side to side.

It is well known also that the liver is situated beyond the muscles that appertain to the peritoneum. Therefore, because three muscles are situated above it, it is impossible to obtain any indication as to its condition by touch, unless the ulcer in it is very large indeed, or unless the muscles outside it have become heavy through phthisis. Nevertheless, the feeling of weight that is in the right side of the epigastric region, and the pain that cometh when patients draw in their breath, when this region riseth, and the pain that extendeth to the clavicle of the shoulder, and the frequent little coughings, and the colour of the tongue, which is first red and at last black, and the great sensitiveness [of the skin], and the great thirst, and the vomiting of a certain amount of pure bile, which is [first] like the yelk of an egg in colour, and finally green like verdigris—all these things indicate that the liver is diseased. And the belly of those who suffer from these ailments is constricted, and if this doth not happen in conjunction with weakness of the liver, there must be an ulcer (or, abscess). Now, like unto these things which have been described are also the attacks that arise through a bloody tumour (or, erysipelas), and these are accompanied by fevers and severe thirst. And we also see that the fevers, to which the ancients gave the name of "burning", follow closely in the track of bilious diseases of Fol. 161 a. the liver and belly. Moreover, burning fevers also arise | through bloody tumours of the lungs, just as they arise in every part of the body through the putrefying of the bilious chymes, which result in excessive inflammation. But fevers which are very violent are wont to arise through the belly and through the liver, and in the majority of cases, through these members, the change of these fevers into phthisis taketh place; phthisis, however, we will treat of separately.

Now the ulcers (or, abscesses) that arise in the lower and deep sides of the liver exceed those that arise in the upper parts thereof in respect of want of sensitiveness, and bilious
Page 335 spewings and vomitings, | and violent thirst, just as those that arise in the upper parts exceed those that arise in the lower

parts in respect of causing patients very great pain, even as
we have said above, and in respect of causing dry coughings,
and in respect of making the patient imagine that the pain
that goeth up as far as the clavicle inclineth to the lower part
of the body. Now, the outside ribs and the parts on each
side cause pain in the majority of cases, and this is common
in both diseases. This, however, doth not happen in all cases
with those who suffer pain, because the livers of all men are
not joined to a rib; on the contrary, we see clearly, not only
from those of the apes that have been dissected, but also from
those of other animals, that the liver in some cases is joined
to a rib, and not in others. And it happeneth sometimes that
an ulcer will arise in the upper part of the liver only, just as
one will arise in the lower part only, but it is not possible for
it to rise up and die down in exactly the same place in which
it was first of all in such a way that this can be demonstrated,
because the whole substance of the member is joined to it in
all its parts. When he whose epigastrium is naturally very
flat is smitten by some disease, the great ulcers that exist in
the liver are seen clearly, for they subsist | by themselves out- Fol. 161*b*.
side the muscles which are situated outside it, and there is a
certain limited region within which the groups of ulcers can
be easily felt. For each one of the muscles, the long and the
short, because the parts thereof are united each with each,
little by little acquires ulcers; so then it is not only the liver,
which is wholly contained within a kind of circular boundary,
that is attacked by ulcers. It is well known that when the
liver hath become hard, the indication of the fact is more
evident, so far as its touch is concerned, than when it becometh
hard through ulcers, and that, for the same reason, the muscles
of the body which are outside it | become heavy (?) through Page 336
diseases of this kind. For after a considerable time, even
though the ulcer that produceth the hardness thereof hath
become very large indeed, it happeneth that it is exceedingly
difficult to obtain an indication of this fact by the touch,
because some effusion of liquid, such as a gathering together
of water, hath taken place therein. For besides the fact that
the liver is diseased, it is impossible for there to be in it any

25

disease of this kind; but no liver existeth that hath always suffered, and because of this fact it happeneth very often. On the contrary, because the liver is the organ of the existence of the blood, it followeth of necessity that this existence perisheth when the liver is smitten with sickness. Therefore it happeneth, as hath been already said, that, when one of the other members becometh exceedingly cold through a change of condition, this coldness extendeth even to the liver. For when well-established changes of condition take place in the spleen, and in the belly, and in all the bowels (or, intestines), and especially in those that are in the upper part of the body, and in those veins that are between the intestines, they cool very easily and quickly with them those veins which are in the lower portions of the liver, and through these there also taketh place a change of condition in the whole body of the liver. Now, in disease of the lungs, and of the membrane of the diaphragm, and of the kidneys, those veins that are in the upper portions of the liver first become affected sympathetically, Fol. 162*a*. after a period which is well known, and the liver | itself also suffereth sympathetically. Now these facts are even as they are described here, and no man can dispute their accuracy; if there be a man who gainsayeth them, he is not a man of learning at all.

As concerning obstructions, the indication of these is not so clear. Now, the veins in the lower portions of the liver come forth from the vein that is called the "door", and terminate in small very fine ramifications. And this fact is well known, just as it is equally well known that to that region where these Page 337 veins terminate | there come other ends of veins from the vein that is called the "cave", which is situated on the upper, rounded side of the liver, and that they are divided and terminate among the ends of the other veins, but the ends of the two groups of veins are not attached one to the other. Moreover there is no man who can dispute this fact, nay, all the physicians confess as with one voice, that all the food when it goeth up into all the body, having passed through the veins that are in the lower sides of the liver, entereth into the veins that are in the. upper side of the liver through

the ends of the veins, which have been already mentioned.
Because this is well known, and because it is also well known
that the inflamed ulcer, or the hard ulcer, obstructeth the
passage of the blood, rightly doth it appear, even as many
physicians have also thought, that that matter in the blood
which is fine and watery passeth through, and goeth up
throughout the body; and that the actual blood itself remaineth,
by reason of its thickness, in the lower portions of the liver,
and becometh the cause of the ailment of overfullness. Now,
when we enquire into this confession (or, statement), it is seen
that certain parts of it agree with the things that take place,
and that certain parts do not. The part of it that saith that
the watery part of the blood cometh to that vein [called] the
"cave", and goeth up into the whole body, seemeth to agree
with the opinions that have been enunciated above, as also
doth the observation about the blood that is thick, for the
blood doth not stand and remain inactive in the veins of the
lower portions of the liver, but it moveth on, and descendeth
to the delicate intestines, and to | the colon, and to the blind Fol. 162b.
bowel (or, intestine) and to that lower one also. And perhaps
some one will say, "also into the belly", but this doth not
appear to take place, either in the case of the ulcers, or of
the stoppages which happen in the liver; on the contrary, the
whole body [of the liver] is seen to be full of blood charged
with phlegm, through the collecting of the liquid that is called
"phlegmatic", which formeth in all the flesh. Now, through
the belly nothing | of the nature of blood is discharged, neither Page 338
in the case of dropsy, nor in the case of the other kind of
dropsy which is called "zekkânâ", nor in the kind [called]
"teblânâ". So then the whole region which is between the
membrane of the peritoneum, and the other bodies which are
beyond it inwards, are filled with a fluid of a watery nature.
Now, it is well known when cold hath attacked the liver that
the food which goeth up into it is not transformed into blood,
and also that the veins of the whole body become cold when
it becometh cold, and especially, as often happeneth, when
there can be no ulcer in the liver. But those who are sufferers
are attacked by a dropsical discharge when the spleen is in an

25*

evil condition (or, state), or the most delicate of the intestines, or the veins which are called "mesenteric", or the lungs, or the kidneys are attacked, or when there taketh place a large discharge of blood, or the flux of women, or the retention of the menstrual flow, or some other disease of the womb. For in all these it is seen that, when there is no ulcer whatsoever in the liver, the body is attacked by the disease of dropsy. Especially may it be seen clearly by those who observe, that, through the drinking of cold water at an unseasonable time, the liver becometh exceedingly cold, and that it immediately beginneth to collect water before there is any ulcer in it. Therefore in those who suffer in this manner a great lust for food taketh place, but this is not worth wondering at, because we have already learned that, when the mouth of the belly becometh very cold, the lust for food, which is called "bestial", attacketh a man. Now, for the purpose of making clear [our] discourse, we will go back and explain the differences that Fol. 163 a. exist between the three | kinds of dropsy, and describe the parts of the body in which they take place, and the causes that produce them.

Of the different kinds of dropsy.

Now, the cause of the disease of dropsy is a defect (or, injury) in the power of transforming [food into] blood, which is in the liver. As concerning the places wherein the disease Page 339 can subsist, it is meet to know that sometimes it | subsisteth in the whole body to such an extent that there is no one part of it, large or small, that is free from its malign operation. Sometimes it existeth [only] in the region between the bowels and the membrane which overlieth them, and which is called the peritoneum. When it existeth in this place the water is sometimes very greatly in excess of the wind, but sometimes the wind is greatly in excess of the water; and it is from this fact that the different names which are given to dropsy originate, for the dropsy that attacketh all the members of the body they call "besrânâ" (*i.e.*, appertaining to the flesh). The dropsy that attacketh the part low down, in the region between the intestines (or, bowels) and the membrane (*i. e.,* peritoneum), in which wind is in excess of water, they call "ṭablânâ", (*i. e.,* appertaining to a drum), because when thou strikest on the epigastrium there cometh forth from it a sound like that caused by striking a drum, and for this reason they call this kind of dropsy by the name "ṭablânâ". The third kind of dropsy, wherein water is in excess of wind, and in which the epigastrium when touched shaketh like a wine skin which is not quite full of wine, they call "zeḳḳânâ", because when it existeth in the body the belly resembleth a wine skin (or, bottle).

The causes, which produce the disease of dropsy, have already been mentioned, but it is meet that they should also be elucidated here. The generic causes of dropsy are excess of cold and immoderate heat, for things which are opposite in their natures are wont to produce [dropsy] in the bodies of the children of men. When the natural heat, which is implanted in the liver, and which is the power that produceth the blood, is destroyed—now it can be destroyed either by cold in excess, or immoderate heat—then dropsy is produced. For as cold, | through the condition of its being, killeth and Fol. 163 b. destroyeth this natural heat, in whatsoever member it may be, so also in like manner doth immoderate heat strangle and destroy it, and bring it completely to an end. For behold, | we Page 340 see that a little flame also perisheth before a great flame of fire and is no more found, and if a large quantity of oil be poured on the flame of a lamp it extinguisheth it, even though oil be the substance by which it is fed; but when it is fed with oil in small quantities, the oil nourisheth the flame, and maketh it to shine brightly. Even so is it with an unnatural excess of heat, for when it attacketh the liver, it strangleth the natural heat which is therein. These are the two common (or, general) causes through which dropsy cometh into being, for they quench the natural heat which is implanted in the liver, and they destroy the power in it which transformeth food into blood, and they do not allow nourishment to be transmitted from the liver to the whole body, for they take away the second digestion from the organ that transformeth food into blood, which taketh place when a man is in a sound state of health.

Further, it is necessary to know that this taketh place not
only when there is in the liver one of these two causes which
kill the natural heat thereof, and which effect the destruction
of the agent that produceth the blood, and which through this
act establish the disease of dropsy, but also when one of these
two causes existeth in one of the other members that are
closely associated with the liver through the veins, or that are
nigh unto it by reason of their position in the body, for in a
similar way these causes then also destroy the producer (or,
source) of the blood, and they produce dropsy when the force
of the cause creepeth onward, and arriveth at the liver, and
injureth the digestive power which is therein, which we call
the producer (or, source) of the blood. For behold, how often
do we see the disease of dropsy supervening, not only when
there is an ulcer, or a hardening, or bloody tumours, or stop-
Fol. 164a. pages, or a change of condition, | of whatsoever kind it may
be, whether simple or compound, in the liver, but also when
one of these diseases subsisteth in the belly, or in the spleen,
Page 341 or in the mesenteric veins, | or in the upper, delicate intestines!
Similarly, we have very often seen this disease arise in connection with the letting of blood in large quantities, and with
a perpetual flow of menstrual fluid in the cases of women,
and with the retention of the menstrual flow also. For this
reason no man will ever be so bold as to say that it is possible
for dropsy to be produced by any one of these diseases without
the intervention of the destruction of the liver. For if the liver
hath become cold beforehand, and all those distinguishing
characteristics, which have been mentioned above, and which
exist through that digestive power, are destroyed, then the
disease of dropsy will arise.

Now this power itself, which is situated in the liver, sometimes suffereth independently through the liver, when there is
any cause that maketh it to suffer, and sometimes in sympathy
with the other members, which are near the liver or which
are associated closely with it, when there is any disease in
them, and when this reacheth the liver it is injured. For it is
right to know that when the liver suffereth sympathetically
with the belly, because of its proximity thereto, [the other

members] are associated with it in the disease. And when the liver suffereth sympathetically with the delicate (or, smaller) intestines, or with the mesenteric veins, because of the association that existeth between it and them through the veins, the power that is in the liver suffereth also, and is injured, and the disease of dropsy cometh into being. And when this disease ariseth through the letting of a large quantity of blood, or through the menstrual flow, then the liver becometh cold through the exhaustion of the heat that is therein, and the great abundance of the blood is destroyed and strangled (or, choked) because the heat of the liver is destroyed. Thus also is it in the case of retention of the menstrual flow, and also in the case of the ulcers and bloody tumours, which arise in the liver and injure | Fol. 164 b. it, and of those which arise in the other members, for their disease extendeth to the liver, and causeth it to become diseased (or, sick); for some of them, because of excessive cold, | and some of them because of immoderate heat, quench Page 342 the power therein. It is, moreover, meet to know why the disease of dropsy sometimes ariseth in all the members of the body, and why sometimes only in the region that is between the intestines and the peritoneum.

Now we have learned that this digestive power in the liver is sometimes injured greatly, and sometimes but slightly. When the liver is injured greatly, and its operation is rendered idle, the food that cometh from the belly to the liver, being then in the form of certain chymes, remaineth there undigested, and is not transformed into blood; thence it goeth up into the whole body, and each member, which ought to be nourished (or, fed) by it, is impeded in its work, because it is not blood. And it is well known that it doth not only not nourish those members, but that little by little it destroyeth their [normal] condition, and it produceth in them a cold and watery substance, and they become swollen, and their whole substance becometh dropsical. Now when this power in the liver is only slightly injured, it digesteth the food partly and it becometh in part blood, and the part that is good the members draw into themselves and nourish themselves thereon. The part that is useless for this purpose returneth to the veins and to the intestines, where it is dissolved into watery vapours of some kind, and these watery vapours are emptied out and go forth through the coats of the intestines and of the veins, to the region between the intestines and the membrane that surroundeth them. But, as we have already said, when the wind is in excess of the water, or when there is more water than wind, those varieties of dropsy which we have mentioned above arise. Since then, so far as it lieth in our power, we have explained clearly the distinguishing characteristics of the three kinds of dropsy, let us return to that part of the subject about which it is necessary to speak, and describe also the

Fol. 165 a. other symptoms of the diseases that arise | in the liver.

Page 343   Since | we know that the work of the liver is the production of the blood, we would say that the direct cause that produceth this effect is the power of the liver alone; this it is that transformeth [the food] into blood, and it is one of the Four Natural Powers, of which mention hath been made very frequently; now ye have learned that they are the substances of all the powers in the natural conditions (or, compositions) of the members. Therefore it happeneth in respect of the liver, when it is in one of the eight changes of condition, that its powers are injured by the injuries that are akin to the change of condition wherein it then is. The hot changes of condition which take place in the liver burn up and consume the chymes which are therein, and also those which ascend into it through those veins that come into it from the region of the intestines. The changes of condition that are cold make thick the chymes which are confined therein, and they move and flow with difficulty; and the living chyme, and that which is half digested, which go up into the liver, these cold changes of condition render phlegmatic. Then as concerning the two other changes of condition, the dry one maketh the chymes dry and thick, and the wet one maketh them to be diluted, and thin, and watery.

When ye see stool that is like the water in which freshly killed meat hath been washed, let this be unto you a sure indication that the disease of the patient is situated in the liver only; the diseases which arise through weakness of the

these is that which we call "mettalĕyânâ" (*i. e.,* "which hath the power to draw"), because it draweth to a member the chymes that are akin to it for the nourishment thereof, and it meaneth "desirer", and "drawer". Next after this cometh the | "mĕsha- Page 346 khĕpânâ" (*i. e.,* "transformer"), which transformeth that which is drawn to a member and digesteth it, as, in the case of the chyme which is drawn to a member, this is transformed and made to be like unto the member that is nourished thereby. | Fol. 166*b.* Now, the other two powers (or, effects?) are servants of these. "Akhûdhà" (*i. e.,* "holder") always referreth to the chyme that cometh to a member to be transformed to its nature and is still undergoing digestion. "Mĕnappesànà" (*i. e.,* excretor) referreth to the expulsion of that substance which is useless, and which remaineth over after digestion is accomplished; this is called "superfluity". I advise, then, the physician always to make an examination of these powers in every member that is diseased, as well as of the disease itself, and to find out what its condition is in weakness, and in strength. Now concerning these things we intend to speak. For when the power which draweth is itself diseased, it leaveth the food in the belly, which is made chylus (χυλός), in order that it may be excreted from below by means of the anus, it being moist, undigested matter. This should be to physicians a sign of disease of this power, for everything that happeneth through something else, as it were, from a cause, is a mark and a symptom of that thing itself. Some have said that this disease is due to the veins that go up from the liver, and therefore they have named the diseases from which men suffer after the names of the veins. Thus they call these veins "mesenteric", and they also call "mesenteric" those who are attacked by the disease. Such men, however, make a mistake, in the same way as do those who think that the hands are diseased when fainting fits come on through the stomach, or that men have heart disease because they are unable to move them as they did formerly. For the veins that are between the liver and the belly act as hands to the liver, because they draw away | Page 347 and take food from the belly and convey it to the liver. And in like manner do they act when the legs (or, feet) suffer because of some disease of the lumbar regions, for they bring assistance to the legs (or, feet), and leave alone the part of the spinal column that is affected. For if these veins were

Fol. 167 a. affected by any kind of ulcer, | or by any kind of ailment whatsoever, one would think, and quite rightly, that this disease appertained to them. Now, if because of disease of the liver, the liver were unable to draw to itself food through the veins that are in it, it would not be the mesenteric veins that would be in need of healing, because it is the liver itself that is diseased. Similarly, the hands and the feet would not require healing if they were paralysed through a disease of the spinal column. But let this be a sure indication to you:—if there be any disease caused by an ulcer or a bloody tumour in those veins, ye will be able to detect it if ye inspect carefully the substances that are excreted from below. Because, not only are the substances excreted by such patients like, in every respect, unto those which men say are excreted when the liver is too weak to draw to it foods of this kind, but they will be seen by you to have mixed with them some foul matter, as it were, which cometh forth from the ulcer. For, from all the members that have ulcers in them, and that are not overlaid by some dense, thick covering, various kinds of thin foul matters exude and flow; when the ulcers begin to get ripe, then foul substances which are thick and pus-like flow. When matter of this kind is seen to come forth in the stool that is from the chylus, there being no ulcer whatsoever in the liver, then it is right to conclude that the mesenteric veins are diseased.

Page 348 So also | when these veins are unable, through weakness, to draw to themselves, and to take what they require from the belly, it is well known that their power to draw is impaired; and in like manner when they are unable to hold anything, it is well known that their power of retention is feeble. For this reason the excretions are in the first instance blood mixed with foetid matter, and later they become thick and full of scum. Sometimes evacuations of another kind take place, and these are not due to weakness but to the strength of the excretory

Fol. 167 b. power. These are | frequently due to excess in the food-supply, the substances which are evacuated being healthy in every particular. Very often these take place through the purification of a diseased liver which is effected by natural means, the liver itself being strong and vigorous; for when the evil chymes that are in the liver are digested (or, dissolved) then there goeth forth an evacuation, the things that are of use being retained, whilst those that are foul are expelled. Certain men call evacuations of this kind "bloody disease of the inside" (*i. e.,* dysentery). These happen in many of those in whom some member hath become severed, and in many of those who, after leading a strenuous life of athleticism, have settled down to a life of ease and idleness. Sometimes this kind of evacuation is seen to take place in women, it being due to the stoppage of the menstrual flow, and the blood floweth in large quantities from the anus, like stool. Now this happeneth to many persons through other causes. In cases like these pure blood is evacuated, which is like unto that of lambs, [the throats of] which have just been cut, and sometimes it cometh from the lower part of their belly, and sometimes from the upper. Blood, which is bad, and hath a scum, and is putrid, is evacuated during the diseases that are due to bloody tumours and ulcers, when they come in the organs of digestion, but it appeareth now and again also in diseases of the liver, when it is strengthened by the helping medicines that are administered to it, for when the liver is being purified the excretions are horrible both in their colour and in their smell. In like manner there goeth forth in cases of this kind | urine that is exceed- Page 349 ingly foul (or, bad), and this hath led astray many physicians so greatly that they imagined that the patient was in very .grave danger. For those who without rational enquiry, and who by experience only acquire a simple training, are very remote from the knowledge of those things that only take place at rare intervals, since they only have in their minds those things that are always happening, and in the same way. Now, in conjunction with the indications of digestion, when the illness hath lasted some time, emptyings out of this kind take place, not once, but very often, as ye have learned in the expositions of the books of Hippokrates.

For, as it sometimes happeneth, when the power of retention

Fol. 168 a. becometh so weak that it is unable to hold | that which is [unduly] heavy upon it, or when the power which expelleth ejecteth it, the power of retention will retain and hold that which doth not weigh very heavily upon it; for each of these powers, in each of the members, is sometimes strong, and sometimes weak. When we come to the understanding of each one of them, [we find] that this taketh place through their vigour of attack in respect of [their] operations. For as in very truth each member performeth its appointed work at all times to the utmost power of the constitution (or, condition) which it possesseth naturally, and although there are occasions when it forbeareth for a long time together, during which periods it holdeth fast that which vexeth it, it proceedeth straightway to the expulsion of the same. Similarly there are times, when a member is greatly tortured, during which it eagerly desireth to cast out from itself that which vexeth it, but is unable to do so because of the sickness which is in it, and because great weakness hath descended upon it. It is meet therefore for you to meditate upon all these things in connection with each of the natural organs, and in this wise, when ye exercise your minds in the symptoms of each one of them, having carefully examined them in this manner, ye will find that the diseases of certain members have never been investigated, in any way whatsoever, by those who were before

Page 350 us. | Now, inasmuch as the operation of the liver in our bodies hath been already described, with all the various kinds of diseases that arise in it, and the art of understanding the symptoms of these same, it is meet for us to proceed at once to the description of the treatment to be employed in healing them.

Of the treatment to be employed in healing diseases of the liver.

If now an ulcer beginneth to grow in the liver, and its growth is accompanied by symptoms that indicate clearly that it is there, which symptoms have been already described, and if it seemeth to us that the body is over-full, and the strength of the patient, and the season of the year, and other things,

are satisfactory, then we must let blood from the lower vein
of the right arm, in the fore-arm, or in the palm of the hand,
from the same vein, in order that we may draw away the
blood from the liver, and may empty the body of its over-
fullness. It hath been said | very often in this treatise that Fol. 168*b*.
every ulcer, no matter of what kind, which cometh in a
member, existeth through a flow of blood thereto; and the
first object to be aimed at in healing it is to draw away the
blood to the other members and to empty it, and the second
object is to empty it by means of a purge, and we receive
the proof of [the correctness of] this [view] from the constitution of the members. So that if there be a hot ulcer in
the mouth, we do not employ medicines that draw away
phlegm by any means that giveth relief, and if the stomach
be swollen (or, thick), it is not well to excite vomiting. So
also if the intestines (or, bowels) be swollen it would not be
the act of a skilful physician if we were to employ a cathartic
to bring down the belly. And even so is it with the liver; if
a hot ulcer beginneth to grow in the upper, rounded sides
thereof, it is not good for us to administer a medicine which
would bring down the belly. Similarly also, if the lower sides
of the liver are diseased through an ulcer of some kind, it is
not meet for us to make use of medicines which produce urine
before the ulcer beginneth to attack the digestion; | if we did Page 351
so this act would constitute a danger to the members. The
object of skilful healing is to draw away that chyme, which
floweth to the member that is swollen, to other parts of the
body, in order that we may empty the body, so that we may
be prepared to heal the member that requireth to be healed.
And in addition to this, from the time when the ulcer beginneth
to grow until the time when it is ripe, it is necessary for us
to use things that check and purify it and assistance both
internal and external. When the ulcer hath attained to its full
strength, we must mix with these things those which quiet and
alleviate it slightly; and then, little by little, we must increase
the things that relieve and quiet it, and reduce those that cool
and check it as the sickness dwindleth, until the helpful means
which we are then employing are the exact opposites of those

that were applied at the beginning. And thus also do we act in the matter of the healing of the liver. At the beginning

Fol. 169 a. of the disease we apply the plasters that | cool and check it, if it be a hot ulcer, and we mix with these medicines that are aromatic, since the nature of the member itself demandeth to be strengthened, such as plasters made of bread soaked in wine, and plasters made of asparagus boiled in wine, and in such plasters we mix spikenard, and absinthe, and other aromatic substances, with fine flour of barley, and warming unguents which possess power of themselves, and anointings with oil of asparagus, or oil of myrtle, and pure oil, and foods that possess refreshing powers and are not difficult to digest, such as extract of husked barley, or gourd soup, or spinach soup, or orach soup, or pulse soup, or sorrel soup, or cabbage soup, or stewed fox-grapes. Use these especially if the ulcer hath the nature of a bloody tumour, and the medicines made with

Page 352 oxymel boiled with root of fennel, and the medicines | made with roses and spikenard, the greater number of which have their names written down in the Chapter on the Diseases of the Stomach. And we must treat the patient in this way until the ulcer beginneth to dwindle (*i. e.*, be absorbed), taking good heed at the beginning to avoid a treatment that checketh the ulcer because of the tender nature of the member, and until the signs of the dwindling of the ulcer are seen. Now when the signs of dwindling appear, we apply the medicines which break through obstructions, and clear the mouths of the veins. And if the flanges of the liver and all its upper side be swollen (or, thick), we must reduce the belly by means of foods, such as those that contain carthamus tinctorius (safflowers?) and the seed of nettles, and *khĕbĕlbĕlĕ*, or polypodium, or epithymum, or some medicine that is stronger than these. And, if necessary, we must reduce the belly by means of clysters, at first using those that are gentle, and later those that are more severe. And if the ulcer be situated in the *kûstâ* (Pers. کوست, "vault") of the liver, or in the lower sides, the medicines that produce urine are useful, especially after the ulcer hath begun to dwindle

Fol. 169 b. away, | as, for example, spikenard, and petroselinum (rock parsley), and ammi Copticum, and other such-like medicines.

And when the ulcer hath dissolved itself still more, the medicines that force a way through obstructions and bring down the urine are useful.   This is the method to be followed in healing the hot ulcers, which break out in the liver, and which are accompanied by fever and other characteristic happenings that indicate excessive heat.

And if there be a hard and obstinate ulcer in the liver, and especially in the flanges thereof, the writer of the present work saith, we have not ourselves been able to heal such, and we have never seen one healed by other physicians.   And according to the evidence of the majority of cases of such ulcers, when the greatest care hath not been used in respect of them, they have, after some considerable time, broken up into corruption and ended in dropsy.   For the protracted hardness of the liver all the medicines that break through obstructions, and open the veins which are therein, are useful, and the plasters which are the most softening and which are good for hardness and for excessive flooding, and those | which are good Page 353 for hardness and inflation of the spleen.   For the weakness that ariseth in the liver, without ulcer and fever, all the course of healing that hath been prescribed for the healing of the stomach is useful.   Sometimes we make infusions in wine of the medicines that are used in hot fomentations, and sponge the patient therewith, but sometimes, when pounded up finely and mixed together with wax, which hath been melted down with oil of nard, we make plasters of them, and lay them over the liver; and we also administer draughts which cause the urine to flow, and strengthen the liver.   Now therefore it is meet for us to write down first of all the medicines that are good for the hot ulcer which breaketh out in the liver and for fever.

Tablets made of roses and eupatorium which are good for hot ulcers of the liver and for fever.

Fresh roses	200	drachms
Extract of eupatorium	20	,,
Dew of honey (manna)	200	,,
Spikenard	8	,,

Pound, and work into a paste with water, and make into

26

tablets, each containing one drachm, and administer in extract

of husked barley, or oxymel. |

Tablets made of ambarberis (امبربريس, berberis), which are good for a burning fever, and for ulcers of the liver, and for excessive thirst.

Berberis	4 drachms
Seed of garden cucumbers	6 ,,
Stacte	6 ,,
Bamboo	6 ,,
Absinthe	40 ,,
Fresh roses	40 ,,
Amylum	3 ,,
Tragacanth	2 ,,
Crocus	2 *denkè*

Make into tablets, each containing one drachm, and administer in oxymel.

Tablets made of oil of | aniseed, which are good for the liver, and for fever, and which break through obstructions.

Oil of aniseed	3 drachms
Absinthe	1 drachm
Asarum	1 ,,
Seed of rock parsley	1 ,,
Bitter almonds	1 ,,
Stacte	1 ,,
Spikenard	1 ,,
Laurus malabathrum	1 ,,
Eupatorium	2 drachms
Aloes	2 ,,

Make into tablets with a thick infusion of absinthe, each containing one and two drachms, and administer in one measure of oxymel.

Tablets made of absinthe, which break through the obstructions that arise in the liver, and in the stomach, and which produce urine.

Absinthe	5 drachms
Asarum	5 ,,
Seed of rock parsley	5 ,,

Oil of aniseed······················5 drachms
Bitter almonds, cleaned·········5 „

Make into tablets of one and two drachms, and administer in some convenient draught.

Tablets made of lichen, which are good for the obstructions that arise in the liver and stomach, and for pain in the spleen, and protracted fevers, and which produce urine.

Take in equal quantities the following:

Lichen (?)
Madder
Oil of aniseed
Rock parsley
Absinthe
Asarum
Bitter almonds, cleaned
Costus
Rhubarb
Aristolochia, long
Extract of eupatorium, or of the root.

Make into tablets of one drachm, and administer in some convenient draught.

An antidote made of lichen, which is good for | hardness Fol. 170b. **of the liver and spleen, and which openeth a way through obstructions, and produceth urine, and mitigateth pains in the kidneys.**

Lichen	8	drachms
Bitter almonds, cleaned	5	„
Cinnamon	5	„
Caryophyllus aromaticus	5	„
Laurus malabathrum	5	„
Meum (bear-wort)	8	„
Valerian	8	„
Chamaepitys (χαμαίπιτος, ground pine)	8	„
Myrrh	8	„
Spikenard	12	„
Gentian	7	„

26*

Aloes	14 drachms
Dûḳôn	8 „
Seed of rock parsley	8 „
Black cummin	8 „
Rock parsley	8 „
Ginger	8 „
Crocus	3 „
Asarum	7 „
Valerian	6 „
Cassia	7 „
Incense berries	7 „
Stacte	7 „
Reed of incense	7 „
Bdellium	7½ „
Aristolochia, long	7 „
Peppercorns, long	7 „
Page 355 | Licorice root	12½ „
Rhubarb	12 „
Teucrium polium	2 „
Pistacia lentiscus	2 „
Peppercorns, round	10 „
Costus	10 „
Oil of balsam	3½ „
Seseli	3½ „

Pound the dry medicines well, and reduce to a powder, dissolve those that can be dissolved in strong-smelling wine, pour in with them as much honey as sufficeth, mix [all] together and use. Dose, a portion as large as a chestnut in some draught which is suitable for the disease.

Another [antidote] made of lichen. It is a most excellent and marvellous medicine, and no other existeth which is to be compared with it.

Lichen	2 *estîrê*
Peppercorns, long	2 „
Aristolochia, round	2 „
Bitter almonds, cleaned	5 drachms
Cinnamon	5 „
Meum (bear-wort)	4 „

Valerian	4	drachms
Chamaepitys (ground pine)	4	,,
Caryophyllus aromaticus	4	,,
Seed of rock parsley	4	,,
Rock parsley	4	,,
Dûķôn	4	,,
Cummin	4	,,
Ķarmânî	4	,,
Ginger	4	,,
Dried hyssop	4	,,
Oil of balsam	4	,,
Oil of sweet almonds	4	,,
Myrrh	1	drachm
Crocus	1	,,
Frankincense	1	,,
Gentian	1	,,
Asarum	1	,,
Cassia	1	,,
Myrtle berries	1	,,
Ķîrpâ (kîrķà = κιρκαία?)	1	,,
Spikenard	1	,,
Rûnd	4	estîrê
Madder	4	,,
Aristolochia, long	3	,,
Licorice root	$4^1/_2$	,,
Flowers of pistacia lentiscus	$4^1/_2$	,,
Peppercorns, white	$4^1/_2$	,,
Peppercorns, round	$4^1/_2$	,,
Costus, white	$4^1/_2$	,,
Seseli	$4^1/_2$	,,
Extract of Abgar (agrimonia eupatorium)	10	,,
Hyssop	8	,,
Seed of parsnip	8	,,
Oil of aniseed	8	,,
Cardamoms	8	,,
Extract of mint	5	,,

| Socotra aloes | 5 *estîrê* |
| Good extract of absinthe | 3 „ |

Pound and reduce to a very fine powder, and work into a paste with honey from which the scum hath been removed. The dose is according to what God shall shew thee.

[A medicine] made of crocus, which is good for debility Page 356 **and hardness of the liver, | and spleen, and stomach, and for protracted vomitings, and for dropsy, and it maketh the face to shine.**

Crocus	2 drachms
Spikenard	2 „
Cassia	2 „
Cinnamon	1 drachm
Costus	1 „
Flowers of pistacia lentiscus	1 „

Honey as much as sufficeth.

Dose, one drachm in an infusion of rock parsley, and aniseed, and stacte; it is a helpful medicine.

[A medicine] made of little(?) lichen, which is good for cold of the stomach and liver, and for hardness of the liver of long standing, and it breaketh through the obstacles that exist in the liver and in the kidneys.

Lichen	2 drachms
Costus	2 „
Laurel berries	2 „
Lupins	2 „
Trigonella	2 „
Peppercorns	2 „
Rhubarb	3 „

Honey as much as sufficeth.

Dose, one drachm in an infusion of absinthe.

[A medicine] made of myrrh, and it is called "Ḳûpâr". It is good for hardness of the liver, and for pleurisy, and for coughs.

Gum of terebinth	4 drachms
Myrrh	4 „
Spikenard	1 drachm
Crocus	1 „

Cinnamon	1 drachm
Cassia	2 drachms
Aspâlitôs (harts-tongue fern?)	2 „
Flowers of pistacia lentiscus	2 „
Flesh of dried grapes, cleaned inside and out	45 „
Honey as much as sufficeth.	

Dose, one drachm. Dissolve the medicines which can be dissolved and the dried grapes in strong-smelling wine, pound the dry ones and clean them, melt the gum of terebinth with the honey, mix them all together, work up well and pour into a vessel, and administer in some drink that is suitable for the particular disease. It may also be burnt like incense before the table, and its smell is very pleasant.

The medicine "Atenasyâ" (ἀθανασία, *i. e.*, "Immortality"), which is made of the liver of a wolf. It is good for pains in the liver, and spleen, and stomach, and for dysentery, and wind, and protracted coughs, and for those who spit blood, and it giveth relief in attacks of such a disease as that of Philo (Pîlôn). | And the present writer beareth wit- Fol. 171*b*. ness to the fact that one dose of it, given to him with extract of chicory and fresh fennel, removed entirely the hot pain in the liver. This medicine is compounded | of Page 357 the following:

Myrrh		
Crocus		
Seed of poppies		
Opium		
Castoreum		
Hyoscyamus		
Costus	in equal
Cardamoms	quantities
Spikenard		
Root of Abgar (agrimonia eupatorium), or the extract		
Wolf's liver		
The right horn of a goat, roasted		

Honey, as much as sufficeth.

Wine for the medicines to be dissolved.

Mix well together, and give as a dose a portion as large as a chick-pea, in some convenient draught.

The Atenasyâ of Asclepiades, which is good for pains of the liver, and for wind, and burning, and pain in the stomach, and for dysentery, and strangury, and blows of reptiles (*i.e.,* **bites of insects), and for those who bring up blood and pus.**

Spikenard	4	drachms
Costus	4	„
Rock parsley	4	„
Dûḳôn	4	„
Asarum	4	„
Peppercorns	4	„
Castoreum	4	„
Opium	8	„
Aniseed	8	„
Seed of rock parsley	8	„
Seed of hyoscyamus	8	„
Myrrh	8	„
Crocus	8	„
Acorus	3	„
Meum (bear-wort)	3	„
Styrax	8	„
Sea-seals	6	„
Bark	6	„
Root of mandragora	6	„

Honey as much as sufficeth.

Pound, and administer as a dose a portion the size of a chick-pea, in some convenient draught.

The Little Atenasyâ, which is good for the diseases. A well-tried remedy.

Styrax	4	drachms
Crocus	4	„
Costus	4	„
Spikenard	4	„
Cassia	4	„

Opium   4 drachms
Extract of eupatorium   8   ,,
Licorice root   200   ,,
Honey as much as sufficeth.

Dose, a portion the size of a walnut, in some convenient draught.

[A medicine] made of gentian, which is good for hardness of the liver, and for obstructions, and for pains in the stomach, and spleen, and kidneys, and for protracted fever.

Gentian   10 drachms
Peppercorns, round   10   ,,
Costus   1 ounce
Malabathrum   1   ,,
Spikenard   1   ,,
Rhubarb   1   ,,
Honey as much as sufficeth.

Dose, one drachm in an infusion of rue.

Têryâḳê (Theriaca) [composed of] four medicines which are | good for pains of the liver and spleen of long stand-Fol. 172 a. ing, and for diseases of the stomach, and for the falling sickness, | and for insanity and delirium, and for the bites Page 358 of noxious reptiles, and for palpitation of the heart.

Gentian   ⎫
Laurel berries   ⎪ in equal
Aristolochia, round or long ⎬ quantities
Myrrh   ⎭
Honey, with the scum removed, as much
      as sufficeth.

Dose, one drachm, in hot water. Others substitute costus for the myrrh, and others add a little saffron.

The Antidote Ariston, which is good for a bad condition of the liver, and for phthisis, and for protracted pains of the stomach; it is helpful in every kind of disease, provided that a man useth it with the skill and understanding which are meet.

Euphorbium   6 drachms
Crocus   6   ,,
Cassia   6   ,,
Amomum (?)   6   ,,

Opium	6	drachms
Acacia	6	„
Costus	6	„
Myrrh	6	„
Nard	6	„
Gum Arabic	6	„
Seed of nettles	6	„
Seed of croton	6	„
Bdellium	6	„
Frankincense	6	„
Sumach	6	„
Gum (or, glue)	6	„
Yellow sulphur	6	„
Dry styrax	6	„
Peppercorns, white	6	„
Dried roses	4	„
Pyrethrum	4	„
Seed of cyclamen	4	„
Seed of rue	4	„
Seed of rock parsley	4	„
Seed of thorns	1	drachm
Seed of lettuce	4	drachms
Wild bitter herbs	4	„
Ammi Copticum	4	„
Seed of hyoscyamus	10	„
Safflower	4	„
Ginger	4	„

Pound the dry medicines very carefully, and steep them in wine for three days. Dissolve the soluble medicines in wine, and melt the glue with honey. Mix them all together, work them into a paste with as much honey as sufficeth, and whilst thou art working them up add to them a measure of oil of balsam. Then pour them all into a vessel and set it on the fire, and let them boil for a little; take and use after six months. Dose, one drachm for every ailment, in some convenient draught.

An Antidote [made of] many kinds of medicine. It is Page 359 good for pains of every kind, and especially | for the hard

liver, and the cold stomach, and it removeth wind from
the belly, and effecteth digestion. |

Castoreum	1	drachm
Opium	1	,,
Cinnamon	1	,,
Meum (bear-wort)	1	,,
Valerian	1	,,
Dûḳôn	1	,,
Asarum	1	,,
Others (*sic!*)	4	drachms
Râmîn	6	,,
Myrrh	6	,,
Peppercorns, round	6	,,
Peppercorns, long	6	,,
Galbanum	6	,,
Costus	6	,,
Crocus	6	,,
Leeks	—	,,

Honey as much as sufficeth.

Dose, a portion the size of a bean, in some draught which is
convenient. Dissolve the soluble medicines in wine, pound the
dry ones and reduce to a powder. Melt the galbanum with
honey, and mix them all together, and administer according
to what is meet.

[A medicine] made of rhubarb, [which is good] for every
ailment of the liver and stomach, and for obstructions (or,
stoppages).

Rhubarb	10	drachms
Greek absinthe	10	,,
Gum Arabic	2	,,
Malabathrum	1	drachm
Seed of parsnips	5	drachms
Crocus	2	,,
Spikenard	2	,,
Aloes	5	,,
Myrrh	1	drachm
Musk	1	,,
Castoreum	4	*denḳê*

Seed of rue	2 drachms
Seed of rock parsley	2 „
Honey as much as sufficeth.	

Dose, one drachm, in an infusion of trigonella.

An Antidote made of gentian, which is good for protracted hardness of the liver and spleen, and for removing obstructions, and it is helpful in coughs of long standing, and in phthisis.

Gentian
Rhubarb
Pistacia lentiscus
Incense berries
Aniseed
Spikenard
Asarum
Castoreum
Crocus
Seed of rock parsley
Cassia
Costus
Ammi Copticum
Rock parsley
Chamaepitys   all in equal
Myrrh   quantities
Fir cones
Licorice root
Madder (parsîôn)
Σκόρδιον (water germander)
Χαμαίδρυς (Chamaedrys,
    wall germander)
Madder (pôthâ)
Stacte
Aristolochia, long
Aristolochia, round
Arsenic
Irises

Honey, with the scum removed, as much as sufficeth.

Dose, one drachm, in infusions of rue and aniseed.

The Antidote Pîkhadôtôs, which is good for cold of the liver and stomach, and spleen, and for coughs, and jaundice, | and twisting of the intestines, and fistular Page 360 pains, and for disease of the womb, and difficulty in passing water, and for dropsy, and for all the diseases of women.

Crocus	5 drachms
| Spikenard	4 ,,
Malabathrum	8 ,,
Seed of hyoscyamus	6 ,,
Castoreum	6 ,,
Peppercorns, white	4 ,,
Ginger	4 ,,
Cassia	2 ,,
Euphorbium	$^1/_2$,,
Fat of yathmâ	$4^1/_2$,,

Fol. 173 a.

Honey as much as sufficeth.
Dose, a portion the size of a chestnut.

A Mixture of rhubarb which is good for pains in the liver and spleen, and which breaketh through obstructions and expelleth jaundice.

Rhubarb	6 drachms
Scamonia	6 ,,
Dried grapes	100 in number
Fir cones	100 ,,
Peppercorns	100 ,,

Honey as much as sufficeth.
Dose, two drachms.

Another Mixture which is good for protracted fevers, and protracted hardness of the liver and spleen.

Acorus	10 drachms
Ammoniac	10 ,,
Rhubarb	8 ,,
Cassia	8 ,,
Lichen	8 ,,
Castoreum	200 ,,
Amomum	3 ,,
Aniseed	6 ,,

Bdellium	200 drachms
Honey as much as sufficeth.	

[Dose,] one drachm.

Another Mixture.

Root of ferula asa foetida	1	drachm
Ginger	1	„
Rhubarb	2	drachms
Honey as much as sufficeth.		

Dose, two drachms.

Another Mixture, which is good for the person who falleth down and injureth his liver, and for him that bringeth up blood.

Rhubarb	10	drachms
Ginger	20	„
Lichen	5	„

Pound, and administer one spoonful dry with some convenient draught. Others add to the above two drachms of yellow arsenic.

Another [Mixture] which is good for those who are smitten, or whose livers are injured, and for those who bring up blood. Take camphor leaf, and smelters' dross, and myrrh in equal quantities, pound them, and administer | dry, or infuse them in water, which is to be given to the patient in draughts.

Another Mixture made of rue, which reduceth the belly and is good for tumidity, and for pains of the liver and stomach.

Rue	10	drachms
Cummin	10	„
Peppercorns, round	8	„
Peppercorns, long	8	„
Amomum	8	„
Talispar	8	„
Seed of ammi Copticum	8	„
Seed of gourds	8	„
Scamonia	5	„
Pulp of colocynth	7	„
Convolvulus	7	„

Page 361

Bark of incense	2 drachms
Seed of rock parsley	7 ,,
White sugar	7 ,,
Honey as much as sufficeth.	

Dose, one drachm, in wine mixed with water.

Gughershân Shahrîrân, which is good for severe cold in the liver and stomach, and for dropsy, and it cleareth out the belly.

Lepidium latifolium	6 drachms
Ginger	6 ,,
Peppercorns, long	6 ,,
Peppercorns, round	6 ,,
" Many feet ", *i. e.,* polypodium	6 ,,
Incense bark	6 ,,
Amomum (shôshmîr)	6 ,,
Caryophyllus aromaticus	6 ,,
Nadh-bûsht (= nagh-bûsht)	6 ,,
Laurus malabathrum	6 ,,
Amylum	6 ,,
Cardamoms	6 ,,
Cinnamon	6 ,,
Stacte	6 ,,
Spikenard	6 ,,
Bark (cinnamon ?)	6 ,,
Seed of rock parsley	6 ,,
Ammi Copticum	6 ,,
Seed of fennel	6 ,,
Aniseed	6 ,,
Castoreum	2 ,,
Epithymum	12 ,,
Convolvulus	16 ,,
Panîd (sugar?)	60 *estîrê*

Melt the loaf sugar (?) with the wine, mix all the other things with it, add as much honey as sufficeth, and administer a portion as large as a walnut to him in whom thou wishest to produce a motion of the bowels. If, through the cold in the liver and in the stomach, it is necessary for thee to administer this medicine daily, then give him a portion as large as a

Pontic nut, either in wine mixed with water, or in an infusion of aniseed and stacte.

A medicine which is made of black ferula asa foetida, and is good for hardness of the liver, and for dropsy, and for cold in the stomach and kidneys.

Black ferula asa foetida	10	drachms
Seed of gourds	8	,,
Seed of leeks	8	,,
Ginger	7	,,
Bĕlilḵê	7	,,
Amɫàgh	7	,,
Ammi Copticum	5	,,
Seed of rock parsley	5	,,
Aniseed	5	,,
Amomum	5	,,
Cardamoms	5	,,
Cummin	5	,,
Cinnamon |	5	,,
Black hellebore	7	,,
Bark of incense	7	,,
Peppercorns, long	4	,,
Peppercorns, round	4	,,
Spikenard	2	,,
Caryophyllus aromaticus	1	drachm
White sugar (?)	10	drachms

Page 362

Honey, with the scum removed, as much as sufficeth.

Dose, two drachms, in wine mixed with an infusion of aniseed, stacte, and spikenard. This medicine will strengthen the liver and stomach, and will make beautiful the colour of the face.

Another, which is good for the liver and spleen which have become hard.

Fol. 174a.

| Dry pitch
Myrrh
Costus   in equal
Trigonella   quantities
Peppercorns

| Dry pitch
Myrrh
Costus } in equal quantities
Trigonella
Peppercorns

Pound them and administer two drachms in a draught of wine mixed with water, but if fever be present administer in pure water.

The simple medicines which are good for protracted hardness of the liver.

Two spoonfuls of the finest oil of nard in an infusion of rue and dill, to be taken in a draught daily for seven days.

Or oil of balsam, in the same way.

Or oil of costus, in the same way.

Or make the patient drink one measure of oil of nuts in an infusion of trigonella, and rue, and spikenard, and cypress berries.

Or pound up two drachms of costus, and let him drink it in wine mixed with water.

Or make an infusion of costus, twigs of agrimonia eupatorium, trigonella, and grapes, and make the patient drink it, in draughts, each of which containeth one measure of oil of walnuts. Let the patient sleep on his right side, and drink a draught daily for three days.

Or make an infusion of absinthe, rhubarb, costus, rue, pistacia lentiscus, peppercorns, and grapes, and make the patient drink draughts of the same [mixed with] lichen, or crocus, or "athanasia" (ἀθανασία), or malabathrum.

If there be fever, with thirst, and an ulcer on the liver, administer draughts containing extract of chicory and extract of fennel.

Or fox-grapes pounded to a fine powder with one of those medicines compounded with roses, which have been written down in the beginning [of the Chapter].

Or infuse a piece of malabathrum about the size of a chickpea in water of absinthe, asarum, lupins, and costus, and make the patient drink draughts of the same, each containing one drachm of myrrh. |

Page 363

Or crush half a drachm of dry pitch, and mix it with any one of these infusions, and let the patient drink it in the form of a draught.

Or take in equal quantities gentian, absinthe, asarum, costus, twigs of agrimonia eupatorium, root of camphor, madder, root

27

of rock parsley, bark of fennel root, rue, pistacia lentiscus, thyme, lentils, trigonella, *ashdahân* (?), and dried grapes, and steep them in water until one half of it hath evaporated, and let the patient drink a draught of two measures daily, [mixed] with one of those medicines which are good for the ailment, Fol. 174*b.* and which have been already mentioned | above.

Or make an infusion in water of spikenard, and stacte, and aniseed, and asarum, and cypress berries, and cinnamon, and kôlîghân, and lichen, and absinthe, and administer two measures as a draught daily for three days.

The foods that are good for those who have fever, and inflammation and disease of the liver, have been written down above, and the patient must abstain from animal food, and from drinking wine.

If there be in the liver hardness and obstructions, without fever and inflammation, medicines that break through the obstructions, and open the veins of the liver, and cleanse them painlessly, are useful, as, for example, aristolochia of both kinds [*i.e.*, round and long], costus, lichen, madder, absinthe, asarum, and all medicines of this kind. In short, all the medicines that have a bitter taste, and are, at the same time, slightly astringent, are good for the hardness and the stoppages, which take place in the liver, and spleen, and stomach. Sometimes we steep them in wine, which we heat and then use to foment the parts of the body that are over the liver. Finally, we anoint these parts of the body with one of the unguents that are good for the purpose, such as oil of nard, or oil of asparagus, or we lay on them one of the well known plasters that have been mentioned in connection with diseases of the liver, such as pôlrakhnôn (?), or one of the others that are good for hardness of the liver. And [we prescribe] food which is easily digested, Page 364 and | soft wine which is not very old, and from time to time baths in sweet (*i.e.*, fresh) water.

If there is in the liver such great weakness, although it is not accompanied by hardness and an ulcer, that, owing to some member which is situated near to it, it falleth into an evil condition, and there ariseth from it a bloody flux of the belly, or [one of those watery, undigested fluids which have been mentioned above, all kinds of aromatic herbs are good for it, as, for example, spikenard, and cassia, and such like, when steeped in water. And in infusions of these herbs must be placed one of the medicines that are good for over-fullness of the liver and stomach. | Sometimes these must be steeped in Fol. 175 a. wine that is slightly astringent, and fomentations must be applied as in cases of hardness of the liver. And sometimes these medicines must be crushed and pounded, and thrown into strong-smelling wine and oil of nard, or into a preparation of jasmine, or of aromatic reeds, and then heated over the coals and boiled; and into this rolls of wool must be dipped and spread out on the body above the liver. After the application of these plasters patients must be made cool (or, refreshed), and as dry food they must eat things made of meat and asparagus, and they must drink wine made of the flowers of the wild vine; and they must eat food that possesseth strength and which will strengthen the liver, and they must drink strong-smelling wine, and they must abstain from everything which is difficult of digestion.

Now if the liver be so much destroyed that dropsy ariseth through it, in attempting to heal it, it is meet to multiply the use of the medicines that will produce urine; and from time to time we must use the medicines that produce evacuations, and that purge gently without being sufficiently strong to injure the liver. And there must be frequent washings and bathings, and anointings with hot and stimulating oils, and sprinklings of natron and sulphur on the epigastrium, and plasters that make the water to run out of the body, such as those composed of the dung of oxen, or the dung of goats. Other things good for this disease are, the | drinking of the fat Page 365 of camels [which have passed] many days without food and drink, or let the patient drink each day three measures of the urine of camels, or of goats, and a spoonful of green hellebore pounded and powdered. Or take in equal quantities dill, and seed of rock parsley, and aniseed, pound them to a powder, and administer each day one spoonful in the urine of a camel. Or let him throw into his mouth one drachm of sagapenum, and drink afterwards the urine of camels, or of goats. All kinds of dropsy are, however, difficult to cure. The kind of dropsy that is called "fleshy" can, perhaps, be cured, provided that the sick man, and the physician who ministereth to him, and the season of the year, and the period of the patient's life, all suit each other, and the physician treateth the case with the proper skill and understanding. It is now meet that Fol. 175 b. the compound | medicines, which empty the belly, and bring down the water, and break through obstacles, be written down.

Pills made of rhubarb which are good for obstructions of the liver, and stomach, and spleen, and for the dropsical, and which purge gently.

Rhubarb	7	drachms
Barbarikôn (a foreign drug)	7	„
Sagapenum	5	„
Bdellium	5	„
Ammoniac	5	„
Lichen	7	„
Opopanax	3	„
Tithymalus (a kind of euphorbium)	4	„
Convolvulus	12	„
Spikenard	5	„
Polypodium	5	„
Aloes	5	„
Stacte	2	„
Green hellebore	10	„
Amlagh	4	*estîrê*

Dissolve the soluble drugs in wine, pound the dry ones to a fine powder and mix with them, make pills the size of peppercorns, and administer as a dose two drachms in wine mixed with water.

Pills made of spikenard, which break through the obstructions that arise in the liver and stomach, and which help the dropsical, and reduce the belly gently (*i. e.,* cause painless evacuations).

Spikenard	3	drachms
Agaricum (larch fungus)	3	„
Rhubarb	3	„

Amomum	3 drachms
Stacte	3 „
Epithymum	3 „
Crocus	3 „
Costus	3 „
Semadrêôs (Chamaedrys?)		3 „
Myrrh	1 drachm
Bark of incense	1 „
Pistacia lentiscus	1 „
Aloes	7 drachms

Page 366

Make into pills with old wine, having [first] anointed thy fingers with oil of jasmine, and administer as a dose two drachms in wine mixed with water, before and after food.

Another well tried medicine which is good for the drop-sical (?), and also for the healthy.

Green hellebore	10 drachms
Bĕlîlk̄ê	2 ounces
Convolvulus	10 drachms
Indian salt	4 „
Mâzrîôn	1 ounce
Peppercorns	5 drachms
Ginger	3 „

Pound, work into a paste with honey, and administer as a dose one drachm, in strong-smelling wine mixed with water.

Pills made of agaricum (larch fungus), which break through the obstructions, and purge the diseased, to whom they must be administered with camels' fat (or, milk). And they help the liver and stomach.

Epithymum	6 drachms
Aloes	6 „
Agaricum (larch fungus)	4 „
Scamony	2 „
Rock parsley	2 „
Nettles	2 „
Aniseed	2 „
Seseli	2 „
Seed of rock parsley	2 „
Dûk̄ôn	2 „

Fol. 176a. | Pound, and administer in the form of a dry powder, a dose being two drachms, or if taken in wine mixed with water, one drachm.

A medicine made of Khamâlîâ (χαμελαία or, dwarf olive?), that is to say, mazriôn, which is good for the dropsical, and which cleareth out the water in large quantities.

Mazriôn, steeped in wine and dried	1	drachm
Convolvulus	1	„
Epithymum	1	„
Aniseed	$^1/_2$	„
Seed of rock parsley	$^1/_2$	„
Green hellebore	$^1/_2$	„

Pound, and administer as a dose two drachms.

Another medicine made of sagapenum.

Sagapenum	2	*estìr è*
Gum ammoniac	1	drachm
Opopanax	1	„
Aristolochia	1	„
Convolvulus	2	drachms
Mazriôn	2	„
Agrimonia eupatorium	1	drachm
Shaṭharagh (?)	1	„
Pulp of wild cucumbers	1	„
Calamint (?)	1	„
Squills	2	drachms

Pound all these, and steep in sugar (syrup?), and set in the sun for seven days; clarify (or, strain) and administer daily a Page 367 dose of three measures. | Let the patient drink it for seven days and he will be better.

A medicine made of figs, which is good for the dropsical, and it will expel the water. Soak fine fleshy figs in sweet oil for one whole night and day. Then take them, and open them, and fill them with either the root or pulp of wild cucumbers well crushed, and administer two or three figs as a dose. Give the rest according to the strength [of the patient], who, after them, must drink tepid water. And afterwards let him be well rubbed with nitre, olive oil, and wine, and let his whole body, from head to food, be bandaged, and he will become well.

Pills, made of sagapenum, which are good for the drop-sical, and for pains in the stomach, and for fistula (?), and for the wind that goeth about in the intestines, and for pains in the back, and they strengthen the liver.

> Sagapenum
> Aloes
> Bdellium — in equal quantities
> Gum Arabic
> Larch fungus (agaricum)

Make into pills with wine, or with extract of ḳaḳang (Pers. کاکنج, solanum halicacabum, or vesicaris, i. e., bladder-wort), and administer as a dose two drachms.

Pills made of Khamelaia (dwarf olive?) which are good for the dropsical, and they expel water from the body freely. Steep khamelaia, that is to say, marzîôn, in strong vinegar for a whole night and day, and then dry it in the shade. Crush it and work it into a paste with vinegar, and make it into thin (small?) pills, and dry them in the shade. Then take some of | these pills, and burnt copper and aniseed Fol. 176 b. in equal quantities, pound them and make into a paste with extract of ḳaḳang, and make into pills. Administer as a dose one or two grammes in dill water.

Pills made of bdellium, which are good for the dropsical, and also for healthy folk, and they expel the water freely.

> White bdellium
> Pulp of colocynth — in equal quantities
> Larch fungus (agaricum)
> Scamony

Make into pills with wine, and administer as a dose two drachms in wine mixed with water, before and after food.

Pills made of rhubarb, which are good for pains in the liver, and stomach, and spleen, and they break through obstacles, and make patients to micturate freely.

> Myrrh   1 ounce
> Bdellium   1   „

Ammoniac	1	ounce
Sagapenum	1	,,
Stacte	1	,,
Frankincense	1	,,
Aloes	1	,,
Castoreum	1	,,
Cinnamon	2	drachms
Crocus	2	,,
Spikenard		2	,,
Peppercorns	2	,,
Seed of rock parsley	1	drachm
Aniseed	1	,,
Cummin	1	,,

Page 368

Make into pills with extract of cabbage, and administer as a dose two drachms.

Pills made of mazrîôn, which are good for the dropsical.

Mazrîôn tablets soaked in vinegar	20	drachms
Green hellebore	20	,,
Peppercorns	2	,,
Ginger	2	,,
Cummin	2	,,
Ammi Copticum	2	,,
Root of fennel	2	,,

Make into pills with extract of cabbage, and administer as a dose four drachms in cold water; if taken in hot water the passage of the water will be prevented.

Pills made of euphorbium, which are good for the dropsical and for pains in the back, and for fistular pains, and for gout, and they help the healthy, and those whose members are sluggish. Now they must be taken before and after food.

Euphorbium	2	drachms
Stacte	2	,,
Scamony	5	,,
Larch fungus	5	,,
Aloes	10	,,
Epithymum	10	,,

Extract or flower of absinthe   5 drachms
Bdellium   5   „
Salt   1¹/₂ „
Ammoniac   1¹/₂ „
Peppercorns, long   2   „
Aniseed   4   „
Spikenard   10   „

Make into pills the size of peppercorns with extract of cabbage,
| and administer [as a dose] ten, and let the patient drink hot Fol. 177 a.
water after taking them.

**Pills made of dates, which are good both for those who
are sick and those who are well. They may be taken in
winter and in summer, and at all seasons of the year, and
they are beneficial in cases of fever.**

Fleshy dates, cleaned inside and out   10 [drachms]
Soak them in strong, pungent vinegar from evening till morning, and then pound them well.   Then take

Fresh rue   9 drachms
Scamony   9   „
Peppercorns   200 in number
Armenian nitre   2 drachms
Ginger   2   „
Seed of rock parsley   1 drachm
Bitter almonds, cleaned   50 in number

and pound them all and mix with the dates, and add as much
honey | as is sufficient to make them into a paste. Administer Page 369
as a dose two or four drachms, but it is a useful medicine
taken in any quantity.

**Another medicine made of cinnamon, which is good for
a bad state of the liver, and for tumidity of the liver, and it
strengtheneth the stomach, and beautifieth the complexion.**

Black hellebore   65 drachms
Stacte   65   „
Ginger   10   „
Cinnamon   10   „
Peppercorns, white   14   „
Ṭalíspar   3   „
Kúlanghán (خولَنّخان, galanga) 10   „

Nad-bûsht (flowers of bûsht) 6 drachms

New wine ⎫
Wine of lilies ⎭ as much as sufficeth.

Make into pills, and administer as a dose two drachms.

Another [medicine], which is good for the dropsical, and it causeth the water to pass freely.

Wild doves' droppings	4 drachms
Euphorbium	4 ,,
Burnt copper	7 ,,
Dry rue	3 ,,

Make into pills with extract of cabbage, and administer as a dose one drachm.

Another [medicine], which is good for the dropsical, and for pains in the back, and for fever.

Aloes	6 drachms
Epithymum	3 ,,
Rhubarb	4 ,,
Asarum	4 ,,
Agaricum (larch fungus)	4 ,,
Crocus	3 ,,
Spikenard	3 ,,
Rock parsley	1 drachm

Make into pills with extract of ḳaḳang; dose, one drachm.

Another [medicine], which is good for the dropsical, and which maketh the water to pass freely.

Burnt copper	4 drachms
Rue	4 ,,
Aniseed	4 ,,
Mazrîôn	3 ,,

Fol. 177 *b*.

Make into pills with boiled honey, and administer as a dose one gramma, or two, in wine mixed with water.

Another [medicine], which maketh the water to pass very freely. Take in equal quantities burnt copper, the droppings of doves, rue, mazrîôn, and euphorbium, and make into pills with wine or boiled honey, and administer as a dose one gramma, or two, in wine mixed with water.

A medicine made of sagapenum, which is good for an evil condition of the liver, and for the dropsical.

Sagapenum	200 (?) —
Ammoniac	4 drachms
Aloes	4 ,,
Bdellium	4 ,,
Myrrh	3 ,,
| Agaricum	5 ,,
Euphorbium	2 ,,

Page 370

Make into pills with extract of ḳaḳang, and administer as a dose two drachms.

An Antidote of Asclepiades, which is good for pain in the liver, and which removeth obstructions in the veins, and it benefiteth the dropsical greatly.

Gentian
Chamaepitys azuga
Madder
Seed of rock parsley
Colocynth pulp
Mustard
Seed of garden cucumber
Σκολοπένδριον (Venus's hair)
Root of ferula opopanax
Sea seals
Madder
Cabbage seed
Aristolochia, long
Peppercorns, white
Spikenard
Costus
Seed of gourds
Rock parsley
Eupatorium
Origanum
Fir cones
Garden nigella
Chamaedrys
Origanum dictamnus (dittany)
Dûḳôn

all of them in equal quantities

Honey as much as sufficeth.

Dose, a portion as large as a bean, in wine mixed with water. This medicine hath been well tried, and Galen said that it was most excellent.

Another medicine of the same kind, for the same diseases.

Rhubarb	1	drachm
Dûk̄ôn	1	,,
Pistacia lentiscus	1	,,
Incense berries	1	,,
Aniseed	1	,,
Spikenard	1	,,
Crocus	1	,,
Cinnamon	1	,,
Cassia	1	,,
Valerian	1	,,
Asarum	1	,,
Rock parsley	1	,,
Chamaepitys	1	,,
Wild mint	1	,,
Fir cones	50	in number
Honey as much as sufficeth.		

[Make into pills] the size of a chestnut, and administer as a dose one measure in honey wine mixed with water. It maketh the water to pass, and relieveth the pain.

An Antidote made with crocus, and prescribed by Galen. It is good for protracted pains of the liver, and is an ex- Fol. 178a. **cellent check | to the diseases [thereof].**

Crocus	200	drachms
Meum (bear-wort)	4	,,
Madder	4	,,
Asarum	4	,,
Dûk̄ôn	4	,,
Rock parsley	4	,,
Myrrh	4	,,
Indian nard	6	,,
Nard k̟liṭàyâ	6	,,
Costus	1	drachm
Cassia	1	,,

Pistacia lentiscus	1 drachm
Incense berries	1 ,,
Madder	2 drachms
Extract of licorice root	3 ,,
Venus's hair	3 ,,
Malabathrum	3 ,,
Eupatorium	3 ,,
Oil | of balsam	6 ,,
Ardekhrôn	5 ,,

Page 371

Honey as much as sufficeth.

Dose, about as much as a chestnut in honey wine.

A medicine made of costus, which is good for protracted hardness of the liver, and for [removing] obstructions in the veins thereof.

Costus
Eupatorium
Lupins   } in equal quantities
Trigonella
Peppercorns

Pound dry and reduce to a powder. Dose, two drachms in kandîkôn, which is mixed with an infusion of aniseed, stacte, and spikenard.

Another medicine for the same disease.

Rhubarb	2 drachms
Costus	4 ,,
Peppercorns	4 ,,
Trigonella	4 ,,
Laurel berries	4 ,,
Lupins	4 ,,

Pound and administer in doses of the same size as of the preceding medicine.

Another.

Costus	2 drachms
Eupatorium	1 drachm
Peppercorns	$^1/_2$,,

Pound and administer in [one of] the infusions which have been mentioned.

Another.

Costus ⎫
Resin ⎬ in equal quantities.
Gentian ⎭

Pound, and administer a spoonful in water to those who have fever, and in wine to those who have no fever.

Another medicine [which is good] for protracted hardness of the liver; it hath been well tried.

Liquid styrax	3 drachms
Costus	4 ,,
Stacte	4 ,,
Gum of almonds	4 ,,
Galbanum	4 ,,
Laurel berries, cleaned	4 ,,
Peppercorns	1 drachm
Opium	4 grammes

Pound, and mix with extract of trigonella, and honey, and administer as a dose one drachm in kandíḳôn, mixed with an infusion of spikenard and stacte.

Another medicine which is good for hardness and obstructions in the liver.

Fennel seed	200 drachms
Rock parsley	6 ,,
Aniseed	6 ,,
Bitter almonds, peeled	6 ,,
Spikenard	6 ,,
Absinthe |	6 ,,

Honey as much as sufficeth.

Dose, two drachms, in kandíḳôn, which is mixed with an infusion of asarum and absinthe.

Another medicine which is very good for the same diseases.

Madder   ⎫
Lichen   ⎪
Origanum   ⎬ all of them
Eupatorium ⎪ in equal
Âlinôn, or resin ⎪ quantities
Malabathrum ⎭

Costus
Caryophyllus aromaticus
Peppercorns
Aristolochia |
Venus's hair   all of them
Oak-earth, that is chamae-   in equal
    drys   quantities
Spikenard
Gentian
Cabbage seed
Gourd seed
Honey as much as sufficeth.

Page 372

Dose, a portion the size of a walnut, in honey wine. [This medicine is] also [good] for jaundice.

Or crush three grammes of aristolochia, round or long, and administer in water to those who have fever, and in wine mixed with water to those who have not fever.

Or make an infusion of blue melilotus flowers in water, and [reduce it by evaporation] to one half, and strain it, and administer as a dose two measures. Let the patient take this dose daily for three days.

Another medicine [prescribed by] Severianus, which is good for the dropsical, and it maketh the water to pass very freely. Crush one drachm of euphorbium, and administer in an egg which is to be sucked.

Another medicine made of squills, which is good for the dropsical.

        Roasted squills   1 *lîṭrâ*
        Strong-smelling wine   4 measures

Boil in an earthenware vessel until only one measure remaineth, then strain, and administer to the patient a spoonful at first, and add to the dose gradually until it becometh five spoonfuls. Then diminish the dose until it becometh one spoonful only.

Another medicine which maketh the water to pass freely, and which is good for the dropsical. Take in equal quantities euphorbium, mâzrîôn, doves' droppings, and rue, pound them, and administer as a dose one drachm in wine, mixed with water.

Another medicine which maketh the water to pass. Take in equal quantities copper rust, costus, mâzrîôn, and aniseed, pound them, and administer as a dose two drachms in hot water.

Another medicine for the same diseases.

Euphorbium	4 grammes
Cummin	4 drachms
Peppercorns	4　　"
Rue	4　　"

Pound, and administer as a dose one drachm in honey wine.

Or take in equal quantities mâzrîôn, calamint, and spikenard, and make an infusion of them | in water, [reduce by evaporation] to one half, and let the patients drink this infusion daily in doses of one measure, and thou wilt marvel.

Or make the patient drink daily for seven days a hot draught containing two measures of fresh extract of calamint, and thou wilt marvel.

Or let patients drink daily for forty-one days a draught containing a portion of "teryâ<u>k</u>ê",[1] about the size of a chick-pea, in an infusion of calamint and rock parsley, | and at the ninth hour let him eat meat stewed with olive oil, or with sesame oil. And make them bathe daily in baths of natron, or sulphur, or some astringent substance [*e. g.*, alum]; and thou shalt throw into the water the roots that reduce the body, and a large quantity of alum, and natron, and sulphur, and bitter salt. Or let the patient go to the baths and have his whole body rubbed with sulphur, and natron, and olive oil. And when he hath sweated profusely, let him bathe in those waters. In the morning let him take teryâkê, and at night let him bathe [again], after he hath digested his food, and at the ninth hour let him partake of a meal, and pass water.

Now if there be hardness and stoppages in the liver, and there is as yet no water in the epigastrium, make an infusion in water of chamomile, dill, flax seed, trigonella, origanum majorana, spikenard, cypress berries, king's crown (melilotus), absinthe, pistacia lentiscus, dried roses, cassia, balsam, and all

---

[1] *l. e., theriake.*

other medicines of this kind, and apply with sponges fomentations of these infusions to the parts of the body that are over the hardened liver and spleen, and anoint the patient with some suitable oil, and lay on the body one of the plasters that have been written down, and that appeareth to thee to be that which should be used. And if there be in the liver weakness as well as hardness, let infusions of the above-mentioned things be made and applied to the body, provided there be no fever. If there be weakness or hardness, but no fever, make infusions in strong-smelling wine of spikenard, root of pistacia lentiscus, dried myrtle, cypress berries, bark of the incense tree, and absinthe, and apply them in the form of fomentations with sponges. And anoint the patient with oil | of nard, or oil of Fol. 179*b*. stacte, or oil of asparagus, and sprinkle over him, when pounded and reduced to powders, stacte, frankincense, absinthe, roses, cypress berries, and any kind of herbs that are used in making fomentations. And thou shalt do this every day until the patient is better, and thou shalt apply to him the following

Plaster, which is good for great weakness | [of the liver] Page 374 **and stomach, and which softeneth the hardness and relieveth the pain in the belly that beginneth through weakness of the liver and stomach.**

Greek bread	4 drachms
Absinthe	4 ,,
Stacte	4 ,,
Frankincense	4 ,,
Myrrh	1 drachm
Aloes	1 ,,
Pargenag	1 ,,
Aromatic reeds	1 ,,
Awlôg	1 ,,
Acacia	1 ,,
Ladhnâ gum	2 drachms
Peeled asparagus boiled in wine	4 ,,
Dates of dactylus	5 in number
Wax	8 drachms
Oil of nard ⎫ Oil of roses ⎭	as much as sufficeth.

28

Steep the dates and bread in wine, crush and peel the asparagus and boil in wine, pound up and mix with the dactylus dates, melt the wax with the oil, pound the dry drugs to a powder, mix everything together, and stir up well and make into a well compacted paste, and lay it as a plaster on the belly, over the liver and stomach.

A plaster made with wine from the flowers of the wild vine, which is good for weakness of the liver, and for the bloody discharge from the belly, which is caused by the rheum that is sent into it from the head, or from any other part of the body.

Oleum omphacinum (green oil)	3	drachms
Wine made from wild vines	3	,,
Dried roses	3	,,
Aloes	3	,,
Alum	3	,,
Dried grapes	3	,,
Acacia	3	,,
Purple βαλαύστιον	3	,,
Stacte	3	,,
Absinthe	3	,,
Spikenard	4	,,
Cypress berries	4	,,
Crocus	4	,,
Flesh of dates	18	,,
Wax	36	,,
Oil of myrtle } Oil of asparagus }	as much as sufficeth.	

Pound, and use as a plaster.

Another plaster, which is good for the same diseases.

Cummin	1	drachm
Purple βαλαύστιον	1	,,
Greek bread	20	drachms
Myrrh	2	,,
Acacia	2	,,
Dross of the smelter's furnace	2	,,
Ladhnâ gum	2	,,

Stacte	2 drachms
Frankincense	2 ,,
Wax	2 ,,

Oil of asparagus | as much as sufficeth.   Fol. 180 a.

[The directions for use are wanting.]

Another medicine, which is good for the hot ulcer of the liver and stomach, and for fever.

Aloes	4 drachms
Frankincense	4 ,,
Dried roses	4 ,,
Stacte	2 ,,
Absinthe		2 ,, Page 375
Spikenard	2 ,,
Crocus	2 ,,
Wax	8 ,,

Oil of roses as much as sufficeth.

[Directions for use are wanting.]

A plaster made of laurel, which is good for hardness of the liver and spleen, and for the dropsical.

Laurel berries	8 drachms
Wax	4 ,,
Tincture of bark (?)	4 ,,
Ox fat	1 measure
Ḳaysê (aromatic woods?)	4 measures
Galbanum (χαλβάνη)	4 ,,
Bdellium	4 ,,
Pig fat	2 ,,
Ammoniac	6 drachms
Aphronitrum	5 measures

Oil of cedar, or oil of olives, as much as sufficeth.

[Directions for use are wanting.]

A plaster made of Glûsḳâ (i. e., the finest flour), which is good for hardness of the liver and spleen, and for the abscesses that come in any part of the body, and it checketh, and scattereth them, or transformeth them into pus.

| Glûskâ, or Greek bread | 20 drachms |
| Cypress gum | 26 ,, |

Ammoniac	10	drachms
Bdellium	10	,,
Oil of olives, or oil of nard	20	,,
Wax	20	,,
Ḳalpûnyâ	20	,,
Dill	2	,,
Chamomile	2	,,
Flax seed	2	,,
Trigonella	2	,,
King's crown (melilotus)	2	,,
Dates	2	,,
Liquid myrrh	2	,,
Fleshy figs	7	,,
Wine	1	ḳesṭâ

Steep all the dry medicines in wine for a whole night and day, and infuse them well, and crush and squeeze dry all the solid ingredients and throw them away. Pound the Greek bread, and soak it in the wine that hath been squeezed out of the medicines, work it up with it, and make it into large tablets, and dry them. Dissolve the bdellium and the ammoniac in the same wine, and mix with them the tablets which thou hast dried. Boil the dates and figs by themselves in new wine, pound the tablets, and melt the wax, and gum, and ḳalpûnyâ in oil, then mix them all together, and work up into a paste, and use a plaster.

A plaster made of wild cucumbers, which is good for
Fol. 180*b*. **the dropsical, and when it is placed | on the belly it produceth a motion of the bowels in the healthy as well as in the sick.**

Page 376

Pulp of wild	cucumbers	8	drachms
Convolvulus	8	,,
Scamony	8	,,
Euphorbium	8	,,
Bdellium	7	,,
Ammoniac	7	,,
Opopanax	7	,,
Sagapenum	7	,,
Seed of dill	22	,,

Indian salt	22 drachms
Aloes	22 ,,
Ox gall	22 ,,
Bitter salt	22 ,,
Garden nigëlla	22 ,,
Mountain grapes, dried	22 ,,
Peppercorns	22 ,,
Ginger	22 ,,
Green hellebore	22 ,,
Mâzrîôn	22 ,,
Bĕlilķê	22 ,,
Epithymum	26 ,,
Natron	26 ,,
Sulphur	26 ,,
Trigonella	10 ,,
Chamomile	10 ,,
Flax seed	10 ,,
Liquid styrax	10 estîrê
Wax	10 ,,

Ox oil ⎫
Honey ⎭ as much as sufficeth.

Dissolve the soluble ingredients in wine, melt with oil those that can be melted, pound to a powder the dry ones, mix all the others with them, work up into a paste, and use as a plaster, and it will make the water to pass. If a man wisheth to have a motion of the bowels, and he cannot take medicine, let him put this plaster on his belly, and he will be purged.

A plaster, which is good for the dropsical, and especially for those whose bellies are swollen tight like drums.

Rust of copper (verdigris)	2 ounces
Ox fat	1 ounce
Droppings of goats	3 ounces

Pound the goats' droppings and the verdigris together, boil them in the urine of goats until the mixture becometh thick, mix the ox fat with them, pour olive oil, honey, and vinegar on them, work up well into a paste, and use as a plaster on the belly.

Another, which is good for pain in the liver and for the dropsical.

Flour of trigonella	2 *ḳesṭê*
Yellow sulphur	2¹/₂ *liṭrê*
Red nitron	2 ,,

Pound all these, and work into a paste with honey, and use as a plaster.

Or take two *liṭrê* of wax, and melt with as much olive oil as is sufficient, and mix with them, and work up into a paste, and use as a plaster.

Another [plaster, which is good] for the dropsical. Take ox-dung, or the droppings of goats that pasture in the mountain Page 377 Fol. 181 a. (*i. e.*, desert), dry and crush, and boil | in vinegar | mixed with water, add to them of yellow sulphur one-third of the quantity of the dung, and mix all together, and work into a paste, and use as a plaster.

Another plaster made of absinthe, which is good for excessive debility of the liver and stomach.

Absinthe	4 drachms
Stacte	4 ,,
Bark	4 ,,
Frankincense	4 ,,
Spikenard	4 ,,
Styrax	4 ,,
Myrrh	1 drachm
Asarum	1 ,,
Incense berries	1 ,,
King's crown (melilotus)	1 ,,
Cassia	1 ,,
Crocus	1 ,,
Wood of balsam trees	1 ,,
Aloes	5 drachms
Bdellium	5 ,,
White wax	5 ounces

Oil of nard as much as sufficeth.

Melt down, mix together, make into a paste, and apply as a plaster.

Another plaster, which is good for the dropsical.

Roots of rape
Natron
Black mustard
Wax } all in equal quantities
Gum of cypress
Stacte

Pound and mix together, and use as a plaster with olive oil and wine.

A plaster made of figs, which is good for hardness of the liver, and spleen, and stomach, and for abscesses.

Figs	40	drachms
Bdellium	8	,,
Ammoniac	8	,,
Flour of beans	3	,,
Flour of chick-peas	3	,,
Flour of trigonella	3	,,
Flour of bitter vetches	3	,,
Flour of lupins	3	,,
Flour of flax seed	3	,,
Flour of King's crown	3	,,
Flour of chamomile	3	,,
Flour of fennel	3	,,

Oil of chamomile
New wine } as much as sufficeth.

Steep the figs in the new wine for a whole night and day, and pound, crush, and boil them thoroughly. Dissolve the soluble ingredients in the new wine wherein the figs have been infused, add the dry ingredients to them, work them all up well together with oil, and apply the mixture as a plaster to the diseased part of the body.

A plaster made of cardamoms, which is good for protracted pains of the liver and stomach, and for a hard spleen, and for the dropsical, and for excessive cold.

Cardamoms	4	drachms
Amomum	4	,,
Spikenard	4	,,
Peppercorns, long	4	,.

Peppercorns, round	4	drachms
Costus	4	,,
Frankincense	4	,,
Cassia	4	,,
Pyrethrum	4	,,
Bdellium	4	,,
Ammoniac	4	,,
Myrrh	4	,,
Aloes	4	,,
Styrax	4	,,
Incense berries	4	,,
Aristolochia, long	4	,,
Aristolochia, round |	4	,,
Cypress berries	4	,,
King's crown (melilotus)	4	,,
| Bark	4	,,
Caryophyllus aromaticus	4	,,
Ladhnâ gum	4	,,
Crocus	1	drachm
Gum of cypress	12	drachms
Wax	12	,,
Irises	5	,,
Stacte	5	,,
Oil of balsam	5	,,
Ox fat, or goose fat	5	,,
Oil of nard, as much as sufficeth.		

Fol. 181 *b*.

Page 378

Melt down the ingredients that can be melted, dissolve those that are soluble, pound the dry ones to a powder, mix all together, and work up into a paste, and apply as a plaster. This plaster is also good for protracted pains in the ribs, and lungs, and for pains in the loins and excretory organs, and for pains in the kidneys, and it healeth obstinate ailments [of all kinds].

A plaster made of euphorbium, which is good for cold diseases in every part of the body, and for the retention of the menstrual flow in women.

Euphorbium	2	ounces
Thapsia Asclepium	1	ounce

Seed of gourds	1 ounce
Cardamoms	1 „
Opopanax	1 „
Ardankîs	3 ounces
Wax	3 „
Pyrethrum	1 ounce
Peppercorns, white	3^{1}/$_{2}$ ounces
Pulp of squills	2 „
Oil of nard, or Oil of roses }	1 *lîṭrâ*

Mix well together, and apply as a plaster to the places that require to be made very hot.

Another plaster, which is good for great weakness of the liver and stomach, and for fever.

Aloes	4 drachms
Stacte	4 „
Juice, or flowers, of absinthe	4 „
Dried roses	3 „
Spikenard	2 „
Crocus	2 „
Oil of nard Wine of lilies } as much as sufficeth.	

Mix well together and use as a plaster.

A plaster made of droppings of goats, which is called "Espúgê", and which is good for the dropsical, and for great tumidity of the liver and stomach.

Moist pitch	2^{1}/$_{2}$ *lîṭrê*
Natron	1^{1}/$_{2}$ „
Pulp of wild cucumbers	3 ounces
Pyrethrum	3 „
Opopanax	3 „
Nettle seed	3 „
Bark of camphor root	6 „
Frankincense	6 „
Irises	6 „
Ox fat	6 „
Rust of copper (verdigris)	6 „
Seed of rock parsley	6 „

　The substance that is between
　　　the bones of sheep, which |
　　　they form in the time of
　　　Nîsàn　　　　　　　　　　6 ounces
　Droppings of goats that feed
　　　in the desert　　　　　　6　　„
　Cardamoms　　　　　　　　6　　„
　Garden nigella　　　　　　　6　　„
　Cypress berries　　　　　　6　　„
　Laurel berries　　　　　　　$^1/_2$ *ḳesṭâ*
　Flour of trigonella　　　　　$^1/_2$　　„
　Oil of cedar　　　　　　　　$^1/_2$　　„
　Juice of flowers of wild cu-
　　　cumbers, or of cucumbers,
　　　which is called Alîṭrîôn　$1^1/_2$ ounces
　Wax　　　　　　　　　　　I *liṭrâ*
　Gum of cedar　　　　　　　2 *liṭrê*

Pound, dissolve, melt, stir up together, and use as a plaster
on the belly. This plaster is also good for great pain in the
excretory organs, and for gouty pains.

　　**A plaster made of castoreum, which is good for protracted
hardness of the liver, and spleen, and for the dropsical, and
for those who emit watery fluids with the urine.**
　Castoreum　　　　　　　　I ounce
　Peppercorns, round, long, and
　　　white　　　　　　　　　$1^1/_2$ ounces each
　Myrrh　　　　　　　　　　3　　„
　Ârdâkhrôn　　　　　　　　3　　„
　Oil of balsam　　　　　　　3　　„
　Incense berries　　　　　　3　　„
　Laurus malabathrum　　　　3　　„
　Costus　　　　　　　　　　3　　„
　Opopanax　　　　　　　　3　　„
　Cypress (or, juniper)　　　　3　　„
　Gum of cedar　　　　　　　3　　„
　Stacte　　　　　　　　　　3　　„
　Sampsuchum marjorana　　3　　„
　Aloes　　　　　　　　　　3　　„

Styrax	3 ounces
Spikenard	3 ,,
Wood of balsam	3 ,,
Cypress berries	6 ,,
Aristolochia	6 ,,
Irises	6 ,,
Ammoniac	6 ,,
Bdellium	6 ,,
Wax	4 *lîṭrê*
Oil of nard	3 ,,

Wine as much as sufficeth.

Pound the dry ingredients to a fine powder, dissolve the soluble ones in wine, melt with oil those that can be melted, mix all together, work up into a paste, and use as a plaster.

A plaster made of goats' droppings, which is good for the dropsical, and it cleareth out the body thoroughly by means of the urine.

Droppings of goats that feed in the desert	1 *lîṭrâ*
Yellow sulphur	20 drachms
Natron	20 ,,
Garden nigella	20 ,,
Ox gall	20 ,,
Doves' droppings	20 ,,
Pulp of the wild, bitter cucumber	20 ,,
Bark	20 ,,
Bitter salt	20 ,,
Alum	10 ,,
Flour of trigonella	10 ,,
Flax seed	10 ,,
Liquid myrrh	10 ,,

Crush all of them and mix together. If the plaster is required over the liver, work them up with wine and oil | of nard; and, if over the spleen, with vinegar and olive oil; and, if on the whole belly to clear out the water, mix with the urine of goats, or with sugar and olive oil. These remarks on the symptoms of all the diseases that attack the liver, and on the means to be employed in healing them, must suffice for this the Sixteenth Chapter.

Page 380
Fol. 182 *b*.

HERE ENDETH THE SIXTEENTH CHAPTER.

CHAPTER XVII.

Of all the diseases of jaundice.

IT hath already been said in the preceding Chapter that there is in the liver a certain fountain (or, source), which produceth all the various chymes whereby our bodies subsist, and that each one of these chymes is absorbed (or, drawn) by the members that are destined to receive it, together with the blood that remaineth in the liver, and that each chyme is sent forth by means of the veins into the whole body, in the proper proportion. Also, as concerning the other chymes, the use of each one of them in the body is seen clearly. Now the blood nourisheth and strengtheneth the members, and beautifieth the colour of the body, and maketh it ruddy. The power of the blood is hot and moist, and its colour is red, and the natural taste thereof is salt.

And the red bile maketh hot the liver and stomach, and produceth digestion, and by means of its pungency (or, acidity) it stimulateth the expulsion of urine and of the faecal matter; and it openeth the pores of the skin, and reduceth the superfluity of the body by means of the sweat thereof. The power of the red bile is hot and dry, and its taste is bitter, and it hath three colours, namely, red, yellow, and green.

The phlegm also is of a moist nature, and it nourisheth the bones, and the nerves, and the membranes, and the tendons, and the ligaments, and the joints, and the lungs, and the brain, Page 381 and it cooleth the heating power (or, flame) | of the red bile. Now its power is cold and moist, its colour is white, and its taste is sweet; if it changeth, it becometh acid and salt.

Fol. 183a. The black | bile is, so to speak, the foundation of these, and it knitteth together firmly the bodies (i. e., organs), and it maketh firm the skin and the members, and it produceth the desire for food in the stomach, and it brighteneth the intellectual (or, mental) powers of the brain, and it maketh the heart to be bold and courageous. In its effect the black bile is cold and dry, and its colour is black, and its taste is acid (or, sour, or, bitter).

It hath been already shewn that all these chymes are separated from the liver as from a single source, and with the destruction of the liver all these chymes are destroyed. Now, it sometimes happeneth that the source of these chymes is destroyed, and that with itself it destroyeth the members that receive them. And it sometimes happeneth, when there existeth any disease whatsoever in them, that they destroy the liver, and this taketh place either when they become enfeebled and are unable to draw [the chymes] from the liver, or when the veins whereby they draw the chymes to them are blocked up. And the chyme doth not destroy the member that receiveth it, which is one of these organs (or, vessels), namely, the gall-bladder which is opposite to the liver, for that it is which receiveth the black bile, and that it is which inflameth the liver in the manner which hath been described, and that it is which draweth away the black bile from the blood by means of the veins which are in it, and clarifieth it. Now, when this vessel (*i. e.,* the gall-bladder) becometh weak, or the veins thereof be stopped up, and this bile remaineth in the blood, fouling the nature of its substance (or, being), and it existeth together with the blood in the whole body, the disease that ariseth through it is called "yarḳânâ" (*i. e.,* jaundice). But it is meet for us to examine very carefully and see whether the liver itself had been diseased in every case when the disease of jaundice appeared, or whether the sickness of jaundice is caused by any other disease, for we can always see when the liver is not in any way diseased. Now, | the discharge of the yellow bile Page 382 into the whole body taketh place at the time of the crisis, just as other discharges take place in other crises. And we also see | that, without any crisis, the whole of the blood can become Fol. 183*b*. bilious through some destroying agent, as, for example, in cases of snake bite. For behold, among the royal slaves whose duty

it was to catch vipers, there was a certain man who was bitten by a viper. Up to a certain time he drank one of the drinks which he had been in the habit of drinking, but when his whole body had become changed, and he had become yellow (or, green), he came to me and related to me everything that had happened to him. Having drunk draughts of "têryâḵê", in a short time his body returned to its proper state. And the physicians debated among themselves whether any particular signs followed [the absorption] of the poison, for we see always that, without poison, the chymes in the bodies are destroyed, like the milk which is in them, by the poison. Therefore it is not a matter for wonder if the chymes that are in us are changed, and assume such a variation of this kind that the whole body becometh yellow through them, and through the change in the natural condition of the liver, and that same kind of evil condition happeneth without the presence of stoppages, or ulcer, or hardness of the liver. For it is clearly seen that sometimes the whole body becometh like grass which becometh white and yellow, and that sometimes its colour becometh like that of lead, or even blacker than lead. Moreover, other mixtures of colours, which are indescribable, occur in men during the other diseases of the liver, even when there are no ulcers at all in it, just as when disease of the spleen is present other kinds of colours, which are blacker than those caused by the liver, appear in the body. These are very difficult indeed to explain (or, describe), but they are very
Page 383 easily recognized by those who have seen them | many times.
Fol. 184a.   Now, when Saṭsînôs was lying sick,—I know not wherefore many of the physicians who used to visit him came to imagine that he had an abscess | on his liver—and much time had passed by and he had made no progress [towards a recovery], he called in me, even me. And having seen him, immediately I entered the chamber (or, house) wherein he was lying, I said this to him:—"Know thou that there is nothing bad in thy "liver. As for the other things, when thou hast uncovered "thy epigastrium, I shall be able to know what they are." What in truth he had was a pustule (or, abscess) in the deep-seated muscles, and pus was already flowing between the

muscles that lie from side to side, and those that go upwards
from below obliquely; but these ye know are placed between
those that are attached to the peritoneum, and that lie from
side to side, and those other external muscles which are below
the skin, and that go downwards in an oblique direction. And
ye know that I have had many other similar cases in which
I have diagnosed the disease from the colour of the patient
only, and that by it alone I have been able to decide immediately when it was the liver which was diseased, and when
it was the spleen, and not by the answers of the patients to
the questions that I had asked them beforehand, and it was
not by touch that I had ascertained what was the disease of
the member. It is for this reason that ye have also heard me
on many occasions curse those men who dared to call themselves pillars of the art of healing, when they had no wish to
help those who were sick; but it is not our intention to speak
of these men on the present occasion. And for this reason
we proceed to discuss the other kinds of jaundice that arise
through injuries of the gall-bladder, which receiveth the bile
which is fixed in the liver. Now, it is proven in our opinion
that the bile is absorbed by this bladder from the bilious liquid,
just as the kidneys absorb the watery part of the blood, and
as the spleen also absorbeth | from the blood that substance Page 384
which is to it as scum is to wine. Therefore, if this bladder
which absorbeth the bile becometh weak, the blood must sometimes become polluted. And there are other kinds of jaundice
besides the three | kinds that have been mentioned above. Now Fol. 184*b*.
it is possible also that the gall-bladder may sometimes become
full (*i. e.*, overfilled), just as that other bladder for the urine
becometh filled, and that it may not be able to empty itself
because of certain obstructions, or because of the weakness of
its emptying power, or again, because of obstructions, or because
of the weakness of the veins that are situated close to it in
the liver, and that it is therefore not permitted to absorb into
itself the bilious moisture. Now it is absolutely necessary to
enquire into these diseases of jaundice, and to examine carefully the substances that are excreted, for through this method
of procedure we shall gain no small benefit in respect of

indications (or, symptoms) of the diseases. As for myself, I have understood this, and I have observed and seen that the stools of certain men, who were attacked by the disease of jaundice, were sufficiently coloured by red bile as to be distinguishable; and the same was the case with the urine of some of them. And I have looked at the water in which some of them have had a bath, and found it to contain bile [which hath come into it through the pores of] their skin; in some of them, however, the bulk of the bile was kept back, [and in the water] only a very little of it [appeared]. Now as we were wishing to obtain exact information on the subject, having anointed the bodies of those who were sick with absinthe, we ordered [them to be bathed, and] their sweat [to be scraped off them with a leathern scraper], for thus have ye seen before all men their sweat collected in the lower part of the scraper (*i. e.,* strigil). And the sweat of some of them was found to be watery, and that of some of them to contain bile. Therefore from these [experiments], which were performed some time

Page 385 ago, there shall be unto you | an obvious indication of the spot that was diseased. And to [the results of] your investigation ye may add the following also.

Behold, a certain man, having been attacked by an acute bilious fever, came out of this on the seventh day, and a very large quantity of the red bile had been discharged into all his skin. And since during the days that followed the jaundice

Fol. 185*a.* remained on him, whilst we were visiting him, | we inspected his stool and his urine, and because both of these were natural (*i. e.,* normal) they shewed that the jaundice was not due to disease of the liver. Now, the thought entered my head that it was possible that the bile, which was poured out into all his skin, was so thick that it remained therein, and guided by this thought I believed that it was possible for me to learn what was the spot that was affected. Having seen that [the water in which he washed was bilious], I understood that the bile was thick and difficult to dissolve. Thereupon I ordered the patient to use in bathing the water that cometh forth hot from the earth, since it is very good for reducing and dissolving [bile], and a course of treatment that was wet and was at the same time moderately reducing, and was capable of diluting the thickness of the chymes. Now, this man was by these means healed of this disease, and the indication thereof being confirmed became a help [to him], and his healing was also established.

And in the case of another man I found a large quantity of bile in the scraper, and I therefore understood that bile was discharged into every part of his body in large quantities, and having made use of the helps which were suitable for his disease I cured him. Now, in all those patients who, though having no fever, had nevertheless a feeling as of some weight resting on the right side of the epigastrium, there was jaundice. When in all these I had broken a way through the obstacles (or, stoppages) which they had by means of foods, and draughts, and helps that reduced them, and had afterwards administered to them a medicine that caused them to have a motion | of Page 386 the bowels, I healed them all in about one day. Now, in the cases of those on whom the [aperient] medicine had no effect, I used to give them in addition a draught which broke with violence a way through the obstacles, and similarly, when it was necessary to do this a second time, I used to give them a medicine which was very much stronger than that which I gave on the first occasion, so that at the conclusion of the purging, the bile that was lighter in colour than the red was expelled with severe griping. | Fol. 185 b.

Now, from these things one would imagine that the bladder that receiveth the bile is suffering from some disease which is like unto that from which the other bladder that collecteth the urine suffereth. For in this also the quantity of the water which is sometimes collected therein is so great, that the bladder becometh greatly stretched, and is unable to empty out that which is contained therein. In 'the case of the bladder that [collecteth] the urine, this taketh place for two reasons, namely, through weakness of the emptying power,—now there are also times when it happeneth through deep sleep—and through the retention of the urine owing to non-emptying. For this reason the bladder becometh distended to so great an extent that it becometh the cause of disease in its voiding power. Now as

29

concerning the bladder which receiveth the gall, since it cannot
in any way be helped to empty itself by the mental powers,
there is one source and cause of its weakness, just as is the
case with the other natural organs, and this weakness taketh
place through a change in the condition of the member (or,
organ). But since it is necessary for us to write about the
means to be employed in healing jaundice, it is meet for us to
collect [the facts] about all the various kinds of jaundice, which
we mentioned a little way back very briefly, in order that the
means to be employed in healing them may be elucidated.

Now, the first kind of jaundice is that which taketh place
Page 387 in the crises of fevers, when the yellow bile is poured out | into
all the body. The second kind of jaundice is that which cometh
through the bite of a viper or of some other noxious reptile,
which ejecteth his bile (or, venom), [into a man] and changeth
the colour of his body. The third kind of jaundice is, as hath
been said, that which ariseth through a change in the condition
of the liver. The fourth kind of jaundice is that which occurreth in sympathy with the spleen, when it doth not absorb the
chyme of the black bile, and it produceth in the body the
Fol. 186a. various colours that have been mentioned above. The fifth |
kind of jaundice is that which ariseth through some injury to
the bladder that receiveth the gall, and when it doth not absorb
the chyme of the red bile from the liver. Now, this happeneth
when the gall-bladder becometh full, just as the bladder for
urine becometh full, and is unable to absorb the bile because
of its weakness, or because of stoppages, or because of some
ulcer (or, abscess) in the veins, which are hollowed out from
it into the liver. And the bile remaineth with the blood, and
it changeth the colour of the body to that of its own substance. These are all the various kinds of jaundice, and these
are also the causes that produce them, and it is now therefore
meet for us to proceed to the description of the means of
healing that are suitable for them.

The means to be employed in healing jaundice.

Now, if we follow the opinion of the writer of this treatise, the means to be employed in healing jaundice are the following: the first kind of jaundice mentioned, namely, that which ariseth during the crisis of a fever, is to be healed by means of constant bathing in baths, and by anointing with some aromatic oil that dissolveth, and by rubbing the body with natron, and by suitable food and drink. All these must be used until the bile, which hath been poured out into the body, becometh dissolved, and the body returneth to its natural state.

The second kind of jaundice, namely, that which cometh through the bite of a reptile, or through some deadly drug, can be healed by "Têryâḳê", and by suitable medicines.

The third kind of jaundice, namely, that which ariseth through a change in the condition of the liver, | can be healed by the Page 388 treatment that hath been described in the Chapter preceding this, and the jaundice that ariseth through some injury to the spleen can also be healed by the means already described in the Chapter that treateth of the spleen.

Therefore it now remaineth for us to describe in the present Chapter the means to be employed in healing the kind of jaundice that ariseth through the injuries that attack the bile that is fixed in the liver; this disease is very serious indeed when it happeneth to exist in conjunction with hardness of the liver. Now, if it should arise in connection with an abscess in the liver, and a burning fever, and the patient's strength is well sustained, it is, first of all, an excellent thing to let blood from the vein | that hath been mentioned in connection with Fol. 186b. disease of the liver, and next we must administer draughts of pure oxymel with extract of husked barley. And if, as it doth happen, there are present much inflammation, and [a raging] thirst, and very hot urine, administer also extract of chicory, and stewed orach (or, spinach), or stewed sorrel, or stewed gourds, and the emptying out of the belly is very useful. And if the disease of jaundice arise, unaccompanied by fever and [a raging] thirst, but with hardness and obstructions [of the liver], then all the drugs that are bitter, and that break through

29*

the obstacles, are useful, and also frequent emptying of the belly, which is produced by agaricum (larch fungus) and scamony, and the application to the parts of the body over the liver of the plasters that have been already described (or, written down), and into which enter largely, as ingredients in their composition, aloes, and absinthe, and stacte, and crocus, and asarum, and everything of the same kind.

A medicine which is good for jaundice, and which breaketh through obstacles of the liver, and produceth a motion of the bowels.

Aloes	6	drachms
Stacte	4	„
Agaricum	2	„
Manyfeet (polypodium)	2	„
Epithymum	2	„
Carthamus tinctorius	2	„
Bark	1	drachm
Laurus malabathrum	1	„
Spikenard	1	„
Honey as much as sufficeth.		

Dose, two drachms, if there be no fever, in wine mixed with water, and if there be fever, in hot water.

Another medicine which is good for the same diseases, and which produceth a motion of the bowels.

Fir cones, cleaned	2 drachms
Peppercorns	1 drachm
Fleshy dried grapes, cleaned	$^1/_2$ „
Scamony	$1^1/_2$ drachms

Page 389 Pound, and administer as | a dose one drachm, in a tincture containing three parts of vinegar and one of myrrh, mixed with an infusion of dill. Roast the scamony in bread, or in a double vessel, and then mix it [with the other medicines].

Another well tried medicine which is good for the diseases of jaundice.

Dried grapes, cleaned,	6	drachms
Cummin	6	„
Peppercorns	4	„

| Roasted scamony | 3 drachms |
| Seed of dill | 2 ,, |

Pound, mix together, and administer as a dose one and a half drachms in vinegar, and *môryâ* (wild myrtle, or basil?), and an infusion of asarum and absinthe.

Another.

Aloes	3 drachms
Sulphur		3 ,,
Bark	2 ,,

Fol. 187 a.

Crush, mix together, and administer two drachms in a *kandîkôn* mixed with an infusion of dill.

Another.

| Sulphur of the sea | 4 drachms |

Crush, and let the patient suck it out of an egg.

Another [prescribed by Galen], which expelleth the jaundice powerfully.

Euphorbium ⎫
Epithymum ⎬ in equal quantities
Aloes ⎪
Rock parsley ⎭

Dose, one or two drachms in an infusion of aniseed, or in cheese water.

Cheese water of milk, which is good for the jaundice that ariseth from heat of the liver. The cheese water of milk is prepared in the following manner:

| Milk of goats | 3 *lîtrê* |
| Orach (or, spinach) | 1 handful |

Pound the orach, add the milk to it, and stir it up with thy hands for a time, strain and squeeze out the juice from the rinds of the orach, and throw them away, and set the milk aside for a whole night. In the morning when thou hast worked it up together as in making cheese, strain off the water, add to it a little honey, and two drachms of Indian salt, and one gramma of roasted scamony. It will expel bile freely. Let the patient drink this for three days.

[A list of] the simple medicines which are good for jaundice.

All the bitter herbs that produce painless motions of the Page 390 bowels, such as absinthe, asarum, aniseed, spikenard, acorus reed, gysophilla struthium (or, saponaria officinalis, ptarmica, or gum ammoniac), agaricum (larch fungus), madder, balsam wood, gentian, costus, aristolochia, centaureum, and every herb of this kind, singly, or together. They are to be pounded, and infusions are to be made from them, and these, when drunk with honey water in hot water, clear away the jaundice, and break through the obstructions in the liver. And, in cases of jaundice, the tablets for breaking through obstructions, which are described in connection with diseases of the liver, are very useful, and those that have roses in their composition are useful in cases of jaundice with fever and in cases where there is excessive heat in the liver, and they must be drunk with extract of orach (or, spinach).

Another medicine which is good for jaundice, and for expelling [the bile].

Scamony	
Aloes	in equal
Agaricum (larch fungus)	quantities
Roasted Armenian nard	

Dose, one drachm, in hot honey water.

Another medicine which breaketh through obstacles.

Bitter cypress berries, cleaned	2 drachms
Absinthe	
Asarum	
Spikenard	1 drachm each
Aniseed	

Fol. 187 b.

Pound, and administer one drachm in honey wine if they have no fever, and, if they have fever, in honey water.

Another.

Asarum	8 drachms
Absinthe	8 „
Seed of dill	4 „
Aniseed	4 „

| Seed of rock parsley | 4 drachms |
| Bitter cypress berries, cleaned | 4 „ |

Make into tablets with water, and dry, and administer as a dose one drachm in an infusion of ἀδίαντον (warrior's hair).

Another. Take "Adianton", that is, hair of the warrior, dry it, pound it up, and administer four drachms of it in an infusion of aniseed. Or steep a very large quantity of adianton in water, and make a strong infusion of it, and let the sick man wash in it, and thou wilt see the jaundice floating on the top of the water.

Another.

| Rinds of pomegranates | 4 drachms |
| Yellow arsenic | 2 „ |

Crush, and administer as much as will lie on thy thumb nail in ass's milk, three measures. Or give the patient to drink two measures of extract of sorrel, with milk. Or give a dose of extract of radishes, two measures. Or half a drachm of natron in one measure of "living" | wine, mixed with water, and Page 391 let the patient drink this when he is seated in his bath, and thou wilt see the jaundice float on the top of the water.

Simple medicines for jaundice.

Take gysophilla struthium and madder in equal quantities, pound them, and administer a spoonful in honey, and insert some of it into the nose.

Or pound ḳandûsh (veratrum album, or white hellebore?), and give as a dose one drachm in honey water, and it will act as an emetic, and will break through the obstructions [in the liver].

Or pound madder, and administer one spoonful in a measure of wine, and half a spoonful separately.

Or take cypress nuts, two drachms, and a drachm of bark of cinnamon (?), pound them, and administer in old wine, and let the patient run about until he getteth rid of the jaundice in his sweat and urine.

Or take in equal quantities myrrh, and peppercorns, and the excrement of a dog, and pound them, and administer one spoonful with wine, and let him wash daily | in a bath, even Fol. 188 a. though he have fever.

Or pound three drachms of orach (or, spinach) seed, and one drachm of ḳandûsh (white hellebore?), and administer the whole as a dose in a draught of honey water, and it will expel the jaundice by vomiting.

Or pour wine into a wild cucumber from which the seeds have been withdrawn, and heat it over the coals until it boileth, and clarify, and give it as a draught. It will produce a motion of the bowels, and will clear out the jaundice; now this medicine is good also for dropsy.

Or take the droppings of a dog which are white, and which contain a large quantity of bones, and they must be collected when the Mighty Man (or, Warrior, *i. e.,* Orion) is in the ascendant, and pound them, and give four drachms to the patient worked up into a paste with honey, and let him use it as a linctus each day until he is cleansed of the jaundice.

Or take ḳandûsh and irises in equal quantities, pound them and work them into a paste with vinegar and honey, and give the patient three drachms, and let him use it as a linctus, and afterwards let him drink hot water, and it will cleanse him by means of vomiting.

Or pound up gysophilla struthium and eupatorium, and administer one spoonful in honey; and let some of it be blown up into his nostrils.

Or dry river calamint, and pound it, and administer four drachms in wine mixed with water, and let the patient drink a dose daily for three or four days. Or make an infusion of it in water when it is in an unpounded state, and let him drink Page 392 the infusion [daily] for | seven days.

Or pound the horn of a stag, which hath been burned, take four grammas, and one gramma of sulphur, and work it into a paste with honey, and administer the whole of it to the patient in a draught.

Or pound a radish, with its leaves, squeeze out the juice, and take two measures of it, and add thereto one measure of wine, and let the patient drink half a drachm. Heat it a little, and let him drink it whilst he is sitting in the basin (?) of the bath. It hath been well tried.

Or pound one drachm of peppercorns, and administer to the

patient in honey water, and let him be sitting in a basin (⸓)
of cold water [when he drinketh it].

Or crush two drachms of crocus, and give it to the patient
to drink in two measures of asses' milk. It is a sure cure.

When the whole body is languid, and the eyes become
yellow, the following medicines are useful.

**A medicine which is ejected into the nose, and which
driveth jaundice out of the eyes.**

[Take] extract of anglîdôs, | (ἀγγελοειδής?) and inject.   Fol. 188*b*.

Or inject extract of the root of leontice leontopetalon
(cyclamen).

Or pound the root thereof, and steep it in the milk of a
woman for a whole night, and squeeze it out in the morning;
then warm it up, and inject it, and it will clear away the jaundice. This is a well tried medicine.

Or dissolve one *ḳertâ* (Arab. قيراط, carat = ¹/₃rd of an obolus)
of alîṭriôn (*i. e.*, elaterium, a preparation of wild cucumbers) in
the milk of women, and inject. Or crush it, and blow it in
the form of a dry powder into the nostrils.

Or warm the extract of the leaves of wild cucumbers, and
inject it.

Or dry the leaves, and pound them, and blow them into the
nostrils. This is a well tried and certain cure.

Or pound up sorrel, and dissolve in the milk of women, and
inject into the nostrils, and it will break through obstructions,
and will remove the jaundice from the eyes.

Or heat up the milk of women and vinegar, and inject.

Or pound the root of clover (*espestâ*), squeeze out the juice,
and mix with it oil of *yasmengh* (jasmine?) and sugar, warm,
and inject.

Or let the patient snuff up into his nostrils strong vinegar,
and he will be freed from jaundice.

Or warm up extract of beetroot (or, plantain), and inject into
the nostrils.

Now, it will help those who have jaundice to drink an infusion of absinthe continually (*i. e.*, regularly), and to eat food
that is pungent, and delicate, and purifying, and to take baths

Page 393 very frequently; | in short, all the medicines that are good for the disease of jaundice are bitter and purifying (or, astringent), and they break through the obstructions in the liver, and it helpeth patients also if the veins which are under the tongue are cut.

Another Gurgarshân, in which the broad-leaved pepperwort is an ingredient. This is good for internal and external fistulas, and for the cold winds which arise in the belly, and it grippeth (or, inundateth) the stomach, and produceth a motion of the bowels.

Hellebore	10	*mathḳâlê*
Bĕlîlḳê	10	„
Amlagh	10	„
Pepperwort	10	„
Seed of rue	10	„
Seed of leeks	10	„
Seed of radishes	10	„
Seed of gourds	10	„
Seed of rock parsley	10	„
Seed of ammi Copticum	10	„
Mustard	10	„
Pepperwort (taḥlê)	10	„
Seed of bâshôshâ	10	„
Wild rue	10	„
Angabûsht (flowers of bûsht)	10	„
Doronicum scorpioides	10	„
Amomum zerumbeth	10	„
Cinnamon	10	„
Peppercorns, long and round	10	„ each
Seed of marzangûsh (origanum majorana)	10	„
Seed	of shablîlta	10	„
Seed of fennel	10	„
Zĕrîwand	10	„
Iron rust (*i. e.*, oxide of iron)	12 + 30 + 30	„
Honey	1	*lîṭrâ*
Cow oil	1	„
Sugar	12 + 30 + 30	*mathḳâlê*

Fol. 189 a.

Pound the dry ingredients, make into a powder, dissolve in honey, mix together, and use. Dose, a portion the size of a walnut, in wine mixed with water. Let him take one *naṭlà* (= two *estîrê* = one *menêkîthâ* = one spoonful) after food and before food. It is a sure cure.

Another Ṭarsûm.

Sugar	12	drachms
Black hellebore	12	,,
Ammi Copticum	12	,,
Seed of rock parsley	12	,,
Ginger	6	,,
Cinnamon	6	,,

Pound to a fine powder. Dose, one spoonful with hot water, and it [will be found] helpful. Or let the patient take seed of ammi Copticum, and seed of wild rue, and when using add a little salt. These things must suffice for the Seventeeth Chapter.

HERE ENDETH THE SEVENTEENTH CHAPTER.

CHAPTER XVIII.

Of the symptoms of the diseases of the spleen.

Now it hath already been shewn above in respect of the spleen, that, by means of its veins, it draweth away the scum (or, impurity) of the blood, namely, the black bile, from the liver, and that it sendeth a little of it into the stomach, as thou mayest see from the desire for food [which is in] the stomach. This organ is situated on the left side, below those ribs called the "alien (*i. e.*, external) ribs", and it is surrounded, with the belly and with the membrane that envelopeth the intestines, by the arteries that nourish it, and with its diseases are associated constantly the liver, and the stomach, and the kidneys. Now, from the things that have already been said about the liver and the gall, it is easy for us to discover by what signs it may be known that the spleen is diseased, and also from the fact that the ulcers that arise therein are easily known by the touch; but because it hath the greater number of them in conjunction with the liver, for this reason they become large, and diminish and change. In the disease caused by the

Fol. 189 *b*. weakness of the spleen, the colour of the whole body | inclineth to black, because its natural function is to draw away and to absorb from the liver the chyme of black bile; for by this, it seemeth, it is nourished. When, then, the absorbing power of the spleen is weakened, because this black bile is not cleared away, it goeth up from the liver into the whole body, and because of this the colour of those who suffer from disease of the spleen becometh black. Moreover, it poureth out ofttimes a superfluity of black bile, just as the liver doth in certain vomitings and retchings when it poureth out the blood of black bile; similarly also it goeth downwards to the bowels, and if the black bile is not emptied out in some such manner, a change of mind (*i.e.,* mental derangement) taketh place, and it produceth sickness and distress. There are occasions when it also produceth a violent desire for food, especially when the excess of black bile that cometh into the belly | is sharply Page 395 acid [in taste]. Sometimes, however, it effecteth the destruction of the craving for meats, and when it happeneth that the blood is destroyed and putrid, the patient loatheth them. When the spleen becometh hard it produceth dropsy, because the liver suffereth in sympathy with it, as we have already said above. Moreover, when both the liver and the spleen are diseased, the diseases of jaundice appear, and the colour of the sufferers inclineth so much to black, that thou mightest think that the red bile had been mixed with a certain quantity of soot. Thus certain physicians have thought that the liver was not diseased in jaundices of this kind, and therefore they enquired how jaundices could arise sometimes through the spleen. Similarly, certain other physicians thought that, without disease of the liver, men could become dropsical, because the spleen alone was hard. Let them, however, keep to their enquiry as to jaundices. And also as concerning the dropsy that ariseth in acute illnesses, in some of these the liver suffered so violently through the hot change of condition, and perhaps also through the dry change of condition, that it was unable to change food into blood, and yet even in such | illnesses these physicians Fol. 190a. would not believe that the liver was diseased. If, because there is no ulcer whatsoever in the liver, they think that it cannot be diseased, let them consider that there is no other member which is diseased in those who have suddenly drunk very cold water, and have been attacked through this by dropsy. Now, that cold water, drunk in this manner, doth from time to time injure other members, and other members of the internal group, is well known; sometimes it injureth one more than its neighbour, and sometimes it attacketh one before it attacketh its neighbour to such an extent because it happeneth that one member is found to be weaker than its neighbour at that time. But if dropsy is about to arise, it is necessary for the cold to

Page 396 go up from the member that is injured | to the liver. Thus sometimes dropsy ariseth because of the spleen itself, or when it becometh chilled in conjunction with an ulcer of some kind, as, for example, when it is exceedingly hard; or also without an ulcer, when it becometh chilled through the drinking of cold water suddenly and unseasonably. Now, disease and derangement of the mind, which are closely connected with insanity, arise when any excess of black bile of this kind is transferred to the mouth of the belly, a fact which hath already been mentioned by us in the section treating of the mouth of the belly. It is therefore not meet to say anything more about it here, because the organic ailments, which take place therein, do not need mental symptoms [to reveal their existence]. As concerning the other diseases, which take place in the spleen through a change of condition, their symptoms are well known from the things that have been said [here], and also from those things that have been said in the sections that treat of the liver. Some of them have been described with exactness and have been mentioned by name, and like unto these some of them have been established, and therefore the symptoms of the latter resemble those of the former. And, as we know well what the function of the spleen is, and as we have learned also what happenings are attached to the diseases thereof, it is necessary for us to approach the [considerations of the]

Fol. 190b. operations thereof | also with great care. Now therefore also all the medicines that are bitter—with some of them those that are acid must be mixed—are good for the spleen, and, as things that are also good for it, we apply to it oil, and plasters, and draughts of oxymel. And to these we add also other kinds of medicinal draughts, which are compounded with drugs that are well known to be good for it, and that do not heat it overmuch, and that do cut and destroy the various kinds of thick chymes that arise in it.

An Antidote compounded with spikenard, which is good for hardness of the spleen and liver, and which breaketh through obstacles, and maketh the urine to flow.

Spikenard	3 drachms
Crocus	2 „

Aristolochia, long	3 drachms	
Ammoniac |	4 ,,	
Bark	4 ,,	
Camphor root	12 ,,	
Costus	12 ,,	

Page 397

Honey as much as sufficeth.

Dose, one drachm, in an infusion of the leaves of the camphor tree and absinthe.

Another compounded with spikenard, which is good for pain in the spleen, and for livers that are tumid, and for dropsy, and for pain in the stomach and kidneys, and for jaundice, and for coughs of long standing, and for the bites of noxious reptiles.

Spikenard	2 drachms
Costus	2 ,,
Bark	2 ,,
Arsenic	3 ,,
Peppercorns	1 drachm
Seed of rock parsley	1 ,,
Acorus reed	1 ,,
Trefoil (blue melilotus?)	1 ,,
Irises	1 ,,
Balsam wood	1 ,,
Camphor root	4 drachms
Dried grapes, cleaned	5 ,,

Honey as much as sufficeth.

Dose, two drachms, in an infusion made from the leaves of the camphor tree, calamint, and oxymel.

An Antidote compounded with nightshade, which is good for obstructions of the liver, and spleen, and for cold in the womb, and for third-[day] fever, and for wet coughs, and shortness of breath, and jaundice, and palsy of the limbs.

Seed of wild rue	1½ *manyâ*
Frankincense	10 drachms
Stacte	10 ,,
Incense berries	10 ,,
Bark	10 ,,

King's crown (melilotus)	10	drachms
Crocus	10	,,
Spikenard	10	,,
Aristolochia	20	,,
Rhubarb	20	,,
Amomum zerumbeth	4	,,
Doronicum scorpioides	4	.,
Ginger	12	,,
Costus	12	,,
Opium	12	,,
Caryophyllus aromaticus	6	,,
White hellebore	6	*estîrê*
Dried roses	6	,,
Cypress berries	10	,,
Aloes	14	,,
Nigella sativa	6	,,
Peppercorns	13	drachms
Honey as much as sufficeth.		

Dose, one drachm, in some convenient draught.

Another Antidote which is good for the spleen, and stomach, and liver.

Aristolochia, long	13	drachms
Root, or leaves, of the camphor tree	13	,,
Pepperwort	8	,,

Page 398 | Honey as much as sufficeth.

Dose, one drachm, in oxymel mixed with water. Or make into tablets with strong vinegar, each containing two drachms, and dry them, and administer [to the patient].

Tablets compounded with madder, which are good for hardness of the spleen, and for pain in the liver, and for protracted fever.

Madder	12	drachms
Bark of camphor root	2	,,
Aristolochia, long	2	,,
Irises	2	,,

Make into tablets with oxymel. Dose, two drachms, in an infusion of absinthe and leaves of the camphor tree.

Tablets compounded with squills, which are good for the same diseases.

Roasted squills	4 drachms
Cardamoms	4 „
Jackals' milk	4 „
Gum ammoniac	2 „
Pyrethrum	1 ounce
Rape seed	2 drachms
Gypsophilla struthium (saponaria officinalis)	2 „

Make into tablets with vinegar; dose, one drachm.

Another.

Ḳandûsh	5 drachms
Roasted squills	5 „
Opopanax	5 „
Camphor root	5 „
Crocus, cleaned	2¹/₂ „

Make into tablets with oxymel, and administer as a dose one drachm.

Another.

Seed of agnôs (the chaste tree, Abraham's balm)	4 drachms
Peppercorns, white	2 „
Spikenard	2 „
Gum ammoniac	2 „
Camphor leaves	2 „

Make into tablets with vinegar. Dose, two drachms in oxymel.

Another.

Ammoniac	8 drachms
Fresh senna twigs	8 „
Camphor root	4 „
Seed of tamarix (μυρίκη)	4 „
Scolopendrium vulgare (Venus's hair?)	4 „
Roasted squills	4 „
Peppercorns, white	4 „

Make into tablets with vinegar; dose, two drachms.

Another.

Myrrh	6	drachms
Costus	6	„
Spikenard	4	„
Camphor root	4	„
Ferula opopanax (panaces heracleum)	4	„
Madder	4	„
Dittany	4	„
Ammoniac	4	„

Make into tablets with vinegar; dose, two drachms.

Fol. 191*b*. | **Shlâbhê, which are good for a spleen which is much thickened, and for fever.**

Irises	6	drachms
Camphor root	8	„
Myrrh	1	drachm
Dried grapes	1	„
Rape seed	1	„

Page 399 | Pound, reduce to a powder, steep in vinegar and water, boil, and administer in a draught two measures daily for three days.

Another.

Seed of garden nigella (شونيز)	10	drachms
Seed of myrica	10	„
Flowers of senna	10	„
Camphor leaves	10	„
Calamint	10	„
Derostho [Aegyptiaca] (cyclamen?)	10	„
Absinthe	10	„
Agrimonia eupatorium	10	„
Strong vinegar	2	*lîṭrè*

Boil down to one half, and let the patient drink daily two measures for seven days.

Or steep terebinth [seeds] in strong vinegar for seven days. Then let the patient eat three spoonfuls of them daily, and afterwards drink the vinegar.

Or pound four drachms of pepperwort and two drachms of garden nigella, and work them into a paste with honey, and administer daily three drachms in vinegar and water; let the patient drink this draught daily for seven days.

Or pound acorus reeds, and administer one spoonful daily, in oxymel with which camphor is compounded.

Or boil in strong vinegar unboiled cypress nuts and juniper berries, until the vinegar is reduced to a third. Administer this to the patient, and let him drink it, according to his strength, and let him sleep on his left side, and with the scum of the vinegar make plasters, and bind one on his left arm for three days.

Or administer a draught containing two drachms of the seed of purslane (Pers. بَرِزَهٔ) and vinegar and water.

Or dry and pound lamb's tongue, and administer one spoonful daily in oxymel for seven days.

Or scrapings of

Bark of camphor root	12	drachms
Fennel	6	„
Irises	6	„
Potter's dross (or, ashes)	2	„
Mustard	2	„
Myrrh	8	„

Pound and work into a paste with vinegar and honey, and administer as a dose two drachms; and use some of it as a liniment to rub over the spleen.

Or pound four drachms of centaureum (κενταύρειον), and administer daily with oxymel for seven days.

Or use aristolochia in the same way.

Or administer two drachms of eupatorium in an infusion of absinthe.

Or dry the rinds of fresh gourds pounded, and administer two drachms | with | oxymel, daily for seven days.

Page 400
Fol. 192 a.

Or kill a bat, dry it, pound it, and administer as much as will lie on the top of the [thumb] nail in hot water.

Or dry aristolochia and yellow hellebore, and pound them, and administer daily one spoonful in the urine of goats or camels for seven days.

Or pound up fennel, and administer daily in a cupful of the urine of camels; it is very beneficial also in dropsy.

Or let the patient drink daily for seven days two measures of water wherein the blacksmiths have cooled their red-hot iron.

Or take in equal quantities camphor leaves, ammoniac, centaureum, eupatorium, cypress leaves, juniper berries, ears of barley, or husked barley, and steep them in vinegar and water, and let the patient drink every day one measure, and let him apply fomentations of the same to the part over his spleen, and let him apply also plasters that soften, and take the foods and drink that are good for him.

And if patients be in a good state of health, the letting of blood from the lower vein of the left fore-arm, or from the palm of the hand of the same arm, and from the same vein is useful. Most useful of all, however, in cases of pain in the spleen is cautery above the lower vein of the left arm. And useful also are fomentations with medicaments which are to be prepared in the following way:—

Spongings which are good for protracted hardness and tumidity of the spleen.

Chamomile	I	hand
Fennel	I	„
Flax seed	I	„
Trigonella	I	„
Bran	I	„
Liquid myrrh	I	„
Camphor leaves	I	„
Calamint	I	„
Juniper berries	I	„
Cypress leaves	I	„
Cabbage leaves, or leaves of beet	I	„
Strong vinegar	2	measures
Water	I	measure

Soak until the liquid is reduced to one half, foment the patient with it, anoint him with oil of fennel, or oil of chamomile, and let him apply the following plaster:—

A plaster which softeneth.

Fleshy figs	4 ounces
Ammoniac	I ounce

Bdellium	1 ounce
Aloes	1 ,,
Opopanax	4 drachms
Sagapenum	4 ,,
Asa foetida	4 ,,
| Flour of trigonella	4 ounces
Droppings of goats	4 ,,

Page 401

Pungent vinegar as much as sufficeth.

\| Boil the figs and vinegar together, dissolve the soluble in- Fol. 192*b*. gredients in the vinegar wherein the figs have been boiled, pound the dry ingredients to a powder, mix everything together, work into a paste, and use as a plaster. Thou must first use fomentations on the patient, and then apply the plaster, which thou must keep on until the spleen is softened and it returneth to its natural state.

Another, which is good for excessive hardness of the spleen.

Ammoniac	12 drachms
Irises	12 ,,
Bdellium	12 ,,
Ox fat	12 ,,
Styrax	8 ,,
Frankincense	8 ,,
Opopanax	8 ,,
Galbanum	8 ,,
Greek ligusticum	8 ,,
Peppercorns	70 in number
Wax	6 *estìrê*
Cypress gum	6 ,,
Oil of irises	
Wine	} as much as sufficeth.

Mix well together, and use as a plaster.

Another, which is good for excessive hardness of the liver and spleen.

Rhatany	58 drachms
Wax	20 ,,
Dry pitch	20 ,,
Smelters' dross	10 ,,

Sulphur	10	drachms
Frankincense	10	,,
Myrrh	10	,,
Pyrethrum	5	,,
Camphor root	5	,,
Gum ammoniac	3	,,
Chamomile	3	,,
Glue (or, gum)	3	,,
Aristolochia	3	,,
Crocus	3	,,
Cardamoms	3	,,
Amomum	3	,,
Oil of nard	3	,,

Vinegar as much as sufficeth.

Mix together well, and use as a plaster.

Another, which is good for protracted hardness of the spleen, and for pain in the excretory organs.

Cardamoms
Amomum
Pyrethrum } [in equal quantities]
Mustard
Flour of trigonella

Pound, work into a paste with vinegar and oil, and use as a plaster, and if a bath be near, let the patient wash daily and then apply the plaster.

Or crush equal quantities of ammoniac, bdellium, myrrh, and frankincense in strong vinegar, dip a piece of cotton-wool into it and wet it well, lay it over the spleen, and let it stay there until it cometh away by itself. When it falleth off, dip another Page 402 piece of cotton-wool in the vinegar and apply it. |

Another, which is good for hardness [of the spleen], and is very powerful.

Rape seed	2	drachms
Camphor root	2	,,
Pyrethrum	2	,,
Cardamoms	1	drachm
Glue, prepared	2	ounces
Gypsum (lime)	4	drachms

Pound the dry ingredients, and mix them with the glue, and finally add the lime (plaster of Paris), work into a paste, | and Fol. 193 *a*. use as a plaster.

Another.

> Trigonella ⎫
> Mustard ⎬ in equal quantities
> Flour of barley ⎭

Pound, boil in vinegar, and use as a plaster.

Another. Take ten large, fleshy figs, soak them in vinegar, boil them, crush them and pound them, add to them a quantity of mustard and camphor root, pounded to a powder, equal in weight to that of the figs, and work up into a paste, and use as a plaster. Let the plaster remain on the patient as long as he is able to bear it, and then let him go into a bath, taking the plaster on him, and let him sit in a vessel of hot water.

Or boil wine with the vinegar, or liquid myrrh, or flax seed, or trigonella, or barley-flour, and use as a plaster. Boil each by itself, with vinegar, and smear him with it.

Or boil fleshy figs with strong vinegar, pound them together, and mix with them equal quantities of ammoniac, and bdellium, and camphor root, and mazrîôn (or, mazerîôn, or, mâzrîôn), and work up into a paste, and use as a plaster.

Another, which is very good, and which is called the "Idiot's medicine".

> Oak glue   2 *lìṭrè*
> Unslaked lime   1 *lìṭrâ*

Pour the glue into a potter's vessel, set it on the fire until it is melted, then sprinkle the lime over it, and mix them together. When it is hot smear it on the skin of the patient, like a plaster, and afterwards let him take a bath.

Another.

> Oak glue   1 *lìṭrâ*
> Unslaked lime   6 ounces
> Flowers of "Asia stone"   4   ,,
> Resin   1 drachm

Mix and work well together, and use.

Another, which is called the "Flower", and which is very good for pain in the spleen, and for dropsy, and for distention of the epigastrium, and for pains in the excretory Page 403 organs, and for gout, and for every | protracted pain.

Wax	2 ounces
Cedar gum	2 ,,
Dry pitch	2 ,,
Dross of the furnace	2 ,,
Arsenicum	2 ,,
Unslaked lime	2 ,,
Myrrh	2 ,,
Glue	2 ,,

Vinegar and oil as much as sufficeth.

Pound up and use as a plaster.

Another, which is exceedingly powerful, and is good for the dropsical, and for fistulas of long standing.

Pepperwort
Trigonella   } in equal quantities
Unslaked lime
Natron in tablets

Fol. 193 b. | Crush, and work into a paste with honey, and smear on the skin which is tender, and use as a plaster (or, bandage). Let the plaster remain on the patient as long as he can bear it, and when he beginneth to suffer great pain let him go into a bath, and having taken off the bandages, let him sit in a vessel of hot water, and stay there for some time. In doing this men suffer a great deal, but they go out of the bath healthy and well.

Another, which is good for hardness of the liver, and for dropsy.

Wax	1	ounce
Ammoniac	1	,,
Gum of terebinth	1	,,
Irises	2	drachms
Sampsuchon	4	,,
Ferula opopanax	4	,,
Ox fat	4	,,
Aniseed	4	,,

Ammi Copticum	4 drachms
Ox gall	4 ,,
Peppercorns	4 ,,
Flax seed	5 ,,
Trigonella	5 ,,
Honey	4 ,,
Wine	1 measure

Old oil as much as sufficeth.

Mix together, and work up, and use as a plaster.

Another, which was prescribed by Galen, and is very useful.

Moist pitch	1 *lîṭrâ*
Wax	40 drachms
Sulphur	3 ounces
Natron	3 ,,
Smelters' dross	3 ,,
Wine } Oil }	as much as sufficeth.

Mix together, and work up, and use as a plaster.

Another, which was prescribed by Asclepiades, and which is exceedingly useful for obstinate pains, and pains of long standing of all kinds.

Moist pitch	40 drachms
Wax	1 *lîṭrâ*
Gum of cedar	1 ,,
Natron	1 ,,
Sulphur of the sea	1 ,,
Wild grapes of the mountain	4 ounces
Pyrethrum	4 ,,
Dregs of burnt vinegar	$\frac{1}{2}$ *ḳesṭâ*
Galbanum	4 ounces
Pepperwort	1 *ḳesṭâ*

Wine and oil as much as suffice.

Mix well together, and work up, and use as a plaster.

Or dissolve ammoniac in vinegar, and smear it over the spleen, and dip cotton-wool in it.

Or dissolve frankincense | in extract of cabbage, and use it Page 404 as a liniment.

Or boil beet in vinegar, crush, pound, and mix with it cow oil, and use as a plaster.

Or boil cabbage in strong vinegar, and bathe the body with the vinegar; fry the cabbage with oil, and use it as a plaster.

Or make an infusion of fresh gourd leaves in vinegar, then fry the leaves, and use as a plaster.

Fol. 194a.   Or treat the leaves of the croton | plant in the same manner.

Now concerning the symptoms of the diseases of the spleen, and the means to be employed in healing the same, the above must suffice for the Eighteenth Chapter.

[HERE ENDETH THE EIGHTEENTH CHAPTER.]

CHAPTER XIX.

Of the symptoms of the diseases which take place in the bowels, and of all the various kinds of looseness of the belly.

SINCE it is meet for us to write in this Chapter concerning the different kinds of diarrhoea which exist, it is necessary first of all to state in what different forms it existeth, and what are the causes that produce them. Diarrhoea ariseth naturally from the stomach, and from the liver, and from the bowels. The diarrhoea that ariseth from the stomach is due either to the weakness of the natural powers that are therein, when they are unable to digest the food, and when they retain it, and it is subsequently cast forth undigested, or to some collection of the chymes in it, and these are emptied out by nature as the result of some critical happening, as, for example, by vomiting, or by a loosening of the belly, or through some evil quality of the food, or through insufficient preparation of it, or through excess in the quantity thereof, when it doth not digest well and is expelled alive (*i. e.,* in the state in which it was eaten), or through some heat in the belly, which destroyeth the food and dissolveth it, or through excessive cold, through which the liver is unable to digest it. Now, diarrhoea ariseth from the liver also. When there are obstructions in the liver, or in the mesenteric veins, and it cannot absorb the juice from the intestines, then this juice goeth down undigested, and produceth diarrhoea. Or when there is feebleness in the liver, | it produceth Page 405 that bloody diarrhoea which hath been mentioned, and which is like unto the water in which meat that hath just been killed hath been washed, or that other kind which is like unto filth.

And diarrhoea ariseth also from the intestines (or, bowels), either from weakness of their natural powers, or when they "cough" through the natural roughness that they possess, and there taketh place in them that which is called a "passage",

Fol. 194*b*. or when the chyme of bile or of phlegm | becometh collected in them and inciteth them to evacuation, or when there is griping of some kind, or when some opening of the veins, or arteries, which are between the intestines, taketh place and produceth a discharge of blood, or when there are boils and fistulas in the anus, and these incite the belly to make a motion. These are, in brief, all the causes that produce a discharge of the belly, and these are the places in which they subsist. Now, their exact symptoms are well known, both from the canons of the art of medicine, which have been written down in the Second Chapter of this treatise, and from the indications that are here laid down.

Now, in the description which hath been given above of the diseases of the belly, we have in our discourse divided the belly into two portions. The upper portion thereof, which is united to the channel (or, tube) that bringeth down the food, they call "stomach", and it is composed of very many sensitive nerves; the other portion, which is attached thereto, descendeth as far as the exit of the bowels (or, intestines). As each of these portions filleth the need of the use [thereof] in animals, and in proportion to its need produceth also happenings, which are different from [those of] its companion, so also all those diseases that arise in the interior, that is to say, from the members that are similar in their parts (*i.e.,* homogeneous), and from the organs, are common to these, and to the others, and to all the intestines. Now, the similarity of the things that happen to the members is the same not only in the case

Page 406 of those that resemble each other, | but also in the cases of the other members which are associated with them, even as it is also in the case of the kidneys in respect of the disease that is called "disease (ἵλαος) of the colon" (*i.e.,* colic). And from many of the happenings (or, attacks) it is easy to diagnose the position of the place that is diseased, even as in the case of the internal pain caused by haemorrhoids. Now, it is meet for you to hearken to a description of a disease of the interior, which is accurately thus named as this appellation indicateth that there is a wound in the intestines.

Of dysentery, that is, ulceration of the bowels.

Now, this disease doth not come into existence suddenly, | Fol. 195 a. like that other disease from which, as we have said, the liver suffereth, and when it beginneth there are associated with it also particular symptoms, such as evacuations of bile in large quantities which cause griping pains and afterwards there come forth pieces scraped off the bowels, and at the same time there cometh forth also a little blood. Now, inasmuch as the pain in the interior is a symptom of the disease, when the pieces scraped off the bowels are evacuated, it is meet to examine [them] and see lest peradventure something that resembleth a kind of oil is evacuated with them. If thou seest that such is the case [then it is clear] that the ulcer is in the thick intestines. When something is evacuated with the portion scraped off the bowels, it is right to examine closely in order to find out if this something is attached to all the matter of the stool, or whether it occurreth in one part of it only. If it be attached [to the whole stool], it sheweth that the ulcer is situated in the upper intestines; if it occurreth in one part only, and is not attached [to the whole] stool, it sheweth that the ulcer is situated in the lower intestines. Now this seemeth to take place also in those evacuations of detached membrane, but if it be known that it is not thus, it resembleth that which taketh place when blood is passed. So likewise is it when skin is evacuated, for it indicateth in which intestines the ulcer is situated, and the fact that it is attached to the things that are evacuated indicateth also some commotion in its substance. | Page 407 If the disease be situated in one [member, or part] only, it will be benefited in no small degree by a course of healing treatment, and this treatment will be indicated to us when we know in which of the intestines the ulcer is situated. For when the ulcer is in the upper intestines, it may be healed by the remedies that are drunk, but, if it be in the lower intestines, it will be benefited by the remedies that are transmitted to it from below by means of washings (or, bathings). Now, internal ailments of this class are defined by the evacuations that are caused by the liver, because, in the initial stages of diseases of the liver, a thin, bloody, putrid fluid is evacuated, and likewise, when the disease hath gained strength, a thick chyme, Fol. 195*b.* which resembleth the scum (or, sediment) | of wine, is evacuated, and also by the fact that during such diseases no pieces of abraded membrane of any kind are evacuated, and by the fact that the emptyings out, which are due to the liver, frequently take place only after intervals of two or three days, and then, when patients have evacuated, the distress returneth to them in a more aggravated form than before.

As concerning ulceration of the bowels, this doth not seem to arise suddenly, moreover, in the evacuations that are seasonable blood is not evacuated. The ulcer that is below, inside the anus, and that is called "haemorrhoid", produceth great straining at stool, and an exceedingly frequent desire to evacuate. In the initial stages of its growth patients evacuate a small quantity of some substance that resembleth phlegm, or a kind of fatty matter; but if the ulcer hath been in existence for a considerable time, pieces of detached membrane appear. Now, all these appear, during the whole period of the disease, not to be mixed in any way with the things that come from above. And certain physicians have written to this effect:—"In cases "where strong haemorrhoids have existed, certain men have "sometimes evacuated from the anus stones, full of holes, and "they resemble those that exist in the bladder." But this I myself have never seen, moreover, I never heard any man say [that he had seen such]. The above remarks on the symptoms Page 408 of ulcers in the bowels must suffice, and we must | append here a description of the course of treatment suitable in each kind of disease that hath been enumerated above.

Of the healing of looseness of the bowels.

Now, we have already said above that emptyings out (or, discharges) of the chyme, which is confined in the stomach or in the intestines, take place during some natural crisis. This kind of discharge it is not meet for the physician to restrain, on the contrary, he should assist nature, and he should make the patient to empty himself through that part of the body to which the matter inclineth. Thus, if he see the patient emptying himself by vomiting, he should administer to the patient such things as will help him to effect the act of vomiting. And if he see that the patient hath looseness of the bowels, in this case likewise he should assist him by administering to him that which appeareth to him to be capable of making him able to empty himself of superfluous matter, either in the form of food or in the form of medicine. | In the case of Fol. 196a. excessive heat, which transmitteth the food too rapidly, we administer the kind of food that is difficult to digest, such as ox flesh, or something similar, in order that the heat may be concentrated upon the difficulty of digesting the food. Now, as concerning the cold and heat of the belly, a description of the treatment to be employed hath been already written in the Chapter on the belly. An unsuitable kind of food we replace with one that is good, and the foods that are irregular we replace by those that are regular, and we diminish the quantities of the food, which hath been given hitherto in excess, and thus healing followeth by the exercise of medical skill. The means to be employed in healing the looseness of the bowels, which ariseth through weakness of the liver, and through obstructions in the veins, hath also been very fully described in the Chapter that treateth of the diseases of the liver. It now remaineth to write down in this Chapter a description of the means of healing that are to be employed in cases of injuries in the intestines, and we will first of all write down the medicines that are to be drunk, and that are beneficial in cases of ulceration of the intestines.

Disparmaṭôn Tablets, which is, being interpreted, tablets compounded of seeds, which are good for ulcers of the bowels, and for those whose food doth not digest, and for violent colic, and for violent straining at stool.

Myrtle berries	2 ounces
Seed of dill	1 ounce

Page 409

Aniseed	1 ounce
Ammi Copticum	1 ,,
Seed of hyoscyamus |	1 ,,
Opium	6 ounces

Pound and work into a paste with wine, and make into tablets, each containing half a drachm, and use after seven months.

Aromatic tablets, which are good for gripings of the belly, and for diarrhoea, and for nausea, and for pain of the stomach.

Cassia	2 ounces
Purple βαλαύστιον	2 ,,
Spikenard	2 ,,
Bark	2 ,,
Cypress berries	3 ,,
Pyrethrum	1 ounce
Ginger	1 ,,
Peppercorns, white	1 ,,
Peppercorns, long	1 ,,
Caryophyllus aromaticus	1 ,,
Frankincense	1 ,,
Crocus	4 drachms
Costus	4 ,,
Dried roses	20 ,,
Styrax	4 ,,

Pound and make into tablets with water. Dose three(?) drachms.
Fol. 196b. Some add to the above ingredients | myrtle berries two drachms, sumach two drachms, and malabathrum half a drachm.

Tablets compounded with opium, which are good for looseness of the bowels, and for straining at stool, and for gripings. They can be swallowed, and when dissolved in tincture of lamb's tongue may be administered in the form of an injection.

Opium	8 drachms
Myrrh	2 ounces
Aristolochia	2 ,,
Crocus	2 ,,
Myrtle berries	2 ,,
Sumach	2 ,,

Hellebore stalks, roasted	2	ounces
Burnt paper	2	,,
Seed of thorns	2	,,
Seed of hyoscyamus	2	,,
Flowers of thorn	4	,,
Frankincense	4	,,
Amomum	4	,,
Garden abrotonum (southern- wood)	4	,,
Oak bark	1	ounce
Acacia	1	,,
Madder	1	,,
Bamboo sugar	1	,,
Ṭarmâgbîn (a kind of manna)	1	,,
Purple βαλαύστιον	1	,,
Cassia	1	,,
Sea seals	1	,,

Make into tablets with strong-smelling wine, each containing a drachm, and administer to every man according to his strength, in some convenient draught.

Tablets compounded with hypocistis (tragopogon pratense, goat's beard) which are good for protracted looseness of the bowels, and for pain in the intestines.

Extract of hypocistis	6	drachms
Purple βαλαύστιον	6	,,
Acacia	6	,,
Pulp of pomegranates	6	,,
Glaucium phoeniceum	4	,,
Myrrh	2	,,
Frankincense	2	,,

Make into tablets with extract of myrtle. Dose, one drachm, in extract of asparagus.

Tablets compounded with | Mùshakṭôn (moschatum, or Page 410 odoramentum Romanum), which are good for diarrhoea and gripings.

Gum Arabic	3	drachms
Purple βαλαύστιον	3	,,
Acacia	3	,,

31

Extract of hypocistis	3 drachms
Glaucium phoeniceum	1 drachm
Styrax	1 „
Moschatum	1 „
Fresh extract of roses	1 measure
Myrtle berries	4 drachms

Make into tablets with extract of roses, each containing a drachm, and administer in an infusion of myrtle berries.

Pills compounded with castoreum, which are good for gripings, and diarrhoea, and severe strainings at stool, and for intermittent fevers.

Castoreum	2 drachms
Myrrh	2 „
Opium	4 „
Aniseed	4 „
Seed of rock parsley	8 „
Seed of white hyoscyamus	3 „

Make into pills about the size of peppercorns with wine mixed with water, and administer seven, or six [as a dose]. And Fol. 197 a. after [taking] them let the patient drink extract of | asparagus, or extract of myrtle berries. If thou givest them for fever, let the patient drink hot water after them, and let him take them one hour before the period of the fever.

Another [medicine which is good for] the same diseases, and which is to be taken in the same way.

Seed of hyoscyamus	4 drachms
Seed of rock parsley	4 „
Aniseed	4 „
Styrax	4 „
Crocus	3 „
Opium	3 „
Myrrh	2 „
Castoreum	2 „

Make into pills with wine, and administer six or seven drachms in an infusion of aniseed and cummin.

Another [medicine which is good] for the same diseases.

| Seed of rock parsley | } All in equal |
| Ammi Copticum | } quantities |

Seed of hyoscyamus ⎫
Cummin
Seed of poppies   ⎬ All in equal
Aniseed   quantities
Myrrh
Opium   ⎭

Pound, and make into small pills, and administer seven or five according to the directions given above.

Another medicine which is compounded with mandragora, and which is good for severe gripings, and for strainings at stool, and for intermittent (or, periodic) fevers, and which restraineth the bowels, and relieveth the pains.

Bark of mandragora root	4 drachms
Seed of hyoscyamus	4 ,,
Crocus	5 ,,
Myrrh	5 ,,
Frankincense	5 ,,
Opium	5 ,,
Castoreum	5 ,,

Pound, and work into a paste with wine, and make into tablets, each containing two *den̲ke̲*, or, make into small pills, and administer according to the directions | given above. Page 411

Tablets compounded with dried grapes, which are good for protracted looseness of the bowels, and for the passing of excremental matter.

Dried green grapes   ⎫
Pulp of small pomegranates   ⎪
Purple βαλαύστιον   ⎪
Pulse   ⎪ in equal
Sumach   ⎬ quantities
Frankincense   ⎪
Myrrh   ⎪
Gum Arabic   ⎪
Crocus   ⎭

Pound, and make into pills with fresh myrtle juice, and administer two drachms [as a dose]; and afterwards let the patient drink two measures of an infusion of myrtle berries.

31*

Pills compounded with myrrh, which are to be placed in the anus, and which will arrest the looseness, and relieve the straining at stool.

Myrrh
Frankincense
Crocus
Opium
} in equal quantities

Pound and work into a paste with the yolks of fresh eggs, and make into pills as long as small acorns. Place inside each a thread, smear them with oil of roses, and place them in the anus, and let them remain there for a whole | night. If thou wishest to remove them, take hold of the threads which are in them and pull them out.

Fol. 197b.

Another [kind], which is to be placed in the hollow beyond the anus in the straight intestine, and which is good for protracted diarrhoea, and for gripping firmly the belly which is loose.

Alum from the furnace
Frankincense
Crocus
Powdered frankincense
Styrax
Castoreum
Indian goat's beard
Opium
Gum Arabic
} in equal quantities

Pound and work into a paste with extract of lamb's tongue, or extract of polygonum. Dip wads of wool in it, and dry them, and place them in the anus, and let them remain in the patient the whole night. Now, to each of these wads, as to each pill, a long thread must be tied, so that if it be necessary for us to draw them out we may be able to do so by means of it.

An antidote for flux of the belly, which is good for looseness of the bowels, and for gripings, and for straining at stool, and for frequent "coughing".

Castoreum
Styrax
} in equal quantities

Opium ⎫
Seed of hyoscyamus ⎪
Crocus ⎬ in equal quantities
Asarum ⎪
Myrrh ⎭
Honey as much as sufficeth.

Dose, a piece about the size of a vetch.

Roasted drugs, which are good for straining at stool, and for the discharge of blood.

White pepperwort, roasted	5	*estîrê*
Cummin steeped in vinegar |		
and roasted	12	drachms
Flax seed	12	„
Seed of leeks, roasted	12	„
Stacte	4	„
Black hellebore steeped in cow oil	8	„

Page 412

Pound and mix together, and administer in the form of a dry powder; dose, one spoonful in extract of asparagus, or hot water.

Tablets compounded with crocus, which are good for pain in the inside, and for gripings, and for those who cough and bring up blood, and for the flow of the menstrual fluid.

Crocus	3	drachms
Castoreum	4	„
Myrrh	4	„
Seed of rock parsley	4	„
Hyoscyamus	4	„
Dûk̩ôn	4	„
Opium	4	„
Styrax	4	„

Make into tablets with water, and administer half a drachm in extract of myrtle berries.

Tablets compounded with βαλαύστιον, which are good for gripings, and for diarrhoea.

Βαλαύστιον	8	drachms
Kardt̩înà	8	„

Sumach	8	drachms
Roasted acorns	8	,,
Polenta of the service tree		8	,,
Cummin steeped in vinegar and roasted	4	,,
Dried grapes, roasted and cooled in vinegar	4	,,
Myrtle berries	8	,,

Make into tablets with extract of Persian apples, or extract of lamb's tongue, each containing a drachm, and administer.

Tablets compounded with roses, which are good for diarrhoea, and for those who bring up blood.

Rose blossoms	1	ounce
Opium	1	,,
Acacia	1	,,
Gum Arabic	1	,,
Extract of hypocistis	1	,,
Purple βαλαύστιον	1	,,
Dried grapes	2	ounces
Fleawort (or, fleabane)	2	,,
Extract of lamb's tongue	2	,,
Indian goat's beard	2	,,
Extract of roses	1	ounce

Make into tablets with extract of roses, and administer in some convenient draught.

Another kind of tablet, which is good for protracted diarrhoea. And I use it myself. Take in equal quantities dried grapes, fruit of "îrâḳânê", and opium, and make into tablets, or pills, each containing one *denḳâ*, and use.

Tablets compounded with purple shell-fish, which are good for diarrhoea, and for pain in the belly, and for those who bring up blood. After one dose the flow is arrested.

Shell-fish, roasted and washed	12	drachms
Dried grapes,	roasted	12	,,
Cedar bark, burnt	12	,,
Extract of sumach	12	,,
Root of mandragora	12	,,
Gesamyâ (γῆ Σαμία, or talc)	12	,,

Stag's horn, roasted	6	drachms
Vine-flower wine	6	,,
Extract of hypocistis	6	,,
Opium, roasted	6	,,
Acacia, roasted	6	,,
Black myrtle berries	15	,,
Sumach	14	,,
Pomegranate rind, roasted	7	,,
Frankincense	8	,,
Seed of rock parsley	14	,,
Seed of hyoscyamus, roasted	4	,,
Sumach of the leather-workers	2	*kestê*
Black wine as much as sufficeth.		

Boil the sumach with the wine until the mixture becometh thick, strain out the sumach, mix up the drugs in the wine, and make into tablets, each containing one drachm, and administer.

Tablets compounded with môrîḳâ (tamarix), which are good for looseness of the bowels, and which will restore them to a proper condition.

Tamarix fruit, that is, *'ârâ*	4	drachms
Edible sumach	4	,,
Black myrtle berries	4	,,
| Acacia	2	,,
Opium	1	drachm

Fol. 198 *b*.

Make into tablets with extract of Persian apples, each containing one drachm, and administer, with extract of asparagus, or with an infusion of senna and an infusion of myrtle, two measures.

The Medicine of Asclepiades, which is good for looseness of the bowels.

Myrtle berry wine	6	*kestê*
Sumach	1	*lîṭrâ*
Honey	1	,,

Boil the sumach with the wine, stir up with a senna stick until the mixture is thick, strain out the sumach, pour in the honey, and boil until the whole obtaineth consistency, and administer as a dose one spoonful fasting.

Another medicine of Asclepiades, which is good for the same diseases. Crack a newly-laid egg, pour the contents thereof into a vessel, add to them an equal quantity of sweet Page 414 oil, an equal quantity | of omphacium,[1] an equal quantity of crushed peppercorns, an equal quantity of pounded raisins, and an equal quantity of wheat-flour, and work into a stiff paste, and make into large pills, and give them to patients to swallow fasting. Let them swallow as many of them as they can, and let them not drink anything after them.

Another medicine for the same diseases.

Green dried grapes
Rind of pomegranate
Sumach } in equal quantities
Peppercorns

Crush these, and work them into a paste with the yolks of eggs, pour the mixture into the shell of a pomegranate, smear the edge thereof with fat, and roast on hot coals. Then crush and pound the shell of the pomegranate, and give to patients as much as they can eat, and let them drink nothing. Very many [sufferers] have been benefited by this medicine.

Pills which are compounded with raisins, and which are good for pain in the belly and diarrhoea. Take in equal quantities green, unslit raisins, and opium, and rock parsley, pound them, make into pills, with water, about the size of peppercorns, and administer seven or ten as a dose.

Another kind which is very powerful indeed. Take rock parsley, fruit of îrânâķê, sumach, and opium, in equal quantities, make into pills with extract of myrtle berries, and administer.

[A medicine] which is good for those who have diarrhoea, and protracted looseness of the bowels. [Take] milk of goats or of cows, drop into it red-hot stones until one-third Fol. 199a. of it hath boiled away, let it | cool a little, but keep it stirred all the time it is cooling, and when it is cold, let the patient drink it, and let him do this for seven days. This will form an excellent medicine for looseness of the bowels if physicians will pour into it the white droppings of dogs which are costive

---

1 Oil made from unripe olives, and mixed with aromatic substances.

in the season of summer, provided they contain many bones. These droppings must be pounded and reduced to a powder, together with the bones in them, and then poured into the milk which hath been boiled by means of hot stones. Dose, one or two spoonfuls. Let the patient take this dose seven, or five, days, and the droppings will assist the rheum of the belly.

Or eggs boiled | in vinegar and eaten. These will be much Page 415 more beneficial if there be mixed with them some astringent drug. The following medicine is good for those who have obstinate (or, protracted) dysentery.

Ashes of roasted shell-fish	4 parts
Raisins	2 ,,
Peppercorns	1 part

Crush and reduce to a powder, and sprinkle some of it over the food of the patient, and make him drink also weak white wine, or wine mixed with water.

For those who have severe gripings and diarrhoea the following medicine is good:

Psyllium	2 drachms
Oil of roses	1 measure
Water	2 measures

Mix together and administer in the form of a draught.

Or administer extract of purslane and let patients drink it, or let them eat it raw instead of vegetables.

Or let them drink two measures of extract of lamb's tongue daily.

Or let them drink extract of polygonum in the same way.

Or let them drink one measure of extract of asparagus, and two drachms of gum Arabic. Dissolve the gum in the extract, and mix with water, and administer.

Or pound the skins of grapes, roasted, and administer two spoonfuls in cold water.

Or let the patient drink two drachms of pepperwort, roasted, one drachm of roasted cummin, and two drachms of sugar. Administer these in one dose with cold water. If it seemeth to thee that there is some kind of bile eating into the bowels, and producing the diarrhoea, mix two parts of yellow hellebore with the pepperwort, and administer one spoonful as a dose. If there be severe straining at stool, and also continual griping, Fol. 199 *b*. mix therewith psyllium seed with | cold water and oil of roses, and administer in the form of a draught. And let the patient anoint his anus with warm oil of roses, and sit in an infusion of fennel, flax-seed, trigonella, and liquid myrrh. If there be dysentery, and it produceth a discharge without blood and griping, an injection of salt water will be very good for the Page 416 patient; the bowels must first of all be washed with honey water, and after this the salt water may be | injected. Or inject salt water in which olives have been soaked, with sea-seals, and it will prevent the bleeding. Now, for severe griping, and straining at stool, and intense gnawing pains, an injection made of husked-barley water, and oil of roses, and the yolks of eggs, is very good.

Or rice [mixed] with the fat of the kidneys of stags. Take the juices of these, and mix with them sea-seals, and inject the mixture.

Or inject the extract of [lamb's] tongue with one of the tablets that are compounded with paper, and that have been written down.

Or mix extract of polygonum in the same way, and inject.

Or infuse in strong wine dried roses, rinds of pomegranates, sumach, raisins, senna leaves, and root of thorns, until the wine becometh thick. Then strain and mix it with oil of myrtle, and one of the tablets that are compounded with paper, and inject it after the patient hath eaten his supper. It is a helpful medicine in cases of protracted diarrhoea, and of pain in the bowels that is due to great putridity.

Tablets compounded with paper, which are good for severe ulceration, and for the putridity that ariseth in the bowels, and for protracted diarrhoea.

Burnt paper	10	drachms
Sandarach (σανδαράκη)	5	,,
Burnt copper	5	,,
Verdigris	5	,,
Smelters' dross	5	,,
Oil of unripe olives (omphacium)	5	,,

Tincture of arsenic	5 drachms
Unslaked lime	5 ,,
Crocus	2 ,,
Opium	2 ,,

Crush thoroughly, mix with extract of lamb's tongue, and make into pills, each containing four drachms, dry in the shade, and use according to the strength of the patient. Thou shalt inject it mixed with one of the extracts that have been described above. To patients of full age (*i. e.,* over forty) give three drachms, | to those of middle age (*i. e.,* about thirty-three), two Fol. 200 *a.* drachms, and to children (or, youths), one drachm.

Another medicine, which is compounded | with paper, Page 417 **and which will assist immediately those who have no fever.**

Burnt paper	10 drachms
Acacia	5 ,,
Hypocistis	5 ,,
Unslaked lime	5 ,,
Sandarach	5 ,,
Tincture of arsenic	5 ,,
Copper rust	5 ,,
Smelters' alum	5 ,,
Oil of unripe olives	5 ,,
Lycium	5 ,,
Opium	2 ,,
Crocus	2 ,,

Pound and work into a paste with myrtle-berry wine, and make into pills of four drachms each. Mix with new wine, and inject according to the strength of the patient. Or mix it with extract of lamb's tongue, or strong wine in which sumach hath been steeped with myrtle, and rinds of pomegranates, and dried roses, and the yolks of eggs, and oil of roses.

Another very useful medicine.

The outside of burnt copper	8 drachms
Burnt paper	16 ,,
Unslaked lime	16 ,,
Acacia	6 ,,
Oil of unripe olives	3 ,,
Sediment of burnt wine	1 drachm

Make into four-drachm tablets with myrtle-berry wine, and inject with strong wine two drachms, either in extract of lamb's tongue, or in extract of polygonum, four drachms.

Another medicine, which is most excellent.

Burnt raisins	16 drachms
Burnt paper	16 ,,
Burnt Greek bread	16 ,,
Sandarach	16 ,, .
Tincture of arsenic	20 ,,
Acacia	5 ,,
Oil of unripe olives	3 ,,
Sediment of burnt wine	5 ,,
White lead	5 ,,
Unslaked lime	10 ,,
Lithargyrum	3 ,,

Pound and make into four-drachm tablets with old wine, and use when the pains are violent, mixed with an infusion of trigonella, or of flax seed, or yolks of eggs, and oil of roses. After the violence of the pains hath passed use the medicine mixed with an infusion of senna and myrtle, with strong wine in which sumach hath been steeped.

Page 418   **Another medicine, which I myself use, | because experience of its use proveth that it is a sure remedy.**

Tincture of arsenic	12 drachms
Sandarach	6 ,,
Unslaked	lime	26 ,,
Burnt paper	30 ,,

Fol. 200*b*.

Make into tablets with extract of lamb's tongue, and use according to the instructions that are written above.

Another medicine, which is exceedingly powerful.

Opium	6 drachms
Crocus	6 ,,
Burnt paper	30 ,,
Smelters' alum	2 ,,
Tincture of arsenic	2 ,,
Sandarach	8 ,,
Verdigris	8 ,,
Oil of unripe olives	8 ,,

> Extract of grape pulp 8 drachms
>
> Unslaked lime 15 ,,

Make into large tablets with myrtle-berry wine, and mix with some useful extract, and administer in quantities according to the strength and age of the patient.

The above medicines may be injected in cases of cancer (or, gangrene) and ulceration of the bowels, sometimes with an infusion of dry myrtle, or with extract of roses, or with extract of husked barley which. hath been melted down with goats' fat, or with extract of lamb's tongue, or with strong wine, wherein astringent substances have been boiled, and oil of roses, and extract of husked barley.

Now, the simple medicines, which may be injected with the extracts that have been mentioned, are, for example, sea-seals, acacia, lithargyrum, white lead, amylum, goat's beard, frankincense, hypocistis, dirty wool which hath been burnt, paper, burnt raisins in vinegar, electrum (*i. e.,* amber), and coralium. Very useful also for those who suffer from looseness of the bowels and diarrhoea are washings in infusions made from roots and twigs of plants that possess astringent properties, now, the infusions must be made with strong wine, and he who suffereth must wash himself therewith. If the patient be very weak let physicians add to them unleavened bread, and strong-smelling wine, and wine of lilies, and all those substances which are sweet-smelling. And on each occasion when the patient is sitting | in the bath, let them pour on his head a Page 419 little cold water. Now, for those patients who are very weak the following plaster will be found to be very beneficial.

A bread-plaster (poultice), which is good for looseness of the bowels, and which will greatly strengthen | those Fol. 201 a. who are weak.

> Greek bread ⎫
> Leaves and berries of myrtle ⎪
> Crocus flowers ⎪
> Asparagus ⎬ in equal
> Apples, cleaned ⎪ quantities
> Cummin ⎪
> Sandal-wood ⎭

Wood of aloes
Aromatic reeds
Crocus
Rind of pomegranates } in equal
Acacia   quantities
Gum of ladanum
Frankincense
Myrrh
Wine of lilies } as much as
Oil of jasmine } sufficeth.

Mix these well together, and apply as a plaster to the belly after bathing.

Another medicine, which is good for diarrhoea, and thirst, and for excessive burning.

Peeled asparagus boiled in vinegar	4 drachms
Myrtle berries and dactyl grapes boiled in strong-smelling wine	4 „
Flowers of crocus	4 „
Origanum majorana	4 „
Myrrh	4 „
Greek bread	8 „
Stacte	3 „
Aromatic reeds	1 drachm
Oil of roses and Vinegar }	as much as sufficeth.

Mix together skilfully and carefully, as is meet, and use as a plaster.

Another medicine, which is good for diarrhoea, and for looseness of the bowels; this medicine hath been well tried and is a sure remedy.

Myrrh	4 drachms
Frankincense	4 „
Stacte	4 „
Acacia	4 „
Alum	4 „
Gum of ladanum	4 „

Aloes	4	drachms
Crocus	2	,,
Fine flour of barley	8	,,
Dried roses	7	,,
Purple βαλαύστιον	7	,,
Extract of hypocistis	7	,,
Yellow raisins	7	,,
Glaucium	7	,,
Lycium (πυξάκανθα)	7	,,
Opium	4	,,
Bark of mandragora root	4	,,
Seed of hyoscyamus	4	,,

Pound all these, and dissolve in vinegar, or in wine, and smear with it all the epigastrium, and the back and loins. And place on the patient a layer of cotton-wool, which hath been soaked in the mixture, and let it remain there until it falleth off of its own accord. It is an exceedingly powerful remedy. Other physicians mix with this medicine wax and oil of roses, and then use it as a plaster. Now, for looseness of the bowels and for diarrhoea, | the plasters that have been prescribed for Page 420 weakness of the liver and stomach are also useful. And this Chapter, which is the nineteenth, must come to an end at this point.

HERE ENDETH THE NINETEENTH CHAPTER.

CHAPTER XX.

Of the symptoms and distinguishing characteristics of the diseases | that subsist in the intestines of the colon.

NOW, it seemeth that there exist in the intestines other diseases, which are the opposites of those that have been mentioned above, as, for example, the disease called "disease of the colon". It hath been demonstrated that, after (*i. e.*, beyond) the lower mouth of the belly which is called "the fingers", there is situated the intestine called "Ṣâwmâ", and it is so called because it appeareth always to be empty. From the mouth of this intestine the mesenteric veins (or, arteries) receive juice and convey it to the liver. After the intestine Ṣâwmâ is the "little intestine", and after the "little intestine" is the intestine called "ʿawîrâ" (*i. e.*, Blind). At the mouth of this [last-named intestine] is situated the thick intestine that is called "Colon", and it extendeth downwards as far as the intestine that is called "Tĕrîṣâ" (*i. e.*, Straight), and after this there is the intestine that is close to the anus.

Now, the casings of the colon extend right and left, to the liver, to the spleen, to the loins, and to the region above the kidneys, and for this reason men have come to call every ailment that attacketh the other intestines, and also the kidneys, and that causeth effects resembling those that take place in the colon itself, a "disease of the colon". For the causes of the ailment which arise in the intestines, and which produce the effects of retention of the belly (*i. e.*, costiveness), and severe pain, are, to speak briefly, the following:—ulcers which are hot and firm, ulcers which are hard and cold, the chyme

of some kind of bile, or phlegm, dense wind, | which produceth inflation [of the belly], some kind of faeces which have become dry in the intestines, and round or flat clots of blood; now, some men have said that a stone is produced in the intestines of the colon. These are the causes that produce constraint in the belly, and severe pain in the intestines, and the other effects that follow diseases of the colon. Now, violent pain appeareth very often in the colon, which is thought by physicians not to arise from the colon, but from the kidneys, just as they have also thought that pain of the kidneys arose | in the colon; Fol. 202a. and certain of these physicians thought that pain of the colon never appeared in the left side. These diseases are, in their initial stages, very hard to diagnose, and there is likewise very little difference between the various kinds of assistance which they require; therefore it is meet to examine carefully the effects that are predominant in each one of them. Now, spewings and vomitings take place in very many cases of disease of the colon, and they are constant in those who suffer therefrom, and they pour out in large quantities some substance which is decayed and full of phlegm. On the other hand, their stool is restrained to such an extent that they cannot expel wind, not even small puffs; so long as they retain it they feel as if the pain were a fire burning in their inside, and as if it were occupying a very large space. And sometimes, when the pain increaseth, it seemeth to be in many other parts [of the belly], but when pains arise in the kidneys they seem to be fixed in one place. When, then, the pain is fixed in one place, above the seat of the kidneys, it is impossible from this to derive any indication (or, symptom) of the disease, but in addition to the directions which I am now going to enumerate, let us look also at the urine, because in the initial stages of diseases of the kidneys the urine is clear and watery, whilst a few days later there appeareth in it a substance of unequal density, and still later grains of sand appear | in it visibly. And even though the bowels may make Page 422 motions during disease of the colon, the stool is windy, and very often the faeces that are expelled have the appearance of having been dipped in water. Moreover, the diseases of the colon are relieved by injections which contain substances

32

that soften, whilst those of the kidneys are aggravated thereby. Now, there are times when if some kind of cold chyme be evacuated, patients are relieved immediately, and in such cases the assistance that giveth refreshing becometh not only a means of refreshing, but also a well recognized healer.

Now, inasmuch as the cold chyme, when it is evacuated, delivereth patients from pain, and indicateth the part that is **Fol. 202 b.** diseased, | so that henceforward we may also occupy ourselves carefully with the part that is not diseased, and inasmuch as sometimes the frequency of the pains [in] the two organs [sheweth] that they require the self-same remedies, so also during the later stages of the diseases these organs require remedies that are different. Therefore we think also that nothing is of use as concerning healing or injuring, since the first difference in their initial stages is difficult to distinguish, because at that time they do not require remedies which are different, neither external nor internal, for the medicines which give relief are alone sufficient.

Since, then, all the intestines are situated after the belly, that is to say, the upper intestine which is called "Sâwmâ", and after that the "Little" intestine, and after that the "Blind" intestine, which separateth from it, and looketh in a downward direction,—now the colon is the intestine which goeth in an upward direction until ofttimes it toucheth the liver and the spleen,—for this reason, I say, it cometh to us to wonder how it is that physicians, as well as simple folk also, have been persuaded to call all the diseases that arise in these intestines, which are parts of the belly, "diseases of the colon". Now **Page 423** it seemed to me also that | I ought to be persuaded of the truth of this opinion, but when I sought out the cause of the stability of the pain, I was not, as it happened, convinced of its truth by the things which had been said. For it is not with the colon as it is with the kidneys, or with the veins that are between the kidneys and the bladder when there is a stone in them, and when an insistent pain ariseth in their exits; and one must think that it is the wind that is confined, or the cold chymes like them that produce the pains. For the texture of the coats of the little intestines, because it is exceedingly light and fine, is incapable of holding for a long time such causes [of pain] as these; therefore it is clear that when a cold, thick, and viscous chyme is generated in a thick, dense body (or, organ), or a wind that produceth swelling, and when it cannot find a place of escape, it causeth a violent pain for these two | reasons, namely, a change of composition (or, constitution), Fol. 203a. and the distention of the bodies (i. e., organs) in which the wind is confined. Moreover, this pain becometh continuous also, because, owing to the fact that it is kept in by the thickness and density of the bodies wherein it is confined, it cannot be evacuated sufficiently.

Of the disease which is called Îleôs (εἰλέος).

Now there arise also at times in the chest pains . . . . sufficiently, and they excite vomitings [which are so severe that the man who is attacked] by this disease can only live with difficulty, and this disease they call "Îleôs". [Now, it ariseth] when there is an ulcer that dischargeth itself outwardly in the region of the Little Intestine, and one thinketh that it is like unto the string of a harp which is fastened to the intestine. And to the physicians who were before us it hath seemed, and rightly, that this disease is due to the stoppages that are caused by the faeces, and that the diseases, which arise in the little intestines, and which are like unto it, are due to a kind of hardening [in the faeces]. Now, the physicians of olden time believed, and rightly, that other diseases arise when all the intestines are diseased together, | Page 424 and also the belly with them, just as arise the diseases of the intestinal canal, and of the entrance to the belly, when they are not of a gnawing character. The diseases that gnaw are closely connected with weakness of the intestines, because they are unable to hold fast that which is confined in them, even for a short time, but they expel it immediately, as if it were some weight that was pressing upon them, even as it is in the disease called strangury. For it appeareth that this happeneth also sometimes when the bladder is emptied continually because of the acidity and the erosive character of

32*

that which entereth it; and there are also times when it cannot bear the weight thereof, and although this be small, it ejecteth it. Now, the causes that produce [disease of] the intestinal canal, and the course of healing treatment that is suitable for them, have been described above. For the pustules, and the ulcers, and the hard sores, and the hard red abscesses, which arise in the various regions of the belly, | are well known, and they make manifest the place that is diseased as well as the disease itself, and these have been sufficiently described in the preceding sections [of this work].

Now, when the round and flat maw-worms become many in the intestines they produce severe pain. These we are able to recognize from that which patients evacuate before us in the time of health, just as the gourd-seeds which come forth with the stool form indications for us that maw-worms actually exist in the intestines. Similarly also people eject round maw-worms in the season of health in the vomit from the mouth, and sometimes in the stool from the anus, either because of the heat of fever, or through some medicine. Let this be a sure symptom to thy mind: maw-worms do not produce serious disease like unto the diseases of the colon, and they do not remain fixed in one place continually. And when thou layest thy hand [on the belly] the position of them will become well known to thee, for when thou hast held thy hand there for some time, and hast pressed it lightly upon it, | the maw-worms will seem to thee to be moving away from under it. They also produce languor of the mind, and vomitings, and spewings, especially before food, and they fill the mouth with saliva. Now, therefore, having described the symptoms of all the diseases that arise in the intestines, it is time to append a description of their treatment, even as the plan of this work demandeth. First of all we will write down a list of the medicines which are good for them, and then, one by one, we will enumerate the other helps (or, remedies) that are useful.

The medicine called "Pîlônyâ Antîdîṭôs" (Φιλονεῖον Φάρμακον), which was prescribed by Philo, and which is good for many ailments, and especially for pain in the colon, and it giveth relief to the pain.

White peppercorns	1 drachm
Seed of hyoscyamus	1 "
Opium	10 drachms
Crocus	5 "
Euphorbium	1 drachm
Spikenard	1 "
Pyrethrum	1 "
Honey as much as sufficeth.	

Dose, a portion as large as a chick-pea.

Greek Antîdîṭôs Pîlônyâ, | which is good for the colon, and for pain in the liver, and in the spleen, and in the kidneys, and for a cough of long standing, and for the uterine asthma of women who are pregnant.

Peppercorns, white	20 drachms
Seed of hyoscyamus	20 "
Opium	1 drachm
Crocus	5 drachms
Rock parsley, or	4 "
Seed of rock parsley	4 "
Spikenard	4 "
Bark	1 drachm
Laurus malabathrum	1 "
Euphorbium	1 "
Pyrethrum	1 "
Incense berries	1 "
Honey as much as sufficeth.	

Dose, a portion as large as a chick-pea. For disease of the colon administer in an infusion of fennel; and for pain in the liver, in an infusion of polium Teucrium; and for pain in the spleen, with oxymel compounded with camphor; and for pain in the kidneys and strangury, in an infusion of dill and rock parsley; and for [uterine] asthma, with the vinegar of squills; and for a hacking cough, in extract of husked barley.

Persian Pîlônyá, which is good | for diseases of the colon, and for looseness of the bowels, and for the menstrual flow, and for pregnant women whose courses still flow, and for winds that are held fast in the womb, and it protecteth the child in the womb and preventeth him from being destroyed.

Peppercorns, white	20 drachms
Seed of hyoscyamus	20 ,,
Sea seals	10 ,,
Opium	10 ,,
Crocus	5 ,,
Euphorbium	4 ,,
Spikenard	4 ,,
Pyrethrum	4 ,,
Castoreum	1 drachm
Zûrbadh (zerumbeth)	$^{1}/_{2}$,,
Drônag	$^{1}/_{2}$,,
Unpierced pearls	$^{1}/_{2}$,,
Musk	$^{1}/_{2}$,,
Camphor	$1^{1}/_{2}$ *den̲k̲ê*
Honey, skimmed and clarified, as much as sufficeth.	

Dose, one drachm, in some draught which is suitable for the disease.

The antidote Pîrôz (Pers. پیروز) -anôsh (*i. e.*, Pîrôz the Good), which is good for severe disease of the colon, and for flatulence and wind, and for diarrhoea, and for women who are held by wind.

Peppercorns, white	20 drachms
Opium	20 ,,
Seed of hyoscyamus	20 ,,
Crocus	6 ,,
Euphorbium	2 ,,
Pyrethrum	2 ,,
Spikenard	2 ,,
Honey as much as sufficeth.	

Dose, a portion as large as a chick-pea. It is good for every kind of pain, not that it doeth away with it entirely, | but it produceth relief from the disease.

Fol. 204 *b*.

Another, which is good when the pains are intense, and which produceth sleep, and affordeth relief.

Seed of rock parsley	8	drachms
Peppercorns, white	8	,,
Crocus	6	,,
Seed of dill	20	,,
Seed of ammi Copticum	20	,,
Opium	20	,,
Stag's horn, burnt	20	,,
Bark of mandragora root	10	,,
Honey as much as sufficeth.		

Dose, a portion as large as a chestnut, in some convenient draught. It will give relief in the disease immediately.

Tablets, which are good for the disease which is called Îlêôs (εἰλέος).

Seed of rock parsley	6	drachms
Aniseed	6	,,
Absinthe	4	,,
Bark	12	,,
Myrrh	2	,,
Peppercorns	2	,,
Opium	2	,,
Castoreum		2	,,

Page 427

Make into tablets with water, and administer as a dose one drachm.

Now these medicines, which are compounded of opium, and hyoscyamus, and mandragora root, and which are prescribed for diseases of the colon, do not put an end to the pain, and they only afford relief in these diseases. For this reason they are useful in diseases of all kinds, provided that a man maketh use of them with the skill that is meet; but if a man maketh use of them in an immoderate degree, and without skill and understanding, they produce very serious ill-effects in the bodies of the children of men, for through the great coldness which opium, and hyoscyamus, and mandragora root possess, they kill the natural heat of the ordinary member of the body, and bring the sick man into great and inconceivable peril. Now, we make use of the medicines only of necessity, and only in cases where the diseases are accompanied by great pain, or in severe attacks of sleeplessness, once every few days, in order that we may have an opportunity of making a plaster [and applying it to the patient] .... Now, often it is as it were caused by some hot wind........... or when diseases arise in one of the member through the hot, wet chymes, these medicines destroy the diseases entirely.[1]

Another medicine, which is good for disease of the colon; it hath been well tried.

Willow | bark	4 drachms
Ginger	3 „
Nuts	6 „
Dates	6 „
Water	1 measure

Boil until the mixture is reduced to one-third, stir up with a branch of rue, and administer as a dose two measures daily for three days. Now when the disease of the colon is due to some bilious chyme, or phlegm, or to faeces which have become hard in the intestines, the following are good:

Hiera píkrâ	2 drachms
Croton oil	7 „

Administer as a dose two measures in an infusion of fennel, galanga, and rue. If, however, thou thinkest that there is some kind of hot ulcer, \| or erosive chymes in the intestines, administer four drachms of psyllium, and half a measure of oil of roses, and two measures of tepid water, mixed together. And use also the pills compounded with sagapenum, which are prescribed in the section on dropsy, and give the patient two drachms in an infusion of fennel. And apply fomentations also to the belly, and use infusions of medicines that are of a quieting and softening character, such as spikenard, chamomile, flax seed, liquid myrrh, and every drug of this kind. And for flatulence and wind in the colon the white droppings of a wolf swallowed with honey, or tied up in the skin of a stag or wolf, and drawn over the belly, are useful. Concerning these droppings it hath been stated by Galen that they assist disease of the colon because of their natural properties. The white droppings of a dog, smeared over with honey, also assist flatulence

Page 428

---

[1] The gaps in the text make it impossible to give a connected rendering.

of the belly and colon. And small gourds when dried and swallowed assist the colon and also the severe gripings that take place in the intestines. An infusion of wild mint, drunk daily for three days, will often remove disease from the colon. Now, it is also useful | for them to take an old cock, and to kill him, and cut him up, and fill his inside with salt, and to boil him thoroughly with fennel, and rue, and ground-ivy; and if the patient be given three or four measures of the water in which the bird hath been boiled to drink, it will remove disease from the colon. Or crush galanga, and steep it in water, and give the infusion to patients and let them drink it. Or kill white, fig-eating wagtails (motacilla ficedula), boil them in salted fennel water, and let patients drink the liquor and eat the flesh of the birds, and he will be helped. Or take the other kind of wagtail, with coloured wings, which is called "cock of the desert", and kill one, and soak it in salted fennel | water, and let the bird be eaten [by the patient], and the liquor be drunk by him. Concerning this medicine, the writer testifieth that the patient is helped naturally, and not by any mixture whatsoever. It is also said that, if the red worms, which are found in the ground wherever there is mud, be dried, and pounded, and swallowed in an infusion of fennel, they remove disease from the colon. Or mix together two drachms of pepperwort and one measure of sesame oil, in seven drachms of tepid water, and let the patient drink it. Or pound four drachms of irises, and administer in the water of curdled milk. Very beneficial also are baths and washings in infusions of drugs that possess quieting and softening properties, and also hot water poured on the epigastrium, for it giveth relief in cases of disease of the colon, for it breaketh up the .... And if thou thinkest that it is worms which produce this disease in the colon, then make use of the medicines which bring down worms.

Now the simple medicines that kill worms are these: Abrotonon, absinthe, | senna, lupins, costus, garden nigella, pepperwort, compound of aromatic spices, bark of mulberry root, bark of the root of the pomegranate tree, both of the bitter and sweet kinds, saponaria officinalis (or, gypsophilla

struthium, or ptarmica, or gum ammoniac), leontice leonto-
petalem (European cyclamen), copper, ferns, ḳambîlâ, coriander
seed, thorns, copal, and the medicines that are called "alterative".
All these medicines will kill worms if they be soaked in water
until two-thirds thereof have been absorbed, and if patients
drink the infusion, and they will also kill worms if the water
in which they have been boiled with salt and natron be injected
through the anus into the belly. Similarly, if they be pounded
and be drunk by patients with vinegar, or with wine, or with
goat's milk, they will expel worms both flat and round. And
Page 430 if they be mixed | with ox gall and laid upon the region of
the navel, and on the whole epigastrium, they will expel the
worms and bring them down in the stool. And especially useful in cases of worms is the following medicine:

Costus	4 drachms
Garden nigella	2 ,,
Inside of copper	3 ,,
Ferns	3 ,,
Copal	3 ,,
Convolvulus	8 ,,

Pound these well, and administer as a dose four or seven
drachms in vinegar and tincture of myrrh, or in pure wine, or
in goat's milk. The night before the medicine is to be taken,
let the patient fast, and in the morning let him drink it, taking
care not to smell the odour thereof.

**Or let the patient drink the following well tried medi-
cine:**

Inside of copper	4 drachms
Convolvulus	4 ,,
Ferns	4 ,,
Indian salt	3 ,,
Costus	6 ,,

Pound these well together, and administer as a dose four or
five drachms. For three days before the patient is to take
the medicine, let him drink one cup of goat's milk each morning, and on the fourth morning let him drink the medicine
in milk, and evacuation of the bowels will follow. In diseases
of the colon injections are useful, the drugs employed being

sometimes aromatic and sometimes bitter; and sometimes the medicines that produce vomiting are useful.

Injections, which are good for pains in the colon. |

Cyrenean ointment	4 ounces
Jackal fat	4 ,,
Stacte	4 ,,
Crocus	4 ,,
Pyrethrum	4 ,,
Galbanum	4 ,,
Sagapenum	4 ,,
Moist pitch (bitumen)	5 ,,
Honey as much as sufficeth.	

Pound all the drugs and mix them with the honey, dissolve the mixture in sweet wine and oil of irises, and inject into the body.

Another.

Aloes	
Myrrh	
Opium	in equal quantities
Cream of goat's milk	
Castoreum	

Crush them, work into a paste with sweet wine, dissolve in four drachms of wine, by weight, and oil of rue, or oil of fennel, and inject into the body.

| **Another.**

Moist pitch (bitumen)	3 drachms
Aromatic herbs (or, incense)	2 ,,
Opium	1 drachm

Pound, mix together, and inject with wine, and oil of rue, and the cream of goat's milk. First give the patient food, and then administer to him the injection.

Another.

Aloes	2 drachms
Opium	2 ,,
Cream of cow's milk	4 ,,
Moist pitch	4 ,,
Aromatic herbs (or, incense)	4 ,,
Castoreum	4 ,,

Mix these together, and dissolve them in wine and oil of rue, a *mathkâl*, four drachms, and transfer to the body in an injection.

Another, which is good for pain in the colon, and for protracted pain in the excretory organs, and for rigidity (paralysis).

Fennel	1	hand
Chamomile	1	,,
Land-caltraps (or, water-caltraps)	1	,,
King's crown	1	,,
Ground-ivy	1	,, .
Cabbage	1	,,
Beet	1	,,
Flax seed	—	
Trigonella	—	
Bran	—	
Liquid myrrh	—	
. . . .	. . . . . .	
Fleshy figs	10	[in number]
Heriôn dates	1	hand
Pulp of colocynth	1	,,
. . . .	. . . .	
Calamint	1	hand
Origanum majorana	1	,,
Sagapenum	4	drachms
Ammoniac	4	,,
Bdellium	4	,,
Cyrenean ointment	4	,,
Opopanax	4	,,
Galbanum	4	,,
Castoreum	3	,,
Fleabane (?)	10	,,
Convolvulus	10	,,
Mâzêriôn	6	,,
Tithymalus (spurge)	6	,,
Seed of rock parsley	8	,,
Lettuce	8	,,

Malabathrum	8 drachms
Rue	8 „
Wild rue	8 „
Licorice	root	8 „

Soak all these in water until they have absorbed one half of it, clarify, pour on them one measure of olive oil, one measure of honey, one measure of sheep oil, or cow oil, two drachms of crushed salt and one drachm of natron, mix these together, and inject into the body. Now the quantity which thou shalt inject at one time is two *lițrê*, and it shall remain in the body of the patient so long as he can bear it, and thou shalt keep his legs well raised in the air. And if no motion of his bowels followeth after the injection hath come out, administer to him a second injection made of the self-same medicine, only add | to the quantity which thou didst give the first time. And if even after this the pain be not removed, take one *dûlķâ* of old sweet wine, and add to it two measures of mulberry water, and three measures of olive oil, and warm them slightly, and inject the whole of it if it be possible, and if the patient can bear it, and the pain will be removed immediately. This medicine is beneficial in cases of pain of the excretory organs, and of sciatica, and it is good for those who suffer from attacks of gout, and from the pain caused by the discharge of thick matter. Now, the suppositories which are placed in the anus, and which are compounded of the following drugs, are very good for the patient.

Pills which are placed in the anus, and which produce motions of the bowels; these are also good in cases of fistula.

Natron	1 drachm
Common salt	2 drachms
Colocynth pulp	1 drachm

Pound these, work them into a paste with honey that hath been boiled solid (or, hard), and make long suppositories of it, and place in the anus.

Another, which is good for the sick and the healthy alike, and for internal fistulas, and disease of the colon, and for the cold winds that are produced in the lower part of the belly.

Sarcocolla (Persian gum)	4 drachms
Fleabane	4 "
Natron	1 drachm
Vegetable alkali	1 "
Fuller's earth	3 drachms
Pulp of colocynth	2 "
Aloes	5 "
Glaucium phoeniceum	5 "
Ammoniac	5 "
Indian salt	5 "

Pound, crush, work into a paste with water, and make into oblong pills about the size of small acorns. | Anoint them with oil of roses, and place them in the anus, and let the patient bear them [as long as he can]. And if they produce a burning sensation in the belly, inject after them goat oil, or oil of roses, and the yolks of eggs. And very good also for those who are attacked by disease of the colon are the softening plasters (or, ointments) that are enumerated in the section on diseases of the stomach and liver; and the plasters that expel the water in cases of dropsy are also good for protracted diseases of the colon.

The medicines that are good for ailments of the anus.

The diseases | that arise in the anus are exceedingly difficult to heal, and this is due, first, to the extreme sensitiveness which this member possesseth, and, secondly, because it is not easy of access. Moreover, owing to the heat and moistness of this member, it requireth medicines that dry it and soothe it at the same time, and it is very difficult to heal disease in it by means of medicines that possess astringent properties, because this member, being extremely sensitive, is unable to bear pain. Therefore it is meet for us to apply to it the medicines that dissipate its diseases, and that are neither very pungent nor drastic in operation. The following are the medicines that effect painlessly what it is meet for them to do.

A medicine that is good for ulcers and cuts (or, tears, cracks, &c.) in the anus.

Aspadhkhâ (white lead)	5 drachms
Lithargyrum	5 ,,
Frankincense	3 ,,
Smelters' dross (or, alum)	3 ,,
Crocus	$^1/_2$ drachm
Wine	
Oil of roses	as much as suffice.

Another, which is good for the boils and itch [that arise in the anus].

. . . .	6 drachms
. . . .	6 ,,
Frankincense	2 ,,
. . . .	2 ,,
Dried hyssop	2 ,,
Oil of roses	2 ,,
Wine as much as sufficeth.	

Another, which is good for the boils and cracks that come in the anus.

Incense	4 drachms
Ammoniac	4 ,,
Blood-stone	4 ,,
Frankincense	4 ,,
Alum, round	4 ,,
Dried grapes	1 drachm
Crocus	1 ,,
Gum of terebinth	4 drachms
Finest wax	2 ,,
Oil of roses	
Wine	as much as suffice.

Mix up well together and use.

Another, which is good for the anus that protrudeth and for the boils in it.

Lead dross	4 drachms
Flowers of roses	4 ,,
Myrrh	2 ,,

Pound these, and, having first washed the anus with strong wine, crush the medicines and sprinkle them on it in the form of a dry powder.

Another, which is good for the anus that protrudeth, and that hath in it boils and ulcers.

Green dried grapes	2 drachms
Rind of pomegranate	2 "
Sumach	2 "
Lead, roasted and washed	4 "
Lithargyrum	8 "
Frankincense |	2 "

Page 434

Crush in wine, and use as a plaster.

Another medicine, which was prescribed by Galen, and which is good for a protruding anus.

Fruit of îrânîķê	
Acacia	
Dried grapes	
White lead	in equal
Extract of hypocistis	quantities
Cedar bark	
Frankincense	
Myrrh	

Pound, and apply in the form of a dry powder, having first of all washed the anus with old wine.

Another, which is good for the ulcers that come in the anus.

Wax	4 drachms
Lithargyrum	4 "
White lead	4 "
Goose fat	4 "
Chicken fat	4 "
Crocus	2 "
Oil of roses |	as much as suffice.
Wine |	

Another.

Lithargyrum	1 drachm
White lead	1 "
Lead powder	1 "
Fresh leaves of hyoscyamus	1 "
Stacte	1 "
Wax	2 drachms

Oil of roses 7 drachms

Wine as much as sufficeth.

Another.

White lead
Roasted lead
Oil of roses } [in equal quantities]
White of egg

Pound in a mortar to a fine powder, and rub on the anus.

Another, which is good for a protruding anus, and for ulceration thereof, and for a relaxed anus.

Break up dried grapes, boil them well in water, crush and pound them, mix oil of roses with them, and use as a plaster. If there be bleeding of the anus, let the grapes be soaked in wine, and wash the anus with this infusion and then apply the plaster.

Or take two parts of aloes, and one part of frankincense, pound them with the white of an egg, and use as an ointment. Or | wash the anus with wine, and sprinkle this medicine on Fol. 208*b*. it in the form of a dry powder, until the ulcers be dried up.

Now, the medicines that have been prescribed for bleeding of the nose, or for bleeding of the organs of respiration, are also good for bleeding of the anus, as is also sitting in infusions of substances that possess astringent properties, such as myrtle, dried grapes, rind of pomegranates, | senna, olives, roses, and Page 435 everything of the same kind.

Or soak rinds of pomegranates, and lentils, and roses in water, crush them, pound them, mix with them oil of roses, and apply to the anus in which there are ulcers, or which is relaxed, or which protrudeth. Now, on the occasions when there is bleeding of the fistulas which are in the anus, the following medicines, which burn and prevent the flow of blood, are also useful.

Sharp waters (*i. e.*, acid lotions), which burn and cut fistulas.

Alkali (vegetable) 2 drachms
Lime 2 „
Arsenic 2 „
Vitriol 2 „
Water 2 *ḳôḳè*

33

Pound the drugs and throw them into a metal pot, and let them steep for a whole night and a day. In the morning boil the mixture down to one half, then strain off the liquor, and keep them in a vessel, [and use] .... necessary, bring unslaked lime, and .... and vegetable alkali, and arsenic, and vitriol, [and ......], three drachms of each. Pound these and work them up in the liquor, and smear the mixture on a piece of cotton-wool, and anoint the fistulas therewith. And when they have been well burnt, take barley flour, and oil of roses, and the yolk of an egg, and mix them together, and work them into a paste, and use it as a plaster on the fistulas until the burns are removed, and then dry them with medicines that possess drying properties. And if there be none of these acid solutions prepared, take unslaked lime, and vegetable alkali, and yellow arsenic, one drachm of each, pound them and reduce them to a powder, work them into a paste with figs hang them in the sun for seven days, and move them once every day, and use. This medicine operateth with very great effect.

Another medicine, which operateth by burning.

Unslaked lime
Vegetable alkali
Cantharides |
Yellow arsenic
Red arsenic
Sal ammoniac

} in equal quantities

Fol. 209a.

Crush, reduce to a powder, work into a paste with water of alkali, and use, but be careful lest thou burn any part un-

Page 436 necessarily. |

Or work up sarcocolla in water, and smear the mixture on strips of wool, and wrap them round the fistulas, and then anoint with the medicine.

Another.

Unslaked lime   5 drachms
Cantharides   4   „

Pound and work into a paste with water of alkali, and keep in a vessel, and use in the same way as the preceding medicine. When the fistulas have been well burnt, roast sesame

seeds, and pound them, and mix them with sheep oil, and use as a plaster until the fistulas are removed, and then dry up the places where they were with koḥl and oil of roses. Or crush and scatter over the places, after the fistulas have been removed, the following medicine:

Burnt oyster-shells ⎫
Purple βαλαύστιον ⎪
Lithargyrum ⎬ in equal quantities
Glaucium pills ⎪
Burnt lead ⎭

Crush to a powder, and scatter it over the anus until [the disease] cometh to an end. And if it be necessary, make the patient sit in wine in which thorns, rinds of pomegranates, dried myrtle, and roses have been steeped; and when he riseth up out of the wine let him wash in an infusion of myrtle, and sprinkle some of the medicine on the sore, and lay a piece of linen over it. Or anoint the fistulas daily with olive oil, and pound fuller's earth, and sprinkle it over them, and if thou doest this for seven days it will remove them.

And now, having written a description of the symptoms of all the ailments that attack the intestines, and of the means to be employed in healing them, it is necessary for us to address ourselves to the consideration of the ailments that attack the kidneys, according as the plan of the present work demandeth, and therefore the above must suffice for this the Twentieth Chapter.

HERE ENDETH THE TWENTIETH CHAPTER.

33*

CHAPTER XXI.

Of the symptoms of disease which appear in the kidneys.

Fol. 209b.

Page 437

Now, it hath been shewn that the kidneys perform two functions in the bodies of the children of men: first, | they draw the urine from the liver | and transmit it to the bladder, and secondly, they stir up the organs of the seed that they may transmit the seed to the testicles. The kidneys are situated on the spinal column, near the loins, and they are enveloped in fat, and they are fixed, one on the right and the other on the left, and they and the inner loins of the spinal column are girt about with the veins that bring the seed, and with those that bring the urine. They are nourished by means of the great vein that is called "Ḥalîlâ", (*i. e.,* "the cavern"), which bringeth the blood from the liver; and their nature is fleshy and nervous, like that of the heart. Now, the kidneys are attacked by three kinds of diseases (or, sicknesses), namely, those which are organic, and those which are homogeneous, and those in which the unity is done away. They suffer organically when there are in them ulcers, or some kinds of obstructions which are due to the chymes or to stones. They are attacked [by diseases] that are common to all of them owing to their homogeneity, when they become hot, or cold, or dry, or wet; [and they suffer from breach of unity] when there is in them either [a rent], or an opening, or something [of this kind]. Now, the symptoms of the diseases that arise in the kidneys may be thus described:—If fever beginneth suddenly, accompanied by pain in the kidneys, when a stone is deposited in one of the kidneys, or in [one of] the veins that extend from it to the bladder, then the pain resembleth the pains that belong to the colon. The diseases of the colon are to be distinguished by the frequency and intensity of the vomitings, and by the fact that that which is vomited is saltish, or bilious, or phlegm-like. Now, there are cases in which some portion of the food which hath been eaten becometh detained in them, and this patients often feel through the region of the upper parts [of the belly], when the disease is in the colon. There are times also when [this may be known] from the fact that the pain is not fixed in one place, but it moveth about, and is spread about in no small degree, and also from the fact that patients do not break wind. Now, all | these symptoms Page 438 are closely associated with the pains in the colon | only, some Fol. 210 a. of them in a greater degree, and some in a lesser; but when they are due to those that subsist in the urine, or to a stone of some kind which hath gone forth into it, this can be defined in a manifest manner, and in that case not a single question about it can be permitted. Moreover, the pains that arise in the bladder are accustomed to arise during the period of early manhood (or, in youth), but those that arise in the kidneys arise during the period that followeth maturity. And stones are produced when the kidneys possess naturally a hot, burning constitution. Now, very many of those who suffer from this ailment (or, pain), feel in a moderate degree some pain in the depth of their body, which is near their epigastrium, at the beginning, when as yet nothing hath been evacuated that is visibly of a dusty nature. In such cases as these, as ye also well know, when I administer certain of the potions that remove pains of the kidneys, I make an exact diagnosis of the disease itself, and of the place that is diseased, at one and the same time, and [I] thus possess the means for the beginning of the curative treatment, through this same diagnosis. For if things that are like unto sandy particles are found in the urine, after the patient hath drunk the remedy, I recognize immediately that the disease ariseth from the kidneys, and I administer afterwards draughts of the same medicine, and I exercise care in other ways [for] the healing [of the disease], for this testifieth that the disease is due to the kidneys. When the pains appear together with the shiverings due to fever,

after irregular intervals of cessation of them, and the attacks
of fever do not come in regular order, we then throw ourself
on the belly of the patient, and afterwards we also turn him
over on the other side, so that the side that is diseased may
be higher [than the other], and besides this we ask him whether
he feeleth any weight suspended above and pressing upon the
kidney that is suffering. For, if it be found that these things
are happening to him, it is meet for us to think that there is
Page 439 a pustule [in the kidney], and | if it be ruptured and pus come
forth therefrom by means of the urine, the man who is afflicted
Fol. 210b. in this manner will be delivered | from the pain. Now, this
bringeth the danger of a tumour upon the kidneys, and therefore it is meet to try in every way which is in our power to
concern ourselves seriously with the closing up of the rupture;
for if we do not close it very quickly indeed, it will be exceedingly difficult to heal it. A very well-known sign whether
the tumour persisteth is derived from the urine.

[here the text is broken.]

sometimes [that which passeth] is like unto slough, and sometimes also it is like unto blood, and this maketh known that
the tumour itself is checked. Moreover, when a vein bursteth
in the kidneys, which sometimes happeneth through excess of
blood, or through a fall, or through a violent blow, people
pass a large quantity of blood in the urine. And sometimes
also a vein will open of its own accord in the kidneys, and
this happeneth suddenly. Now, of the tumours that arise in
the kidneys a sure symptom is the small thin particles of flesh
which come forth in the urine, and these are some portions
of the substance of the kidneys themselves. And in the case
of cancer, which very often indeed is due to tumours, other
very fine (or, thin) particles of flesh, which are like unto strings
(or, filaments), come forth [in the urine]. Now, Hippokrates
observed them coming forth in the urine, even as he himself
wrote in [his] book. Both he and we ourselves have seen them,
sometimes as long as a span, and sometimes longer than a
span, and on another occasion when seen by us they were so
long that we wondered that the kidneys were large enough
for anything of the kind to subsist in them. Therefore it seemed

to us that the place where they were produced must most
certainly be in the veins. And they are like unto the things
which are "dragons", and which, they say, exist in the legs in
one of | the countries of Arabia; they resemble the nerves in Page 440
their nature, but in their colour and thickness they are like
unto maw-worms. Many people say they have seen them. | I Fol. 211a.
have heard of these things, but because I have never seen
them it is not my affair to think anything definite about their
place of origin, or about their nature. As concerning the
[worms which are like unto] strings, and which, as I said, go
forth with the urine, as I have seen them and know their substance and colour, I am convinced of the correctness of the
opinion of those who say that they consist of thick and viscous
chyme, and that they are transmitted to the kidneys, and subsist there. For their length, I have not sufficient knowledge
to assign a cause, but from the first time I saw them, I thought
that their cure could be effected by means of the medicines
that are called "combative", and even so hath it happened to
all those who have suffered from this disease. Now, no pain
of the kidneys recurred to those who were cured by the
urinary remedies, either at first or subsequently. But although
the other noxious chyme was cleared away from the vein by
means of the urine, I never saw that a man suffered through
this, either in his kidneys, or in his bladder, or in those
veins which bring the urine, moreover, I have never seen
that he was injured in any of these even if he discharged
a large quantity of pus by means of the urine. Thus then
are all these things, and they resemble the things which
happen in the intestines. For neither do the intestines suffer
in any way through the discharges that take place from the
liver, although they are injured by the acidity of the bile, just
as the bladder is injured also by the acidity of the urine, when
the exit therefrom remaineth in this state for a long time, for
then a tumour breaketh out.

Page 441

| Of debility of the kidneys.

Now there is a disease [of the kidneys] ....
[Here a large section is wanting.]

Fol. 211b. **Forecast [based] on the test days of the month of Tammûz, from the nineteenth day of the month, on which the Dog (Sirius) of the Giant (Orion) riseth, until the twenty-sixth day at daybreak.**

The nineteenth day of Tammûz, in the evening. If thou seest a cloud in the heavens at the beginning of the First Teshrî there will be rain. If it be in the middle of the night, in the middle of the month the rain will come. And if it be towards morning, the rain will come at the beginning of the month. And if the cloud appeareth in the middle of the day, the rain will come in the middle of the month. If it appeareth towards the evening, the rain will come at the end of the month. And if there be no cloud [at the times] mentioned above, there will be no rain at all in the month. And if there be cloud scattered over all the heavens that month will be lacking in rain. And if there be frequent cloud in the north, whether in the evening, or at midnight, or at dawn, or in the day-time, [at the periods] which are written above, then understand that the winter will be severe. Whenever there happeneth to be cloud in the north, according as God wisheth, thus will be these days, in each month one day, from the First Teshrî until the month of Iyyâr.

Forecast [based] on shooting stars.

If a star goeth from the east to the west, the king of Persia will die in Bêth Ḥûzâyê (i. e., Ḥûzistân, or Ahwâz, near Al-
Page 442 Baṣrah), and the children | of men will become sick and diseased.

And if the star goeth from the west to the east, there will be great wrath, and all men will be in tribulation, and evil words will be spoken, but finally [all] will be well.

And if the star goeth from the north to the south, the king will see evil, and women will not become pregnant, and a father will kill his son, and a brother his brother, [for a period of] three years, and then there will be peace.

And if the star goeth from the south to the north, and shall be red in colour, or if it fall from the heavens to the earth, there will be a pestilence in that year. And the king will go to a far country, and war and darkness will prevail, and there will be great sickness, and [leading away into] captivity, but in towns there will be peace.

And if the star goeth from the west to the south, and breaketh up into fragments, towns will attack towns, and there will be pestilence and slaughter.

And if the star goeth from the north | to the south, and Fol. 212a. breaketh up into fragments, there will be pestilence .... and men will increase in their habitation.

And if the star descendeth from the upper heaven to the earth, there will be tranquillity and peace, and glory to God.

And if the star bursteth forth from the east, and is not broken up into pieces, the progeny of animals will flourish, and flowers will bloom, and the cities which have been laid waste will be rebuilt.

Another, derived from the shooting star in the heavens.

When the star shooteth from the west to the east, there will be wrath and tribulation to men, but finally there will be happiness.

When the star shooteth from the east to the west, the king of Persia will wage war on Bêth Hûzâyê, even as it is written in the Book of Andronicus.

When the star shooteth from the north to the south, the kingdom will be disturbed and shaken, and the women who are with child will utter wailing lamentations. And a father will rise up against his son, and a brother against his brother, and there will be division (dissension) for three years, and then there will be peace.

When the star shooteth from the east to the west, and becometh divided into fragments on the south and on the west, it indicateth the many thrones which have been set up on the earth. | And many will go in and will seize principalities, Page 443 and there will be many wars, and they will be in number as is the number of the portions into which the star hath been divided by the force of the atmosphere. Similarly [men] will lay violent hands on principalities, and powers, and dominions, and thrones.

When the star shooteth from the north to the south, and is divided into portions, the towns will fall upon each other, and there will be pestilence and slaughter, and great countries will be laid waste and desolate.

And when a star shooteth from the heavens, and it is seen to fall upon the earth, man will dwell in peace and prosperity. And God [only] knoweth the truth!

Forecast about the winter derived from the appearance of the sun.

When the sun riseth, and there is splendour (or, a clear sky) before him, and his rays spread abroad a little, and a cloud appeareth in the firmament, then understand that it indicateth dampness in the atmosphere, especially when before he riseth
Fol. 212*b.* he maketh the firmament red, | and when whilst he is still below the earth (*i. e.,* horizon) his light falleth on the stars and maketh them dark (*i. e.,* invisible). This sheweth, and it is the same in the case of his setting, if a cloud be seen before him, that there will be rain.

And if the year be born in the sign of the Ram, the rain will cut the early seed (crop).

And if the year be born in the sign of the Bull, sow with seed first of all the plain and it will yield good crops.

And if the year be born in the sign of the Twins, sow the dry land.

And if the year be born in the sign of the Crab, the rain will cut first of all the seed [sown] in the high ground.

And if the year be born in the sign of the Lion, the latter crop will not prosper in the plain or yield good crops.

And if the year be born in the sign of the Ear of Corn (Virgo), the beginning of the winter will be hot, and the middle thereof temperate, and at the end there will be heavy snow and ice, and the [spring] rain will be late.

And if the year be born in the sign of the Scales, the early seed and the latter crop will not prosper.

And if the year be born in the sign of the Scorpion, the early seed (crop) and the wheat will be destroyed.

And if the year be born in the sign of the Great Image (Sagittarius), the early seed (crop) in the plain will prosper, and there will be damage to the wheat.

And if the year be born in the sign of the Goat (Capricornus), the early seed (crop) will lack rain sorely, but the plain will have abundant crops. | Page 444

And if the year be born in the sign of the Water-carrier (Aquarius), sow the high land first of all, and the early seed (crop) will prosper.

And if the year be born in the sign of the Fishes, do not sow seed too early, and the crops will prosper on the mountains, and the early seed (crop) will prosper.

Of the Malwâshê of the Year.

The Sign of the Ram. The rivers and streams will be filled with water.

The Sign of the Twins. There will be languor and scattering abroad among men; one shall go against them, and they shall be put in subjection. Now it is meet for us to sow seed while the earth (or, ground) is dry.

The Sign of the Crab. And there will be damage, and the locust, and multitudes of mice, in [the earth]. And man will not threaten (or, attack) man. And the fruit will be destroyed in the country of India, but there will be good crops and great joy in Ṭibaryâ. At the close of the year there will be misery, and in Media the winter will be severe. It is meet to sow seed from the beginning of the month of the First Teshrî until the end of the period, fifty days. If Bêl be associated with that year, there will be great peace, and if Ârîs (Mars) be associated therewith, there will be blood and war.

The Sign of the Twins. Great disturbance and rebellion will take place.

The Sign of the Lion. That year will be good, and there will be great abundance, and vineyards, and wheat, and barley will thrive. And if Bêl | be associated with the year there Fol. 213a.

will be great rest, and abundance, from the beginning of the First Teshrî.

The Sign of the Ear of Corn (Virgo). It is meet for us to sow seed from the beginning of the First Teshrî, up to fifty days.

From the "towers".

The Sign of the Scales—and those which are useless to him, the Bull, the Ear of Corn (Virgo), and the Goat (Capricornus); and those which are very useful to him—.

The Sign of the Crab—and those which are useful to him of the towers, the Crab and the Fishes—and those which are not useful to him—the Ram, the Lion, and the Bowman (Sagittarius), and those which are useful to him of the stars, and Mercury.

The Sign of the Bowman—and those which are useful to him of the towers—the Crab, the Scorpion, and the Fishes.

The Sign of the Goat—and those which are not useful to him of the towers—the Twins, the Scales, and the Water-carrier (Aquarius).

The Sign of the Water-carrier—and of the months, Shebât.

Of the marrying of wives.
B (2), G (3), D (4), H (5), W (6), Z (7).

Page 445 **B (2)** Haste to thy house, | to the woman whom thou lovest.

G (3) It will be stupendous if thou marriest now.

D (4) There is no work for thee with women, on the contrary, thou art useless, and canst do nothing.

H (5) Abandon [the idea of] marriage, for thou shalt not attain it for two years.

W (6) Know that everything is from God, and will take place in its season.

Z (7) Thou shalt prosper with the women who are known [to thee], and shalt fulfil thy affair (or, act) at the end.

If thou wishest to know which hand hath something clasped in it, reckon the [letters of the] name of the man, and [those in] that of his mother, and make all thy reckoning one, and as it ascendeth, cast out every sixth number. If pairs remain (*i.e.,* if the number is an even one), the thing is clasped in his right hand, and if the number remaining is an odd one, it is in his left.

Again, if anything be stolen from thee. Write the [letters of the] name of him whom they accuse of the theft one after the other, and the name of Kronos at the top of them all, and also the name of the thing which is lost [at the bottom of them], and the name of its owner in the middle [of them]. Then reject according to their number, however many they may be, and watch the number that remaineth so that thou mayest not succeed in rejecting them all. Begin with the first letter and assign one to each man. He to whom the lot cometh, and he agreeth with it, he it is who hath taken the object. | And if the lot cometh to Kronos, and he agreeth with Fol. 213*b.* it, the man is not guilty; but if the lot cometh to the thing stolen, [all] are free from blame. [This means of discovery of the thief] hath been well tried, and is sure, unless, perhaps, he who employeth it maketh a mistake in his calculations.

A List of the numerical signs employed in the prescriptions. Page 446

9 =	1	**oo** =	40
9ₐ =	2	**7oo** =	50
9ₐ9 =	3	**ooo** =	60
9ₐ9ₐ =	4	**7ooo** =	70
ᴧ =	5	**oooo** =	80
9ᴧ =	6	**7oooo** =	90
9ₐᴧ =	7	**7₉** =	100
9ₐᴧ9₉ =	8	**7ₐ** =	200
9ₐ9ₐᴧ =	9	**7ₐ9** =	300
7 =	10	**7₉ₐ9ₐ** =	400
o =	20	**7ᴧ** =	500
7o =	30	**7₉ᴧ** =	600

The method of reckoning by the Lîṭrâ and Mathḳâlâ.

The *lîṭrâ* containeth twenty *estîrê* (*i. e.*, 100 drachms).

The *estîrâ* containeth five drachms (*zûzê*).

$\bar{\varsigma}$ = five drachms.

The *ḳesṭâ* containeth one *lîṭrâ* and a half (*i.e.*, 150 drachms).

The *menḳîthâ* (or, *mĕnêḳîthâ*) containeth half a *ḳesṭâ* (*i.e.*, 75 drachms).

The exact equivalent of the *mĕnêḳîthâ* is one spoonful, and the spoonful containeth four *mathḳâlê*. They say also that Page 447 one *mĕnêḳîthâ* containeth five spoonfuls, and that one spoonful | containeth two *mathḳâlê*.

The *mathḳâlâ* of the sanctuary containeth twenty drachms.

The *mathḳâlâ* of the sanctuary and the *dînârâ* are the same.

The *mathḳâlâ* which they use in weighing *zûzê* (*i. e.*, money), and *dînârê*, and pearls, is the equivalent of the *dirham*.

One *garmâ* (*i. e.*, gramma) = a quarter of a *zûzâ* (*i.e.*, drachm).

One *ḳerṭâ* (carat) = one half, or one quarter of a *denḳâ*.

Eighteen *ḳerṭîn* (carats) are equivalent to one half of a *mathḳâlâ* of the sanctuary.

One *ḳesṭâ*, that is to say, the wine measure, containeth one *manyâ* and a half.

The *manyâ*, which is called *shâmûnâ*, weigheth eighteen *ḳerṭê* (carats).

The *ḳerṭâ* is four barleycorns, or six.

Fol. 214*a*.   **Again, if thou wishest to know whether it is to good or to evil that they are calling thee.** Stand in the sun, and measure [the length of] the shadow of thy person by paces made by thy legs, and mix (*i. e.*, add) thy name and the name of thy mother to the number, and count them out two at a time. If thou hast a remainder of two, sit where thou art, and if the remainder be one, go.

Again, if thou wishest to know, in the case of a man and a woman, which of them will die first, count up the names of the two together, and reject the numbers one by one. If the number which remaineth with thee be even, the woman will live, and if it be odd, the man will live. Or divide the numbers three by three; if three remain in thy hand, the man will live, and if two or one remain, the woman will live.

| In the name of our Lord I write down a computation arranged in a tabular form of lunar months. If thou wishest to know on how many days of a solar month the moon will rise, take as many years of Alexander as thou hast in the period which thou art seeking for, and cast out from them one thousand, and seven hundred, and sixty and eight, and what thou hast remaining over, count out (*i. e.*, divide) by nineteen at a time. And that which remaineth in thy hand hold, and add to the number which is written in alphabetical order before thee, in the column with nineteen above it. Then go downwards, opposite to the solar month which thou seekest, opposite the number which goeth forth for thee, [will be found] how often the moon riseth in a solar month. If the year be one with an intercalary month, know that the month of Shebhât passeth with less than the number which goeth forth for thee by one day. And know that in the months which have two numbers, now they are divided by two red dots, in those months, I say, the moon riseth twice, at the beginning and at the end thereof. And that door is empty, in that year the moon doth not rise in Shebhât; but if there be one hour intercalated, it will rise on the twenty-ninth day. When thou hast counted out (*i. e.*, divided) the years of Alexander by nineteens, thou shalt deduct from them as followeth: from twenty, one; and from one hundred, five; and from four hundred, one, and from one thousand, twelve.

Page 448
Fol. 214*b*.

Number	1	2	3	4	5	6	7	8	9	10	11	12	13	14	15	16	17	18	19
Îlûl	9	27	17	6	24	13	2	21	10	29	18	7	25	15	4	22	12	1:31	19
Âbh	10	29	18	7	26	16	4	22	12	1:30	20	8	27	16	6	24	13	5	21
Tammûz	12	1:30	20	9	27	16	5	23	13	2	21	10	28	18	7	26	15	4	22
Khazîrân	12	1	20	9	27	17	6	24	14	3	22	10	1:30	18	7	26	15	4	23
Iyyâr	14	3	22	11	29	18	7	26	15	4	23	13	1	20	9	27	17	6	25
Nîsân	14	3	22	11	30	19	8	26	15	5	24	12	1	20	9	28	17	6	25
Âdhâr	16	5	14	13	1:31	20	9	28	17	6	25	14	3	22	11	1:31	19	8	26
Shebhât	14	3	25	11	—	19	8	26	16	5	24	12	1	21	9	23	17	6	25
Kanôn (1)	16	5	24	13	1:31	20	9	28	17	6	25	14	3	22	11	23	19	8	26
Kanôn (2)	17	6	25	14	3	22	11	29	19	8	27	15	2	24	12	1:31	20	9	28
Teshrî (1)	18	7	26	15	4	23	12	1:31	20	9	28	16	6	26	13	2	21	10	28
Teshrî (2)	19	8	27	16	5	24	13	2	21	10	28	17	22	14	3	2	22	11	29

When [this table] endeth, go back to the beginning, and pray for me.

The signs of the months of the Arabs	Number of days	Names of the months of the Arabs. Muharram is the Second Kânôn	Foundations of the years of the Arabs
7	30	Muḥarram	z S 1 4 L ε 9
2	29	Ṣafar	4 L ε 9 z S 1
3	30	Rabi‘ al-Awwal	S 1 4 L ε 9 z
5	29	Rabi‘ at-Tânî	L ε 9 z S 1 4
6	30	Gamâdah al-Awwal	1 4 L ε 9 z S
1	29	Gamâdah at-Tânî	ε 9 z S 1 4 L
2	30	Ragab	4 L ε 9 z S 1
4	29	Sha‘bân	9 z S 1 4 L ε
5	30	Ramaḍân	L ε 9 z S 1 4
7	29	Shawwâl	z S 1 4 L ε 9
1	30	Dhu’l-Ḳa‘dah	ε 9 z S 1 4 L
3	29	Dhu’l-Haggah	S 1 4 L ε 9 z

That thou mayest know the intercalary years of the moon by a second inversion, that is to say, *baigô*, and by five, that is *hayag*, and by seven, that is *zeklîd*, and by ten, that is *khahâ*, and by thirteen, that is *yagîbah*, and by sixteen, that is *yôtab*, and by eighteen, that is *khayînazag*.

Now, if thou wishest to know on what day of the Page 451 week the beginning of the lunar month is born, take the sign of the month, and add to the foundation in red one of the seven [numbers] which are written in the table before thee, and see how much they amount to. If their total be more than seven, cast out seven from them, and the number which remaineth in thy hand is the [number of the day] of the beginning of the solar month in the week. If seven remaineth in thy hand, that is the [day of] the beginning of the month. If one remaineth the month beginneth on the first day of the week; and if two, on the second day of the week; and if three, on the third day of the week; and if four, on the fourth day of the week; and if five, on the fifth day of the week; and if six, on the Eve of the Sabbath; and if seven, on the Sabbath.

And if thou makest a mistake in the foundation of the year of the Arabs, or if thou dost not know which it is, add of the years of Alexander nine hundred and fourteen, and what remaineth over with thee cast out, seven by seven (?), that is to say, from thirty take two, from one hundred take two (?), and from two hundred take four, and from four hundred take eight, and from eight hundred take sixteen, and from one Fol. 215 b. thousand take twenty. And when thou hast cast out | those years of Alexander which are in excess of nine hundred and fourteen seven by seven, see how many remain with thee. If seven remain, that number is the foundation. If the total is more than seven, strike out seven. Then what remaineth in thy hand double four times, and see what their total is. Then reject seven by seven and see how many remain with thee. If the remainder be one, then one is the foundation, and if two, two is the foundation, and if three, three is the foundation, and if four, four is the foundation, and if five, five is the foundation, and if six, six is the foundation, and if seven, seven · is the foundation.

That thou mayest know in what hour the moon is born, (*i. e.*, the new moon will appear) by night or by day. Consider how many moons there have been from Nìsân until the moon under which thou standest, and for every month reckon Page 452 one moon; count them out six at a time, | and in the number that doth not complete a group of six the moon will be born. If one remaineth over, the moon will be born in the first hour of the night. If two remain over, the moon will be born in the fifth hour of the night. If three remain over, the moon will be born in the ninth hour of the night. If four remain over, the moon will be born in the first hour of the day. If five remain over, the moon will be born in the fifth hour of the day. And if six remain over, the moon will be born in the ninth hour of the day.

In like manner also take the days of the months from the First Teshrî till the day wherein thou standest, count them out six at a time, and see how many remain in thy hand. If one remaineth, the moon will be born in the first hour of the night.

And if two, the moon will be born in the fifth hour of the night. And if three, the moon will be born in the ninth hour of the night. And if four, the moon will be born in the first hour of the day. And if five, the moon will be born in the fifth hour of the day, and if six, the moon will be born in the ninth hour of the day.

And if thou wishest to know how long the moon will shine, and when it will set. Know thou that every day of the waxing moon addeth four minutes (or, measures (ܡܕܝܢ̈ܬܐ = κέντημα) to its light until it is full, when it immediately beginneth to diminish, and each day its light diminisheth four minutes (or, measures). If it hath gone one day, or two, or Fol. 216 a. however many days it be, double the number four times, and then see how many times four are contained in it, and the number of times will be the number of the minutes (or, measures) which the moon will shine. Every six minutes (or, measures) represent an hour. And if thou hast half a day, reckon to it two minutes (or, measures). And if the moon hath gone up more than fifteen days, then what remaineth to thee double in the same manner, and thus thou shalt find [the answer]. And when thou hast counted fifteen minutes (or, measures), take away from them one minute (or, measure), because on the fifteenth day [the light of] the moon diminisheth one minute (or, measure). Now, an hour containeth thirty se'aryâthâ,[1] and the ḳanṭîmâ[1] six. Each day the sun addeth one sĕ'arĕthâ, and when diminishing, he taketh away one sĕ'arĕthâ [daily]; now, the pôdîthâ (= Gr. ποῦς) is one sĕ'arĕthâ.

Again a calculation | concerning those who are sick. Page 453 First of all reckon up the [numerical values of the letters of the] name of the sick man, and of those of the name of his mother, and cast them out nine by nine (*i. e.*, divide them by nine). If one remaineth to thee, his sickness is from God. And if two, it is caused by the Evil Eye. And if three, it is caused by sorcery (or, witchcraft), but if it be a child who is

---

[1] They are measures of time.

sick, he is too young for the disease to be a punishment (or, revenge). And if four, the sickness is caused by an evil spirit. And if five, there is a *shîdhâ* (*i e.*, a devil) in it. And if six, the sickness is from heaven. And if seven, it is due to trembling caused by fear. And if eight, the sickness is caused by a blow of Sâṭân. And if nine, it is from his mother's womb (*i. e.*, congenital), or is due to vengeance, or disturbing dreams.

Another calculation concerning those who are sick. Reckon up [the numerical values of the letters] of the name of the sick man, and of those of the name of his mother, and cast them out nine by nine (*i. e.*, divide them by nine). And if one remaineth, and [if his sickness] fell [upon him] on the first day of the week, the day of the sun, his sickness is from his shoulders and head and neck, and he will remain sick for nine days. And if two remain, and if his sickness fell upon him on the second day of the week, the day of the moon, his sickness is from his belly, and his whole body is sick. And if three remain, and if his sickness fell upon him on the third day of the week, the day of Ârîs (Mars), his disease is hot and dry; he bathed in water, and the air smote him. Make Fol. 216b. thou for him three lamp-wicks | out of his garments. Set one at his head, and one by his right side, and one by his left side, and he will be sick for fifteen days. And if four remain, and if his sickness fell upon him on the fourth day of the week, the day of Harmîs (Mercury), he passed through (or, over) water, and did not call upon the name of God. Let him give alms. He will be ill for seventeen days. And if five remain, and if his sickness fell upon him on the fifth day of the week, the day of Zeus, his sickness is caused by over-eating, and his heart is mad. Let him make an offering (or, give alms) of a hen. He will be ill for sixteen days. And if six remain, and if his sickness fell upon him on the Eve of the Sabbath, the day of Aphrodite, his sickness is from his head and from his body, and is also due to the turning of two periods (?) of his life. He will be ill for twelve days. And if seven remain, and if his sickness fell upon him on the day of the Sabbath, Page 454 the day of Kronos, | his sickness ariseth from his liver and from his heart, and trembling fell upon him. Bring dust from seven roads, and from seven cemeteries, and from four fountains of water, and repeat over him . . . . And if eight remain, he saw a bad dream on the fourth day of the week. Twenty days he will remain [sick]. And if nine remain, he sat upon the ground on the night of the Eve of the Sabbath and did not call upon the Name of the Living God. He will remain sick for eighteen days, and his illness is due to vengeance.

Thus shalt thou cast out from the letters of the name of a man and from those of the name of his mother, having cast out nine by nine. Take to thee from *yôdh* (10) one, and from *kâph* (20) two, and from *lômadh* (30) three, and from *mîm* (40) four, and from *nôn* (50) five, and from *semkath* (60) six, and from *'ain* (70) seven, and from *pê* (80) eight, and from *ṣâdhê* (90) nine, and from *ḳôph* (100) one, and from *rêsh* (200) two, and from *shîn* (300) three, and from *tâw* (400) four.

And if thou wishest to cast them out by seven, do so thus: Take from *khêth* (8) one, and from *ṭêth* (9) two, and from *yôdh* (10) three, and from *kâph* (20) six, and from *lômadh* (30) two, and from *mîm* (40) five, and from *nôn* (50) one, and from *semkath* (60) four, and from *'ain* (70) nine, and from *pê* (80) three, | and from *ṣâdhê* (90) six, and from *ḳôph* Fol. 217 a. (100) two, and from *rêsh* (200) four, and from *shîn* (300) six, and from *tâw* (400) one.

Another calculation: If thou wishest to know whether a sick man will die or recover: Reckon up the numerical values of the letters in the name of the sick man, and those of the letters in the name of the day in which he became sick, take the total of thy calculation in thy hand, and divide it into three parts, two [representing] the day, and one the sick man. And if one remaineth to thee, the sick man will live; and if two, his disease will prolong itself; and if three, he will die. Now this is a very sure method, but besides dying when three remain of the number made by the letters of his own name, he will die whenever he falleth sick on the day of the number made by the letters of which two remain.

Page 455   **When thou hast counted out three by three the number**
made by the letters of the names of the days, thus shalt
thou reckon: From the first day of the week, two; from the
second day of the week, two; from the third day of the week,
one; from the fourth day of the week, two; from the fifth day
of the week, two; from the Eve of the Sabbath, one; and from
the Sabbath, one.

Another indication concerning a sick man. Reckon up
the numerical values of the letters of the name of the sick
man, and those of the letters of the name of his mother, on
the day wherein he cometh to thee, and see how much they
make, and then count them out three by three. If he came
on the first day of the week, and one remaineth, he will rise
up (*i. e.,* recover) quickly, and if two remain, his sickness will
last a long time, and if three remain, he is nigh unto death.
And if he came on the second day of the week, and two
remain, he will recover [quickly], and if three remain, he will
be sick a long time, and if one remaineth, he will recover
[quickly]. And if he came on the third day of the week, and
three remain, he will recover [quickly], and if two remain,
he is nigh unto death, and if one remaineth, he will be sick
a long time. And if he came on the fourth day of the week,
and one remaineth, he will recover immediately, and if two
remain, he will be sick a long time, and if three remain, he is
nigh unto death. And if he came on the fifth day of the
week, and two remain, he will recover [quickly], and if three
remain, he will be sick for a long time, and if one remaineth,
he will not recover. And if he came on the Eve of the Sabbath,
and three remain, he will recover, and if two remain, he is
nigh unto death, and if one remaineth, he will be sick a long
Fol. 217 b. time. And if | he came on the Sabbath, reckon in the same
way as thou dost for the Eve of the Sabbath.

Another forecast about the sick. Take some bread, and
put it under the head of the sick man, and let it stay there
from the evening until the morning, and then cast it to a dog.
If the dog eateth it the man will not die, and if the dog will

not eat it, the man will die. This is a sure and well-tried indication.

Or give the sick man water to drink, and that which remaineth after he hath drunk sprinkle on the dog of the sick man's house; if the water runneth off the dog the man will die, and if it doth not, he will live.

Or smear the lower parts of the feet of the sick man [in the evening] with goat's fat, and in the morning throw some of it to the dog; if he eateth it the man will live, and if he doth not eat it the man will die.

Or cut off a nail-paring from the foot of the sick man, and throw it into water in a cup; if it sinketh in the water, the sick man will live, and if it | floateth, he will die.   Page 456

Another calculation concerning a sick man, and something which is lost, and a man who hath fled. Take the days from the seventeenth day of Shebâṭ until the day when the sick man took to his bed, or the day when the thing was lost, or the day when the man fled, and see how many they make, and then cast them out by thirty-fives, and then see in which table is the number which doth not go out in thirty-five. If it be in the first table, the sick man will live, and the thing lost will be found, and the runaway slave will return. And if it be in the second table, the patient will remain sick for a long time, but will be healed eventually, although he will be in danger; and the lost thing will only reappear after much delay, and the runaway slave will fall. And if the number be in the last table, the sick man will die, and that which is lost will never be found, and the runaway slave will never return.

Table I. Of life.	Table II. Of danger.	Table III. Of death.
1. 4. 7. 13.	5. 8. 11.	3. 6. 2.
16. 19.	14. 14. 18.	9. 10. 12.
22. 23.	26. 23.	15. 21.
28.	25. 29.	24. 27. 30.
31.	32. 35.	33. 34.

Fol. 218a.   **Another way.** Every one who is attacked by sickness on the first day of the month recovereth on the second day. And if he be attacked on the second day of the month, he will be ill for the whole month. If he be attacked on the third of the month, he will be ill equally the whole month, but he will not die. If he be attacked on the fourth day of the month, he will most certainly die. If he be attacked on the fifth day of the month, he will fall into the hands of the physicians. If he be attacked on the sixth day of the month, he will be quickly healed. If he be attacked on the seventh day of the month, he will not die. If he be attacked on the eighth day of the month, he will not die. If he be attacked on the ninth day of the month, the fever will not leave him. If he be

Page 457   attacked on the tenth day of the month, | he will die in seven days. If he be attacked on the eleventh day of the month, he will fall into the hands of the physicians, and will live. If he be attacked on the twelfth day of the month, his illness will be slight, and he will not die. If he be attacked on the thirteenth day of the month, he will die. If he be attacked on the fourteenth day of the month, he will die in twenty days. If he be attacked on the fifteenth day of the month, he will fall into the hands of the physicians. If he be attacked on the sixteenth day of the month, he will live. If he be attacked on the seventeenth day of the month, the fever will leave him. If he be attacked on the eighteenth day of the month, he will die in twenty days. If he be attacked on the nineteenth day of the month, he will die. If he be attacked on the twentieth day of the month, he will live. If he be attacked on the twenty-first day of the month, he will live. If he be attacked on the twenty-second day of the month, the fever will depart from him, and he will live. If he be attacked on the twenty-third day of the month, he will die immediately. If he be attacked on the twenty-fourth day of the month, he will die. If he be attacked on the twenty-fifth day of the month, he will live. If he be attacked on the twenty-sixth day of the month, he will die. If he be attacked on the twenty-seventh day of the month, he will subsist (or, recover ?). If he be attacked on the twenty-eighth day of the month, he will fall into the hands of the physicians. If he be attacked on the twenty-ninth day of the month, he will live. If he be attacked on the thirtieth day of the month, he will live.

Again, a forecast concerning a sick man, derived from the days of the moon, whether he will live, or die.

First day of the moon. If he is breathing a little, the son of a day, he will live; and if he is altogether sick, he will not die.

Second day of the moon. He is not very ill.

Third day of the moon. It is manifest that the whole day is bad.

Fourth day of the moon. He will fall into the hands of the physicians.

Fifth day of the moon. He will fall into the hands of the physicians, and will be healed.

Sixth day of the moon. He will live two days.

Seventh day of the moon. Change the place, and he will live.

Eighth day of the moon. The sickness abateth with difficulty, and he will live.

Ninth day of the moon. If he is ill from the first hour of the day to the fifth, he will fall into the hands of the physicians, and will live.

Tenth day of the moon. He will be ill until the evening of the day, and will live.

Eleventh day of the moon. He will fall into the hands of the physicians, and will live.

Twelfth day of the moon. If fever precedeth the sickness, he will die.

Thirteenth day of the moon. He will die.

Fourteenth day of the moon. Change the place, and he will live.

Fifteenth day of the moon. He will die.

Sixteenth day of the moon. Whether he eateth, or whether he eateth not, he will live.

Seventeenth day of the moon. If he passeth (*i. e.,* surviveth) fifteen | days, he will fall into the hands of the physicians, and Fol. 218*b*. will live.

Eighteenth day of the moon. He will die.

Nineteenth day of the moon. [He will be sick] seven days, and will live.

Twentieth day of the moon. [He will be sick] nine days, and will live.

Twenty-first day of the moon. He will die.

Twenty-second day of the moon. In twelve days he will be healed.

Page 458 Twenty-third | day of the moon. He will die.

Twenty-fourth day of the moon. In twelve days, he will be healed.

Twenty-fifth day of the moon. He will die.

Twenty-sixth day of the moon. He will live.

Twenty-seventh day of the moon. He will live.

Twenty-eighth day of the moon. After sickness for eight days, he will live.

Twenty-ninth day of the moon. He will fall into the hands of the physicians, and be healed.

Thirtieth day of the moon. Whether he eateth, or whether he eateth not, he will most assuredly die.

Another way. Enquire when the man was taken ill, and see how many days there were of the moon until the day when he was taken ill. Reckon up the numerical values of the letters of the name of the sick man, and add to them the number of the days that have passed since the new moon, and cast them out nineteen by nineteen. What remaineth over above the nineteens, keep in thy mind, and then enter into the circle which is before thee, and thou wilt find the number that remaineth to thee. [To find out] whether he will die or live. One. He will live. Two. He will die. Three. He will live. Four. He will either live or die. Five. He will live. Six. He will either live or die. Seven. He will live. Eight. He will die. Nine. He will live. Ten. He will either live or die. Eleven. He will die. Twelve. He will either live or die. Thirteen. He will die. Fourteen. If he be left for a little, he will certainly live, but another codex maketh known that he will die. Fifteen. He will die. Sixteen. He will live. Seventeen. He is dead already, but in another place it saith that he will live. Eighteen. He will either live or die. Nineteen. He will live.

Another way. See on what day he was taken ill, and how many days have elapsed since the new moon. And thou must learn the name by which the sick man was called when he was born. Then count up the numerical values of the letters of his name, and keep the sum of them in thy mind; add to them the days of the moon's age, which have been mentioned above, and add to these twenty days. When thou hast added them all together, and made one number, then cast it out thirty by thirty, and see how many are left over in thy hands. | And come to the two tables which are drawn up before thee Fol. 219a. here, and if the number that remaineth in thy hand is in the first one the sick man will live, and if it is in the second one, he will die.

Page 459

I. Of life.			II. Of death		
1.	2.	3.	5.	6.	12.
17.	8.	7.	15.	18.	21.
9.	10.	11.	27.	25.	
13.	14.	19.	29.	30.	
16.	29.	20.		4.	
22.	26.	28.		24.	

Another calculation whereby the physician may know which are the days when the moon hath light, and which are those when it hath no light, and whether the sick man will live or die. The days when the moon hath light are the odd days, *viz.*, one, three, five, seven, nine, eleven, thirteen, fifteen, seventeen, nineteen, twenty-one, twenty-three, twenty-five, twenty-seven, and twenty-nine; and the days when the moon is dark are the even days, *viz.*, two, four, six, eight, ten, twelve, fourteen, sixteen, eighteen, twenty, twenty-two, twenty-four, twenty-six, twenty-eight, and thirty. Now, if a man falleth sick when the moon hath light, he will most certainly exist in a weak state for a long time. See then, (*i.e.*, find out) the name of the sick man, and reckon up the numerical values of the letters thereof, and find out the number of the days since the new moon, and the day wherein the sick man took to his bed; and examine the number that remaineth over with thee, and see whether it is found in [the number of] the day on which the moon hath light, and also [the number of] the day wherein he took to his bed. If the moon hath no light on that day, he will subsist for a long time; but if he took to his bed on a day when the moon had light, and the reckoning as to [his] name remaineth to the day wherein it had no light, even though he may suffer greatly, and pass through much affliction, he will be saved. If, however, the

Fol. 219*b.* two calculations | are seen to apply to the day when the moon had no light, he will most certainly die. And thou mayest use this calculation in making a forecast about any fugitive, or about any thing that is lost, and about everything else.

Page 460   **Another calculation whereby a man is able to know beforehand which of two men who are striving together (or, fighting against each other) will die. It was made by Aristotle for King Alexander, his royal disciple, when he was waging war against Darius the Mede, and Alexander conquered Darius. This calculation is sure, and hath been well tried. It is useful to every one who wageth war against his neighbour, and striveth in respect of matters of business, and to kings, and to all men, both little and great. Observe when thou wishest to know when [two] men are waging war against each other, which will conquer. Take the numerical values of the letters of the name of each by itself, and take from the sum of each as many nines as there are in each, and see how many remain to thee in each case, and bear them [in thy mind]. Then come to the following letters (or, numbers), and from them thou wilt learn which will conquer. One conquereth three, five, seven, and nine. Two conquereth one, four, six, and eight. Three conquereth two, five, seven, and nine. Four conquereth one, three, six, and eight. Five conquereth two, four, seven, and nine. Six conquereth one, three, five, and eight. Seven conquereth two, four, six, and nine. Eight conquereth one, three, five, and seven. Nine conquereth two, four, six, and eight. And if there be two in one name, or the numbers in the two names** are equal, the combatant, that is, the elder, will conquer the younger. This calculation is a very sure one, and it is mentioned by the philosophers. When they made the calculation about Alexander's name there remained eight, and when they made the calculation about the name of Darius, there remained seven; and observe that eight conquered seven, so it cometh in one case after another, with the thing that was lost and was found, and with the man who died, and with the man who was healed, | and so on, when thou knowest well the name Fol. 220a. of the sick man and the name of the day in which he [first] perceived his sickness.

Another calculation about a man who is sick. In this manner shalt thou make thy calculation. Reckon up | the Page 461 numerical values of the letters in the name of the star that ruled the day on which the man fell sick, and those of the letters of the name of the sick man; reckon up both sets of numbers, according to the instructions given above. And in like manner make thy calculation about the thing that is lost (or, stolen), and the man who hath betaken himself to flight. Reckon up the numbers of the name of the thing that is lost, and those of the name of him who hath lost anything; and if the owner's name conquereth, his loss will be discovered, and the thing itself. And if he who is sick, or he who hath lost something, conquereth the star that ruleth the day on which he fell sick, he will rise up from his sickness, and be healed, and the thing that is lost will be found. This calculation is sure, and it hath been well tried.

The Days of the Planets.

Hermes ruleth the night of the first day of the week, and the Sun the day-time.

Zeus ruleth the night of the second day of the week, and the Moon the day-time.

Aphrodite ruleth the night of the third day of the week, and Âris the day-time.

Kronos ruleth the night of the fourth day of the week, and Hermes the day-time.

The Sun ruleth the night of the fifth day of the week, and Zeus the day-time.

The Moon ruleth the night of the Eve of the Sabbath, and Aphrodite the day-time.

Âris ruleth the night of the Sabbath, and Kronos the day-time.

And if thou wishest to know how many will remain over, after thou hast divided the sum of the numerical values of the letters of each of the names of the stars and of the days by nine, the following are the numbers:—

Of the Sun two, and of the Moon four. Of Âris one, and of Hermes nine. Of Zeus one, and of Aphrodite five. Of Kronos eight. Of the first day of the week two. Of the second day of the week two. Of the third day of the week one. Of the fourth day of the week two. Of the fifth day of the week five. Of the Eve of the Sabbath four. Of the Sabbath one.

Another forecast concerning the sick. Now, it is right for the wise physician to ask this question:—On what day of the week did the patient take to his bed? | Having made this enquiry | he must then make a calculation by means of the following Twelve Houses (*i. e.*, Signs of the Zodiac), and he must find out clearly in each month, in what day the man became sick, and what its relation was in respect of the moon; and it must also be known in what House the moon is. And from the House, or Sign of the Zodiac, in which the moon is, thou wilt learn clearly about those who are sick, whether they will live, or whether they will die; for in this calculation death and life are written clearly. But, whatsoever the Lord wisheth He doeth, in the heavens and on the earth, and He it is Who trieth secret things, and every man is a liar, even like myself, the writer.

Of the months of the whole year in one.

In Îlûl, when the moon is in the Lion. He who is sick will recover.

In the Virgin. His sickness will afflict him a little, and afterwards he will recover.

In the Balance. After fifteen days he will recover.

Fol. 220*b*.
Page 462

In the Scorpion. Up to twenty days he will die.

In the Bowman. He will be ill for thirty days, and then will die.

In the Goat. . . . .

In the Water-carrier. Seven days he will be sick, and then recover.

In the Fishes. He will recover.

In the Ram. After nine days he will die.

In the Bull. After a few days he will recover.

In the Twins. After some days he will recover.

In the Crab. Even after a long illness he will not recover.

In the First Teshrî, when the moon is in the Virgin. After a few days he will recover.

In the Balance. He will be ill for twenty-five days, and then recover.

In the Scorpion. On the twenty-ninth day he will recover.

In the Bowman, *i. e.,* the Great Image, he will recover.

In the Goat. He will suffer, and after a few days will die.

In the Water-carrier and the Fishes. He will recover.

In the Ram, the Bull, and the Twins. He will be sick for forty days, and then he will die.

In the Crab and in the Lion. He will recover.

In the Second Teshrî, when the moon is in Bĕḳâ Shĕlâmâ, *i. e.,* the Balance, he will be sick for | ten | days, and then he will recover. Page 463
Fol. 221 a.

In the Scorpion. He will recover.

In the Bowman. He will remain sick for a long time, and then recover.

In the Goat. . . . .

In the Water-carrier. . . . .

In the Fishes. . . . .

In the Ram. . . . .

In the Bull. He will recover.

In the Twins. He will be sick for eight days, and will afterwards recover.

In the Crab. He will be sick for twelve days, and will recover.

In the Lion. He will be violently ill for nine days and then die.

In the Virgin. He will do badly for thirty days, and he will then rise up from his sickness.

In the First Kânôn. If a man fall sick when the moon is In the Scorpion, he will recover after a few days.

In the Bowman. He will suffer and then obtain relief.

In the Goat. After a few days he will die.

In the Water-carrier. He will suffer and then have relief from his sickness.

In the Fishes. He will be healed.

In the Ram. He will die, unless he remaineth for thirty days, when he will recover.

In the Bull. After eleven days he will recover.

In the Twins, the Crab, and the Lion. He will die after a few days.

In the Virgin. He will recover.

In the Balance. He will suffer, and then find relief.

In the Second Kânôn. If a man fall ill when the moon is in the Great Image (Sagittarius), he will lie on a sick bed for twenty days, and then die.

In the Goat. He will live.

In the Water-carrier. He will live.

In the Fishes and in the Ram. He will die.

In the Bull, and in the Twins, and in the Crab . . . .

In the Lion. He will die.

In the Virgin. He will recover.

In the Balance. He will recover.

Page 464   In the Scorpion. After a few days he will die. |

In Shebât. If a man fall sick when the moon is In the Goat, he will recover.

In the Water-carrier, the Fishes, the Ram, the Bull, and the Twins, he will recover after a few days.

In the Crab. After a sickness of fifteen days he will die.

In the Lion, and in the Virgin. He will die.

Fol. 221 *b*.   In the Balance. After a few days | he will recover from the sickness.

In the Scorpion. He will die after a short illness.

In the Goat. . . . .

In Âdhâr. If a man fall sick when the moon is

In the Water-carrier. He will recover.

In the Fishes and in the Ram. He will recover from his sickness.

In the Bull. He will die.

In the Twins, and in the Crab, and in the Lion. He will recover.

In the Virgin. After a few days he will die.

In the Balance and in the Scorpion. He will live.

In the Bowman and in the Goat. After a short illness he will die.

In Nisân. If a man fall sick when the moon is

In the Fishes. He will be healed of the sickness which is in him.

In the Ram, and the Bull, and the Two Images (Twins). He will be ·healed.

In the Crab, and Lion, and Virgin, and Balance. He will die.

In the Scorpion, and the Bowman. He will recover.

In the Goat and the Water-carrier. He will recover from his sickness.

In Îyyâr. If a man fall sick when the moon is

In the Ram, he will recover.

In the Bull. . . . .

In the Twins. . . . .

In the Crab, and in the Lion. He will die.

In the Virgin. He will recover.

In the Balance, and the Scorpion, and the Bowman, he will die.

In the Goat. He will recover.

In the Water-carrier. He will die.

In the Fishes. He will be healed.

| **In Khăzîrân.** If a man fall sick when the moon is   Page 465
In the Bull, he will die.

In the Twins, and the Crab, and the Lion, and the Virgin, and the Balance, and the Scorpion. He will die after a short illness.

In the Bowman. He will recover.

In the Goat. He will die.

In the Water-carrier. He will live.

In the Fishes. He will die.

In the Bull. He will recover from his illness.

In Tammûz. If a man fall sick when the moon is

In the Two Images, and the Crab, and the Lion, he will recover from his sickness.

In the Virgin and in the Balance. He will die.

In the Scorpion. He will suffer greatly for a long time from an obstinate sickness, but will afterwards recover.

In Ṣarrâbhâ, that is to say, the Bowman. He will die.

In the Goat. He will die.

Fol. 222 a. | In the Water-carrier, and the Fishes, and the Ram, and the Bull, he will recover.

In Âbh. If a man fall sick when the moon is

In the Crab, he will fall into a sickness for eight days, and afterwards he will recover.

In the Lion, and the Virgin, and in the Balance. He will die.

In the Bowman, and the Goat, [and the Water-carrier], and the Fishes, and the Ram. He will live.

In the Bull. He will die.

In the Two Images. He will recover from the illness which is on him.

An indication whereby thou mayest know how long the moon abideth in each Sign of the Zodiac.

Now the moon abideth in each House two days and a half, from the first [day] of Îlûl, when the moon is in the Lion, until the end of Âbh, when it is in the Two Images.

And this also know:—When the moon ascendeth newly, it travelleth through the width of the Circle of the Signs of the Zodiac, through the Towers (?), in a backward direction, oblique-

Page 466 ly, | until it reacheth its fullness, and then it beginneth to wane, little by little, until it draweth nigh unto the sun, wherever it may be, in the north, or in the south, and from that place it riseth afresh.

And know the following [also]:—Another door (*i. e.,* section) concerning the sick and afflicted.

He who is smitten in the Ram will be sick for seven days, and if he perceiveth it on the seventh day of the moon, he will be healed. And if he perceiveth it on the seventh [day]

of Kronos, there will be great oppression upon him. Of this
man the following are the signs:—His back giveth him pain,
one hour he is hot, and the next he is cold. And if the sun
be there, he will not be healed.

He who is smitten in the Bull. Of this man the following
are the signs:—His arms give him pain, and his left side, and
his neck, and the soles of his feet, and he is smitten in a
secret place, and his eyes are "hard" to him. And if the sun
be in the House in its going down, the pain will never move
from him.

**He who is smitten in the Two Images, that is to say, the
Twins.** Of this man the following are the signs:—His shoulders
and bowels cause him pain, and fever will have dominion over
him in the hour wherein the flux smiteth | him. And if the Fol. 222b.
moon be there on the day when it smiteth him, cure is nigh
unto him, and the atmosphere will be changed about him, and
he will be healed.

He who is smitten in the Crab. Of this man the following are
the signs:—His heart is stirred up against him. And he is smitten
by the water, whether he batheth in water, or is naked by the side
of water. A phantom and a spirit seat themselves on his heart.
If he passeth seven days, and is not healed, he will pass ninety
and four days. If he be smitten in Ârîs (Mars) or in Aphrodite, see that there be not in him some evil and secret blemish.
And if he passeth thirty days, and is not healed, he will die.

He who is smitten in the Lion. Of this man the following
are the signs: - His loins and his heart cause him pain, and if
he be smitten in Aphrodite there will be grievous (or, deep)
sickness to him, and . . . . . . will go forth to him.

He who is smitten in the Virgin. Of this man the following are the signs:—His bowels and shoulders cause him pain,
| he standeth up in a dejected attitude, and winds strike within Page 467
him. If he is not healed in seven days, he will be healed in
twelve or fifteen.

He who is smitten in the Balance. Of this man the
following are the signs:—He hath severe illness. And if he be
smitten on the first day and on the second, he will pass six
months as a sick man, and he will come to the point of death,

35*

but will not die. And if he be smitten on the third day of the Sign of the Zodiac, he will die; and if Kronos be there, and he be smitten at noon of that day, he will be in very grave danger.

He who is smitten in the Scorpion. Of this man the following are the signs:—His bowels, back, and loins cause him pain. He is smitten from a place which is wet, and he becometh a sign thereof.

He who is smitten in the Bowman, Zeus being there. His strength becometh slackness. And if he be smitten on the first day of the Sign of the Zodiac, he will not die; and if he be smitten on the third day, he will die, because he is smitten in the hour of Hermes. Of this man the following are the signs:—His bowels and his legs cause him pain. If the air be changed about him, he will be healed.

He who is smitten in the Goat. Now, he will be smitten with a severe stroke. He will be smitten with heavy eyes. Fol. 223a. And if he be smitten on the first day | of the Sign of the Zodiac, he will die, and if he be smitten on the second day, his sickness will be severe (or, deep), and if he be smitten on the third day, his tears will be poured out in abundance from the air, and he will be healed.

He who is smitten in the Water-carrier. Of this man the following are the signs:—His legs will cause him pain. If he be smitten in the first hour of the Sign of the Zodiac, he will be in very grave danger, and if he be smitten at noon of the same day, there will be bile in his bowels, and it will be poured out into his whole body, and through it he will be sick. If Âris is there, he will be able to move his arms, but if the moon ministereth there, he will die after a short interval.

He who is smitten in the Fishes. Of this man the following are the signs:—His ribs, and his chest, and his members, and his whole body will cause him pain, and his strength will fail utterly. If it be a woman who hath conceived a child in the Fishes, she will discharge much blood, and she will not Page 468 bring forth the child unless thou treat her with either | *rûbâ*, or frankincense and galbanum (styrax). And there are also seven (*sic!*) other roots: wild mint, that is to say, *kharmal*, and *yad'în* (?), and male rice, and *bardâ*, and willow leaves, and leaves of the reed, and artemisia (chamaecyparissus squarrosa), and *beldhâ*; make for her a medicine (?) [of these].

Another kind, concerning the sick.

In the Ram. His sickness will be sore, and his head will cause him pain.

In the Bull. His palate will cause him pain; he cannot be healed.

In the Twins. He will suffer torture; his shoulders will cause him pain.

In the Crab. His belly will cause him pain.

In the Lion. He will find relief immediately. His back will cause him pain.

In the Virgin. He will be afflicted.

In the Balance. He will suffer great torture.

In the Scorpion. [He will suffer] very great [torture] indeed.

In the Bowman. He will live.

In the Goat. He will come to the point of death.

In the Water-carrier and in the Fishes. He will not suffer affliction if a good star be found there. Kronos or Zeus will direct the sick man. If it be Âris, he will become silly; if it be the Sun, he will become blind; and if it be Baltî (Venus), he will fall through love; if it be Hermes, he will be attacked by sickness; and if it be the moon, his sickness will be light.

Forecast concerning a sick man, from the Book of the Signs of the Zodiac.

If a man was born under the Ram, and becometh sick under Cancer, and he surviveth for thirty days, he will not die.

And if a man was born under the Twins, | and becometh Fol. 223*b.* sick under the Virgin, and surviveth for twenty-one days, he will not die.

And if a man was born under the Bull, and becometh sick under the Lion, and surviveth for twenty-seven days, he will not die.

And if a man was born under the Crab, and becometh sick under the Balance, he will die.

And if a man was born under the Lion, and becometh sick under the Scorpion [he will die].

And if a man was born under the Virgin, and becometh sick under the Bowman, and surviveth for seven days, he will not die.

He who was born under the Balance, and becometh sick under the Goat, will die.

He who was born under the Scorpion, and becometh sick under the Water-carrier, if he surviveth for twenty-one days, he will not die.

He who was born under the Bowman, and becometh sick under the Fishes, if he surviveth for eight days, he will not die.

He who was born under the Goat, and becometh sick under the Ram [will die].

Page 469   He who was born under the Water-carrier, | and becometh sick under the Bull, if he surviveth for twenty-two days, he will not die.

He who was born under the Fishes, and falleth sick under the Twins, if he surviveth for sixteen days, he will not die.

Of the stars which are propitious in conjunction with the Signs of the Zodiac. The Sun and Moon are propitious.

In conjunction with the Ram the Moon and Hermes are propitious.

In conjunction with the Bull, the Sun, and Zeus, and the Moon are propitious.

In conjunction with the Twins and the Lion, the Sun, and the Moon, and Kronos are propitious.

In conjunction with the Crab, the Sun, and Aphrodite, and the Moon are propitious.

In conjunction with the Virgin, the Sun, and Zeus, and Hermes are propitious.

In conjunction with the Balance, the Moon, and Hermes, and Zeus are propitious.

In conjunction with the Scorpion, the Sun, and Hermes, and Kronos are propitious.

In conjunction with the Bowman, Zeus, and Aphrodite, and Kronos are propitious.

In conjunction with the Goat, the Sun, and Moon, and Kronos are propitious.

In conjunction with the Water-carrier, the Sun, and Moon, and Aphrodite are propitious.

In conjunction with the Fishes . . . .

Again, I will distinguish and make known concerning the days of the moon, and how they are placed as concerning the Eight Kings.

The first day and the ninth day belong to Baltî (Venus).

The second day and the eleventh day belong to Ârîs (Mars).

The third day and the twelfth day belong to Âthîlyâ (?).

The fourth day and the thirteenth day belong to Zeus.

The fifth day and the fourteenth day belong to Kronos.

The sixth day and the fifteenth day belong to Hermes.

The seventh day and the sixteenth day | belong to the Moon. Fol. 224a.

The eighth day and the seventeenth day belong to the Sun.

That thou mayest know what sufferings each of the stars (planets) produceth in the children of men, and in their bodies:—

Kronos produceth headache, gout, cold (chill), coughs, shortness of breath, and dropsy.

Zeus produceth ulceration of the throat, stoppages of the bowels, scrofula, inflammation of the throat and gums, and pain in the liver.

Ârîs produceth lameness, distortions of the legs, burns, and falling of the mind.

The **Sun** produceth pain in the belly, pain in the intestines, and pain in the heart.

Aphrodite produceth scabies, | leprosy, warts, eruptions of Page 470 sores, paralysis, and blotches and pimples.

Hermes produceth tongue-tied people, dumbness, and sudden death.

The **Moon** produceth thickness of the eyelids, obscurity of vision, blindness, lunatics.

If this section be summarized it decideth concerning them thus:—

Kronos. Those who are drowned in the waters.

Zeus. By means of earthquake, or thunderbolt(?).

Âris. By means of iron, and through rheum.

Sun. By means of rheum.

Aphrodite. By means of food and lust.

Hermes. By means of rigidity, and blows of some kind.

Moon. By means of silver(?).

They arrange (or, assign positions to) the Seven Governors thus:—They say that Kronos, that is to say, Zûkhâl, is above them all in the upper zone of the air, and that he is in the outer circle of the sphere. And he is in every Sign of the Zodiac for thirty months, and he performeth his revolution through the Twelve Houses in thirty years.

After him is placed Zeus, who is surnamed Bêl, who is Mûshtarî, and who is likewise in the zone of the air, and is in the second circle of the sphere. He remaineth in each House twelve months, and he performeth his revolution through all the Houses in twelve years.

And after him is placed Âris, who is Marîk, and he is in the zone of fire, and is in the third circle of the sphere. He remaineth in each House one month and a half, and he performeth his revolution in one year and a half.

And after him is placed the Sun, who is also in the zone of fire, and is in the fourth circle of the sphere. He remaineth in each Fol. 224 b. House thirty days, | and he performeth his revolution in a year.

And after him is placed Baltî, who is Zûhrah, who is in the zone of water, and is in the fifth circle of the sphere. She remaineth in each House twenty-five days, and performeth her revolution in ten months.

And after her is placed Hermes, who is surnamed Nâbô, who is 'Ûtrâdh, and who is also in the zone of water, and is Page 471 in the sixth circle of the sphere. | He remaineth in each House fifteen days, and performeth his course in six months.

And after him is placed the Moon, in the zone of water, and in the seventh circle of the sphere. He remaineth in each House two days and a half, and he performeth his course in twenty-nine days and a half.

This is the order of their positions.   Each one of them hath two Houses, and the Sun and the Moon have one each.   The Goat and the Water-carrier are the Houses of Kêwân, that is to say, Kronos.

Ṣalmâ Rabbâ (Great Image) and the Fishes are the two Houses of Zeus.

The Ram and the Scorpion are the two Houses of Ârîs.

The Lion is the House of the Sun.

The Bull and the Balance are the two Houses of Baltî.

The Two Images and the Virgin are the two Houses of Hermes.

The Crab is the House of the Moon.

Concerning the revolution of each one of the stars (planets).

Two revolutions of Kronos [occupy] sixty years.

Five revolutions of Zeus [occupy] sixty years.

Four revolutions of Ârîs [occupy] sixty years.

Sixty revolutions of the Sun [occupy] sixty years.

One hundred and twenty revolutions of Hermes [occupy] sixty years.

Seven hundred and twenty revolutions of the Moon [occupy] sixty years.

Information concerning the kingdoms of the stars (planets) of the Arabs, and their Houses.

The kingdom of the Sun is the Ram, and his setting place is the Balance.

The kingdom of Aphrodite is the Fishes, and his (*sic!*) setting place is the Virgin.

The kingdom of Hermes is the Virgin, and his setting place is the Fishes.

The kingdom of the Moon is the Bull, and his setting place is the Scorpion.

The kingdom of Kronos is the Balance, and his setting place is the Ram.

The kingdom of Zeus is the Crab, and his setting place is
| the Goat.   Fol. 225 *a*·

The kingdom of Âris is the Goat, and his setting place is the Crab.

Whensoever thou wishest to travel to the east or to the south, do so by the direction of each of the following stars, Aphrodite, the Moon, and Âris. And if thou wishest to go to the west or to the north, do so by means of the Sun, and Page 472 Zeus, and Hermes. |

Another, concerning the road.

If the moon is in the Ram, and a man setteth out on a journey, let him return immediately to his house.

And if the moon is in the Bull, he will return with great profit.

And if the moon is in the Twins, he will go to a far country, and will tarry [there] a long time.

And if the moon is in the Crab, he will return to his house in joy.

And if the moon is in the Lion, be very careful about misfortune, for misfortune will make itself known on the first day, or on the second, or on the third. The misfortunes may be trivial, but the Sign indicateth a bad journey.

And if the moon is in the Virgin, thou wilt be absent a long time from thy house, but no evil will happen unto thee.

And if the moon is in the Balance, thou shalt not depart on thy journey, because thou wilt have no pleasure in it, and it will shew no profit, for it is equally balanced, and will yield neither joy nor sorrow.

And if the moon is in the Scorpion, there shall be both business and profit in thy journey.

And if the moon is in the Bowman, after a very long time thou wilt return happily to thy house.

And if the moon is in the Goat, thy journey will be delayed, and thou wilt not succeed in doing any business, and thy journey will cause thee the most intense grief possible. And if thou returnest under the Sign of the Goat, no evil will befall thee.

And if the moon is in the Water-carrier, be very careful on

the first and on the second day, and on the third day good will come unto thee, and thou wilt find profit.

And if the moon is in the Fishes, and thou goest forth to war, or to anything else which is evil, or to hunt wild animals, or to catch fish, thou wilt hunt and fish well and successfully, but if thou goest forth on any other enterprise it will end un-luckily. "Except the Lord build the house, | the builders Fol. 225 *b.* "thereof weary themselves vainly." (Psalm CXXVII. 1.)

Concerning the days of the moon, and how all the hours thereof are called "working agents." From the twenty-seventh day of the moon to the half of the fourth day of the moon, it is meet for thee to do nothing. From the half of the fourth day of the moon | to the eleventh day of the moon it Page 473 is meet for thee to do anything. From the twelfth day of the moon until the twenty-seventh day of the moon, do every-thing.

When the moon is in the Ram on the fifteenth day, in the early parts [of the day] it is good to go on a journey.

When the moon is in the Bull on the fifteenth day, in the early parts [of the day] it is good to go on a journey. And thou shalt plough the ground, and plant thy vineyard, and sow seed, and thou mayest devote attention to all [thine] affairs.

When the moon is in the Twins on the tenth day, in the early parts [of the day] flee from work of every kind.

When the moon is in the Crab . . . .

When the moon is in the Lion on the fifteenth day, in the early parts [of the day] it is good to set out on a journey.

When the moon is in the Virgin on the fifteenth day, in the early parts [of the day] buy and sell, and sow seed, and plant vineyards.

When the moon is in the Balance on the fifteenth day, in the early parts of the day continue thy journey.

When the moon is in the Scorpion on the fifteenth day, in the early parts of the day continue thy journey, and from every undertaking whereto thou settest thy hand thou shalt derive advantage.

When the moon is in the Bowman on the fifteenth day, in

the early parts [of the day] it is not good for thee to go anywhere at all.

When the moon is in the Goat . . . .

[several lines wanting].

Another forecast as to the day when a man should set out on a journey.

First day of the moon. The whole day is propitious for going on a journey.

Second day of the moon. [Set out] at the turn of the day.

Third day of the moon. Unpropitious.

Fourth day of the moon. The whole day [is propitious].

Fol. 226 a. | Fifth day of the moon. Unpropitious.

[Sixth and seventh days omitted.]

Eighth day of the moon. Unpropitious.

Ninth day of the moon. The morning of this day [is propitious].

Eleventh day of the moon. Unpropitious.

Twelfth day of the moon. It must not be.

Thirteenth day of the moon. The whole day [is propitious].

Fourteenth day of the moon. Unpropitious.

Fifteenth day of the moon. Not good.

Sixteenth day of the moon. The morning of the day [is propitious].

Seventeenth day of the moon. The whole day [is propitious].

Eighteenth day of the moon. [Set out] at the turn of the day.

Nineteenth day of the moon. The whole day [is propitious].

Twentieth day of the moon. The whole day [is propitious].

Page 474 | Twenty-first day of the moon. The morning of the day [is propitious].

Twenty-second day of the moon. The whole day [is propitious].

Twenty-third day of the moon. Unpropitious.

Twenty-fourth day of the moon. [Set out] at the turn of the day.

Twenty-fifth day of the moon. [Set out] at the turn of the day.

Twenty-sixth day of the moon. [Set out] at the ninth hour.

Twenty-seventh day of the moon. The whole day [is propitious].

Twenty-eighth day of the moon. The whole day [is propitious].

Twenty-ninth day of the moon. The whole day [is propitious].

Thirtieth day of the moon. The whole day [is propitious].

And the evil (*i.e.*, unlucky) days of the months.
Nîsân. The third day.
Iyyâr. The sixth and twentieth days.
Khazîrân. The third and eighteenth days.
Tammûz. The sixth and twentieth days.
Âbh. The first, fourth, and fifteenth days.
Îlûl. The third, tenth, and twentieth days.
First Teshrî. The third, sixth, and twentieth days.
Second Teshrî. The third, fifth, and eleventh days.
First Kânôn. The third and twentieth days.
Second Kânôn. The second, third, eleventh, and fourteenth days.
Shebâṭ. The seventh, eleventh, twentieth, and twenty-first days.
Âdhâr. The fourth, fifth, twentieth, and twenty-first days.

If a man falleth sick on any one of these days, he will not live. And if a man be born on any one of them, he will not live. And if he maketh a feast on any one of them, he will have no pleasure therein. And if he setteth out on a journey on any one of them, or if he goeth forth to war, he will be conquered by his enemies. And if he planteth a vineyard on any one of them, it will not thrive. And if he goeth on a journey, evil will befall him. And if he getteth married, he will be destroyed, for these days are evil and are lacking [in good], even as the Living God commanded the children of Israel by the hands of Moses. Understand, then, and take good heed to thyself, even as God counselled the ancients. Nevertheless, whatsoever God wisheth that doth He do.

Another. The good (*i.e.*, lucky) days of the months.
Observe on what day the first day of the month beginneth.

First day. Propitious for everything.

Second day. Propitious for everything.

Fol. 226 b. Third day. Evil and hard. Beware of the | governor.

Fourth day. Bad. On it a man goeth not forth on a journey.

Fifth day. Like the preceding.

Sixth day. Good for betrothal to a wife.

Seventh day. Good. A day of joy and gladness. Do whatsoever thou pleasest therein. Ask the governor and all men for what thou requirest, for God, in His goodness, will multiply thy works.

Page 475 | Eighth day. Like the preceding, but unpropitious for travelling.

Ninth day. Very good.

Tenth day. Like the preceding.

Eleventh day. Like the preceding, and good, and beautiful.

Twelfth day. Like the preceding.

Thirteenth day. Take heed to ask nothing of any man, for this day is evil.

Fourteenth day. Good. He who is born on this day becometh prosperous, and he will possess instruction (or, doctrine).

Fifteenth day. Good. He who is born on this day will become dumb, and restricted in speech (literally, guarded of tongue).

Sixteenth day. Good.

Seventeenth day. Good.

Eighteenth day. Good, but useless for travelling: set not out on a journey.

Nineteenth day. Fair and good.

Twentieth day. Good for those who travel on roads.

Twenty-first day. Bad and cursed.

Twenty-second day. Good.

Twenty-third day. Good.

Twenty-fourth day. Bad, for on this day Pharaoh was born. He who is born on this day will live life but be slain.

Twenty-fifth day. Dark, and there is no light in it. He who falleth sick on this day will die.

Twenty-sixth day. Good for him that goeth on a journey.

Twenty-seventh day. Good.

Twenty-eighth day. Good.

Twenty-ninth day. Good.

Thirtieth day. Good for everything.

A forecast concerning the chances of the road.

If thou goest forth from thy house, and a pig meeteth thee, and he cometh towards thee, it is a good sign.

And if a dog meeteth thee, and he cometh after thee, and barketh at thee, thou shalt not set out on thy journey, for it is not a good sign.

And if it happeneth that thou seest a dog coming out of a house, and he is pleased to meet thee, go on thy journey, for there shall be advantage therein.

And if two men, or two married women, meet thee, go, for it is a good sign.

And if unmarried youths or maidens meet thee, go back, | Fol. 227 *a*. for there will be much weariness on the road.

And if camels meet thee, and they be loaded, and come towards thee, good filleth the road.

And if bulls meet thee, turn back, for there are wrath and vexation (or, trouble) on the road.

And if thou seest a man scattering water, and sitting down, go on thy way, for good filleth the road. | Page 476

And if thou seest a woman, or a maiden, casting forth filth, it is not a good sign.

And if thou goest forth, and seest a black man, it is not a good sign; thou shalt not go forth from thy house.

Of buying and selling. When thou wishest to buy anything, be wise in this manner. Let the thing which is bought be to thee the sun, and let him that selleth be to thee the moon, [in] the place of judgement, and let him that buyeth be he to whom the moon cleaveth. And if the seller doeth wickedness, the moon cometh, and he who selleth is the loser, and no good cometh to him. And if he doeth good, the operation bringeth profit to him that selleth, and the buyer himself is the loser.

Again, another set of forecasts concerning the days of the moon, which will inform thee on which days it is right to do work.

First day of the Moon. This day is good for business of all kinds, and for a man to write his testament (*i. e.*, make his will) in, and to send [his son] to school, and for a man to clarify (or, strain) wine, and to lay the foundation of a house, and to find what hath been lost, and for a sick man to find relief, and for the child who hath just been born to grow, and for a man to be a good father, and the dream which a man seeth on this day will be fulfilled in two days' time.

Second day of the Moon. This day is good for all those who do works of charity, and also for a man to approach great personages, and to put back the wine into the vat. And the slave who flieth to a great personage will be discovered. And the sickness of him that is sick will become severe. And he who is just born will grow up. And he who hath been sent on a journey will come [back] very slowly. And a dream dreamed on this day will be fulfilled in three days.

Third day of the Moon. Winds and whirlwinds (*i. e.*, gales Fol. 227 *b*. or storms) are produced | on this day. Keep thyself from trafficking of every kind, and, as far as it is possible, do nothing whatsoever, but take the greatest care of thy body and of thy house. And travel not, neither by sea nor by land. And do not borrow, and do not lend. And sow no seed of any kind, and give no judicial decision on this day. He who became a fugitive will be found, and the sickness of him that is sick will become severe. And he who is born on this day, if he liveth will become a fool, and the vision which a man seeth on this day is a lying one.

Page 477   **Fourth· day of the Moon.** This day is good | for buying and selling, and for a man to plough in, and to deal in merchandise in, and for those who build houses, and for those who seek to enter into judgement [with their adversaries]. He who hath been a fugitive, and he who hath been lost for a long time, will be found with a great personage. And the thief will be captured, and he who setteth out on a journey will lack a beast [of burden], and he who is sick will be made whole gradually. The tribulations of him that is born on this day shall be many, and he will not be a compassionate man, and his luck (or, destiny) is evil, and his dream will be that of a day only.

Fifth day of the Moon. Keep thyself from buying a slave, and, if thou dost buy one, he will run away and be no more found. Take heed that thou dost not lie, neither shalt thou take an oath, or make another to take an oath; and know that if thou doest these things thou wilt enter into tribulation. And the slave who is lost, and the slave who hath run away, will be found. And he who is born on this day will become a crafty man. And a dream dreamed on this day will be fulfilled in two days.

Sixth day of the Moon. This day is good for those who go a hunting, and for the feast chamber, and for selling property, and for buying. He who is born on this day will live, and he who is taken ill on this day will be healed. And the slave who runneth away, and the slave who is lost, will be found, and a dream dreamed on this day is the son of a day (*i. e.*, ephemeral).

Seventh day of the Moon. This day is good for everything whereto a man putteth his hands, and for planting groves and gardens, and for carrying some one to the house of books, and for buying a slave. And the slave who runneth away, and the slave who is lost, will be found, and he who is taken ill will live, and he who is born on this day will become a great man and a prince of shepherds, and a dream dreamed on this day is the son of its day (*i. e.*, ephemeral).

Eighth day of the Moon. On this day thou shalt not set out on a journey. It is a fair day for those who go out to war. He who is taken sick on this day | will suffer sorely, Fol. 228*a*. and he who is born on this day will be troubled, and a dream dreamed on this day will be fulfilled in two days.

Ninth day of the Moon. Whatsoever thou doest on this day will succeed, and profit will come unto thee therefrom. It is a lucky day for him that borroweth and for him that lendeth, and for him that planteth groves and gardens, and for him that standeth before the king. And the slave who is lost will be found. And he who is born on this day will live, and he will become a great benefactor. And he who falleth sick on this day will find relief, and a dream dreamed on this day is the son of a day.

Tenth day of the Moon. On this day keep thyself from everything. Buy not, and lend not, [if thou dost] they will not Page 478 repay thee. And he who is born on this day | in exile will live in a miserable condition. The slave who runneth away will come [back] in fetters, and a dream dreamed on this day will be fulfilled in six days.

Eleventh day of the Moon. This day is propitious for everything unto which thou shalt put thy hand, and for those who buy slaves, and for those who have cases in law (?), and for trafficking, and for planting groves and gardens, and for gathering fruit, and for a man to set out on a journey, and for the man who is travelling, and for a man to be in the house of the Ruler. And it is propitious for making a feast, and for building buildings, and for a man to become a governor. The slave who runneth away will come [back] very soon. He who is born on this day will become great, and his luck will be good, and he will rejoice. And a dream dreamed on this day will be assuredly fulfilled in four days.

Twelfth day of the Moon. Keep thyself from strife. And shave not thy head and beard, and thou shalt not wash. It is a good day for those who gather grapes in their vineyards, and who put back the wine into the vats. The slave who runneth away will be lost. He who is taken ill on this day will be sick for a long time, but he will live. He who is born on this day will have many illnesses, and he will be a pugnacious man. And a dream dreamed on this day is the son of a day.

Thirteenth day of the Moon. This day is propitious for everything which thou doest, and for making a feast, and for thy going on a journey, and for trafficking, and for meeting and entertaining friends. The slave who runneth away, and the slave who is lost, will be found. And he who falleth sick on this day will be healed. And he who is born on this day will live, and will have good luck. And a dream dreamed on this day will come true within fifteen days.

Fourteenth day of the Moon. Do not go to an election, and do not buy an animal. He who falleth sick on this day will find relief. And the slave who is lost will remain concealed, and he who is born | on this day will be held in contempt. Fol. 228b. And a dream dreamed on this day will come true within six days.

Fifteenth day of the Moon. This day is propitious for laying foundations, and for sowing seed. Take no oath, nor make any one take an oath, therein. The slave who is lost and the slave who runneth away will be found in another district. He who is born on this day will prosper. He who falleth sick on this day will remain ill for a long time. And a dream dreamed on this day will be proved to be untrue within two days.

Sixteenth day of the Moon. It is propitious for everything. It is good for a man going out to hunt and to war, for gradually he will find [prey or, booty]. The slave who is lost will be found within two days. He who falleth sick on this day will obtain relief. He who is born on this day will grow up, and a dream dreamed on this day will come to pass within three days.

Seventeenth | day of the Moon. Keep thyself from trafficking, | and from the swearing of oaths. Shave not thy head and Page 479 beard. The slave who is lost will not be found. He who falleth sick on this day will rise up in a miserable condition. He who is born on this day will be a wise man. And a dream dreamed on this day will be the son of a day, but good.

Eighteenth day of the Moon. This day is good for everything that thou doest, for buying and selling, and for setting out on a journey, and for companionship [with women?], and for beginning to sow crops, and for gathering in fruit, and for planting groves and gardens. The slave who is lost, and the slave who runneth away will be found quickly. He who is born on this day will grow up, and a dream dreamed on this day will be the son of a day, but good.

Nineteenth day of the Moon. On this day it is well for thee to lay foundations, and to sow seed, and to gather in the fruit crops, and to hold [one's] position. The slave who runneth away, and the slave who is lost will be found. He who is born on this day will become a chief. He who is taken ill on this

36*

day will suffer severely. And a dream dreamed on this day will come true within twenty days.

Twentieth day of the Moon. From thy house go not forth, and do no work of any kind. The slave who runneth away, and the slave who is lost will not be found. And he who is born on this day will suffer from many illnesses. And the dream which a man shall dream on this day shall be a lying dream.

Twenty-first day of the Moon. This day is good for hunting. And it is good for a man to keep himself from everything, and from the companionship [of women?], and from buying and selling. He who being bound runneth away will be found, and he who falleth sick on this day will be in tribulation, and will find relief. And the sicknesses of him that is born on this day will be many, and a dream dreamed on this day will come true within two days.

Twenty-second day of the Moon. He who is born on this
Fol. 229a. day will live a long time, and will become | a liar. He who falleth sick on this day will die. On this day it is good to guide the plough, and to marry a wife, and he who is born thereon will die through the blow of an iron weapon. And the dreams of this day are many and seductive.

Twenty-third day of the Moon. It is good to sell and to buy on this day, and to make a feast, and to enjoy women (?), and to hunt. The slave who is lost, and the slave who runneth away will be found. Him who is born on this day the wild beast will eat, and the dream dreamed thereon is a lying dream.

Twenty-fourth day of the Moon. This day is good for those who plough, and for those who traffic therein, and he who falleth sick thereon will be ill for a [long] time. And he who is born on this day, with his life is his power, and he is
Page 480 nigh unto | his end. And the dream dreamed on this day is a lying dream.

Twenty-fifth day of the Moon. Keep thyself from everything which thou wishest to do. He who is born on this day will not live. The slave who runneth away, and the slave who is lost will not be found, and he who is born on this day will die [at once]. It is a day good for the chase. And the dream which is dreamed on this day will become true on the second day.

Twenty-sixth day of the Moon. Traffic in every kind of merchandise. Do not shave thy head and beard. Every one who borroweth from thee will not consume what he borroweth, for with praise he will pay thee back. And the day is propitious for those who make a feast, and if thou plantest a vineyard it will thrive, and whatsoever thou doest therein will be excellent. He who is born on this day will live, and the slave who runneth away, and the slave who is lost will be found. He who falleth sick on this day will find relief. And a dream dreamed on this day is the son of a day, but good.

Twenty-seventh day of the Moon. This day is propitious for trafficking, and for going about from one place to another, and for weaning children. He who is born on this day will live, and he who is sick will be healed. The slave who is about to run away will be observed. And a dream dreamed on this day will come true within . . . . . days.

Twenty-eighth day of the Moon. This day is propitious for planting groves and gardens, and for feasting, and for everything. The illnesses of him that is born on this day will be many. The slave who runneth away, and the slave who is lost will remain hidden. And a dream dreamed on this day is the son of a day and bad.

Twenty-ninth day of the Moon. This day is propitious for buying and selling, and for going on an embassy, and for receiving counsel. He who is born on this day will only live with the greatest difficulty. The slave who is lost, and the slave who runneth away will not be found. He who falleth sick on this day will find relief, and he who is born on this day will live and become a rich man; but for feasting the day is unsuitable. A dream dreamed on this day will come true within seven or fourteen days.

Thirtieth day of the Moon. This day is good for everything, and for taking and giving. | He who falleth sick on this day will Fol. 229*b*. live. The slave who is lost, and the slave who runneth away will be found. He who is born on this day will become rich, and he will be compassionate to the children of men, by the will of God. And the dreams dreamed on this day are bad. Keep thyself from them.

HERE END THE FORECASTS, AND TO GOD BE PRAISE!

From the Discourse of Basil. Concerning the course of the circle of heaven.

Since thou hast asked me, O lover of learning, of what kind is the course of the heavens, according to my ability I will tell thee. It is meet to know that heaven is round on all **Page 481** its sides, and that it travelleth continually and unceasingly | from east to west. And the earth is situated in the middle of it like an egg. And according to the measure of the distance of heaven from the earth, so is it remote (*i.e.*, distant) from all the sides of the earth. The Circle of the Malwâshê (*i. e.*, Signs of the Zodiac), that is to say, the months of the year, embraceth the length of the sphere of heaven obliquely, one half of it being on the south side, in such a position that Nîsân and the First Teshrî ascend from the east, opposite to the earth continually; and the rest of the winter months go up from the south to the earth, that is to say, the First and the Second Teshrî, and the First and the Second Kânôn, and Shĕbât and Âdhâr; and the summer months ascend to all the earth, that is to say, Nîsân, and Îyyâr, and Khazîrân, and Tammûz, and Âbh, and Îlûl. And each one of those months is divided into three parts, and the whole of heaven is divided into three hundred and sixty and five parts, and each part is one day. And one of these parts travelleth, more or less, each day, from west to east, so that the circle performeth its [complete] course of travelling in three hundred and sixty and five days.

Now, the Sun and the Six Stars (*i. e.*, Planets) move through these Signs of the Zodiac from west to east, and the Sun, **Fol. 230a.** through the rapidity of his course, changeth from the east | each day into the west. And these Seven Stars are not situated equally in one place so that their motions (or, courses) can be equal. Kêwân, which is called Kronos, is placed in the outermost part of the sphere, and he fulfilleth his course in three years; [this time is necessary] for him to arrive at the part [of the sphere] wherefrom he went forth, and the region wherein he dwelleth is watery. The star of Bêl (Jupiter), which is called Zeus, is placed in a region of water, and in the second circle of the sphere, and he fulfilleth his course in twelve years. And below him is placed Âris, in a region of fire, and in the third circle of the sphere, and he fulfilleth his course in one year and a half. And below him is placed the Sun, in a region | of fire, and in the fourth circle of the sphere, **Page 482** and he fulfilleth his course in one year. And below this is placed the star of Baltî, which is called Aphrodite, in a region of wind, and in the fifth circle of the sphere; and she fulfilleth her course in one year less two months. And below her is placed [the star of Hermes, who is called "Ûtrâdh", in a [region of water, and in the sixth circle of the sphere, and he [fulfilleth his course in six years. And after him is placed the [Moon],[1] in a region of wind, and in the seventh circle of the sphere, and he fulfilleth his course in twenty-nine days and a half. For ..... the Moon, they are composed of these four elements, earth, water, fire, and air. For one half of heaven is above the earth, and one half is below it; and heaven bringeth back the luminaries each day. During the night the sun dwelleth below the earth, on the northern side, and during the day, above the earth, on the southern side. And because the sun entereth in among the Signs of the Zodiac that are remote from the earth, towards the south, it becometh cold in the winter; and because he returneth to the Signs of the Zodiac that are near the earth in the summer, it becometh hot, and the earth burneth. And the earth itself burneth, because on two days of the year, | at the rising of Tammûz, **Fol. 230b.** [the sun] is [directly] above the earth, and for this reason the children of men cannot dwell in that region. And if the sun [always] travelled directly above the earth, it would burn it all up, because of its great size—now the sun is eighteen times as large as the earth,—and because of the fire thereof which is unquenchable.

Now there are times when the sun and the moon appear to us, as they say, to "become dark", and the reason of this is, according to what those who are skilled [in such matters] say, as follows:—The moon shineth [with light] from the sun, and

---

[1] Restored from the text on pp. 470, 471.

giveth us light. And when these two luminaries happen to be directly opposite to each other, or, as they say, when they balance each other, one on this side of the earth, and one on that, the earth preventeth the [light of the] sun from passing
Page 483 beneath the rays of light that arise from the moon, | because it is impossible for the moon to become dark except when it is fifteen days old, according to the ancient custom which is thus, and the sun also cannot become dark except after the moon is born. Now they say that the moon becometh dark twice in the year; when it becometh dark, being above the earth, we can see it, and when it becometh dark, being below the earth [we cannot see it].

The names of the Malwâshê (*i. e.*, Signs of the Zodiac].

1. Nîsàn = Ram	7. Teshrî, First = Balance
2. Îyyâr = Bull	(Kĕnâi Shêlâmà)
3. Khazîrân = Twins	8. Teshrî, Second = Scorpion
4. Tammûz = Crab	9. Kânôn, First = Bowman
5. Âbh = Lion	10. Kânôn, Second = Goat
6. Îlûl = Virgin	11. Shĕbâṭ = Water-carrier

12. Âdhâr = Fishes

These are the forms that are depicted in the heavens in the Malwâshê. Besides these there are Orion, who is called the "Giant", and who is situated to the south of the crown of the Malwâshê, and the "dog of the Giant" and the "cross of doctrine (?)", and they are situated to the north of the crown of the Malwâshê.

And the seven days of the week are named after these seven luminaries.

Fol. 231a. | The first day of the week is named after the sun.
The second day of the week is named after the moon.
The third day of the week is named after Âris.
The fourth day of the week is named after Hermes.
The fifth day of the week is named after Zeus.
The Eve of the Sabbath is named after Aphrodite.
The Sabbath is named after Kronos.
To each one of these luminaries the first hour of the day named after him belongeth, and after it cometh that of its

neighbour, and thus, little by little, all the hours of the night-time and day-time throughout the entire year are distributed among these luminaries. Now the hours in number amount to eight thousand, seven hundred, and sixty and six.

Of eclipses. A portent which is indicated by the upper part of the Zodiac and the lower part thereof, and the government of the world.

This is the eclipse which maketh the moon to become dark, when it is situated a long way from the sun in the seventh *bûrghâ*, and this is the cause thereof. When the head of the Serpent, or his tail, is in the seventh *bûrghâ* of the sun, and the moon is in the *bûrghâ* wherein is the sun, there taketh place | an eclipse of the moon on the fourteenth day of the Page 484 moon. As concerning the signs which take place therein when this happeneth, observe the colour of the moon. If it be black, there will be gloom, and manifold winds in the air, and sicknesses, and the flocks will die, and exceedingly great fear will fall upon the children of men. And if its colour be black, with redness therein, there will fall upon the children of men pestilence, and slaughter, and the sickness that is caused by the decay of unburied bodies, and famine. And if its colour be yellow, there will be disease of the liver, and the fruits of the trees and the crops will perish, and birds and animals will die. And if its colour be like unto that of dust, in that year there will take place want, and snow, and ice, and frost, and the animals (or, cattle) will die. According to the position of the head of the Serpent there will be good, and according to the position of its tail there will be evil.

And when the eclipse taketh place near the middle of the heavens, above the earth, [it portendeth] . . . . . of temples and of kings . . . . and of blood. And when it taketh place towards the east, | [it portendeth the destruction] of fruit, and Fol.231*b*[1].

---

[1] On the margin is written:—Now Athêlyâ remaineth in each *bûrghâ* seventeen days and six hours, and the total of these is two hundred and seven days, for the twelve Houses make this amount (*i. e.*, 17 1/4 days × 12 1/4 days = 207 days).

of children, and of foundations (*sic!*). And when it taketh
place towards the west, [it portendeth the . . . .] of old men
and of the dead. And when the whole face of the moon
becometh dark, [it portendeth] that what is to come will concern the whole world. And if only a part of the face of the
moon becometh dark, the things which take place will concern
only the part of the world that correspondeth to the part of
the moon that becometh dark. If there be in the moon marks
which are yellow, or black, or of a bright colour, they are due
to the nature of Kêwân, that is to say, Kronos. And if they
be white in colour they are due to the nature of Zeus. And
if they be red in colour they are due to the nature of Ârîs.
And if they be of the colour of wax, they are due to the
nature of Aphrodite. And if they be of a variegated colour
they are due to the nature of Hermes. And as they are, so
is each in his region.

Page 485   **Of Eclipse | of the sun and moon.** If it taketh place in
the Balance, or the Water-carrier, or the Twins, it portendeth
pestilence among men throughout the world. If the eclipse
taketh place in the Rain, or in the Lion, or in Sagittarius, it
will produce far more serious evils, for it portendeth famine,
and wars throughout the world, and the uprooting of cities, and
pestilence among animals, and cattle, and birds, and especially
among camels and flocks of sheep and goats. It produceth
want and scarcity, and especially in the climes [or, regions]
that are near these Signs of the Zodiac. If the eclipse taketh
place in the Scorpion, or in the Fishes, or in the Crab, it will
produce want of water and of everything among the peoples
of the climes that are nigh unto these Signs of the Zodiac.
And men will flee from place to place through want of water
and sustenance, and there will also be scarcity of fish and of
everything which liveth in [or through] moisture. And if the
eclipse taketh place in the Virgin, or in Capricorn, or in the
Bull, there will be scarcity among men of fruit, and of sheep,
and of all kinds of crops. And the earth itself will suffer want,
for fruits and roots will not grow, and there will be no rain,
especially in the climes that are nigh unto these Signs of the
Zodiac.

Of the eclipse, that thou mayest know on what day the moon | will become dark. The eclipse when it taketh place Fol. 232a. on the two hundred and seventh day. Reckon from the day wherein the eclipse took place, and where the two hundred and seven days agree. Direct thy attention, whilst thou art enquiring into the matter, to the fifteenth day of the moon, that is to say, to the day when the moon is full, and to what portion of the upper part of the Zodiac the eclipse can take place. Now, if it be more than twelve parts distant, the eclipse cannot take place. Moreover, observe in what hour the fifteenth day cometh to an end, and the sixteenth day beginneth. And let this be kept by thee [in thy mind] with very great care, that is to say, if the fifteenth day cometh to an end in the day-time, the eclipse is not visible, because it taketh place below the earth, but if the fifteenth day cometh to an end in the | night-time, and beginneth in the sixteenth day, know that Page 486 the eclipse is taking place above the earth, and it will therefore be plainly visible.

If thou wishest to know the House of the upper and lower parts of the Zodiac, divide the years of Alexander one thousand eight hundred and thirty and five, and of those which remain to thee exclude (i. e., divide) them by two hundred and forty and eight.[1] Now, I myself divided them all according to the written instructions, and there remained from two thousand twenty, one hundred and eighty and five. Then these which remain over to thee double twelve times, and see to how many they amount, and then add to them one-third of the number which is over. And when thou hast counted it out (i. e., divided) by two hundred and forty and eight, see unto how much all the numbers amount. Again, take all the days, from the first day of Îlûl to the day that thou wishest to take, and double (i. e., multiply) them three times, and afterwards add to them one-sixth of their number, and then divide the whole number which is known unto thee by three hundred and sixty, and of

---

[1] On the margin of this page is written:—When thou dividest this number by two hundred and forty and eight, from five hundred take four, and from one thousand eight.

Fol. 232b. that which remaineth over to thee, and which doth not | reach three hundred and sixty, assign to each House thirty,[1] beginning to count from the Balance. And where the number endeth, and thou canst not divide it by thirty, there is the upper part of the Zodiac;[2] and count seven Houses, and there is the lower part of the Zodiac.[3]

That thou mayest know the five stars (*i. e.*, planets), and how to make the calculation concerning them. Kronos, that is Zûkhâl. Take the years of Alexander the Greek at the time when thou makest thy enquiry, however many they may be, and divide them by three, though some say divide them by twenty, and the number which remaineth divide into thirties. And of the number which remaineth and which doth not

Page 487 amount to thirty, allot to each House | two and a half. And [begin] to count with the Water-carrier, and, where thy number stoppeth, there is Kronos.

That thou mayest know where Zeus, that is to say, Mûshtarî, standeth. Take the number of years of Alexander, at the time when thou makest thy enquiry, and add to them five, though some say add twenty and one, and divide them by thirty. And of the number that remaineth in thy hand, give to each House one. Then begin to count with the House of the Fishes, and, where thy number stoppeth, there is Zeus.

That thou mayest know where is Ârîs, that is to say, Marîk. Take the number of the years of Alexander, at the time when thou makest thy enquiry, divide them by eight, though some say by three, and divide what remaineth into thirties. And of the number that remaineth to thee, and that doth not make up thirty, give to each House one. Then count from the *Bûrghâ* (*i.e.* House) of the Ram, and, where thy number cometh to an end, there is Ârîs.

That thou mayest know where is Hermes, that is to say, 'Ûtrâdh. Take the number of the years of Alexander, at the

---

[1] On the margin of this page is written:—When thou dividest the number by thirty, thou shalt take from one hundred ten, and from one thousand ten.

[2] *Ânâbîbâzôn*, that is, the head of the serpent, and its form is ☊.

[3] *Kâtâbîbâzôn*, that is, the tail of the serpent, and its form is ☋.

time when thou makest thy enquiry, and divide them by fifteen, and the number that remaineth to thee divide into thirties. And of what remaineth over to thee, give to each House one. Then begin to count with the Two Images, and, where thy number cometh to an end, there is Hermes.

That thou mayest know where is | Aphrodite, that is to Fol. 233 *a*. **say, Zûhrah.** Take the number of the years of Alexander, at the time when thou makest thy enquiry, and divide them by thirty. And of what remaineth unto thee, give to each House one. Begin to count with the Bull, and, where thy number cometh to an end, there is Aphrodite.

The number (*i. e.*, a list) of the Malwâshê.

1. Ram.
2. Bull.
3. Twins.
4. Crab.
5. Lion.
6. Virgo.

7. Balance.
8. Scorpion.
9. Bowman.
10. Goat.
11. Water-carrier.
12. Fishes.

The Characteristics of the Seven Zones.   Page 488

The zone of **Kronos**, the first and the highest, is cold, and it is situated in ice. The planet **Zeus** is in the second zone, wherein there is a moderate temperature, which is nourishing and refreshing. **Ârîs** is in the third zone, where there is the heat that destroyeth. The **Sun** is in the fourth zone, which is the middle zone, and it possesseth the aether, which is warm and refreshing. **Balti** is in the fifth zone, and she is the lady of sowing, and begetting, and the bringing forth of children, and of everything. **Hermes** is in the sixth zone . . . . The **Moon** is in the seventh zone. Now the moon itself is cold and moist, and it is near the earth, and is strong (?) and beautiful (?), when it is remote from the stars, and it waxeth and waneth, and is a symbol of the lives of the children of men.

Again, the characteristics of the Malwâshê.

Kronos is the lord of evil dispositions, and he ruleth Capricorn and the Water-carrier; he is never good.

Zeus is red, and a little bald (?), and he ruleth the Fishes and the Bowman; he is beneficent in everything.

Âris is little, and is a lover of blood, and he ruleth over the Ram and the Scorpion; he is the destroyer of everything.

The **Sun** is red and is long, and he ruleth over the Lion; he is beneficent in everything.

Aphrodite is beautiful in forms, according to her changes. Fol. 233*b*. She ruleth the Bull and the Reed, | that is, the Balance; she maketh [men] to commit follies, but she destroyeth nothing. She is predisposed to adultery and harlotry.

Hermes is swarthy, but beautiful in form, and he ruleth the Twins and the Virgin; [he doeth] neither good deeds nor bad, but he is predisposed to trafficking and to learning. Bind him not to keep a secret, for assuredly he will reveal it. He causeth commotion (or, tumult) in the world.

The **Moon** is bald, and ruleth over the Crab, and over every change.

The appearances (or, similitudes) of the Planets. The Sun Page 489 is like | gold, and the Moon like silver, and Âris like iron, and Aphrodite like brass, and Kêwân like lead, and Hermes like electrum, that is, stamped gold, and Zeus like tin.

The colours of the Planets.

Kronos is very dark, a deep black. **Zeus** is dark, like a good beryl. **Âris** is like a bad agate, a poor red. The **Sun** is of a dull colour, a moderate yellow. **Aphrodite** is white, a good green (*sic!*). **Hermes** is of a greenish colour, like the emerald. . . The **Moon** is white, a moderate blue (*sic!*).

Again, the courses of the Twelve Malwâshê and their changes; here is a list of them.

The **Sun** entereth the Ram on the fourteenth day of Âdhâr, and the Ram riseth with the sun in the middle of the east, ascending in a straight line, and the Balance beginneth to set opposite to him, in the middle of the west, in a straight line. And he riseth above the earth one measure of the six parts of the heavens, which constituteth two hours of the day-time. And at the beginning of the third hour the Bull beginneth to rise from the place where the Ram riseth, towards the north, the measure of one of the six parts of the heavens, and the Scorpion beginneth to set opposite to him, towards the south, where the Balance setteth, according to the measure mentioned [above]. And the Bull riseth up [a distance] of | one of the Fol. 234a. six parts of the heavens, and four hours of the day come to an end. And at the beginning of the fifth hour the Twins rise from the place where the Bull riseth, and towards the north, the same measure, and the Bowman beginneth to set opposite to them in the west, where the Scorpion setteth, and towards the south, and six hours of the day come to an end, which make one half of the day-time. The Ram standeth with the sun in mid-heaven, and the Balance is opposite to him in the middle of the earth. And at the beginning of the seventh hour the Crab riseth in the north-east obliquely, and the Goat beginneth to set | opposite to it in 'the south-west obliquely, Page 490 and eight hours of the day come to an end. And at the beginning of the ninth hour the Lion riseth from where the Twins rise, and the Water-carrier setteth opposite to him, where the Bowman setteth, and ten hours of the day come to an end. And at the beginning of the eleventh hour the Virgin riseth where the Bull riseth, and the Fishes set opposite to her, where the Scorpion setteth in the west, and twelve hours of the day, which form the six parts of the heavens, come to an end, and it becometh evening.

And at the beginning of the first hour of the night the Balance riseth, and the Ram setteth opposite to it, and two hours of the night come to an end. And at the beginning of the third hour of the night the Scorpion riseth, and the Bull setteth opposite to it, and four hours of the night come to an end. And at the beginning of the fifth hour of the night the Bowman riseth, and the Twins set opposite to him, [and six hours of the night come to an end], and it is midnight. The back part of the Bowman and the head of the Goat are in the east, and the back parts of the Twins, and the head of the Crab, are in the west. And the Balance is in mid-heaven, and the Ram is under the earth. And at the beginning of the seventh hour of the night the Goat riseth, and the Crab setteth

opposite to him, and eight | hours of the night come to an end. And at the beginning of the ninth hour of the night the Water-carrier riseth, and the Lion setteth opposite to him, and ten hours of the night come to an end. And at the beginning of the eleventh hour of the night the Fishes rise, and the Virgin setteth opposite to them, and twelve hours of the night come to an end. This is the equality of day-time and night-time on this day.

And know that from this point the Ram beginneth to move up by degrees, and to rise during the night in each day, one-thirtieth part of one-sixth of one Malwàshâ, until the end of the night; now the Ram setteth in the middle of Nîsân. And on the fourteenth day of Nîsân the Ram riseth with the sun, and the Scorpion setteth opposite to him, in the manner that hath been described. And here by [the word] "day", under-

stand midday, | and by [the word] "evening" understand midnight.

And on the fifteenth day of Îyyâr the Twins rise in the morning, and the Bowman setteth opposite to them in the evening, that is, in the west.

And on the sixteenth day of Khazîrân the Crab riseth in the morning, and the Goat setteth opposite to him in the evening.

And on the seventeenth day of Tammûz the Lion riseth in the morning, and the Water-carrier setteth opposite to him in the evening.

And on the seventeenth day of Âbh the Virgin riseth in the morning, and the Fishes set opposite to her in the evening.

And on the seventeenth day of Îlûl the Balance riseth in the morning, and the Ram setteth opposite to him in the evening.

And on the seventeenth day of the First Teshrî the Scorpion riseth in the morning, and the Bull setteth opposite to him in the evening.

And on the seventeenth day of the Second Teshrî the Bowman riseth in the morning, and the Twins set opposite to him in the evening.

And on the thirteenth day of the First Kânôn the Goat riseth in the morning, and the Crab setteth opposite to him in the evening.

And on the thirteenth day of the Second Kânôn the Water-carrier riseth in the morning, and the Lion setteth opposite to him in the evening.

And on the twelfth day of Shĕbâṭ the Fishes rise in the morning, and the Virgin setteth opposite to them in the evening.

And understand | that they all do thus, hour by hour, and Fol. 235 a. day by day, and month by month.

And behold, I will not write them down in a wheel (i. e., circle), in order that thou mayest know their places, as they are in the heavens. And we write down here each group of three Malwâshê together in one picture (i. e., diagram), according to their operation, and according to what the astronomers say; and they are divided into four quarters, corresponding to the Four Elements, and according to the Four Quarters of the heavens, as one might say:—The Twins, and the Balance, and the Water-carrier, these three are hot and moist, according to the nature of air, and according to the eastern Quarter of the heavens [wherein they are]. And the Bull, and the Virgin, and the Goat, these three are cold and dry, according to | the Page 492 nature of the earth, and according to the western Quarter of the heavens [wherein they are]. And the Crab, and the Scorpion, and the Fishes, these three are cold and moist, according to the nature of water, and according to the northern Quarter of the heavens [wherein they are]. And the Ram, and the Lion, and the Great Image (i. e., the Bowman), these three are hot and dry, according to the nature of fire, and according to the southern Quarter of the heavens [wherein they are].

Behold, these are arranged before thee, my lord, in tabular form, on that paper (i. e., page), in another kind of represent-ation, which cometh [before thee].

37

The Course of the Twelve Signs of the Zodiac in tabular form.

The Ram. I.

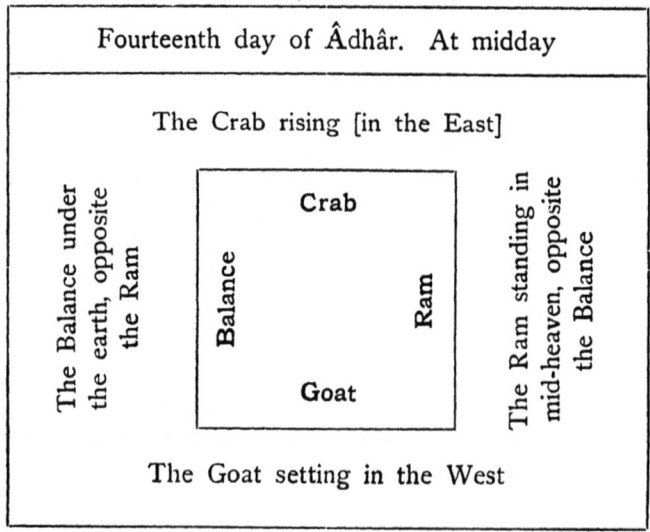

Page 494

Fourteenth day of Âdhâr. In the morning

: 21 : : 11 :

The Ram rising in the East

The Crab under the earth, opposite the Goat

The Goat in mid-heaven, opposite the Crab

Ram

Crab   Goat

Balance

The Balance setting in the West

The Ram. II.

Fourteenth day of Âdhâr. At midday

: 21 : : 11 :

The Crab rising [in the East]

The Balance under the earth, opposite the Ram

The Ram standing in mid-heaven, opposite the Balance

Crab

Balance   Ram

Goat

The Goat setting in the West

The Ram. III.

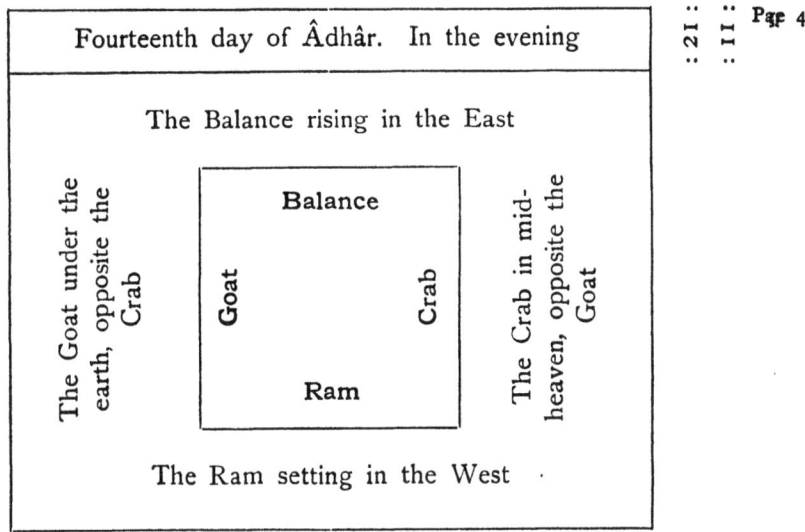

Fourteenth day of Âdhâr.   In the evening

The Balance rising in the East

The Goat under the earth, opposite the Crab

Balance

Goat

Crab

Ram

The Crab in mid-heaven, opposite the Goat

The Ram setting in the West

The Ram. IV.

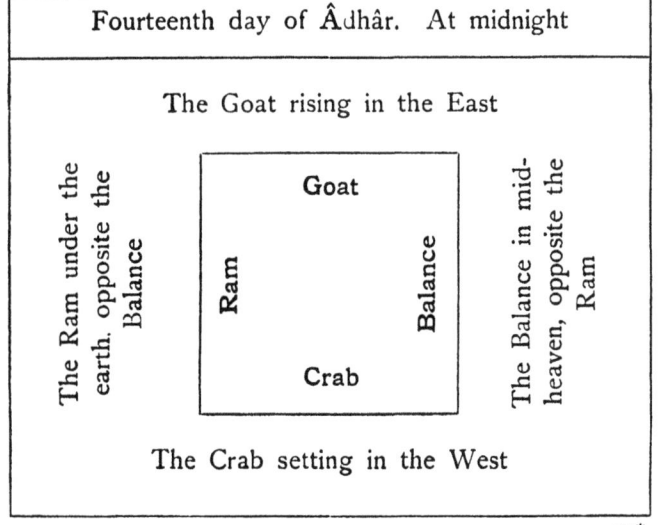

Fourteenth day of Âdhâr.   At midnight

The Goat rising in the East

The Ram under the earth, opposite the Balance

Goat

Ram

Balance

Crab

The Balance in mid-heaven, opposite the Ram

The Crab setting in the West

37*

The Bull. I.

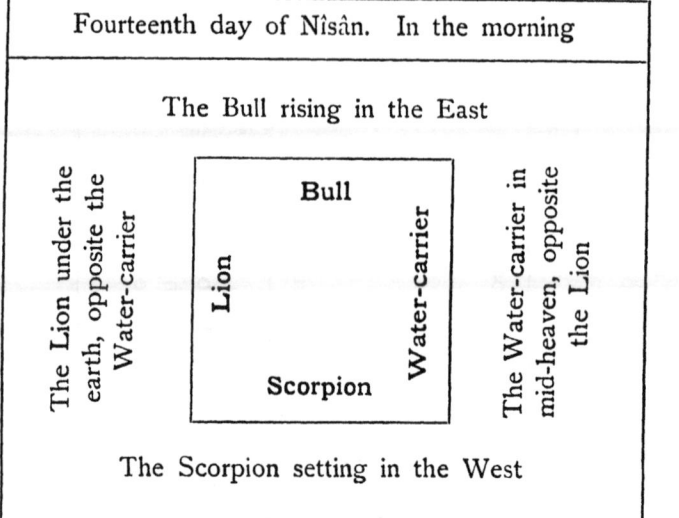

The Bull. II.

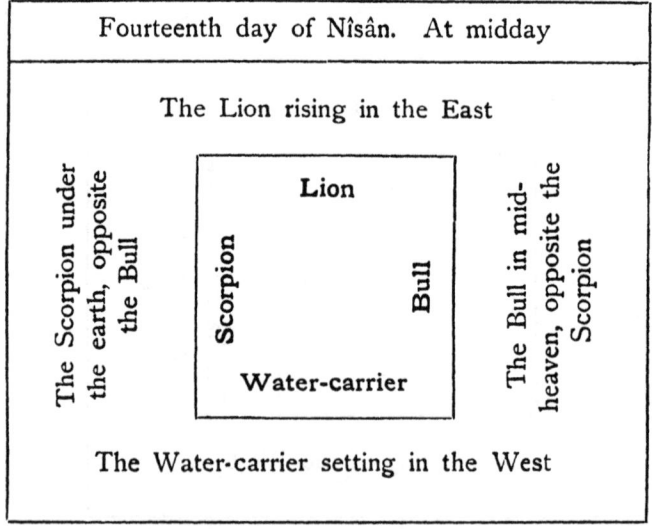

The Bull. III.

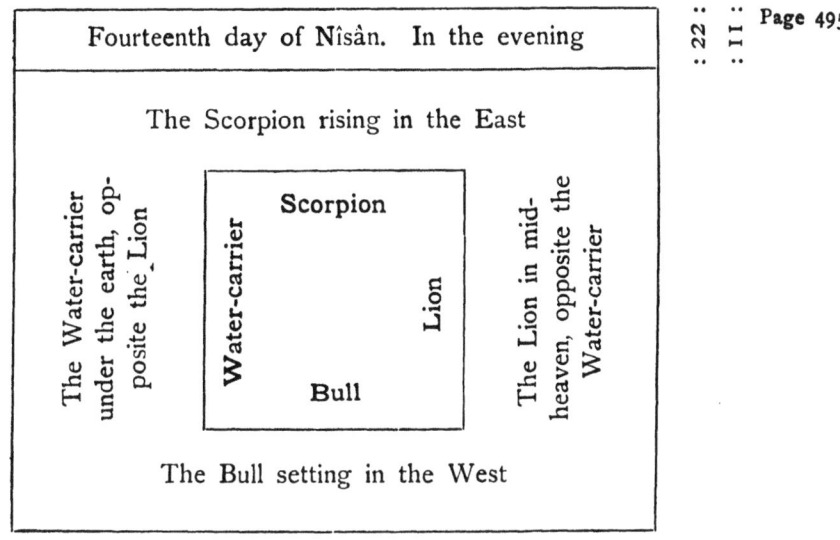

Fourteenth day of Nîsân. In the evening

: 22 : : 11 : Page 495

The Scorpion rising in the East

Scorpion

The Water-carrier under the earth, opposite the Lion

Water-carrier

Lion

The Lion in mid-heaven, opposite the Water-carrier

Bull

The Bull setting in the West

The Bull. IV.

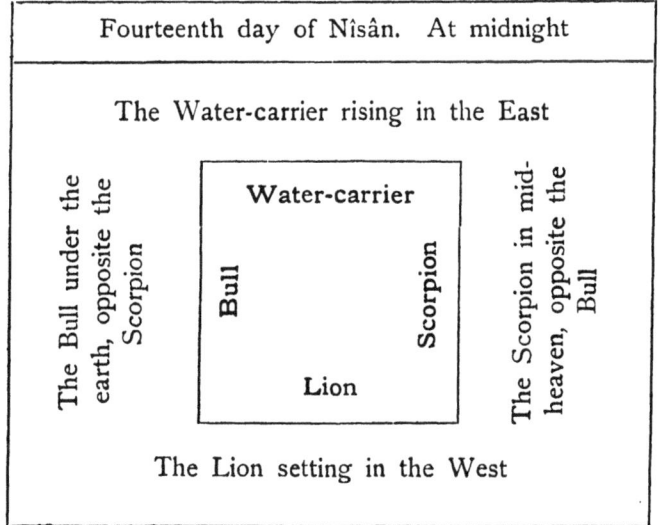

Fourteenth day of Nîsân. At midnight

: 22 : : 11 :

The Water-carrier rising in the East

Water-carrier

The Bull under the earth, opposite the Scorpion

Bull

Scorpion

The Scorpion in mid-heaven, opposite the Bull

Lion

The Lion setting in the West

The Twins. I.

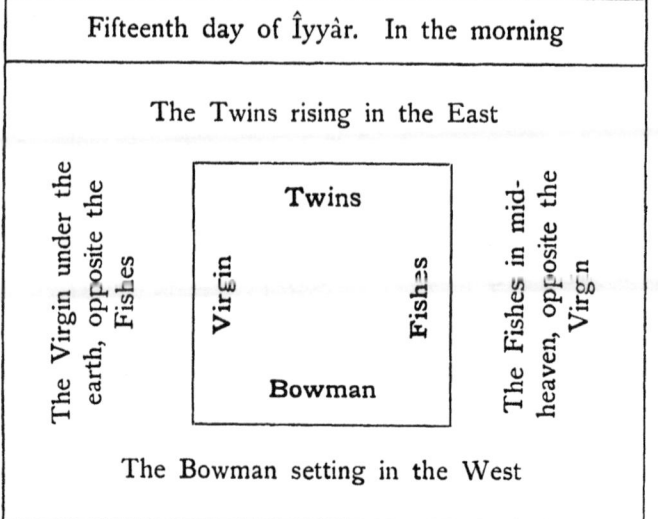

The Twins. II.

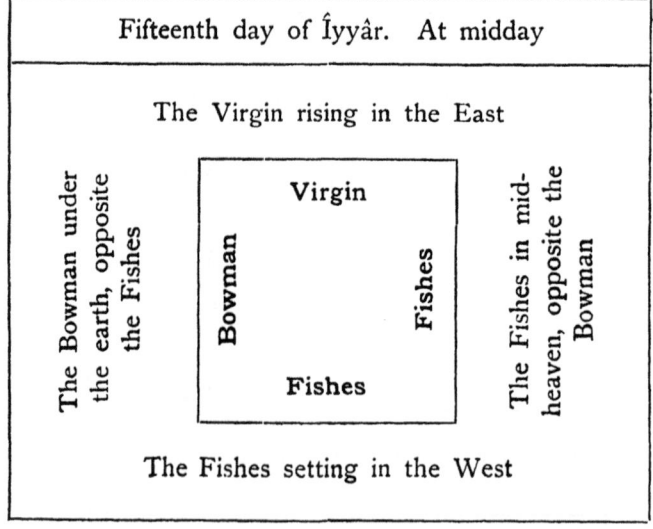

The Twins. III.

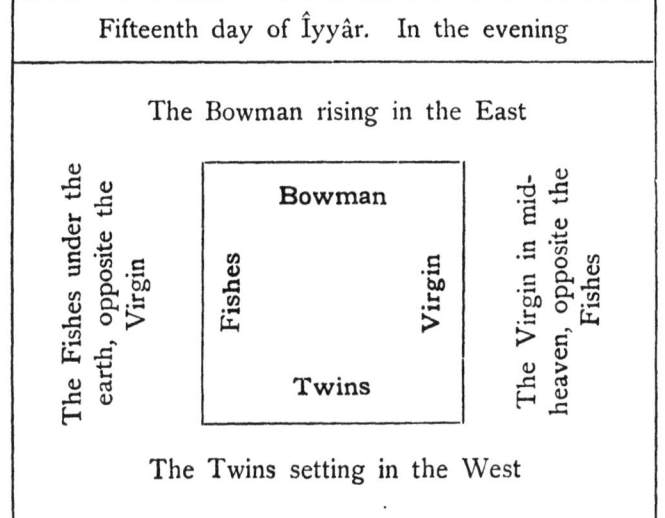

Fifteenth day of Îyyâr.   In the evening

The Bowman rising in the East

The Fishes under the earth, opposite the Virgin

Bowman

Fishes

Virgin

Twins

The Virgin in mid-heaven, opposite the Fishes

The Twins setting in the West

: 22 :   : 12 :   Page 495

The Twins. IV.

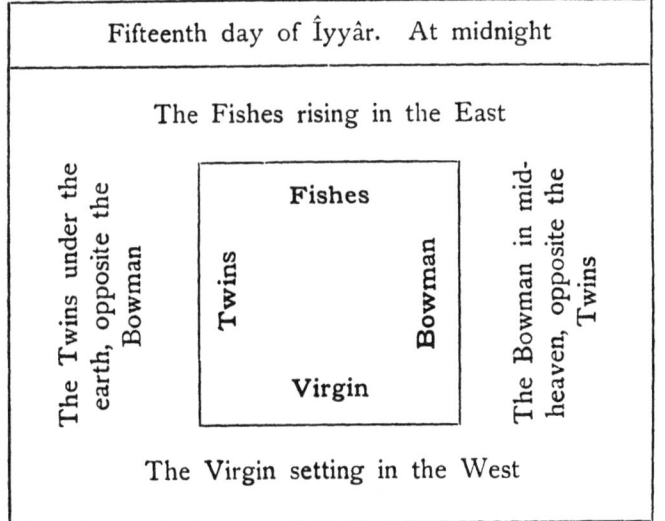

Fifteenth day of Îyyâr.   At midnight

The Fishes rising in the East

The Twins under the earth, opposite the Bowman

Fishes

Twins

Bowman

Virgin

The Bowman in mid-heaven, opposite the Twins

The Virgin setting in the West

: 22 :   : 12 :

The Crab. I.

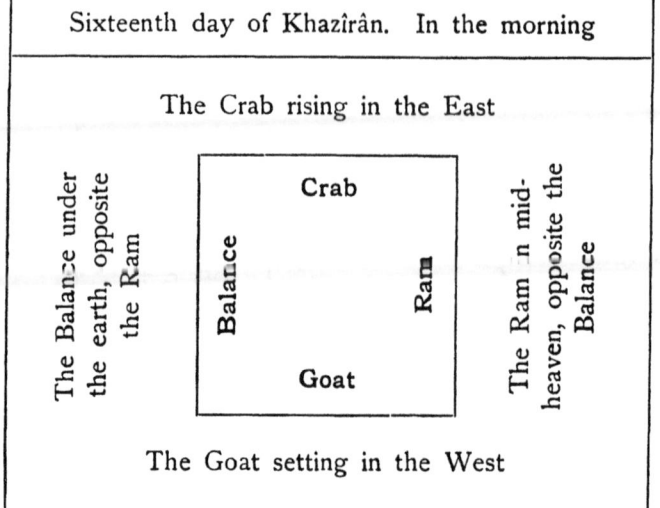

The Crab. II.

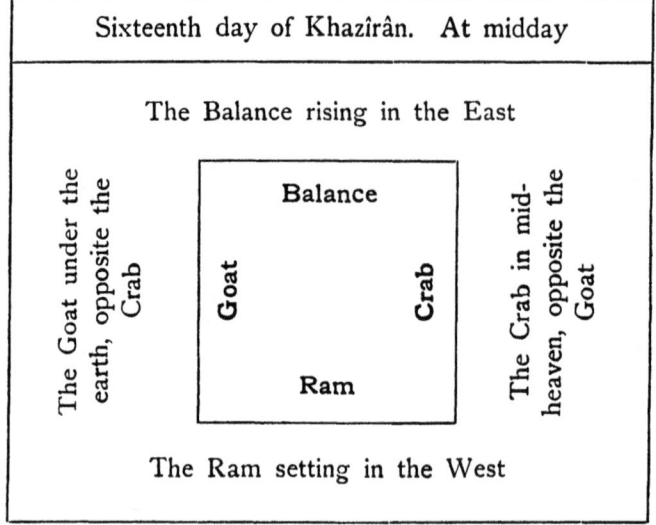

The Crab. III.

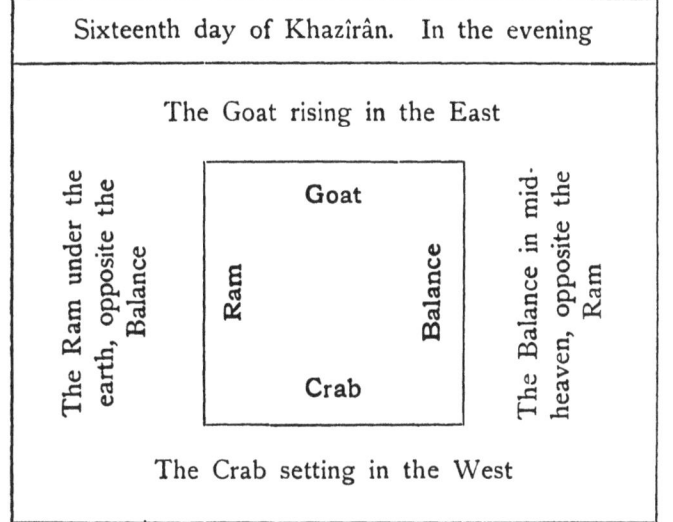

The Crab. IV.

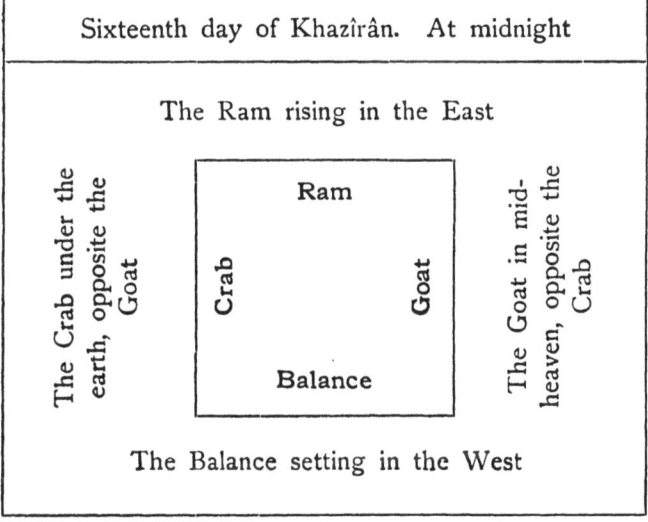

The Lion. I.

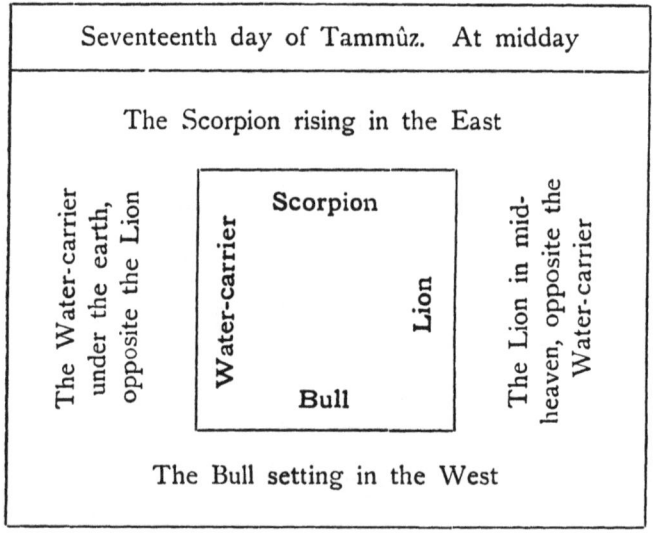

Page 496

Seventeenth day of Tammûz. In the morning   : 23 :   : 13 :

The Lion rising in the East

The Scorpion under the earth, opposite the Bull

Lion

Scorpion   Bull

Water-carrier

The Bull in mid-heaven, opposite the Scorpion

The Water-carrier setting in the West

The Lion. II.

Seventeenth day of Tammûz. At midday   : 23 :   : 13 :

The Scorpion rising in the East

The Water-carrier under the earth, opposite the Lion

Scorpion

Water-carrier   Lion

Bull

The Lion in mid-heaven, opposite the Water-carrier

The Bull setting in the West

The Lion. III.

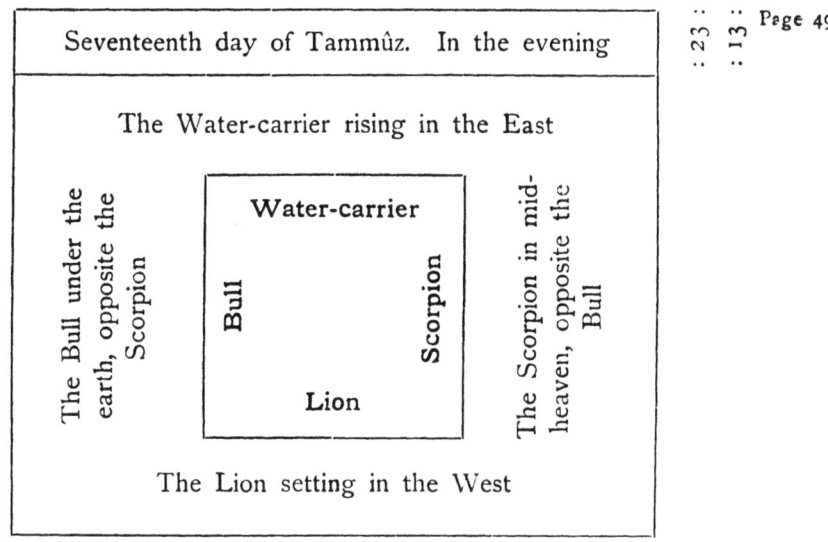

Seventeenth day of Tammûz. In the evening

23 :
: 13 : Page 497

The Water-carrier rising in the East

The Bull under the earth, opposite the Scorpion

Water-carrier

Bull

Scorpion

Lion

The Scorpion in mid-heaven, opposite the Bull

The Lion setting in the West

The Lion. IV.

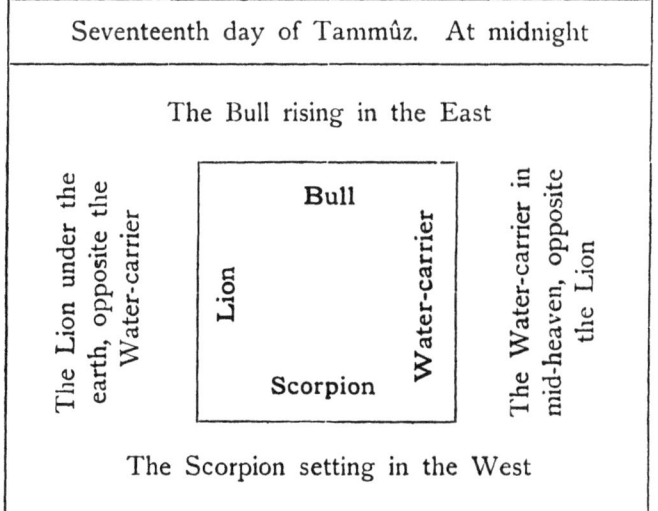

Seventeenth day of Tammûz. At midnight

23 :
: 13 :

The Bull rising in the East

The Lion under the earth, opposite the Water-carrier

Bull

Lion

Water-carrier

Scorpion

The Water-carrier in mid-heaven, opposite the Lion

The Scorpion setting in the West

The Virgin. I.

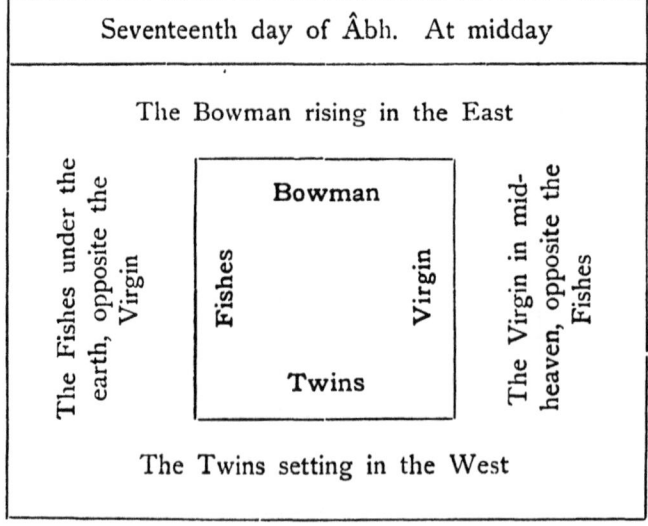

The Virgin. II.

The Virgin. III.

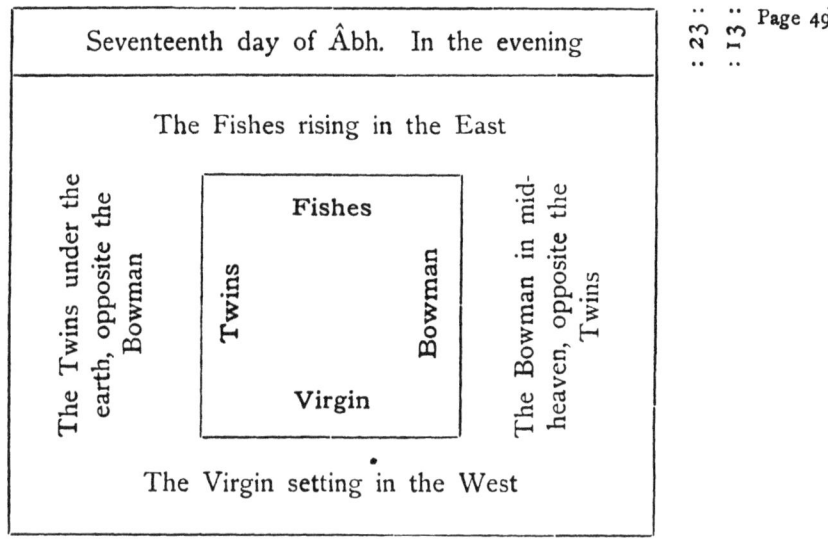

Seventeenth day of Âbh. In the evening

: 23 :
: 13 : Page 497

The Fishes rising in the East

The Twins under the earth, opposite the Bowman

Fishes

Twins

Bowman

The Bowman in mid-heaven, opposite the Twins

Virgin

The Virgin setting in the West

The Virgin. IV.

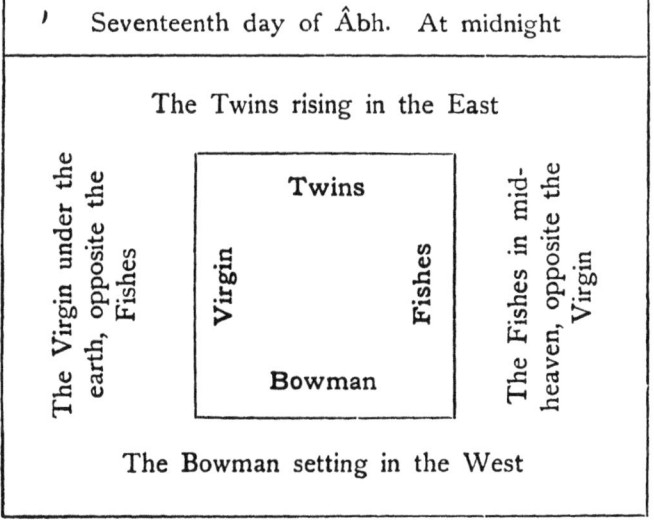

Seventeenth day of Âbh. At midnight

: 23 :
: 13 :

The Twins rising in the East

The Virgin under the earth, opposite the Fishes

Twins

Virgin

Fishes

The Fishes in mid-heaven, opposite the Virgin

Bowman

The Bowman setting in the West

The Balance. I.

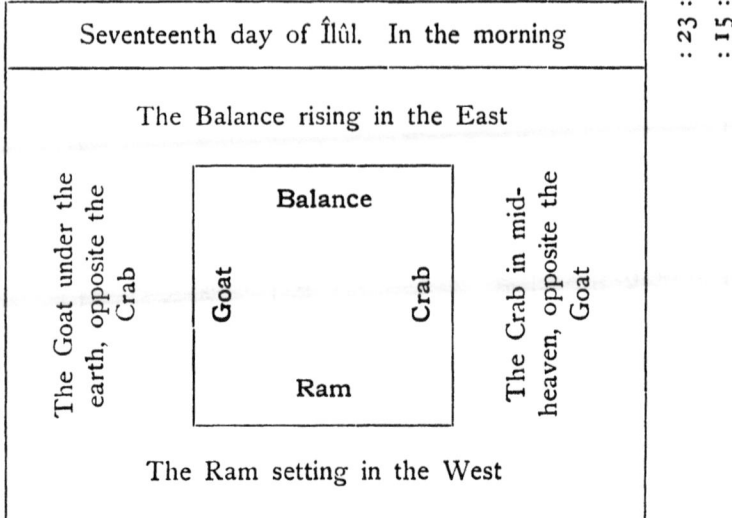

The Balance. II.

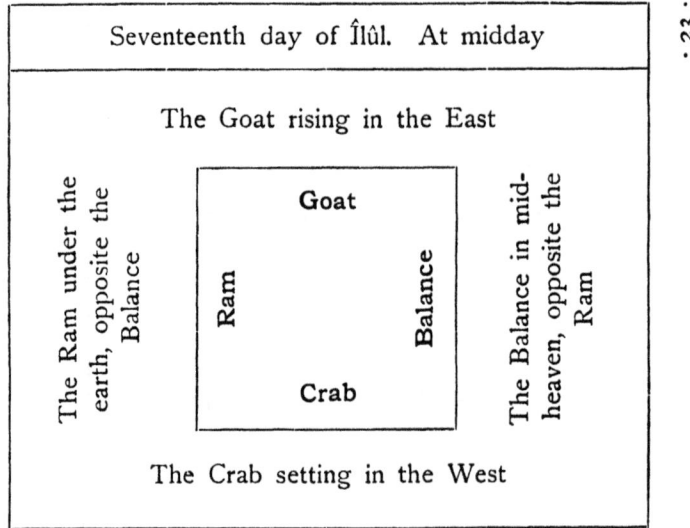

The Balance. III.

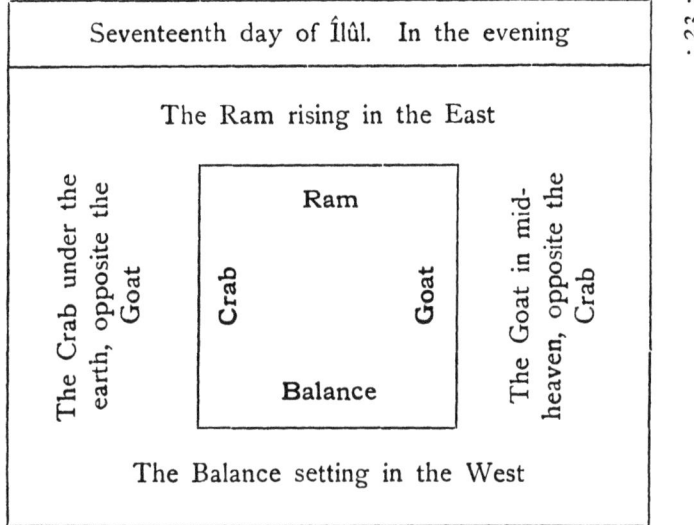

Page 499

The Balance. IV.

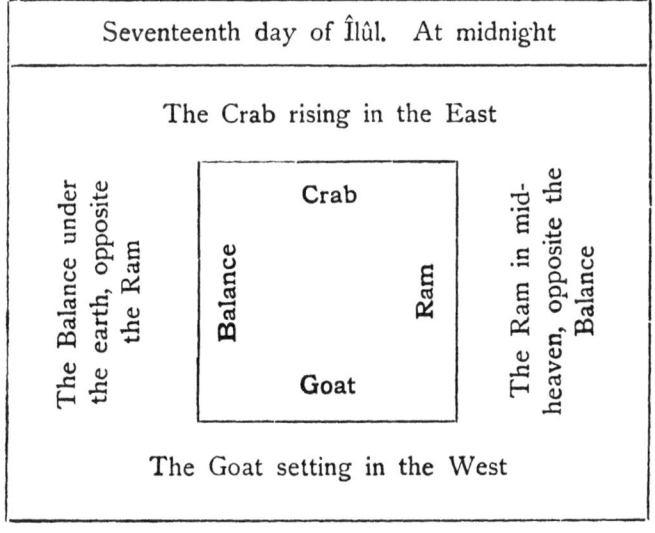

The Scorpion. I.

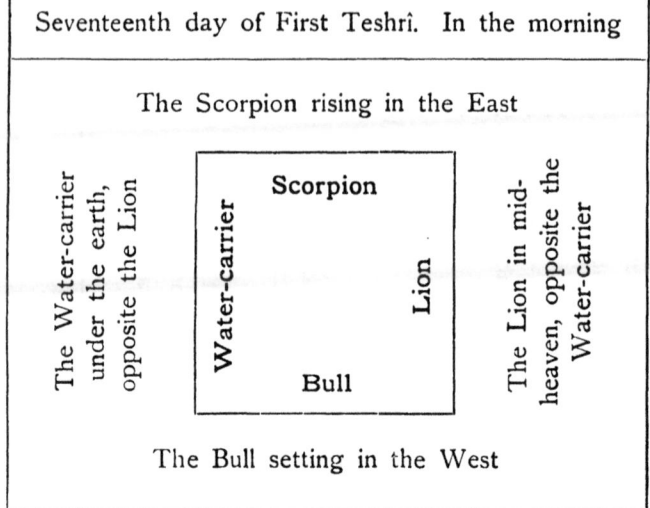

The Scorpion. II.

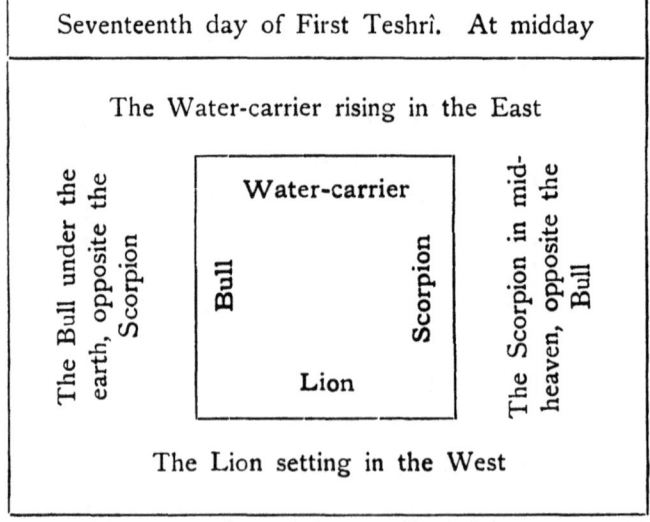

The Scorpion. III.

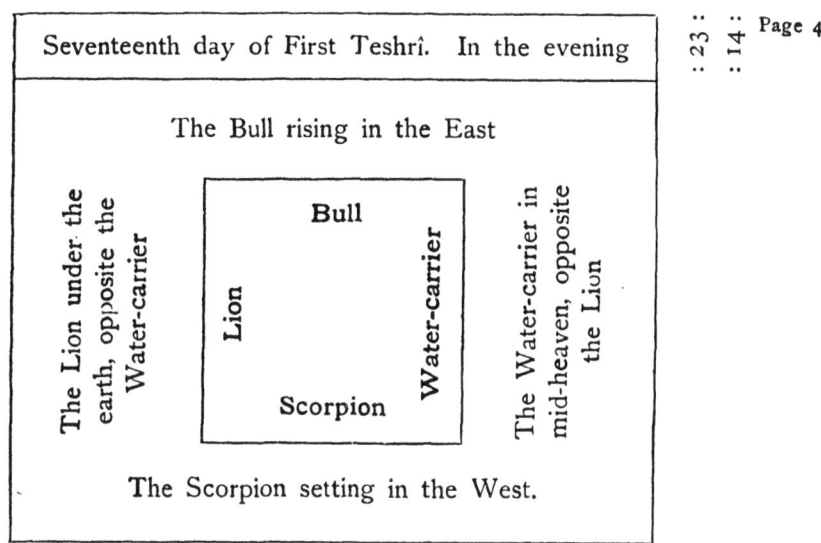

Seventeenth day of First Teshrî. In the evening

: 23 : : 14 : Page 499

The Bull rising in the East

Bull

Lion

Water-carrier

Scorpion

The Lion under the earth, opposite the Water-carrier

The Water-carrier in mid-heaven, opposite the Lion

The Scorpion setting in the West.

The Scorpion. IV.

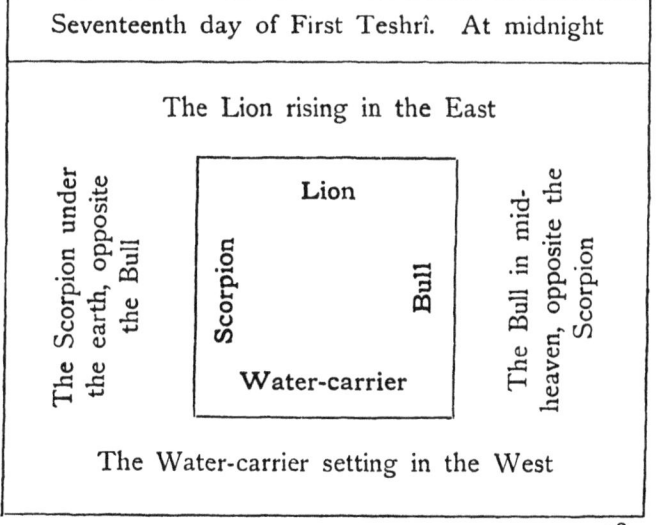

Seventeenth day of First Teshrî. At midnight

: 23 : : 14 :

The Lion rising in the East

Lion

Scorpion

Bull

Water-carrier

The Scorpion under the earth, opposite the Bull

The Bull in mid-heaven, opposite the Scorpion

The Water-carrier setting in the West

38

The Bowman. I.

Page 498

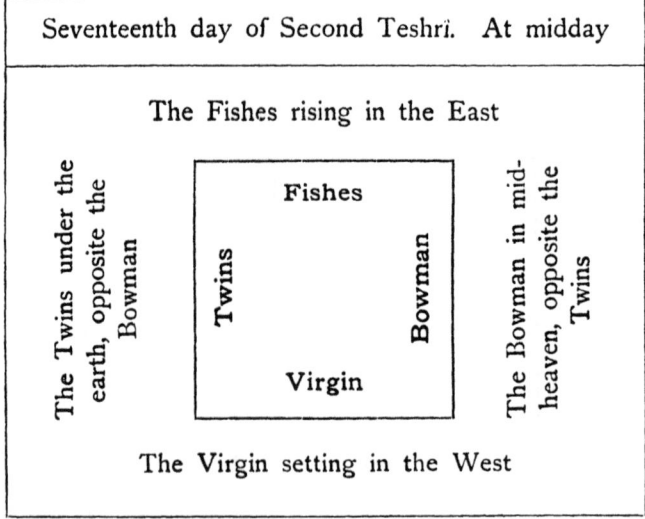

The Bowman. II.

The Bowman. III.

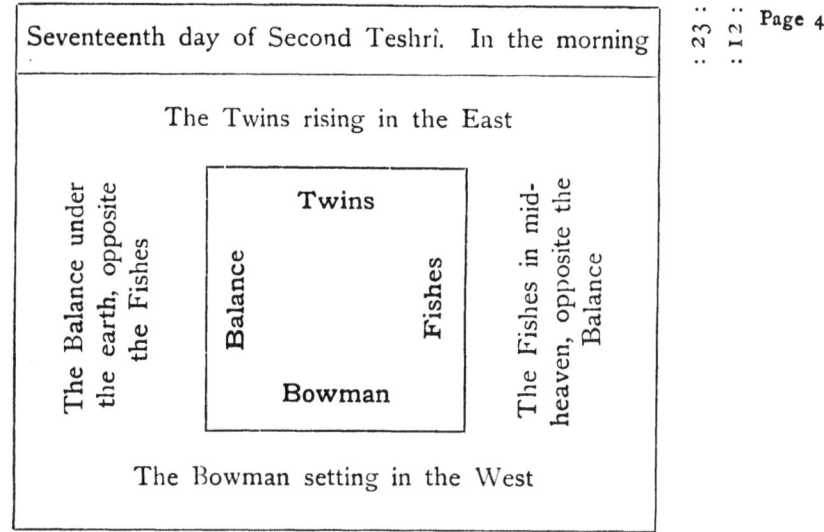

Seventeenth day of Second Teshrî. In the morning

The Twins rising in the East

The Balance under the earth, opposite the Fishes

Balance

Twins

Fishes

Bowman

The Fishes in mid-heaven, opposite the Balance

The Bowman setting in the West

: 23 :
: 12 :
Page 499

The Bowman. IV.

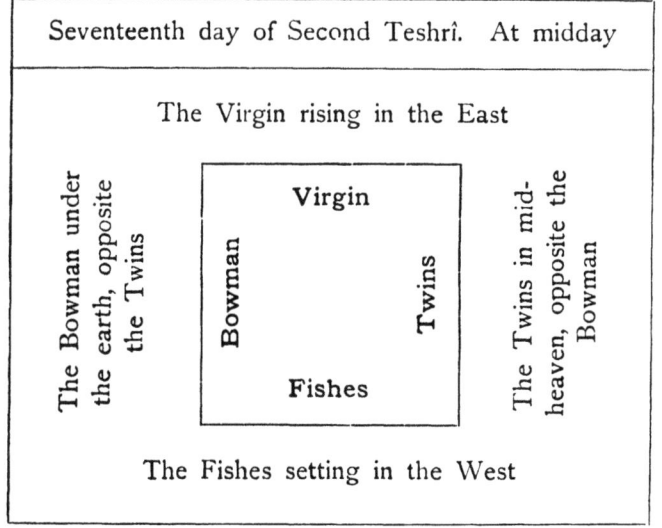

Seventeenth day of Second Teshrî. At midday

The Virgin rising in the East

The Bowman under the earth, opposite the Twins

Bowman

Virgin

Twins

Fishes

The Twins in mid-heaven, opposite the Bowman

The Fishes setting in the West

: 23 :
: 12 :

The Goat. I.

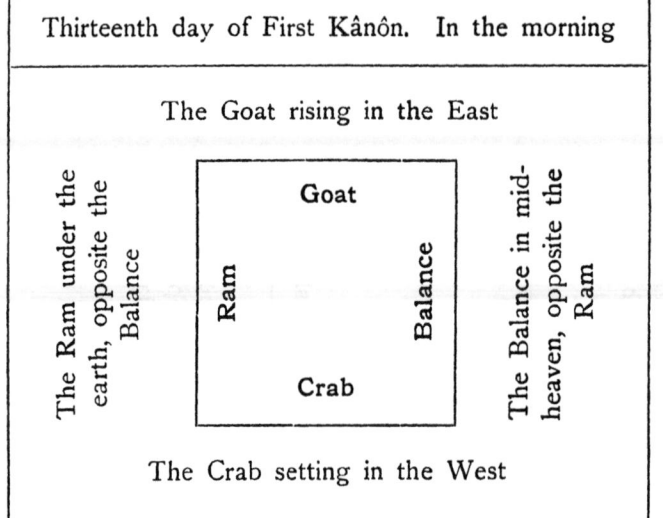

The Goat. II.

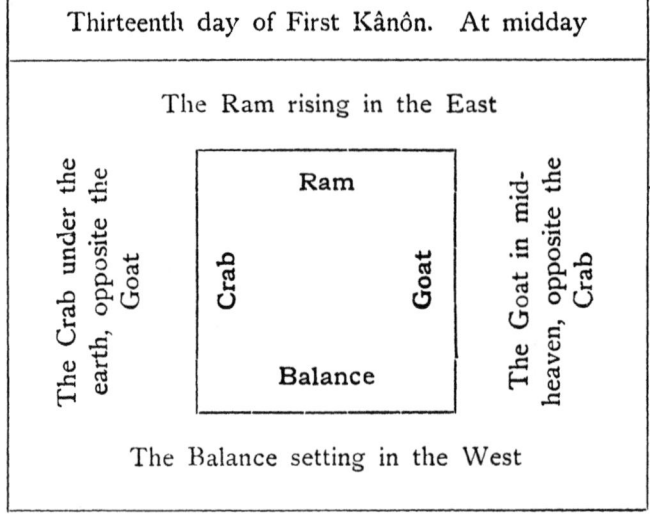

The Goat. III.

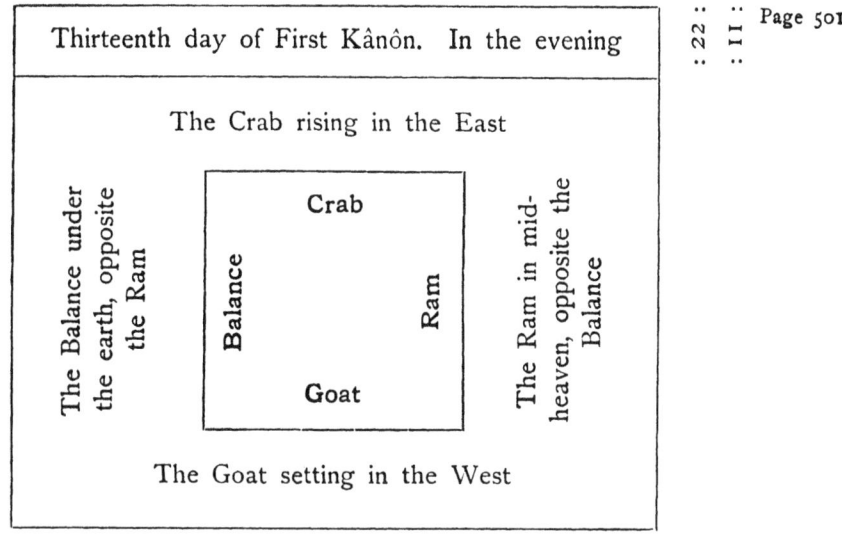

Thirteenth day of First Kânôn. In the evening

: 22 : : 11 : Page 501

The Crab rising in the East

The Balance under the earth, opposite the Ram

Balance

Crab

Ram

Goat

The Ram in mid-heaven, opposite the Balance

The Goat setting in the West

The Goat. IV.

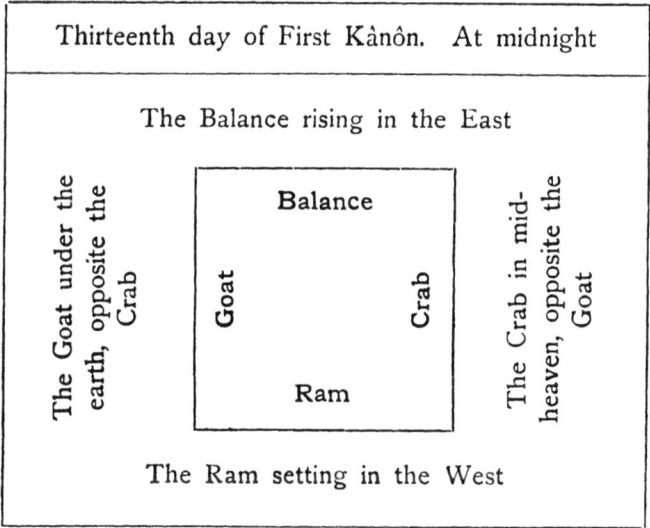

Thirteenth day of First Kânôn. At midnight

: 22 : : 11 :

The Balance rising in the East

The Goat under the earth, opposite the Crab

Goat

Balance

Crab

Ram

The Crab in mid-heaven, opposite the Goat

The Ram setting in the West

The Water-carrier. I.

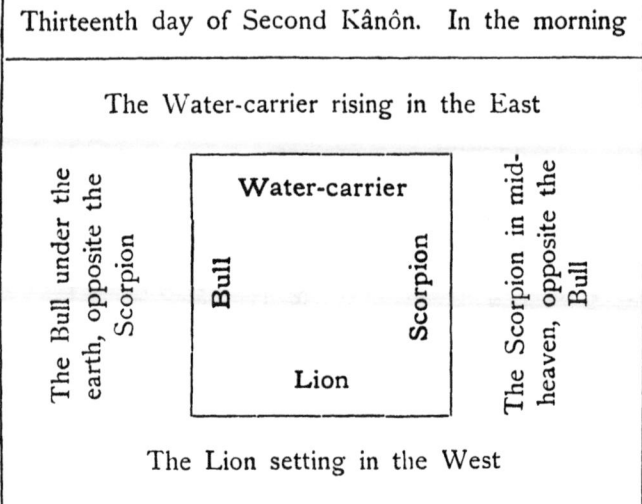

The Water-carrier. II.

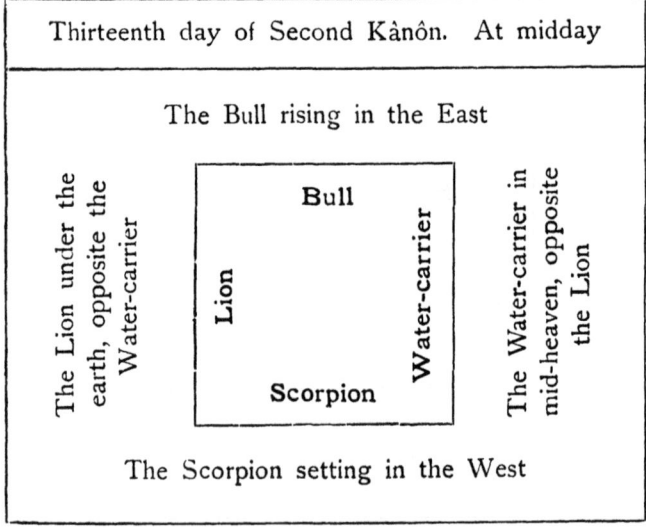

The Water-carrier. III.

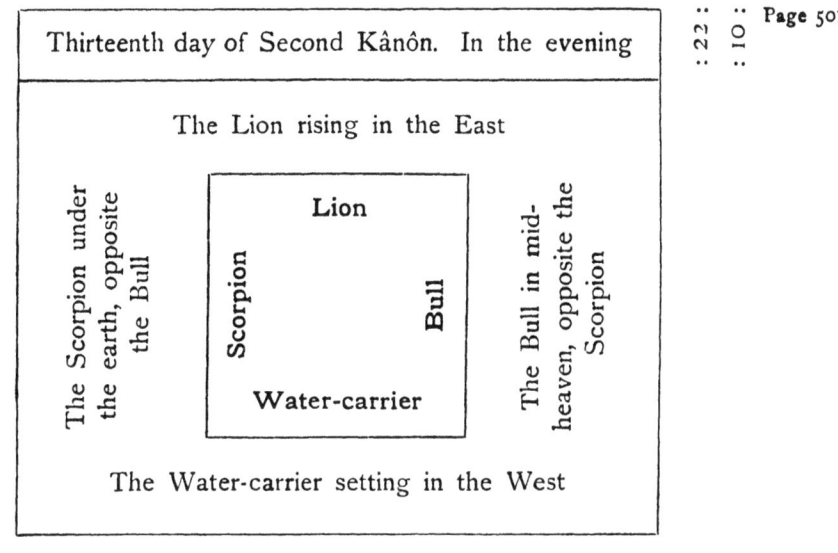

Thirteenth day of Second Kânon. In the evening

: 22 : : 10 : Page 501

The Lion rising in the East

The Scorpion under the earth, opposite the Bull

Lion

Scorpion

Bull

Water-carrier

The Bull in mid-heaven, opposite the Scorpion

The Water-carrier setting in the West

The Water-carrier. IV.

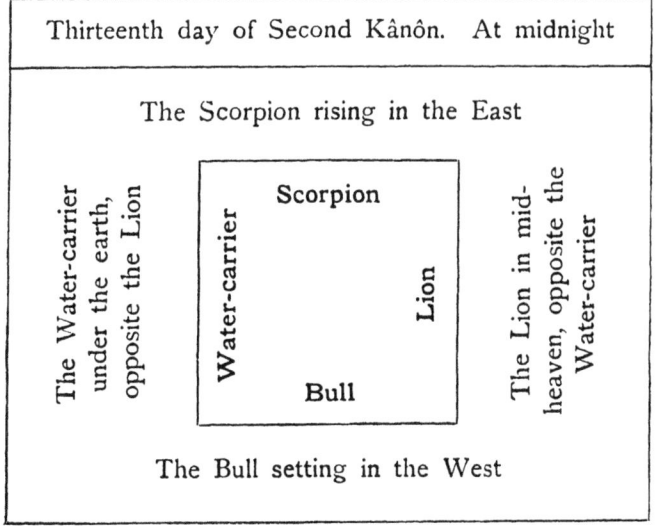

Thirteenth day of Second Kânon. At midnight

: 22 : : 10 :

The Scorpion rising in the East

The Water-carrier under the earth, opposite the Lion

Water-carrier

Scorpion

Lion

Bull

The Lion in mid-heaven, opposite the Water-carrier

The Bull setting in the West

The Fishes.　I.

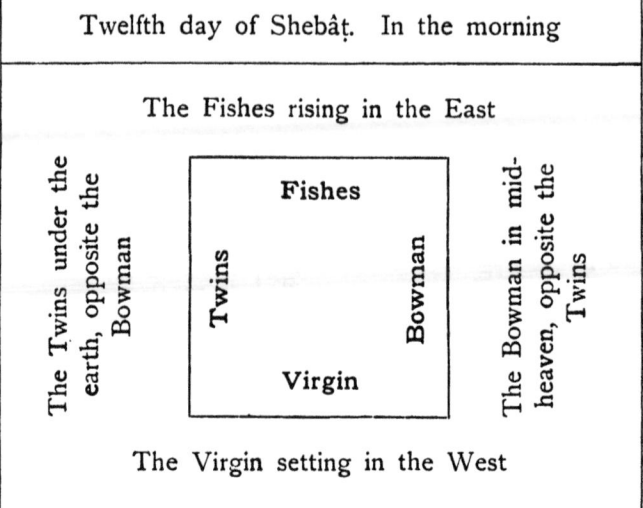

The Fishes.　II.

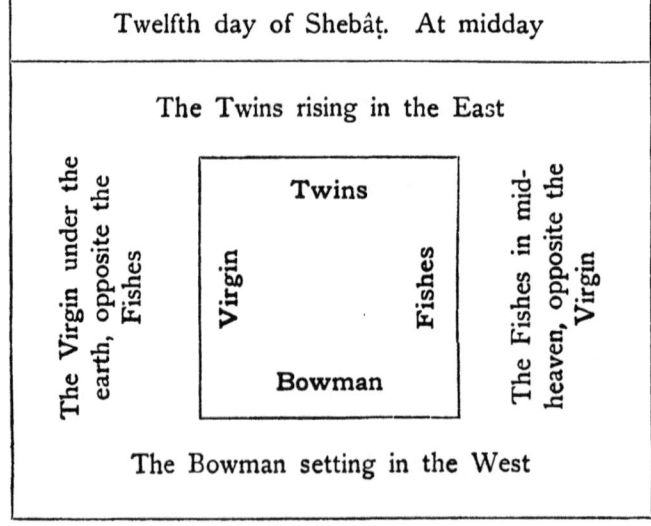

The Fishes. III.

The Virgin rising in the East

The Bowman under the earth, opposite the Twins

Virgin

Bowman

Twins

Fishes

The Twins in mid-heaven, opposite the Bowman

The Fishes setting in the West

Twelfth day of Shebât. In the evening

Page 501

The Fishes. IV.

Twelfth day of Shebât. At midnight

The Bowman rising in the East

The Fishes under the earth, opposite the Virgin

Bowman

Fishes

Virgin

Twins

The Virgin in mid-heaven, opposite the Fishes

The Twins setting in the West

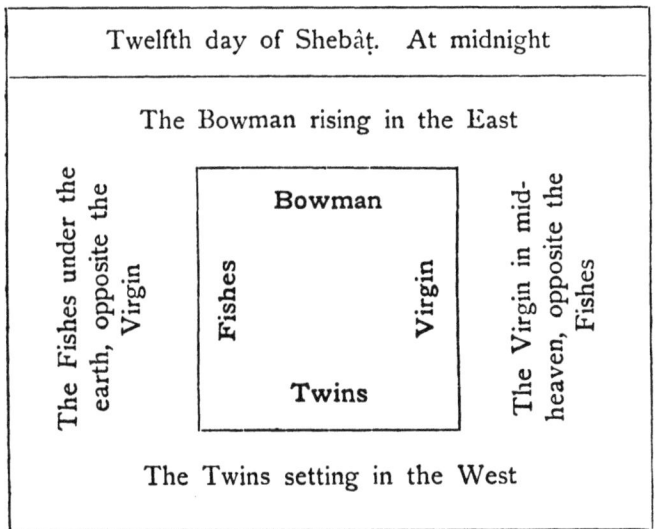

Another List.

Page 502   Twenty-first day of Âdhâr, which is the rising of the Sun, for the Ram riseth, and the Balance setteth, which is its setting (*i. e.*, the Sun's).

Twenty-second day of Îlûl, which is the setting of the Sun, for the Balance riseth, and the Ram setteth, which is its (*i. e.*, the Sun's) rising.

Fol. 236 *b.*   Twentieth day of **Nîsân**. The Bull riseth, which is the rising of the Moon, and the Scorpion setteth, which is the setting of the Moon.

Twenty-second day of the **First Teshrî**. The Scorpion riseth, which is the setting of the Moon, and the Bull setteth, which is its rising.

Twentieth day of **Îyyâr**. The Twins set, and the Bowman riseth.

Twenty-first day of the **Second Teshrî**. The Bowman riseth, and the Twins set.

Sixteenth day of **Khazîrân**. The Crab riseth, which is the rising of Zeus (Jupiter), and the Goat setteth, which is its setting.

Twenty-first day of the **First Kânôn**. The Goat riseth, which is the rising of Âris, and the Crab setteth, which is its setting.

Sixteenth day of **Tammûz**. The Lion riseth, and the Water-carrier setteth.

Twenty-first day of the **Second Kânôn**. The Water-carrier riseth, and the Lion setteth.

Eighteenth day of **Âbh**. The Virgin riseth, which is the rising of Hermes, and the Fishes set, which is its setting.

Nineteenth day of **Shebât**. The Fishes rise, which is the rising of Aphrodite, and the Virgin setteth, which is its setting.

Twenty-second day of **Îlûl**. The Balance riseth, which is the rising of Kronos, and the Ram setteth, which is its setting.

On the same subject.

In the month wherein the Sun goeth up (or, riseth), Kronos goeth down; and in the month wherein Kronos goeth up, the Sun goeth down.

In the month wherein Zeus goeth up, Ârîs goeth down; and in the month wherein Ârîs goeth up, Zeus goeth down.

In the month wherein Hermes goeth up, Aphrodite goeth down; and in the month wherein Aphrodite goeth up, Hermes goeth down.

In this way they were created, each the opposite of the other, and the one the opposite of his fellow.

Again. Kronos four days, | Zeus sixteen days, Ârîs five Page 503 days, Aphrodite seven days, Hermes eleven days, the Sun fourteen days, the Moon twelve days.

That thou mayest know the Planets of the day-time and night-time. The Sun is the planet | of the day. Zeus and Fol. 237 a. Kronos are the planets of the night-time. The Moon, and Aphrodite, and Hermes are the planets of the morning.

That thou mayest know which are the Malwâshê that go down, and which are they that are equal (_i. e._, neither go up or down), and those which go up. The Ram, the Bull, the Twins; these go up. The Crab, the Lion, and the Virgin; these are equal. The Balance, the Scorpion, and the Bowman; these go down. The Goat, the Water-carrier, and the Fishes; these are equal.

That thou mayest know how long the Sun stayeth in each one of the Malwâshê.

In the **Ram**, thirty days and one hour.

In the **Bull**, and in the **Twins**, and in the **Crab**: in each of these he stayeth thirty-one days and twelve hours.

In the **Lion**, and in the **Virgin**, and in the **Balance**: in each of these he stayeth thirty-nine days and eighteen hours.

In the **Scorpion**, and in the **Bowman**, and in the **Goat**: in each of these he stayeth thirty-nine days and six hours.

In the **Water-carrier**, thirty days.

In the **Fishes**, thirty days and six hours.

Total of all these days: three hundred and sixty-five days and six hours.

That thou mayest know the colours and the symbols of the Malwâshê.

The Ram Ⲉ Red.

The Bull ⵏⴹⵏⴹ White.

The Twins ⚌ Reddish.

The Crab ⊂⊃ White.

The Lion ⌐ Red.

The Virgin ⴹⴹ Dark-coloured.

The Balance ⌐⌐ White.

The Scorpion ∿ⱳ Dirty white.

The Bowman ⱪⱬ Red.

The Goat ℧ℏ Dark-coloured.

The Water-carrier ⱳⱳⱳ Dark-coloured.

The Fishes ƆƆ White-black.

The countries of the Malwâshê. The Ram to Persia. Page 504 The Bull to Babylon. The Twins to Cappadocia. | The Crab to Armenia. The Lion to Asia. The Virgin to Alâdâ (Hellas, Greece). The Balance to Libya. The Scorpion to Italy. The Fol. 237*b*. Bowman to Crete and Cilicia. The Goat to Syria. The | Pourer out of Water, *i. e.*, the Water-carrier, to Egypt. The Fishes to India.

The Countries which are included in the courses of the Governors.

The **Sun** travelleth from the Euphrates to Cush, and to Shebhâ, and to the Ocean.

The **Moon** ruleth from the Euphrates to Athens.

Ârîs ruleth from Mâgôgh to Havilah, and Cyrenea, and the occupation [of the peoples thereof is] war.

Hermes ruleth from Athenchôs (?) to the Ocean.

Bêl ruleth Byzantium, and the land of Havilah, and Inner (*i. e.*, Further) India, and as far as the border of the Inner Thebaïd.

Balti ruleth from Ṭarôḳôs to Bêth Gelîsâyâ and the country of the Barbarians.

And **Mashtâwtaphâ** also ruleth from the border of Antioch to Bêth Ḥamîrâyê, the border of which is Further India; [now the people of this district] wage wars continually.

Kêwân ruleth from Ḥalab (Aleppo) to Alexandria, and from Alexandria to Thebes, and to the ends of the South.

Another Door (*i. e.*, Section) on the Malwâshê which hear one another, and see one another, and which give orders to each other, and these are conducive to every kind of companionship.

When thou wishest to see (*i. e.*, find out) about two people who are closely associated, or about the love of a man for his wife, or about a slave and his lord, or about the love between two friends, take the first letters of the names of both of them, and observe to which of the Malwâshê and to which of the Planets they belong. If they belong to the Malwâshê which have friendship with each other, and if they belong also to the Malwâshê which see and hear each other, the friendship [between any two of the persons mentioned above] prospereth (or, is excellent). And if they (*i. e.*, the letters) belong to the Malwâshê which are not associated in friendship, | and which Page 505 do not hear each other, or are not seen by each other, [the persons mentioned above] are separated in friendship from each other.

These are the Malwâshê which do not see each other: | The Bull, the Virgin, the Scorpion, and the Fishes. And Fol. 238 a. Russpôs (*sic!*) seeth the Twins; the Crab, the Lion, the Balance, the Bowman, the Goat, the Water-carrier, **are those which see each other.** The Twins see the Lion, and the Lion seeth the Twins; the Bull seeth the Virgin, and the Virgin seeth the Bull; the Ram seeth the Balance, and the Balance seeth the Ram; the Fishes see the Scorpion, and the Scorpion seeth the Fishes; the Bowman seeth the Water-carrier, and the Water-carrier seeth the Bowman.

The Malwâshê which command each other are the following: — The Bull commandeth the Fishes, and they obey him. The Twins command the Water-carrier, the Crab commandeth the Goat, the Lion commandeth the Bowman, the Virgin commandeth the Scorpion.

The Diastases (*i. e.,* separations) in respect of each other are as follows:—The Ram from the Scorpion, and the Scorpion from the Ram; the Bull from the Balance, and the Balance from the Bull; the Goat from the Water-carrier, and the Water-carrier from the Goat; the Ram from the Fishes, the Crab from the Bowman, the Virgin from the Balance, and the Goat from the Water-carrier. These are the Malwàshê which change, and their friendship is transient.

Again, another treatment of the subject. If Kronos, and Hermes, and Aphrodite be together and if they be propitious, it is very, very good. Kronos and the Sun in opposition to each other [are bad]. Kronos, and Nâbhô (Mercury), and the Moon are good. Kronos, and Nâbhô, and Arìs are good. Kronos, and Nâbhô, and Bîl are good in everything. Kronos and the Sun stir up all evils. Kronos, and Sînà, and Baltî are good, [and produce] all kinds of good things. Kronos and Baltî and Yarîgh are moderately bad. Kronos, and Baltî, and the Sun are moderately bad. Kronos, and Bîl, and Baltî—all good things. Kêwân, and Bîl, and Sînà are good, and enrich [men], but God is the Director of all of them.

[Some lines wanting]

Page 506 and these | are they: Bîl is one among all his companions, and he helpeth, and he hath no enemies: Kronos and Zeus are

Fol. 238b. like a father | and his son. Kronos and Yarîgh are enemies. Kronos and Baltî are like a man and his wife. Bîl and Yarîgh are divided, and they contend against each other. Bîl and Sînà are good and true kings. Bîl and the Sun are good in everything, and they perform good things.

That thou mayest know the day when the day and night are equal. The day and the night together consist of twenty-four hours, and all the hours in the year are eight thousand, seven hundred, and sixty-six. Now, the reckoning [of time] is made by means of the hours; and years, and months, and days, as well as day-time and night-time, are expressed in hours. And the hours themselves are divided into "parts", and "ḳantîmê" (minutes), and "rûphsê" of "rĕthûphê" and of "rĕphâphê". One hour, then, containeth thirty "parts" [of two minutes each]; thirty "parts" contain sixty "ḳanṭîmê" [of one minute each]; five "ḳaṭṭinàthà" contain fifty "rûphsê", and one "part" containeth thirty-five "rûphsê".

Again:—On the twenty-fourth day of the First Kànôn at the sixth hour of the night, the day beginneth to take [time] from the night. On the twenty-fourth day of Âdhàr, at the sixth hour [of the night], the durations of day and night are equal. On the twenty-fourth day of Khazîràn, at the sixth hour of the night, the night beginneth to take time from the day. On the twenty-fourth day of Îlûl, at the sixth hour of the night, the durations of the day and night are equal. And from the twenty-fourth day of Khazîràn to the twenty-fourth day of Îlûl, the night of which taketh [time] from the day, the hours of the morning are longer than those of the turning. And from the twenty-fourth day of the First Kànôn to the twenty-fourth day of Âdhâr, the day of which taketh [time] from the night, the hours of the turning are longer than the hours of the morning. And on the first day wherein the light beginneth to take from the darkness, it taketh twelve "parts", | (i.e., twenty-four minutes) from it in one night. Now twenty- Page 507 four "parts" | are equal to four hundred and sixty "rûphàs". Fol. 239a. Now between Kànôn and Kànôn one hour more is added to the day-time, and the day lengthens [by this amount]. And in Shebâṭ there are added to the day-time forty ḳaṭṭinàthà, which make two "parts" of an hour (i.e., four minutes). In Nisàn the day-time is fourteen hours [long]. And in Îyyàr, when the Sun cometh to the Twins, during the month fifty ḳaṭṭinàthà are added to the day-time, which is then but a little less than fifteen hours, and the night containeth nine hours [only]. And because the Elements seize them, it becometh very hot indeed.

The Number of the degrees of the Sun.

In the First Kànôn, when the Sun standeth in the lower degrees, and the measure (i.e., length) of [thy] shadow when thou standest up increaseth, at the first hour of the First

Kânôn, thou wilt find thy shadow, from thy head to thy feet, which falleth on the ground, when thou standest upright in a flat and level place wherein there is no rising ground or hollow, and there is nothing to obstruct the shadow or to turn it aside or deflect it, thou wilt find thy shadow [I say.] to measure twenty-eight "pôdhyâthâ" (i. e., feet). In the Second Kânôn the length will be twenty-seven feet, in Shebât twenty-six, in Âdhâr twenty-five, in Nîsân twenty-four, in Îyyâr twenty-three, and in Khazîrân twenty-two.

And up to this point the sun is rising, but at this point and henceforth the sun returneth, and he addeth to the length of thy shadow that which was taken therefrom [as he rose]. In Tammûz the length is twenty-three feet, in Âbh twenty-four, Page 508 in Îlûl twenty-five, in the First Teshrî twenty-six, | in the Second Teshrî twenty-seven, and in the First Kânôn twenty-eight.

When thou wishest to know how much thou art to add to the hours, and how much thou art to subtract, do as followeth in all the months:—In the First Kânôn [the length of thy Fol. 239b. shadow is] twenty-|eight feet; and in the second hour eighteen; and in the third hour, and in the fourth hour, and in the fifth hour nine feet; and in the sixth hour eight feet. And again, reverse the process and add that which thou didst take away; in the seventh hour nine feet; and in the eighth hour eleven feet; and in the ninth hour thirteen feet; and in the tenth hour eighteen feet; and in the eleventh hour twenty-eight feet. And count in the same way with all the months. In the Second Kânôn, in the first hour, from twenty-seven feet subtract ten feet. And in the second hour subtract three feet, and in the third hour five feet, and in the fourth hour two feet, and in the fifth hour two feet, and in the sixth hour one foot. Now, in respect of midday, add to these what thou hast subtracted, beginning to add according as thou hast finished. In the seventh hour one foot, and in the eighth hour two feet, and in the ninth hour two feet, and in the tenth hour five feet, and in the eleventh hour twenty-seven feet, and in the twelfth hour the amount is immeasurable.

Prognostications (or, portents) from the Treatise of Hermes on the changes of the days and nights.

The Hours of the day of the First Day of the Week.

First Hour. Of the Sun. Favourable for sailing in a boat and for setting out on a journey.

Second Hour. Of Aphrodite. There is sincerity in it. It is favourable for approaching the Government, and for going on a journey.

Third Hour. Of Hermes. Good for ruling and for all work.

Fourth Hour. Of the Moon. Good for all truth.

Fifth Hour. Of Kronos. Bad and productive of loss.

Sixth Hour. Of Zeus. Favourable for everything. With it is associated upright dealing, and it is good for thought and calculation.

Seventh Hour. Of Âris. Unfavourable for everything, and very full of disturbances and enmity.

Eighth Hour. Of the Sun. No anxiety attacheth thereto, for it is | favourable for everything. Page 509

Ninth Hour. Of Balti. Good and favourable for everything.

Tenth Hour. Of Hermes. | Ordinary, but favourable for Fol. 240 a. calculation.|

Eleventh Hour. Of the Moon. Good and favourable for every work.

Twelfth Hour. Of Kronos. Bad for everything.

The Hours of the night of the Second Day of the Week.

First Hour. Of Zeus. Good for ruling and sailing in a boat.

Second Hour. Of Âris. Bad for everything.

Third Hour. Of the Sun. Good.

Fourth Hour. Of Aphrodite. Favourable.

Fifth Hour. Of Hermes. Ordinary.

Sixth Hour. Of the Moon. Favourable for everything.

Seventh Hour. Of Kronos. Associated with tribulation.

Eighth Hour. Of Zeus. Favourable for going on a journey, and for making a feast.

39

Ninth Hour. **Of Âris.** Bad for everything.

Tenth Hour. **Of the Moon.** Ordinary.

Eleventh Hour. **Of Aphrodite.** Favourable for everything.

Twelfth Hour. **Of Hermes.** Good for going on a journey.

The Hours of the day of the Second Day of the Week.

First Hour. **Of the Moon.** Good for going on a journey.

Second Hour. **Of Kronos.** Full of grief and trouble.

Third Hour. **Of Zeus.** Good for doing work.

Fourth Hour. **Of Âris.** Bad and productive of affliction and trouble.

Fifth Hour. **Of the Sun.** Propitious for setting out on a journey, and for all that a man wisheth [to do].

Sixth Hour. **Of Aphrodite.** Propitious for everything thou doest.

Seventh Hour. **Of Hermes.** [Propitious] for buying and selling.

Eighth Hour. **Of the Moon.** Ordinary.

Ninth Hour. **Of Kronos.** Bad and productive of trouble.

Tenth Hour. **Of Zeus.** Good for trafficking of all kinds.

Eleventh Hour. **Of Âris.** Bad.

Twelfth Hour. **Of the Sun.** Good for everything.

The Hours of the night of the Third Day of the Week.

First Hour. **Of Aphrodite.** Good and propitious for everything.

Second Hour. **Of Hermes.** Propitious for sailing in a boat.

Page 510   Third Hour. **Of the Moon.** Is without grief, | and is propitious.

Fourth Hour. **Of Kronos.** Bad and troublesome.

Fifth Hour. **Of Zeus.** Propitious, but go not on a journey.

Sixth Hour. **Of Âris.** Very bad.

Seventh Hour. **Of the Sun.** Ordinary, and not very propitious.

Eighth Hour. **Of Aphrodite.** Propitious for everything.

Fol. 240b.   Ninth Hour. **Of Hermes.** Ordinary, | but good for going on a journey.

Tenth Hour. **Of the Moon.** Good and propitious.

Eleventh Hour. **Of Kronos.** Bad and productive of loss.
Twelfth Hour. **Of Zeus.** Propitious for everything.

The Hours of the day of the Third Day of the Week.

First Hour. **Of Âris.** Bad for everything.
Second Hour. **Of the Sun.** [Propitious] for all actions.
Third Hour. **Of Aphrodite.** Good for everything.
Fourth Hour. **Of Hermes.** There is much rest therein, and it is good.
Fifth Hour. **Of the Moon.** Propitious for everything.
Sixth Hour. **Of Kronos.** Bad.
Seventh Hour. **Of Zeus.** Good and propitious for everything.
Eighth Hour. **Of Âris.** Very full of disturbances.
Ninth Hour. **Of the Sun.** Propitious for everything.
Tenth Hour. **Of Aphrodite.** Very propitious and good.
Eleventh Hour. **Of Hermes.** Ordinary, and not propitious.
Twelfth Hour. **Of the Moon.** Productive of loss in everything.

The Hours of the night of the Fourth Day of the Week.

First Hour. **Of Kronos.** Bad.
Second Hour. **Of Zeus.** Good.
Third Hour. **Of Âris.** Bad.
Fourth Hour. **Of the Sun.** Good for everything.
Fifth Hour. **Of Aphrodite.** Propitious.
Sixth Hour. **Of Hermes.** Propitious for going on a journey.
Seventh Hour. **Of the Moon.** Propitious.
Eighth Hour. **Of Kronos.** Bad.
Ninth Hour. **Of Zeus.** Good.
Tenth Hour. **Of Âris.** Ordinary.
Eleventh Hour. **Of the Sun.** Propitious for everything.
Twelfth Hour. **Of Aphrodite.** Propitious.

The Hours of the day of the Fourth Day of the Week. Page 511
First Hour. **Of Hermes.** Good for everything.
Second Hour. **Of the Moon.** Propitious for everything.
Third Hour. **Of Kronos.** Bad for everything.

39*

Fourth Hour. **Of Zeus.** Propitious for everything.

Fifth Hour. **Of Àris.** Bad.

Sixth Hour. **Of the Sun.** Bad.

Seventh Hour. **Of Aphrodite.** Propitious for everything.

Eighth Hour. **Of Hermes.** For everything propitious and good.

Ninth Hour. **Of the Moon.** For everything propitious and good.

Tenth Hour. **Of Kronos.** Bad.

Eleventh Hour. **Of Zeus.** Very good.

Fol. 241*a*. | Twelfth Hour. **Of Âris.** Bad for everything,

The Hours of the night of the Fifth Day of the Week.

First Hour. **Of the Sun.** Propitious for sailing in a boat.

Second Hour. **Of Aphrodite.** Propitious.

Third Hour. **Of Hermes.** Ordinary.

Fourth Hour. **Of the Moon.** Good for everything.

Fifth Hour. **Of Kronos.** Unpropitious.

Sixth Hour. **Of Zeus.** Doeth good to no enterprise.

Seventh Hour. **Of Âris.** Bad.

Eighth Hour. **Of the Sun.** Ordinary.

Ninth Hour. **Of Aphrodite.** Propitious for setting out on a journey, and for sailing in a boat.

Tenth Hour. **Of Hermes.** Ordinary.

Eleventh Hour. **Of the Moon.** Propitious.

Twelfth Hour. **Of Kronos.** Bad.

The Hours of the day of the Fifth Day of the Week.

First Hour. **Of Zeus.** Propitious for everything.

Second Hour. **Of Âris.** Propitious for everything.

Third Hour. **Of the Sun.** Driveth away thieves and unprosperous work.

Fourth Hour. **Of Aphrodite.** Good.

Fifth Hour. **Of Hermes.** Propitious for going on a journey.

Sixth Hour. **Of the Moon.** Propitious.

Seventh Hour. **Of Kronos.** Bad and hateful.

Eighth Hour. **Of Zeus.** Good and propitious.

Ninth Hour. **Of Âris.** Bad and full of trouble.

Tenth Hour. **Of the Sun.** Good for everything.
Eleventh Hour. **Of Aphrodite.** Very good.
Twelfth Hour. **Of Hermes.** Bad for everything.

The Hours of the night of the Eve of the Sabbath.   Page 512
Fol. 241 *b*.

First Hour. **Of the Moon.** Good for everything.
Second Hour. **Of Kronos.** Very bad.
Third Hour. **Of Zeus.** Propitious for sailing in a boat.
Fourth Hour. **Of Âris.** Propitious for nothing.
Fifth Hour. **Of the Sun.** Ordinary.
Sixth Hour. **Of Aphrodite.** Good for everything.
Seventh Hour. **Of Hermes.** Ordinary.
Eighth Hour. **Of the Moon.** Propitious for everything.
Ninth Hour. **Of Kronos.** Very bad.
Tenth Hour. **Of Zeus.** Propitious for sailing in a boat.
Eleventh Hour. **Of Âris.** Propitious for nothing.
Twelfth Hour. **Of the Sun.** Ordinary.

The Hours of the day of the Eve of the Sabbath.

First Hour. **Of Aphrodite.** Propitious for everything.
Second Hour. **Of Hermes.** Profitable for everything.
Third Hour. **Of the Moon.** Good for everything.
Fourth Hour. **Of Kronos.** Bad for everything.
Fifth Hour. **Of Zeus.** Good for everything.
Sixth Hour. **Of Âris.** Productive of great loss.
Seventh Hour. **Of the Sun.** Propitious for everything.
Eighth Hour. **Of Aphrodite.** Propitious for everything.
Ninth Hour. **Of Hermes.** Profitable in no respect.
Tenth Hour. **Of the Moon.** Propitious for everything.
Eleventh Hour. **Of Kronos.** Troublesome for everything.
Twelfth Hour. **Of Zeus.** Good.

The Hours of the Night of the Sabbath.

First Hour. **Of Âris.** Full of sicknesses.
Second Hour. **Of the Sun.** Good for everything.
Third Hour. **Of Aphrodite.** Good for everything.
Fourth Hour. **Of Hermes.** Good for everything.
Fifth Hour. **Of the Moon.** Not very good.

Sixth Hour. **Of Kronos.** Troublesome for everything.

Seventh Hour. **Of Zeus.** Propitious for going on a journey.

Eighth Hour. **Of Âris.** Very troublesome.

Ninth Hour. **Of the Sun.** Driveth away the thief.

Tenth Hour. **Of Balti.** Propitious.

Eleventh Hour. **Of Hermes.** Good for everything.

Twelfth Hour. **Of the Moon.** Propitious for going on a journey.

Page 513   **The Hours of the Day of the Sabbath.**

First Hour. **Of Kronos.** Propitious for nothing. He who is sick will not live, and he who is lost will not be found.

Second Hour. **Of Zeus.** Propitious for everything.

Third Hour. **Of Âris.** Propitious for everything.

Fourth Hour. **Of the Sun.** Ordinary.

Fifth Hour. **Of Balti.** Good for everything.

Sixth Hour. **Of Hermes.** Good for everything.

Seventh Hour. **Of the Moon.** Bad and oppressive.

Eighth Hour. **Of Kronos.** Productive of trouble, and loss, and anxiety.

Ninth Hour. **Of Zeus.** Good for everything connected with business.

Tenth Hour. **Of Âris.** Bad.

Eleventh Hour. **Of the Moon.** Good.

Twelfth Hour. **Of Aphrodite.** For thieves lamentable. He who is sick will be healed.

The Hours of the night of the First Day of the Week.

First Hour. **Of Hermes.** Joyful and propitious.

Fol. 242 a.   Second Hour. **Of the Moon.** | Full of falsehood.

Third Hour. **Of Kronos.** Bad.

Fourth Hour. **Of Zeus.** Good.

Fifth Hour. **Of Âris.** Bad.

Sixth Hour. **Of the Sun.** Ordinary.

Seventh Hour. **Of Aphrodite.** Unpropitious for the road.

Eighth Hour. **Of Hermes.** Bad.

Ninth Hour. **Of the Moon.** Ordinary.

Tenth Hour. **Of Kronos.** Propitious for the road.

Eleventh Hour. **Of Zeus.** Propitious for the road.
Twelfth Hour. **Of Âris.** Good for everything.

HERE END THE FORECASTS; AND TO GOD BE
EVERLASTING PRAISE! AMEN.

Of the days of the week, which wise men and men of understanding use.

The night of the first day of the week is the dominion of Hermes. It is good for buying, and selling, and for setting out on a journey. He who is born therein will become a wise man.

The day of the first day of the week. The sun ruleth it, and it is good for everything. He who is born thereon will become a prince. But let Christians take heed to do nothing thereon except to make prayers, and such like things.

The night of the second day of the week. Zeus ruleth it. It is good for buying, and selling, and for starting on a journey. He who is born therein will become a man of peace.

The day of the second day of the week. The Moon ruleth it. It is good for starting on a journey, | and for buying, and Page 514 selling, and for sowing seed, and for sexual intercourse. And he who is born thereon will suffer from many sicknesses.

The night of the third day of the week. Aphrodite ruleth it. The wise should beware of it. It is good for sexual intercourse. For drinking medicine it is useless, and it is not good for travelling. He who is born therein will be beautiful in form, and will be beloved of men.

The day of the third day of the week. Âris ruleth it. It is good for healing. A man must beware of quarrels therein. And he who is born therein will become a physician.

The night of the fourth day of the week. Kronos ruleth it. It is good for travelling, but not for sexual intercourse. He who is born therein will become a very old man. It is good for seeking after learning, and for settling the young in schools. He who is born therein will become a very old man, and he will be wise, and a person of knowledge.

The day of the fourth day of the week. Hermes ruleth it. It is not good. It is propitious for seeking after learning, Fol. 242*b*. but not | for sudden application to letters, nor for settling youths in schools. He who is born therein will become a man of knowledge.

The night of the fifth day of the week. The Sun ruleth it. It is good for sexual intercourse, and for building, and for starting on a journey. He who is born therein will become a prince, and he will go into the presence of the Governor, and he will become an adulterer.

The day of the fifth day of the week. Zeus ruleth it. It is good for buying, and selling, and travelling, and for going into the presence of the Governor, and for letters (*i. e.*, correspondence), and for intercourse with a woman.

The night of the Eve of the Sabbath. The Moon ruleth it. It is good for buying, and selling, and travelling, and for intercourse. He who is born therein will possess a beautiful person, and he will live in poverty.

The day of the Eve of the Sabbath. Aphrodite ruleth it. It is good for buying and selling, but not for travelling. He who is born therein will become beautiful.

The night of the Sabbath. Its governor is Ârîs, and it is good for adulterers and thieves, and he who is born therein will live a long time.

The day of the Sabbath. Kronos ruleth it. It is good for the man who is completing a piece of work, but not for one who is beginning a task. He who is born therein will live a long time.

Page 515   ### Forecasts derived from the Planets.

He who is born under the **Moon** will be a son of guile and wickedness, and thou wilt not be able to stand (*i. e.*, depend) upon his word, and he will be a fornicator. His exterior will be better than his interior.

He who is born under **Ârîs** will have blue and red eyes, and will be a lover of war and blood.

He who is born under **Hermes** will be dark, and lacking in strength, but he will have knowledge and understanding. And he will be a lover of the learning which is in books.

He who is born under **Aphrodite** will possess beauty of person, and will be ruddy, and a lover of wine, women, and pleasure.

He who is born under **Zeus** will be a wretched and poverty-stricken man, and will have a hard heart. He will be a lover of wine and women, and he will overcome every obstacle that cometh in his way.

He who is born under the **Sun** will be dark in colour, and good of soul (*i. e.*, of a happy disposition), and pleasant. His eyes will be small, his wickedness little, and he will love all those who love him.

He who is born under **Kronos**, ...Fol. 243*a.*
[two or three lines wanting]

Forecasts which describe the nativities of the children of men.

First Teshrî. The Balance. He who is born under this Sign will be at peace with all men, and he will be upright (or, straight), and pleasant, and gentle; he will be a husbandman, and of the fruits of his labour he will lend to every man.

Second Teshrî. The Scorpion. He who is born under this Sign will be bad, and bitter (or, obstinate), and warlike; he will have many children, but they will not live.

First Kânôn. The Great Image. He who is born under this Sign will overcome in judgement (*i. e.*, succeed in litigation), and if any man seeketh to do him harm he will not be able to do it, because his eyes do not look at any man. And he will be true, and will possess a good angel.

Second Kânôn. The Goat. He who is born under this Sign will be gentle, and pleasant, and cheerful, and lucky, and he will have many children. He will commit no act of folly, and no evil shall come nigh unto him, for the angel of peace directeth him.

Shebâṭ. The Water-carrier. He who is born under this Sign will be in trouble, | and sores will break out in his body, Page 516 and he will be neither rich nor poor.

Âdhâr. **The Fishes.** He who is born under this Sign will be just and honourable. And swellings will appear in his body, and he will beget many children. His countenance will be bright, and he will become a priest. And he will fall into the hands of thieves.

Nîsân. The Ram. He who is born under this Sign will be just and silent, and no evil will befall him. And his eyes will give him pain, because [his] angel will be unmindful of him.

Îyyâr. **The Bull.** He who is born under this Sign will be a husbandman. His eyes will be large, and he will be bitter (or, taciturn). From the result of his toil he will lend to every man. When he sleepeth water will come from his mouth, and he will have many children.

Khazîrân. The Twins. He who is born under this Sign will be cold, 'and hot, and upright, and merciful. When he looketh to heaven, everything which he asketh from God will
Fol. 243 *b*. be given to him, | because [his] angel is with him and the Sun and Moon direct him.

Tammûz. The Crab. He who is born under this Sign will be a bad man. He will beget many children, and he will not tremble before any man, and he is better than his word.

Âbh. **The Lion.** He who is born under this Sign will be obstinate, and bitter (*i. e.*, taciturn), and strong.

Îlûl. **The Virgin.** He who is born under this Sign will be good, and blessed, and his revenues will increase greatly.

Other forecasts concerning the nativities of men.

Nîsân. The Ram. Whosoever is born at the beginning of this month, will have lepra in his head; he who is born in the middle of the month will have a pleasant odour with him; and he who is born at the end of the month will have an odour of bones (?) with him.

Îyyâr. **The Bull.** The smell of the feet of those who are born in this month is foetid.

Khazîrân. The Twins. He who is born in this month will have a sweet and pleasant odour.

Tammûz. The Crab. He who is born in this month will have lepra, and his mouth and body will possess an odour.

Âbh. The Lion. The mouth of him that is born in the beginning of this month will have an odour, and the odour of him that is born in the middle or end of it will be pleasant, because he is near to the Virgin, and because Bîl is the governor.

Îlûl. | The Virgin. The smell of him that is born in this Page 517 month will be pleasant.

First Teshrî. The Balance. As before.

Second Teshrî. The Scorpion. To him that is born in this month will be attached the smell of his genital organs, and of his hinder parts.

First Kânôn. The Bowman. The smell of him that is born in this month will be pleasant.

Second Kânôn. The Goat. A man born in this month will stink.

Shebâṭ. Water-carrier. The smell of him that is born in the beginning of the month will be very pleasant, and he who is born in the middle, or at the end of it, will have lepra on his head.

Âdhâr. The Fishes. The smell of him that is born in this month will be pleasant.

Another kind of forecasts concerning the person of a man.

The head of a man is interpreted by **Ram.** And his neck and his shoulders by **Bull.** And his two arms by **Twins.** And his breast by **Crab.** And his belly and heart by **Lion.** And his spinal column, and sides, and loins by **Virgin.** And his loins and flanks, that is to say, his sides, by **Balance.** And his bladder, and member, and testicles by **Scorpion.** His thighs by **Great Image.** | His knees by **Goat.** His legs by Fol. 244a. **Water-carrier.** The soles of his feet by **Fishes.** His brain by **Sun.** His skin by **Moon.** His blood by **Âris.** His sinews and veins by **Hermes.** His bones by **Bîl.** His flesh by **Baltî.** His hair by **Kêwân.**

Certain men have said that every man who is born existeth as a mingling and a mixture of the Seven [Governors], and that as the component parts of his members are formed and compounded into one body, so also all his internal operations are mixtures of these Seven Governors. Wickedness cometh from Kêwân, goodness from Bîl, anger and wrath from Ârîs, graciousness and a peaceful disposition come from the Sun, lust and fornication from Aphrodite, knowledge and wisdom from Hermes, and greediness and rapacity from Sînâ. The views were held by the Chaldeans, but now the Christians do not hold them.

That thou mayest know under which of these Seven Planets a man is born, and on what day:—Reckon up the numerical values of the [letters in] the name of the man by themselves. Then divide each of the values of the letters of the alphabet | by seven, and thus shalt thou divide [them]. Seven is *zĕ̆kûth.*[1] From eight take one, from nine take two, from ten take three, and from twenty take six, and from thirty take two, and from forty take five, and from fifty take one, and from sixty take four, seventy is *zĕ̆kûth,*[2] and from eighty take three, and from ninety take six, and from one hundred take two, and from two hundred take four, and from three hundred take six, and from four hundred take one. Divide the total of the numerical values of the letters by seven. **If one remaineth to thee,** the day of his birth is the First Day of the Week, and his Planet is the Sun, and he will "live life", and if he passeth the age of forty-two years, he will live eighty years, and, if God willeth, on the death of his soul he will die. **If two remain to thee,** the day of his birth is the Second Day of the Week, and his Planet is the Moon (?). If he attaineth to the age of thirty years, he will live to the age of seventy-five years, for by the Will of God everything [happeneth]. **If three remain to thee,** the day of his birth is the Third Day of the Week, and his Planet is Ârîs. If he passeth sixty years he will live to the age of seventy-two years. And by the Nod of God everything [happeneth]. **If four remain to thee,** the day of his birth is the Fourth Day of the Week,

Page 518

---

[1] Seven divided by seven has no remainder.

[2] *I.e*, seven divides seventy and there is no remainder.

and his Star is Hermes. If he passeth twenty-eight years he will live to the age of seventy-two years. And whatsoever God willeth He doeth. **If five remain to thee,** the day of Fol. 244*b*. his birth is the Fifth Day of the Week, and his Planet is Zeus. If he passeth forty years he will attain to the age of eighty-four years. And if God willeth to put him to death [before this], the man will die. **If six remain to thee,** the day of his birth is the Eve of the Sabbath, and his Planet is Aphrodite. If he passeth thirty-four years he will add to the years of his life; and if he passeth seventy years he will attain to his eightieth year. And everything is in the Power of God. **If seven remain to thee,** the day of his birth is the Sabbath, and his Planet is Kronos. If he passeth fifty years he will attain to the age of seventy-seven years. And whatsoever God willeth He doeth.

Another kind of forecast for a sick man. When thou Page 519 goest in [to visit] the sick man, observe which of the Signs of the Zodiac goeth in with thee, and reckon up the numerical values of the letters of the name of the sick man, and those of the letters of the name of the Sign, and divide them by seven. If an even number remaineth to thee, the sick man will live, and if an odd number, he will most certainly die.

That thou mayest know from what thing a man will be able to obtain profit and luck. Thou shalt divide the numerical values of the letters of the alphabet by eight in the following way:—Take from nine one, from ten two, from twenty four, from thirty six, from forty *zĕḳûth* (nothing), from fifty two, from sixty four, from seventy six, from eighty *zĕḳûth*, from ninety two, from one hundred four, from two hundred nothing, from three hundred four, from four hundred nothing. Then reckon up the [letters of the] name of the man, and those of name of his mother, and divide each of them by seven. If one remaineth to thee, his fate is from the Government. If two remain to thee, it is from men; and there will be mercy [in it]. If three remain to thee, it is from the Books and Orders of the Church. If four remain to thee, it is from the roads. If five remain to thee, it is from the tillage of the earth. If six remain to thee, it is from buying and selling. And if seven remain to thee, it is from his handicraft.

That thou mayest know whether thy friend loveth or hateth thee. Reckon up the numerical values of the letters in the name of thy friend, and those of the letters in thine own name, and make them one (*i. e.*, add up each group). Then divide each of them by seven. If the figure remaining be in the first table, thy friend loveth thee, and if it be in the second, he hateth thee.

First Table	1. 3. 5. 7.

Second Table	2. 4. 6.

For him that wisheth to betroth a woman to him. Reckon up the numerical values of the letters in the name of the man, and those of the letters in the name of the woman, and divide each of them by eight. If the remainders be odd numbers Fol. 245*a.* | the woman will belong to him, and if they be even numbers she will not.

Another computation for him that wisheth to betroth a Page 520 **woman to him. If thou wishest to know whether | they are suitable, and whether they will agree together or not, divide the numerical values of the letters of the alphabet which are in their names by nine, thus:**—Take one from ten, and two from twenty, and three from thirty, and four from forty, and five from fifty, and six from sixty, and seven from seventy, and eight from eighty, and nothing from ninety, and one from one hundred, and two from two hundred, and three from three hundred, and four from four hundred. Reckon up [the numbers] of the name of the man and those of the name of the woman, and keep each group by itself, and when thou hast added up each group, and arrived at the total of each, divide each total by nine, and the number that remaineth over after the division representeth their fate. And from *ôlaph* (one) to *ṭith* (nine), [take] two numbers at a time, one of [the name of] the man, and one of that of the woman.

:₤: I + I   They will have intercourse, and there will be peace between them.

I + 2   They will live in peace.

I + 3   They will have intercourse, and there will be war between them.

I + 4   They will have intercourse, and there will be hatred between them.

I + 5   They will not have intercourse, and if they do, they will be separated.

I + 6   They will have intercourse, and there will be peace between them.

I + 7   They will have intercourse, and there will be hatred between them.

I + 8   They will enjoy loving intercourse, and will live in peace.

I + 9   They will have intercourse for ever in peace.

ı כ ı   2 + 2   They will have intercourse, and be of one mind.

2 + 3   They will have intercourse, and then will separate.

2 + 4   They will have intercourse for ever with strife.

2 + 5   They will have intercourse, and will live in love, and will cast aside contention.

2 + 6   They will not have intercourse, and if they do a rumour will go forth concerning them, and they will separate.

2 + 7   They will have intercourse, and then will separate.

2 + 8   They will not have intercourse, and if they do, they will separate.

2 + 9   They will have intercourse. and will live in love.

ıﻝı   3 + 3   They will have intercourse, and a rumour will go forth concerning one of them, and they will separate.

3 + 4   Like the preceding.

3 + 5   They will have intercourse, and will live in love.

3 + 6   They will have intercourse, and will live in strife.

3 + 7   They will have intercourse, and will live in peace.

3 + 8   They will have intercourse, and will rear | children. Page 521

3 + 9   They will not have intercourse, and if they do, they will hate each other.

ıꝃ ı 4 + 4 They will not have intercourse because of the bad name which will go forth concerning them. |

4 + 5 They will have intercourse, and then will separate.

4 + 6 They will not have intercourse, and if they do, they will hate each other.

4 + 7 They will have intercourse, and will love each other.

4 + 8 They will not have intercourse, and if they do, they will separate.

4 + 9 Evil men will not permit them [to agree], there is no peace between them.

ı ꜱ ı 5 + 5 They will have intercourse, and will love each other, and then will separate.

5 + 6 They will have intercourse, and will love each other for ever.

[5 + 7] [Omitted.]

5 + 8 They will have intercourse in love and peace.

5 + 9 They will not have intercourse, and if they do, they will separate.

ıꝏı 6 + 6 They will be opponents, and will never be married.

6 + 7 They will never have intercourse, for the woman hath an evil tongue.

6 + 8 They will have intercourse, and will live in peace.

6 + 9 They will have intercourse, and will be separated through an evil tongue.

ıꞁıı 7 + 7 They will never have intercourse, and if they do, there will be war between them.

7 + 8 They will have intercourse and will love each other, and he who would separate them shall be prevented from doing so.

7 + 9 They will never have intercourse, and if they do, they will be separated.

ıꞁꞁꞁı 8 + 8 They will have intercourse, and will love each other, and will rear children.

8 + 9 They will have intercourse, and will love each other, and then will separate.

ıꞭı 9 + 9 They will have intercourse, but will quarrel daily.

That thou mayest know whether a woman is with child or not. Reckon up the numerical values of the letters of the names of the woman and man, and divide each by six; if odd numbers remain to thee the woman is with child, and if even numbers remain to thee, she is not. In dividing the letters of the alphabet each by six, proceed in this manner:—Take one from seven, two from eight, three from nine, four from ten, two from twenty, nothing from thirty, four from forty, two from fifty, nothing from sixty, | four from seventy, two from Page 522 eighty, nothing from ninety, four from one hundred, two from two hundred, nothing from three hundred, and four from four hundred.

That thou mayest know whether a woman hath conceived a boy or a girl. Find out what day of the moon it is, [reckon up the numerical values of the letters in the name thereof,] and reckon up the numerical values of the letters in the name of the woman, and add to it twenty-eight, and then divide each by two. If the remainder be one, the child is a boy, and if it be two, the child is a girl.

Concerning barren folk. Reckon up the numerical values of the letters in the names of the man and the woman, and divide each of them by two. If the remainder be one, both are barren, and they will have no children, and if the remainder be two, they will have children.

That thou mayest know whether a child who hath just Fol. 246a. **been born will live or die.** Reckon up the numerical values of the letters in the names of the father and mother, and of the day on which the child was born, add to them three hundred, and add them all up together, and divide them by seven. If odd numbers remain, the child will live for years, and if even numbers, he will die. If the child be a girl, and even numbers remain, she will live for years, and if odd numbers remain, she will die.

If thou wishest to know which eye of a man is blind. Reckon up the numerical values of the letters of the names of the man and his mother, and make thy reckoning one, and divide it by three. If even numbers remain his right eye is blind, and if odd numbers remain his left eye is blind.

If thou wishest to know whether something good [is hidden] in the place of a man or not. Reckon up the numerical values of the letters in his name, and find out the total, add to them fifteen from thyself, and also a number equal to that of the total of the letters, and divide them by two, and if one remaineth the thing is [hidden] there, and if two, it is not [hidden].

When something is stolen from thee. Find out who is Page 523 the lord of the hour, | and reckon up the numerical values of the letters of his name and of that of the man whom thou seekest, and divide them by three. If one remaineth to thee, good, and if two remain, good also, but if three remain the man whom thou seekest is not there. In dividing the letters of the alphabet by twelve, proceed thus:—[Take] eight from twenty, and six from thirty, and four from forty, and two from fifty, and nothing from sixty, and ten from seventy, and eight from eighty, and six from ninety, and four from one hundred, and eight from two hundred, and nothing from three hundred, and four from four hundred.

Again, concerning a man who goeth forth from his house, and travelleth from place to place (or, from country to country). Reckon up the numerical values of the letters in the names of the man and his mother, and divide them by twelve. If one remaineth to thee he is dead, or is a very long way off; if two, he is in prison; if three, he is near, and is returning from his travels; if four, loss, and evil, and calamity are meeting him on his road; if five, he will find great profit; if six, he will remain a little while longer on his journey; if seven, evil will overtake him; if eight, he hath fallen ill, and is in trouble; if nine, he is in the desert, and is engaged in Fol. 246b. war against | enemies; if ten, he is finding joy, and merriment, and profit; if eleven, he will remain away a little while longer and will then return; if twelve, he will remain away for a very long time on his journey, and finally success and rejoicing will come to him.

Again, if thou wishest to know whether he who is afar off on a journey is in evil case or is happy, and whether he is alive or dead:—Stand up on a piece of level ground, and measure the length of thy shadow in lengths of thy feet, add twelve to them, and divide the total by nine. If one remaineth over to thee, he is in his place; and if two, he is dead; and if three, he is dead; and if four, he is sick; and if five, he is detained by some one, and will not come; and if six, he is in captivity (or, prison); | and if seven, he is travelling Page 524 on his way; and if eight, he is afar off; and if nine, he will come shortly.

Concerning those who fall into prison, whether they will return immediately, or whether they will remain there. If the Moon is in the Ram, or the Twins, or the Crab, or the Lion, or in the Goat, when a man falleth into prison, he will tarry there a time, and will come back. And if the Moon is in the Water-carrier, or the Bull, or the Virgin, or the Balance, or the Scorpion, or the Bowman, he will return immediately. And if the Moon standeth in the Fishes, he will 'not return before he dieth, and he will never go forth from that place. This forecast hath been well tried, and is certain.

Physicians most earnestly warn the sick folk who are under their care to be in a sound state of health during the winter. The sickness which taketh place under the Crab is incurable, and if it be cured it is not due to the Crab (*i.e.*, it is not cancer), because, like the Crab which grippeth and holdeth fast, this disease grippeth the veins, and nerves, and arteries, by means of which the black bile is destroyed, and this collecteth and increaseth until it bringeth about the destruction of the body, and death.

For him that wisheth to wage war against his neighbour. If the Moon is in the Ram, or in the Bull, go not forth to war; and if thou goest to wage war against thy neighbour thou wilt fall. And if the Moon is in the Twins, or in the Crab, he who goeth forth to war will conquer. And if the Moon | is in the Lion, or in the Virgin, or in the Scorpion, or Fol. 247 a. in the Bowman, or in the Water-carrier, or in the Fishes, he who goeth forth to war will conquer. If Ârîs or Kronos be found in one of the Malwâshê do not fight, for there will be no good in thy fighting, but only evil. If the Moon be found in one of these Twelve Malwâshê by itself, there being none of the other six Planets with it, go into battle, and thou shalt conquer. And if the Moon is in the Ram, or in the Crab, or

in the Lion, or in the Water-carrier, or in the Virgin, whenever
_{Page 525} the Moon is found | to be in one of these five Malwâshê, and
none of the other luminaries is with it, enter not into war. Be
warned!

**Again, an admonition as to whether a man shall make
a petition to the king, or anything else.** If the Moon is in
the Ram, or in the Crab, or in the Bull, or in the Virgin, or
in the Balance, that which thou shalt ask from the governor
thou shalt receive, and thou shalt prosper. And if the Moon
is in the Scorpion, or in the Great Image, or in the Goat,
thou shalt not ask for anything. But if Zeus be there, or Baltî,
or Hermes, thy petition shall be answered [favourably], and if
none of these be there, it shall not. And if the Moon is in
the Lion, or in the Bowman, or in the Water-carrier, or in
the Fishes, and thou shalt enter into judgement with a man,
thou shalt not prosper, and thine adversary shall overcome
thee. And if thou makest a petition to the governor, he shall
turn away his face from thee.

Another kind of divination—about marriage. If the Moon
is in the Ram or in the Bull, he who getteth married will
destroy the marriage. And if the Moon is in the Twins, it
will be well with him that is married on the second day; and
if he be married on the third day of the Malwâshâ, the woman
will be to her husband. And if the Moon is in the Crab, the
woman will flee immediately from her husband. And if the
Moon is in the Lion, the woman will be a destroyer. And if
the Moon is in the Virgin, it is better for a man to marry a
_{Fol. 247 b.} widow than a | virgin. And if the Moon is in the Balance, the
woman will be true to her husband. And if the Moon is in
the Scorpion, on the first day of the Malwâshâ, the woman
will be good to her husband, and obedient unto him. And if
the Moon is in the Bowman, the marriage will be very happy.
And if the Moon is in the Goat, the woman will play the
harlot, and commit adultery. And if the Moon is in the Water-
carrier, it will be very good for the marriage, and so also if
the Moon is in the Fishes.

A change, which giveth information about the sub-
_{Page 526} **terranean sea to those who possess understanding.** | Beneath

the earth there is an awful sea of many waters, and beneath
the sea of waters is a sea of fire, and beneath the sea of fire
is a sea of wind, and beneath the sea of wind is a sea of
darkness. And in the days of summer, when the sun riseth
up to that upper region, and to the wind of this firmament,
and his sphere burneth with fierce, blazing heat, and maketh
hot the earth, immediately that fire which is below is quenched,
like a fiery furnace, and the waters of that lower sea become
cold, and the wind of coldness bloweth upon them, and the
coldness thereof riseth up and passeth over the earth, and it
passeth through the veins of the trees, and through the arteries
of the rocks, and through the roots of plants, so that the sun
may not burn up the trees, and the plants, and the seeds.
And if it were not for that cold which ascendeth from within
the earth, the sun would not leave on the earth a single green
thing unconsumed by it, and the children of men would not
be able to walk upon the earth, because of the heat of that
fire which is inside the earth. And it would, moreover, destroy
the flow of the rivers and springs, and the ministration of the
cold would not bring them back again, and in the days of
summer, where there was no water, the sun would burn up
[everything].

Of the sun, and of how he travelleth below this firmament.

During one complete course of his journey the sun | shutteth Fol. 248 a.
[twelve doors], now, he passeth through all these twelve doors
when performing his course. And these twelve doors are
marked on the sundial, in order that the course of his journey
may be equal (?). Each door is distant from its neighbour a
space of one day (*sic!*) only, and every hour hath one degree.
And four winds make the sphere of the sun to travel on its
way. Because that wind which is above is mighty—now it is
the wind wherein the clouds travel—if that celestial blast were
to move a little downwards, and towards the earth, | there Page 527
would be nothing left on it which was not destroyed by it.
Four winds embrace the sphere of the sun on all its sides;

there is one in front of it, and one behind it, and in like manner
there are in all the quarters winds which keep it in its proper
place, lest it should fall, or travel in an irregular manner. If
this were not so it would fall on the boundaries of the earth.
Similarly there is one wind under it, and one wind above it,
and from time to time one of these four storehouses, which
minister to that wind which is above, is opened. Now the
wind that cometh forth from one of [these four storehouses]
is mighty, [and it striketh] the sun, and it turneth back the
sun under the degree (?) of its passage, and the light of the
sun is cut off until the wind that is behind it returneth, and
it stablisheth itself upon its place on the path of its course.
For when the sun riseth up to one of the Great Doors, it
ascendeth from the dry land that they may take it and may
make it to pass beyond the habitation of men. If it ascendeth
to the height, mighty clouds make it to travel along in the
air, and one of these treasure houses is opened, and with earth-
quakes and lightnings he riseth up, and taketh his place in
the mountains of the north. And also in these seas great
serpents (or, Leviathan?) are produced, but not with us; serpents
are only produced in the sea that is called "Demâstâḳâbôs".
Fol. 248ᵇ. Now, when the sun riseth to that fiery wind, it drieth up with
its light all the ends of the earth (?) that are beneath the
firmament, and its sphere blazeth with the heat that is from
above. And the sun itself burneth up the whole earth, because
he riseth up to the wind that bloweth from above.

Now understand ye, O lovers of wisdom, that under this
earth there is an awful sea of many waters, and under these
is a blazing fire, and under this is the Nod, whereby the world
Page 528 standeth. And in the days | of summer, when the sun mounteth
up to that high region of fiery aether, and when the sphere
of the sun blazeth with fire, and when it maketh the whole
earth to glow with fervent heat like an oven of fire, then God
commandeth the winds that blow under that lower sea of fire
to ascend mightily, and they diminish the heat of the lower
fire, and they rise up and pass over the lower sea of waters
and cool them. And that coldness riseth up and ascendeth
from the earth, and it passeth into the veins of the trees and

shrubs, and into the arteries of the rocks, and the dust of the
earth becometh cooled to such an extent that the sun cannot
destroy with its heat the earth and the children of men who
are upon it, and the trees are not destroyed, and the whole
of creation doth not suffer torture. Now, the surface of the
earth is made like a sponge, and the inside of the earth is
[made] in the form of canals and tubes for the winds and the
waters in connection with the service of heat and cold. And
thou mayest learn this fact by experience from the dust of
the earth; the further down thou diggest, the colder doth it
become. Moreover, when in the summer animals are afflicted
by thirst, they dig very deep holes in the earth, and lie down
in them, and find relief. And the men also who dwell in the
southern countries, in the land of Cush, and in the land of
Shĕbhâ, dig holes in their sandy ground, and take up their
abodes therein, and very many of them find refreshing.

| And a different kind of section, which giveth information Fol. 249 *a*.
concerning the winter. Now, in the days of winter, the
sphere of the sun is cast towards the south, for in this direction
is the course of the degrees of its operation. And the heat
thereof is remote from the north, and cold winds rise and
blow strongly, and because the sphere of the sun is lowered
from the heat of the fire of the celestial aether, the sun is
unable to warm the coldness of the winds of snow and ice
which blow and cool all the face of the earth, and they | also Page 529
cool and reduce the heat of the sun. And in order that the
cold itself may not destroy men, and trees, and animals, and
bring to an end everything which is on the earth, the sub-
terranean winds are stilled. And God commandeth the fire
which is beneath them, and it blazeth forth, and the heat
thereof riseth up, and ascendeth through the channels and
tubes which are in the earth, and it passeth into the veins of
the trees, and it melteth the cold which abideth in the trees,
and in the waters, and in the dust of the ground, and through
that fervent heat which ascendeth from the interior of the earth,
the waters of the springs are melted. And instead of an icy
blast there ascendeth from the interior of them the vapour of
heat, and that heat goeth up from below, and warmeth the

severity of the cold. And the animals find refreshing when they dig holes in the lower dust, and lie down in the warmth thereof; and the birds also hide themselves at night-time above the waters of the springs and rivers, and are warmed; and the men who dwell in the north, within the mountains of the children of the north [are also warmed]. Now, the stones of those mountains are of crystal. And within (*i.e.,* beyond) these mountains there is no human habitation, for beyond that river

Fol. 249*b*. which is called the "river . . . ." there is nothing at all | except the Ocean, that terrible sea which surroundeth the whole earth. And as for this great sea, there is no creature whatsoever that moveth in it, and no bird is able to fly over it. And like the wall that surroundeth a city, even so doth this sea surround the whole earth. For what reason are the store-houses of the wind opened? There is a terrible sea beyond the region inhabited by men, and herein are produced serpents and mighty dragons. And when one of these becometh strong, he goeth up and cometh to the great intermediate sea, so that through the children of men the treasure house of the wind may be opened upon him. And the clouds assume form, and pile themselves up upon each other, and they carry him

Page 530 along with thunders and lightnings, | and hurl him on one of the northern mountains beyond the region of human habitation. And there stones of crystal, that is to say ice, descend upon him, and kill him.

A description of the measure of the earth.

The earth consisteth of five divisions, and each division containeth twelve parts, that is to say, in all sixty parts. Three divisions, or thirty-six parts, are water, that is to say, thirty-six days. Of the sixty parts there remain four and twenty, that is to say, two divisions: [of these] fifteen parts may be travelled over, one is uninhabitable because of the heat of the sun, one is uninhabitable because of the cold of winter, one is uninhabitable because of the height of the mountains, and the six parts which remain of the sixty parts into which the whole earth is divided, that is to say, one-tenth of the whole earth, cannot be described.

Again, the measure of the earth from east to west. From the earth to the heavens is [a distance of] seven times, even as it is from | the east to the west. **Again**, from the Fol. 250a. east to the west is [a distance of] eleven thousand and sixty-four parasangs. And from the earth to the heavens is [a distance of] seventy circles, and each circle is one hundred and thirty-eight parasangs, and the total [distance] of the circles is one thousand, six hundred and seventeen parasangs.

Calculation of weight. Ninety-six barleycorns make one *mathḳâlâ* of gold, or silver. The weight of a barleycorn is ten mustard seeds, and the whole *mathḳâlâ* weigheth nine hundred and sixty seeds of wild mustard, according to the opinion of Mâr Êlîyâ, Metropolitan of Nisibis.[1] Seven barleycorns make one finger. Twenty-four fingers make one cubit. | Four thousand arms (*i. e.*, cubits) make one mile. The parasang Page 531 is twelve thousand cubits. Sixty-seven barleycorns make one cubit exactly.

Of the parasang. The parasang of the road is three miles. The mile is ten stadia. The stadium is three *sedê*. The *sedâ* is one hundred paces. The pace is half a reed. The reed is ten cubits.

Of the measure of the Malwâshê. The measure of each of the Malwâshê containeth thirty parts. Each part is seven hundred stadia. A stadium is two hundred feet. The foot is two cubits. The cubit is two spans. The span is twelve fingers.

Of the measures of the hour. The *pôdhîthâ* (or "foot") is one measure of the sole of the foot. The hour is thirty *pôdh-yâthâ* (of two minutes each?). The *pôdhîthâ* is ten *sesê*. The *sesâ* is twenty *rûpsê*. The *rûpsâ* is two thousand twinklings of the eye.

Moreover, one *pôdhîthâ* is two *ḳaṭînâthâ*. One *ḳaṭîntâ* is two *rûpsê*. One *rûpsâ* is one thousand twinklings of the eye. Six *pôdhyôn* are reckoned as one *ewneḳyâ*, and five *ewneḳyâs* are one hour. Twenty-four hours | make one night and one Fol. 250b. day.

---

[1] He succeeded Bar-Ṣaumâ, who died A.D. 458.

Moreover, the *pôdhyâthâ* of the whole year are forty-three thousand, and eight hundred and fifteen. The *pôdhyâthâ* of one day are one hundred and twenty. The *se'ârâthâ* of one day are one hundred and twenty. The *se'ârâthâ* of one day are three hundred (*sic!*). The *se'ârâthâ* of the whole year are two hundred and sixty-two thousand, eight hundred and ninety. An hour is five *pôdhyâthâ*, and a *pôdhîthâ* is six *se'ârê*. Five *ḳanṭîmê* are one hour. The weeks of the whole year are fifty and two, and one day.

Concerning the constitutions of the Malwâshê. The Ram, the Lion, and the Bowman are hot and dry like fire, and like the region of the South, and they are masters of the four-legged creatures (*i. e.,* quadrupeds). The Bull, the Virgin, and Page 532 the Goat | are cold and dry like the earth, and like the region of the West, and they are masters of the trees and grass (or, herbs). The Twins, the Balance, and the Water-carrier are hot and moist like the air, and like the region of the East; and they are masters of men and birds. The Crab, the Scorpion, and the Fishes are cold and wet like water, and like the region of the North, and to them belong all things that emit water.

Concerning the Seasons which thou seekest, Spring, Summer, the period of the Teshrî months, and Winter. In the Spring moisture increaseth because of the persistence of cold, but the beginning of the heat melteth the cold. In the Summer the heat increaseth, because of the descent of the sun. In the period of the Teshrî months (Autumn) the dryness is great, because of the ending of the heat which burneth up the moisture. Now the Winter is very cold because of the remoteness of the sun.

Information concerning the winds. There are four winds which minister to mankind and to everything else, and to them Fol. 251a. are due | the changes in the atmosphere [in] summer and winter. Two of them, those of the east and south, minister to the winter, and when they blow they fill the springs to overflowing, and the rain falleth, and the crops are strengthened, and by means of them the trees flourish and become strong, and bear fruit. But if these two winds, the east and the south,

blow during the summer, they are exceedingly harmful. The
north and west winds minister to the summer, for by means of
them the springs of water are replenished, and they begin to
flow, and their volume is increased; and through them the
cedars and the [other] trees put forth their fruits. Now if
these winds blow during the winter, they produce no benefit
whatsoever, and they do harm to the springs of water.

Concerning the winds. There are four true winds, and
they are the east wind, the west wind, the north wind, and
the south wind. From the east cometh the wind called
"Aplîtôs"; this hath two winds, which blow from the right
thereof and from the left thereof respectively, and they put it
in the middle. From the right side of it cometh | the wind Page 533
called "Ḳîs", and from the left side of it cometh the wind
called "Ôzôs". The name of the wind that bloweth from the
west is "Zôphrôs", and it hath two winds which blow one on
each side of it. The wind on the right is called "Aphrôs".[1]
The name of the wind that bloweth from the north is "Bûrâs",
and it hath two winds which blow one on each side of it.
The wind on the right is called "Tarḳîs", and that on the
left is called "Aprîtôs". The wind that bloweth from the
south hath two winds which blow one on each side of it. The
wind on the right side is called "Lîbîṭôs", and that on the
left is called "Ôrîṭôs". The winds are twelve in number, and
the quarters whence they come are four.

**Concerning the air (*i. e.*, climate) of the [four] quarters
and of the seasons of the year.** The air of the east, and
of the season of Spring, is hot and wet. The air of the south,
| and of the season of Summer, is hot and dry. The air of Fol. 251 b.
the west, and of the season of Autumn, is cold and dry. The
air of the north, and of the season of Winter, is cold and wet.

Concerning the beginning of the Four Seasons. Spring
beginneth on the twenty-fourth day of the month Âdhâr, and
lasteth until the twenty-fourth day of the month Khazîrân.
Summer beginneth on the twenty-fourth day of the month
Khazîrân, and lasteth until the twenty-fourth day of the month

---

[1] The scribe has omitted a line here.

Îlûl. **Autumn** beginneth on the twenty-fourth day of the month Îlûl, and lasteth until the twenty-fourth day of the month First Kânôn. **Winter** beginneth on the twenty-fourth day of the month First Kânôn, and lasteth until the twenty-fourth day of the month of Âdhâr. Now many have proclaimed that the beginning of the growth [of plants] taketh place on the twenty-second day of the month Âdhàr, when the sun entereth the Sign of the Ram, and others have said that trees begin to put forth buds on the seventh day of the month of Shebàt, and others in the middle of Shebât.

A forecast concerning the year, and the airs,[1] and the times (or, seasons).

When the course of the year through the airs is proper and good, everything cometh forth in its season, | and know also that sicknesses will be slight therein. When the course of the year is disturbed and variable, sicknesses which are severe and troublesome, and which cannot be cured quickly, also arise. The changes that take place produce sicknesses which are opposite in character to those of the airs; if they be cold, or hot, or dry, or wet, sicknesses which are different in their natures make themselves visible. When the South Wind bloweth, it maketh the ears heavy, and it tormenteth the eyes, and head, and members; and when the force of this wind increaseth, sicknesses [in these members] arise. When the North Wind bloweth it produceth much coughing, and headache, and sore-throat, and restriction of the bowels, and strangury, and feverish pains in the bones, and pains in the sides and | chest. And if the air of Summer changeth, and becometh like unto that of Nîsân, much shivering and discomfort arise, and it produceth severe pain, and excessive sweating; and fever breaketh out at this season, and men require to abstain from food, and they need purging. Now, if the air of summer becometh soft, and humid, and exhausting, there is severe fever; and if during the greater part of the

Page 534

Fol. 252a.

---

[1] *i. e.*, temperatures, or weather.

year this air continueth, then know that acute illnesses will
arise. And when the air of Winter becometh soft and humid,
and the North Wind increaseth, and the rains of Nîsân set in,
and there is of necessity an excess of rainy weather, then
know that, as is the case in the Summer, there will be acute
fever, and pains in the eyes, and diarrhoea. And women will
suffer in a far greater degree than men, because the nature of
their bodies is more delicate. When the Winter is wet, or
when the month of Nîsân is very mild and soft, then know
that the women who are with child will be afflicted by various
kinds of sicknesses which will attack them, and that towards
the month of Nîsân, through some cause or other, they will
suffer from miscarriages. The woman who bringeth forth her
child will continue to suffer, and the child will suffer; some
women will die immediately, and those who escape [from death],
will be destroyed by sicknesses. People who possess a fresh,
dry constitution will suffer from pain | in the belly, and from Page 535
diarrhoea, and will die forthwith. And if the air of Winter be
soft and humid, and if the air of Nîsân be fair and constant,
there will be a pestilence among the aged. Those who are
young in days will become very weak, and the growth of their
stature will be impeded. When the Summer is soft and humid,
and the North Wind increaseth therein, and the Autumn is
rainy, then know that in the Winter there will be throbbing of
the head forthwith, and pain in the throat, and diarrhoea, and
men will stand in danger [of their lives], and a great pestilence
will take place. If the months of Autumn be hot, and the
airs thereof be variable, then know that many recoveries and
| healings will take place among women and children, whose Fol. 252b.
bodies are moist, and soft, and white. Other folk will suffer
from dry rheum, and acute fever, and black bile, and the
greater number of them will die. As concerning the difference
between the airs, dryness and softness help men more than
moisture, and few men die of these, for they destroy the cause
of the sicknesses.

The sicknesses which take place during a year which is
damp and rainy are very difficult [to cure], and mankind
standeth in danger of attacks from protracted and acute fevers,

and from looseness of the bowels, and from pains in the bones, and from troublesome pains, and numbness (or, stupor), and *âpahlôs* (?), and wasting away of the body, and headaches, and pains in the ligatures of the members, and strangury, and pains in the bowels and stomach. And the food loseth its power, and becometh bitter and cold, and returneth (*i.e.*, is vomited), and during a year of this kind men stand in danger [of their lives]. And the sicknesses which take place during the months of autumn are very alarming, for pestilence breaketh out immediately among the sick, and old cases of consumption carry them off; for the year in which there are months of this kind is one of many changes, and the air (*i. e.*, weather) thereof varieth continually, and its course is marked by manifold changes and by the vicissitudes of perpetual variations. And if the North Wind bloweth, then know that delicate persons will become invigorated, and renew their strength, and improve in health. Their ears will become soft, and their bellies will Page 536 stand (*i. e.*, be healthy), | and their eyes will become moist, and if they have any pain in the chest it will be straightway removed. And when the air of the year is moist, the body languisheth and becometh sluggish, and the ears throb and become heavy, and the head becometh hot and heavy, and the eyes become dark and dull, and the temples throb, and all the arteries cry out in pain, and the body languisheth, and becometh sluggish, and dull, and stupid, and heavy. And in the month of Nîsân of such a year mankind standeth in great danger, and a great pestilence taketh place.

When the East Wind bloweth out of its season, persons become heavy, and the eyes run with water, and the temples Fol. 253*a*. throb, | and the stomach is affected and becometh cold, and the blood loseth its heat and becometh cold, and it diminisheth and arriveth at a state of coldness; if this wind increaseth sicknesses take place. And when the South Wind bloweth the following sicknesses appear: diarrhoea, and the black bile is stirred up, and dry blearedness of the eyes, and rheum in excess, and the brain and the bones are sorely affected. In the year wherein a large number of changes take place in the heights of the air, mankind standeth in a state of great trepidation, because the sicknesses which appear during its course are difficult [to heal], and a great pestilence taketh place. During the winter of such a year men will suffer from pain in the interior, and pain of mind (*i. e.*, depression), and sore-throat and coughs, and pain in the chest, and headache, and madness of heart, and pain in the kidneys, and there will be great tribulation, and mankind will stand in danger. During the month of Nîsân of such a year there will appear black bile, and diarrhoea, and diseases of the intestines, and madness of heart, and pains in the interior, and the flow of blood, both from above and from below, and strangles, and coughs, and scabies, and boils and sores, and pustules, and mankind will stand in danger. During the summer of such a year as this there will be acute fever, and drought, and blight (or, mildew), and intermittent fever, and vomiting, and diarrhoea, and dry blearedness of the eyes, and pain in the temples, and Turkish fever, and consumption, | and the sores of gangrene (or, cancer), Page 537 and affliction, and mankind will stand in trepidation.

And during the autumn months of such a year there will be fever, and pains in the excretory organs, and asthma, and pains in the bones, and terrors, and madness of heart, and dropsy, and rheum, and thick spittle, and diseases of the intestines, and black bile, and mankind will be afflicted and there will be a great pestilence. Of all the airs of the year, the air of Nîsân is the keenest (or, severest), and that of the autumn months sweepeth away mankind; it is only beneficial for cattle, Fol. 253*b*. | wild goats, and elephants. At the beginning of the summer children become strong, and their bodies active; and this is the case with the old, and with elderly persons during the summer, and during the autumn months, and with the young and middle-aged during the winter. Some children, however, become weak and fall sick during the summer, as do also some young and middle-aged persons during the autumn months. Some persons enjoy good health during the winter, because they possess an ardent nature, and there are some who enjoy good health during the summer because their nature is cold, as do also some during Nîsân and the autumn months because of the natural constitution of their bodies. We set for purposes of comparison years against times, and airs against times, and airs against countries (or, regions), and water against countries, and water against food, and food against nature, and nature against them all.

Of the natural dispositions of men during the various periods of their lives.

Children are hot and wet. Between early manhood and old age they are cold and dry. Old men are cold and wet. Strong men are hot and dry. Women are cold and wet.

Of the change of sicknesses. The greater number of sicknesses take place during the month of Nîsân. In this month black bile appeareth, and men lose their senses through it, and tribulation cometh upon them, and flowings of blood break forth, and there appear diseases of the throat, and the skin of the body cracketh, and festereth, and peeleth off, and [men suffer from] the itch, and inflammation, and spasms, and sores, and pustules. And in summer there come fevers, both slight and acute, and intense thirst, and intermittent fever, and vomit-Page 538 ing on the ground, | and pain in the ears, and pains in the genital organs, and consumption, and sores in the mouth. And during the months of autumn there appear fever, and pain in the spleen, and dropsy, and rheum, and emissions of water, and strangury, and pain in the excretory channel, and asthma, and něsôṭanyâthâ (?), and running of the eyes, and ulcers, and black bile. And in the winter there appear pains in the interior, and pain in the lungs, and palsy of the members, and every kind of trying ailment.

Fol. 254a. | **Whosoever feeleth his sickness,** and hath black bile, either from above or below, which is not a cause for medicines, know that he will die. Whosoever hath a sharp attack of sickness with bile, which is either like blood or is black, and which cometh either from above or below, know that he will die. When a sick man hath a sweat for three days, or for five, or for seven, or for nine, [or any] odd number [of days], know that he who sweateth is in a most serious condition. When a cold sweat cometh with fever, know that it portendeth death. Whosoever is sick, if he sweateth a hot sweat copiously, know that the sick man will be healed quickly. Whosoever hath fever, and sweateth not at all, and suffereth from wakefulness (*i. e.,* insomnia), know that he will die.

Again, information concerning touch, or feeling. The feel of blood is great (or, full), and sticky, and thick. The feel of black bile is flat and hard. The feel of phlegm is narrow and soft. The feel of venom and bile is great (or, full) and dense. The feel of winds confined in the belly is like that of leaping animals (?). The feel of the stomach and lungs when thou pressest thy fingers under the middle [thereof] is that they palpitate. The feel of death is terrifying and disturbing. The feel of the veins:—Place thy fingers by the side of the thumb of the left hand | of the sick man, and touch the veins; if he Page 539 draweth himself upwards and leapeth up, like a man who wisheth to flee at all costs, he is suffering great agony, and his fever is strong; and if he is slow to move, and leapeth up, know that his sickness is not acute; and if he leapeth downwards, his bowels are loose.

Concerning the drinking of water. As for the water which cometh from wells, and that which standeth in pools and in hollows, and that which floweth gently, and softly, and quietly, and that which floweth along and runneth slowly | in rivers Fol. 254 b. and brooks, know that all these waters are heavy and injurious. As for the water which floweth swiftly in rivers and in brooks (or, streams), know that it is healthful and light. As for every kind of water which cometh from heaven, that is to say, rain water, and the water which standeth in springs, and in fountains, and in the mountains and hills, know that such water is healthful and light. Therefore it is not meet for the sick, and those who are consumptive, and the feeble, to drink the water which standeth in pits and pools, or which is in the rivers and streams that flow gently, and quietly, and lazily, or to make use of them in any other way whatsoever.

Again, concerning the drinking of water.
The night of the first day of the week. At midnight drink not water.

The night of the second day of the week. Up to midnight drink water, and let it suffice.

The night of the third day of the week. At the tenth hour do not drink water.

The night of the fourth day of the week. Drink water at any time during the night, and have no fear.

The night of the fifth day of the week. Drink water at any time during the night, and have no fear.

The night of the Eve of the Sabbath. At the second hour drink. Observe, and beware the whole night long.˙

The night of the Sabbath. Drink the whole night long, and fear not.

[Of the drinking] of milk. Milk is for him that hath headache, and pains in the eyes, and fever, but to him that is thirsty thou shalt not give milk, because it is very harmful [for him], and it weigheth heavily [on him].

Page 540 | Of the course of dieting which is to be followed during the various seasons of the year. The winter course, which is to be followed from the seventeenth day of the First Kânôn to the eighteenth day of Âdhâr, that is to say, for ninety-two days. During this season sweat and moisture increase until the change which cometh after the winter, therefore use for food things that are warming, and drink wine that hath strength in it, and let a man labour, and become fatigued.

After the winter there are forty-nine days, that is to say, from the eighteenth day of Âdhâr to the sixth day of Îyyâr.

Fol. 255a.| During this season the saliva increaseth, therefore use for food things that possess strong odours, and everything that is pungent (or, acid), and work, and become fatigued.

And from the sixth day of Îyyâr, which is the day of the rising of the Pleiades, until the eighteenth day of Khazîrân, that is to say, for forty-two days, follow the course of dieting which cometh before that of the summer. During this season the yellow bile increaseth, and fever also, and thou must use for food everything that is sweet, and thou must labour in moderation.

The course of dieting for the summer, that is to say, from the eighteenth day of Khazîrân to the eighteenth day of Îlûl, which is a period of ninety-two days. During this season the black, dry bile increaseth. And thou shalt use weak (*i. e.*, light) wines, and shalt eat delicate food, and take everything that is acid (or, sour), and do very little work.

After the summer come forty-nine days, that is, from the eighteenth day of Îlûl to the seventh day of the Second Teshrî. During this season the bile and rheum increase, and therefore use as food everything that is pleasant and acid; and either abstain from work entirely, or work in strict moderation.

From the seventh day of the Second Teshrî, whereon the Pleiades set, to the seventeenth day of the First Kânôn, which is a period of forty-one days, the course of dieting which precedeth that of the winter must be followed. During this season blood increaseth in men, and it is therefore meet for a man to eat sparingly, and to drink wine in strict moderation. And he must work, and become fatigued, and he must continue to work until he beginneth the course of dieting suitable for the winter.

Again, on the different kinds of foods which are to be Page 541 **eaten during the various months of the year.**

First Teshrî. Eat and use what is pungent (or, acid).

Second Teshrî. On no account shalt thou wash with water.

First Kânôn. Partake of as much sweet food as thou art able.

Second Kânôn. Drink "Aḳṭirôn", and guard thyself carefully against the cold.

Shebâṭ. Eat no vegetables of any kind, because of phlegm and cold.

Âdhâr. Eat sweet fruits, and drink infusions | of the seeds Fol. 255 *b.* thereof, and make use of any pungent (or, acid) food.

Nisân. Eat not radishes, or beetroot, or any stale vegetable, and abstain from the flesh of goats and bulls.

Îyyâr. Eat not the heads and milk of goats and bulls, and abstain from the flesh of the same.

Khazîrân. Eat nothing which is boiling, or which hath been heated over the fire, for it injureth the mouth and the interior. And drink cold water fasting (or, on an empty stomach).

Tammûz. Draw not nigh unto a woman, and eat neither pork nor eggs.

Âbh. Eat no vegetable which hath been cooked by fire, and no vegetable soup.

Îlûl. [Drink] milk, and eat flesh continually, and thou shalt find satisfaction (or, rest) for thy body.

Another series of directions which sheweth what a man should and should not eat during each month of the year.

First Teshrî. It is meet to avoid water melons, and cucumbers, and ox-flesh, and fatty foods, and acid things, and honey, and everything sweet, and dry foods made of meat, that is to say, cold meat. In this month eat *estaplîn* that is to say, carrots and all kinds of spring vegetables after food, and new wine, without water, in the morning, fasting, and visit the baths, but do not drink cold water there; partake of olive oil therein. Enjoy thy wife, and perfume thyself with scents, and do not labour overmuch.

Second Teshrî. It is right to avoid honey and butter (or, Page 542 cream). Eat | during this month the flesh of birds of all kinds, and meat cooked with garlic, and drink wine. And eat no vegetable except rock parsley, and gourds, and leeks, and carrots, and *bâbhûlâ* with honey, after food. And drink the following medicine each day: take one drachm of aniseed and one drachm of peppercorns, crush them, and drink them in wine and honey water. This medicine is good for stabbing pains and pains in the knees, and it increaseth seed in men, and it hath been well tried, and is a sure remedy. And also get married in this month, as it is written.

First Kânôn. Follow these instructions which I am now Fol. 256a. going to give thee, and avoid | the flesh of the cock. Get married in this month, for all the children who are begotten in these three months are more beautiful than those who are begotten during the other months of the year.

Second Kânôn. Avoid cold water in the morning, and avoid onions and the flesh of oxen, and goats, and camels, and lentils. And eat in this month honey, and butter (or, cream), and thick soup, and fat, and the flesh of sheep boiled with garlic, and drink all kinds of wines. And after a meal eat a little honey.

Shebât. This is the month of troubles, and it is among the months as was Judas among the disciples. Avoid copulation, and the flesh of goats, and fat, and baths, and butter (or, cream), and fatty substances, and oils. In short, avoid every kind of food that is cold and heavy for the digestion. . . . . .¹ and mix in it hot milk, [and] divide the . . . . . . and the seed of spinach (or, orach) roasted, and nuts and almonds.

Âdhâr. Avoid chick-peas and unboiled onions, and fish, and vegetables (?), and fatty substances. Moderate the appetite, and moderate dry (?) labours.² Get married in this month. Go to the baths on the first day of the month, and on the third day of the last week in it, and keep thyself free ¦from any injury which [draweth] blood.

Nisân. Avoid everything | which is sweet and salt, and Page 543 *gûnyâ* (?), and gourds (or, cucumbers), and chicken. And use [as a medicine] . . . . and pound them up and mix them with vinegar and mustard. And avoid the drinking of wine, and marry in this month of Nîsân, and go to the baths, and anoint thyself with oil of jasmine, and scent thyself with sweet scents.

Îyyâr. Eat no roast meat, and avoid olive oil, and eggs, and fatty substances, and all kinds of wine, and [butter, or cream], and fat, and do not eat gourds (?).

Khazîrân. Eat butter, and honey, | and roast meat, and Fol. 256*b*. fresh (or, raw ?) fish, and walk during this month. And avoid the flesh of sheep, and honey, ...

[Several lines wanting.]

Again, of the proper ordering of foods. Satiety and glut-tony in respect of meats of all kinds produce divers sicknesses, and cause pains in the bones, and thicken the feet (or, legs), and make certain the descent of the body to a state wherein the members become destroyed. One of the philosophers said, "I wonder how a man who doth not over-eat himself can die." Those men who possess blood in excess must eat foods that

---

¹ The scribe seems to have omitted a line or so here.
² The text must be corrupt here.

are cold and dry. And those who lack blood, but who possess red bile in excess, must eat foods that are moist and cool. Those who possess black bile in excess must eat foods that are hot and dry.

The things which make dry are .... and mustard, and everything that is cooked in a frying pan, and that is roasted, and is then eaten. The things that are boiled in water produce Page 544 moist growth. Everything that is sweet | is injurious to the liver and spleen, and heateth the blood. All kinds of polenta food are easily digested by the blood. All field vegetables must be dried, and all garden vegetables must be wetted; everything that is cold thou must roast, for it will then destroy the moisture which is in the stomach. Peppercorns, origanum, hyssop, mint, and fresh mustard, when dried, are not food, but medicine.

The things which cleanse, and cut, and open are:—tri-Fol. 257a. gonella, sweet | grapes, dried beans, chick-peas, especially the black kind. Moreover, the medicine which reduceth the stones that grow in the kidneys is camphor, especially when it is crushed in vinegar and oil, or in vinegar and honey. This is more efficacious than any kind of food. Vinegar and honey reduce the phlegm, and cleanse the internal organs gently, and they afford relief to the chest, and to cases of pain in the lungs. And dry drugs cut up very finely, and almonds chopped up and pounded, do good to the internal organs, and soften the chest and the lungs, and bring up the moisture. And to cure [the evil effects of] *sĕdhâb* and [other] deadly drugs, make an infusion of rock parsley and beetroot [in water], and add to them a little honey and myrrh, and drink draughts of it; if a man maketh an infusion of fresh leaves in water, and mixeth wine therewith, and drinketh of this infusion, it will loosen his bowels. And if he dissolveth sugar therein instead of wine, the mixture will afford relief in cases of hot pains. Water drunk fasting (*i. e.*, when the stomach is empty) reduceth the stomach, and the intestines, and the kidneys, and bladder, and benefiteth slightly the liver during the summer. The drinking of water benefiteth persons who possess in excess red bile and blood, but it is injurious to those who possess in excess black bile and phlegm. Water [drunk] after eating a meal is much more beneficial than when drunk with food; water cleanseth and cooleth. Those who drink cold water in the summer, and hot water in the winter, make a mistake. A very red complexion is due to blood, a yellow complexion is due to red bile, a dark complexion is due to coldness of the black bile, a white complexion is due | to coldness and thick- Page 545 ness of the phlegm. The right side of the body is strong and hot, and the left side is cold and weak. The persons are divided, the natures are confounded, the faces agree, the race is one, the species are three, operations are manifold, the objects of worship are two, the throne is one, the . . . . . . are six, the . . . . . are spacious, the sheep are few, the shepherds | are many, the house is one, the essence (or, person) is divided Fol. 257 b. and turned into three orders, one doth not suit the other, [yet] are they bound together. The subject is ended, and the bile of the dragon hath been expelled.

Again, concerning the forecasting of the year.

Of the indication which taketh place on the nineteenth day of the month of Tammûz, the month of the Sun. On this day the Dog of the Giant (*i. e.*, the Dog-star) riseth, and from it thou canst obtain forecasts about the whole year, [and] about the winter. If thou seest in the region of the sunrise a mist (or, fog) and a black darkness, there will be rain in the first part of the month of the First Teshrî. And if these appear at midday, they indicate that there will be rain during the latter part of Teshrî. And if there appeareth rain with the mist, it indicateth that there will be frost in Teshrî instead of rain. And if there be dew in the night, it sheweth that there will be much rain during the First Teshrî, and also on the second and third days of the Second Teshrî, and on the fourth day of the First Kânôn, and on the fifth and sixth days of the Second Kânôn, and on the eighth day of Shebâṭ, and on the ninth day of Âdhâr, and on the tenth and eleventh days of Nîsân, and on the twelfth day of Îyyâr. And this is true (or, certain).

And if thou seekest to know the truth of this [forecast], during the eighteen days in Tammûz, on the night of the twelfth, after sunset, bring (*i. e.*, take) fig leaves, or olive leaves, or vine leaves, and write upon them the names of the months of winter, each name upon one leaf, and carry them up to Page 546 the roof | of the house, and place the leaves under a vessel until the morning of the next day, and then take them out. And from each leaf that is moist and green thou shalt know that the whole of the month, of which the name is written upon it, shall be rainy and have dew, and from each leaf that is faded in colour and hath no moisture in it, know that during the month the name of which is written upon it there shall Fol. 258*a*. be no rain at all. | This is a well tried and certain [forecast].

Another kind. Observe (*i. e.*, select) one of the days from the fourth to the tenth day, (which is a period of seven days) of the First Teshrî, and let each day represent a month from the beginning of the First Teshrî to the month of Nîsân. Then observe on which day there is rain, and thou shalt know that during the month that is represented by that day there will be rain.

Another kind. Observe during the Festival of the Cross. If there be wind and mist (or, fog) [on that day], the year will be a good one.

Concerning the changes of the clouds. If on the eighteenth day of Îlûl clouds come up from the beginning of the day, there will be rain during the First Teshrî. If they come up at midday, the middle part of the year will be good, and if they come up at sunset, the year will turn to hardness (*i. e.*, the weather will be bad). If clouds are seen in the east, the crops in the east will be good. If clouds are seen in the west, the crops in the west will be good. And this will be the case also if clouds are seen in the north and south. If clouds are seen in all the earth, the crops will be good everywhere. If the clouds are black, the year will be rainy. And if the clouds are white, there will be much snow and ice. And if half the clouds are black, and the other half white, snow and ice will be abundant in that quarter of the world, and the latter part of the year will be cold.

Another kind: concerning the blowing of the winds. If the east wind bloweth on the eighteenth day of Tammûz, and the west wind doth not blow against it, the year will be good. If all the winds blow, and the east wind doth not, the year will be a bad one, and the crops will be deficient. And if the north and south winds do not blow against the east wind, in that | country [wherein this happeneth] the year will be good, Page 547 and there will be abundance of everything. If the north wind bloweth, and the south wind doth not blow against it, there will be evil and a deficiency of the crops in that year.

Another kind: concerning the blowings [of the wind]. If the north wind bloweth | between Nîsân and Îyyâr, and there Fol. 258 *b*. be lightnings and clouds, there will be a great famine in all the earth. If the south wind bloweth, and there be lightnings and clouds, there ,will be a lack of water in all the springs (or, wells). If the east wind bloweth, and there be lightnings and clouds, there will be abundance in all the earth. If the west wind bloweth, and there be lightnings and clouds, there will be dew, and the year will be a good one.

Prognostications about the atmosphere, which are to be derived from the moon. If after three days the moon appeareth small and fair, it indicateth clearness of the air (*i. e.*, fine weather), and this also is the case when the moon appeareth small and fair after four days. If the moon be fair when it is full, it also indicateth clearness of the air.

Prognostications about the winter, which are derived from the appearance of the moon. When the moon is seen on the third day, and the horns thereof are thick and blurred, this maketh known that there will be rain; and the same is the case when it appeareth on the fourth day with thick, blurred horns. When the whole disk of the moon is seen to be red and fiery, it portendeth a hard winter. And if when the moon is full its whole disk appeareth to be dark, it portendeth rain. And if two or three different kinds of darkness be observed in the whole disk, it portendeth a great (*i. e.*, long, severe) winter.

Again, the appearances which take place in the moon. Whensoever the moon appeareth to turn to a dark colour, there will be a pestilence in the country [wherein this is seen]. When the colour of the moon is bright and luminous, there will be wind. When the moon turneth to a blue colour Page 548 (yellowish-green?), the fountains | and springs will dry up. When the moon turneth red, the flocks will perish.

Forecasts concerning winter, which are derived from the appearance of the sun.

When the sun riseth, and the redness thereof is black, it portendeth rain. Similarly also when it setteth: if there be found on its left side black clouds that touch it, it is right to Fol. 259 a. conclude immediately | that there will be rain. And if a thick cloud standeth opposite to its rays when it riseth, it portendeth rain. Now, if there be thunders and lightnings at a distance, they make known [the approach of] winter. And if there be lightnings sometimes from the south, and sometimes from the north, and sometimes from the east, it is right to know that winds are coming from afar, and they make known [the approach of] winter. And also when the birds of the sea are continually flying up out of the water and dipping themselves into it again, and when the bow which appeareth in the clouds is double, they portend rain. And by analogy also when the stag belloweth, and when the crane standeth on the edge of a lake with its head submerged, or diving under water, they indicate [the approach of] winter. And when oxen, having eaten their fill, turn their heads to the south continually; and when the birds of the house flutter about, and move up and down, and cry out frequently; and when flies bite persistently; and when bats fall down into the lamp; and when thou seest sparks in the cooking pot; and when zâmê, that is to say, flies, run greedily towards the food; and when the cock croweth in the middle of the night; and when the spider's web falleth down of its own accord, there being no wind blowing; and when the rays of the sun are black; and when the fire will only burn with the greatest difficulty; and when sheep frisk about; then know that after three days there will be rain. Similarly also when oxen lick their hoofs, and go to their

manger lowing; | and when wolves wax bold and approach Page 549
close to the goats and houses; and when mice squeak a great
deal; all these things portend a very severe winter. And when
the oak trees produce a very much larger number of acorns
than usual, they portend a hard, long winter. And when goats Fol. 259 b.
and sheep ride each other excessively, they indicate [the
approach of] a severe, long winter. And when animals (*i. e.*,
cattle) and dogs dig holes in the ground, and stretch their
heads towards the north, they make known [the approach of]
a severe winter. And when the rooks, that is to say, crows,
come early, they indicate [the opproach of] a severe winter;
but if they delay in coming and only come a few at a time,
they indicate the change of the winter and the mildness
thereof. And when clouds are seen which are low down, and
when the *ḳârônâ* (crane?) uttereth soft, plaintive cries, and
when the crows and ravens collect in groups like stallions, and
cry out in gladness, they make known that the winter is past.

How a man may know whether the winter will be early or late.

When we know this we know when it is right to sow seed,
and whether we should do it early or delay in doing so. If
the winter is going to be a little late in any given year the
following is a convincing way of finding it out:—If after the
harvest there be rain before the Pleiades set, the winter will
be early; but if there be rain at the same time as the Pleiades
set, the winter will be late.

Information concerning the Pleiades.

The Pleiades rise on the eighteenth day of Îyyâr, and the
Giant (Orion) riseth on the eighteenth day of Khazîrân, and
the Dog (Sirius) of the Giant riseth on the nineteenth day of
Tammûz, and Zûphrâ (Zûhrah, Venus?) on the twenty-eighth
day of Âbh, and the Little Waggon on the nineteenth day of
Shebâṭ.

Again, concerning the forms which are seen in the mists in the heavens.

When the form of a man is seen in the heavens, a pestilence will take place in that month.

Page 550   When the form | of a bull is seen in the heavens, war and slaughter will take place in that month, but there will be abundance.

When the form of a horse is seen in the heavens, there will be abundance during that month; but there will be a pestilence among children, and for a long time women will not go with child.

Fol. 260a. When the form of a mule is seen in the heavens, there will be a pestilence among babes in that | country, and women will not go with child.

When the form of a lion is seen in the heavens, there will be a famine in that country.

When the form of a panther (or, leopard) is seen in the heavens, the wild beasts of the mountains will die.

When the form of a wolf is seen in the heavens, father and son will kill each other.

When the appearance of fire is seen in the heavens, there will be a great famine and war in that country wherein the fire is seen, and there will be a pestilence among men in that year, and wheresoever that fire goeth pestilence will appear.

Concerning the stars which have tails, the names of which are as follows: Ḳarithâ, Bûbhâ, and Lakĕthâ.

Now, the operation of these is like unto the nature of the airs. And they cause wars, and bitter envy, and other such like things; and this we know from the Signs of the Zodiac wherein they are, and from the quarter of the heavens to which the tail is directed. For when they are in the east, and their appearance lasteth for a long time, that which is to happen happeneth quickly; and when they are in the west, that which is to happen happeneth after some delay.

Portents derived from these [stars].

The star with a tail portendeth famine and pestilence.

The star with a spear portendeth wars and strifes.

The star with a beard portendeth a change of kingdoms and the downfall of the Government.

And know that in that quarter of the earth to which the tail of the star is directed there will be anger, and that which is to be will be.

Another portent derived from a star with a tail. | If thou Page 551 seest a star with a tail looking towards the east, know that companies of armed men will enter into Babylon, and will destroy all the people [thereof].

And if a star with a tail riseth, and becometh dark, know that there will be much war, and that destroying kings will rise up.

And if it cometh forth in the west or in the north, know that there will be war and rebellion in Bêth Rhômâyê (Byzantium).

Again, portents derived from shooting | stars. If a star Fol. 260*b*. shooteth from the east to the west, the king of Persia will die in Bêth Hûzâyê (Huzistân), and sicknesses and diseases will be many among men.

If a star shooteth from the west to the east, it sheweth that God is wroth with men, and that they will die in strongholds and cities, and there will be many terrifying rumours.

And if a star shooteth from the south to the north, it sheweth that a pestilence will fall upon that city, and on the districts round about it.

And if a star shooteth from the north to the south, it sheweth that the behaviour of men in that country pleaseth [God].

And if a star falleth from the north to the earth, it sheweth that the king of that country will die, and that his general will die, and that there will be war and contention in that country.

Portents [derived] from shooting stars. When a star goeth from the east to the west, the king of Persia will die in Bêth Hûzâyê, and there will be sicknesses and diseases among men.

When the star goeth from the west to the east, there will be great wrath, and all mankind will suffer tribulation, and evil words will be spoken, but finally it shall be well.

When the star goeth from the north to the south, the king will see evil, and women will not go with child, and a father will slay his son, and a brother his brother for [a period of] three years, and then there shall be peace.

When the star goeth from the south to the north, and is red, or if it falleth from the heavens to the earth, there will be a pestilence in that year, and the king will go to a far country, and war and darkness will prevail, and there will be much sickness and [leading away into] captivity, but in the Page 552 towns | there shall be peace.

When the star goeth from the west to the east, and bursteth (?), there will be pestilence among the exalted ones, and men will increase in their habitations (*sic!*).

When the star descendeth from the highest heaven to the earth, there will be tranquillity and peace, and we shall praise God.

When the star bursteth forth from the east, and is like unto fire, and doth not burst (?), the offspring of beasts (or, cattle) Fol. 261a. will flourish, and flowers will bloom, and the towns | which have been laid waste will be [re]built.

Again, portents derived from stars which shoot in the heavens. When a star shooteth from the west to the east, it indicateth the wrath and the tribulation which are about to come upon men, but it also indicateth that, ultimately, there shall be happiness.

And when a star shooteth from the east to the west, the king of Persia shall wage war in Bêth Hûzâyê, as it is written in the Book of Andronicus the Sage.

And when a star shooteth from the north to the south, the Government shall be in a disturbed state, and women who are with child will wail, and a father will rise up against his son, and a brother against his brother, and the . . . . will be divided for three years, and then there shall be peace.

And when a star shooteth from the south to the north, and becometh split into parts on the south side, on the left, it maketh known concerning the many thrones which are established on the earth. And many shall come in, and they shall seize Governments, and there shall be as many wars as there are parts into which the star shall be split by the force of the atmosphere. And with like force shall those who come in seize upon the Principalities, and Powers, and Thrones, and Seats.

And when the star is seen falling from the heavens to the earth, men shall live in tranquillity and prosperity. But God knoweth the truth!

And when the star shooteth from the north to the south, and is divided into parts, kingdoms shall fall upon each other, and there shall be slaughter and pestilence, and great cities shall be laid waste.

A portent derived from the moon. | When the disk of the Page 553 moon is of a fiery red, and its circle is wide (or, broad), it portendeth wind and trouble.

By THE HELP OF GOD WE WILL WRITE DOWN A PORTION OF THE BOOK OF MEDICINES OF THE COUNTRY (*I. E.*, NATIVE MEDICINES). HELP US, O OUR LORD, IN THY MERCY.

For throbbing of the head, and half-throbbing.

Pound *khadwê*, and apply to the head. Or inject tincture of *ḳa'ûrê* roots into the nostrils of the patient on the side where the throbbing is. Or mix vinegar and honey together, and smear the mixture over the patient's head, and wait a little, and then wash it with water. Or work up ashes and vinegar together into a paste, and smear the mixture on the head. Or roast walnuts and terebinth nuts, pour vinegar on them, and mix them together, and smear the mixture on the head. Or pound one measure of dried grapes, and work up into a paste with aromatic tincture, and smear the head with the mixture. Or steep barley in hot water, and put it in a cloth, and wrap it round the head of the patient whilst hot, and he will find relief. Or roast the dung of an ass and the bones of a crab, and work them up into a paste with oil and a little *ḥennâ*, and smear it on the head of the patient. Or take the marrow of the right leg of a sheep, unsalted, and warm it a little over the fire, and pour into the nostrils of the patient. Or inject the juice of a beetroot into the nose. Or inject three drops of radish water into the nose.

For a cut (or, wound) in the head.

Pound frankincense and crocus, and work up into a paste with the white of an egg. Smear the mixture on a strip of linen, or on paper, and apply to the side of the head (or, temple) which throbbeth. Or pound madder and cummin, work up into a paste with the white of an egg, and smear the mixture on paper, and apply it to the cheeks.

A medicine .... for the head which goeth round (vertigo ?).

Take ten drachms | of *hendabba*, and five drachms of seed Page 554 of rock parsley, and five drachms of aniseed, and two and a half drachms of anethum foeniculum, and two and a half drachms of *na'na'*, and soak them for one day and one night in a mixture containing two parts vinegar and one part water. Then boil the mixture, and clarify it, and add hyoscyamus to this tincture of drugs until it becometh thicker than honey. And each night, when the patient is going to bed, | let him Fol. 262 a. swallow as a dose one drachm, or more.

For scabies and for insects in the head.

Take seed of *sagîrâ*, pound it, and work it up into a paste with sheep oil, and smear it on the head.

For lice in the head.

Pound dried grapes in myrtle oil, and smear on the head. Or wash thy head with salt water.

For sores in the head.

Roast some barley, and work up into a paste with the white of eggs, and smear on the head.

For the sores which come in the head, and from which foetid water cometh.

Bring (*i. e.,* take) some pieces of the old soles of sandals, and rub them down like *kohl*, and add the result to oil of pitch, and smear the head therewith.

For sores in the head.

Take the scrapings of *gĕrôd* stone, set them in the sun and let them dry, rub them down, and boil, and anoint the head therewith.

42

For a diseased head.

Burn the bones of partridges (?) until they become a fine powder, pour this powder into olive oil, and warm the mixture, and smear it on the head; and sprinkle vinegar on the head. Or take soft soap and dill, pound the dill and mix it and the soap with old sheep oil, and set it under the stars during the night of the fourth day in three consecutive weeks, and afterwards rub the mixture on the head.

For sores and scabies (or, itch) in the head.

Take frankincense, and *ṣôrâ*, and bread crusts, crush them up together, and make into a paste with olive oil, and smear on the sore part of the head. Afterwards warm an egg slightly over a fire, and smear it over the medicine. Or smear the blood of a man over the diseased part of the head. Or crush camphor

Page 555 | leaves in vinegar, and lay them over this medicine, but beware lest this mixture doeth harm [to the patient]. Or pound root of pyrethrum, and make it into a paste with honey, and smear it [on the head of the patient].

Also for the scabies (or, itch) which cometh in the head, and in the other members.

Heat cypress [flowers], and dried roses, and doves' dung, and sheep oil, over a fire, and smear the mixture on the head of the patient. Or make vitriol (bluestone?) from the stone into a paste with vinegar, and smear on the head of the patient.

Another, for sores in the head.

Pound the insides of walnuts with garlic, and apply to the head.

Fol. 262*b*.

For ailments of the eyes.

Take thirty drachms of water of anethum foeniculum, and ten drachms of run honey, and thirty drachms of water of sweet pomegranates, and mix them all together. Then set the mixture on a slow fire, and boil it until one half hath been boiled away. Take it off the fire, and keep it in a glass vessel, and drop it into the eyes when the belly is empty.

For eyelashes which penetrate the eyes.

Burn the cast skin of a serpent under the eyes [of the patient]. Or make *kohl* from the gall of a stork, or from the gall of a turbot, and apply. Or make *kohl* from the gall of an eagle, [and apply, and the eyelashes] will not enter [the eyes].

For the eye which is wakeful and will not sleep.

Boil the leaves of thorns in wine and vinegar, and apply [the mixture]. Or pound cummin small, and mix it with the white of an egg, and apply to the eyes externally.

For the eyes which have flesh in them.

Beat up eggs with the whites thereof, spread them on paper, and apply to the eyes externally. Or rub down crocus in new wine, and apply it as antimony is applied.

For the hairs which grow on the eyes (eyelids?).

Root out the hairs from the eye, and smear the places where they grew with the blood of bugs, or with the blood of the ticks of a dog. Or mix together the gall of an owl with a little sal ammoniac, | and smear the mixture on the place from Page 556 which thou hast pulled out the hair.

For eyes from which the eyelashes fall out.

Pound, crush, and reduce to a powder the dry dung of a hare and the dry dung of an antelope, and use as stibium is used.

For eyes which have been blinded by snow and frost.

Boil clean wheat straw chopped up in water, strain the liquor, and pour it into the eyes many times. Or roast some wheat, and crush it and reduce it to a powder, and use as stibium is used. Or pound garlic, squeeze out the juice, and smear the eyes therewith.

For eyes in which dust remaineth.

Crush the rinds of sweet pomegranates, and throw into water for a day. Then strain the liquor through a new linen

42*

rag, and pour it into the eyes six times, and after that apply extract of hyoscyamus.

Fol. 263 a.
For the pain of eyes which are dry.

Take menstrual fluid and the semen of a man, | mix them together, and pour into the eyes. And if they continue to be dry, press them together, and smear with *kohl* or put it in the eyes.

For eyes which are weak and are being eaten away.

Rub down the dung of a lizard (or, salamander) in water and old olive oil, which is as thick as honey, and use as stibium. Or apply like stibium the warm milk of a she-ass, or the milk of a bitch.

For eyes which have gangrene.

Boil dried grapes in vinegar and wine, and smear on the eyes externally. Or pound the kernels of almonds and cummin, and work up into a paste with wine, and smear the eyes therewith. Or pound the testicles of a fox, and boil in water, strain off the liquor, and apply to the eyes.

For the eyes of children which are weak and will not open.

Pound black dried herbs with honey, and apply to the eyes.

For red eyes.

Boil well the seeds of sweet pomegranates with chick-peas until the mixture becometh thick like honey, and use like *kohl*.

For a blow and effusion of the eyes.

Pour into them the blood of doves. Or pour into them the warm white of an egg.

For the eyes which weep (or, stream).

Apply the juice of bitter pomegranates as *kohl*. Or apply the blood of white doves daily. Or apply the juice of black thorns as *kohl*. Or pound together the seed of asparagus and the inside of roasted lentils, and mix together in wine and

| smear the eyes therewith. Or boil well red thorns, and mix Page 557 a little vinegar with the liquor, and wash thy face in it. Or make an infusion of anethum foeniculum, and pour it over thy head for seven days. Or squeeze out the juice of grass (*nânḥê*), and rub it on the eyes. This is a well tried and sure remedy.

For inflammation (?) of the eyes.

Take leaves of the *pĕ'êmâ* plant, chew them, and apply them to the eyes.

For throbbing in the eyes.

Smear the eyes with olive oil and with *koḥl* made of olives. Or work up olive wood chips with the white of an egg, and apply to the eyes.

For great and frequent pains in the eyes.

Mix the juice of *tâwlê*, and crocus, and salt with the milk of a woman, and pour it on the eyes. Or dry the flesh of a swallow, and pound it, and mix it with Persian gum, and smear the eyes therewith. And let the patient drink no wine. And an application of the *kûbḥyâ* plants which are found in sponges, when dipped in hot water, is verÿ good for such pains, and also the letting of blood.

For the sore and the blast in the eyes.

| Pound purslane, squeeze out the juice, and mix with it flour Fol. 263b. of barley, and *ḥennâ*, and the whites of eggs, and smear on the eyes externally. Or pound dried grapes, and hellebore, and the rinds of sweet pomegranates, and sumach, and black thorns, boil them together, strain off the liquor, and rub the eyes therewith. These medicines are good for watering of the eyes.

The blight in the eyes.

Pound the rinds of sweet pomegranates in olive oil, and smear round about the eyes. Or let the patient drink hot cow's milk, and it will destroy the yellowness and sore appearance in the eyes.

For attacks of dimness in the eyes.

Pour into the eyes narcissus water. Or smear them with the blood of a fox. This is good also for want of clearness. Or smear them with the blood that droppeth from the liver of an antelope or with the blood of a black sheep made hot. Or roast the liver of an antelope in the fire, and smear the eyes with the juice which boileth out from the liver whilst it is hot, and give the patient the liver to eat. Or smear the eyes with the blood of fish. Or boil the liver of an antelope in a cooking pot, and order patients to keep close to the pot, and let them receive the fumes that arise from the water in the eyes. Or smear the eyes with human excrement. Or Page 558 | smear the eyes with the juice of leeks mixed with the urine of young children. Or make an infusion of sour grapes, and strain off the liquor, and rub the eyes therewith. Or pound gourds, and squeeze out the juice, and pour it into the nostrils. Or mix the juice of the vine with the urine of young children and smear the eyes therewith. Or mix together the gall of a goat and honey from the comb, and smear the eyes therewith.

For darkness of the eyes.

Take a young dove, and dig out its eyes, and tie a mark to it and let it remain in its nest for three days. When its mother cometh and seeth that it is blind, it will go and bring a certain root, which it will lay upon its eyes and they will be opened. If thou canst find that root which the bird hath revealed (?), guard it carefully. Cut off the head of the young bird, and burn it well, and smear with the ashes the eyes of him that doth not see, and he will see.

For him whose sight is weak.

Fol. 264a. Burn the shell of a crab, pound it, and work it up into a paste with the juice of bitter roots, and smear the eyes therewith. Or burn the heads of young swallows in the fire, pound them up and mix them with honey, and smear the eyes with the mixture. Or burn the hoof of an ass, and rub it down in the milk of a she-ass, and smear the eyes with the mixture, and it will benefit the patient greatly. Or melt down the fat of fish, and mix with honey, and smear the eyes with the mixture. Or melt down the fat of a partridge, and mix it with the milk of a mare, and smear the eyes with the mixture. Or smear the eyes with the gall of a partridge. Or pound crocus, and add it to the white of an egg, and lay the mixture on the eyes. Or pour the juice of sweet pomegranates into a glass bottle, set it in the sun until it becometh thick like honey, and mix with it an equal quantity of honey; smear thine eyes with the mixture when thou art going to bed, and thou wilt derive great benefit therefrom.

For the man who cannot see in the night.

Smear the eyes with the marrow from the leg of a mule. Or cut up the liver of a horse, roast it in the fire, chop it up fine, and pour over it one measure (*raknâ*) of oil of musk, work them up well together and smear the eyes with the mixture.

For the eyes over which descendeth sweet, black water.

Smear the eyes with the gall of a vulture [mixed with] honey. Or get a yellow frog, and take some of its blood, | and smear the eyes with it when the sickness is of long Page 559 standing.

For white deposit in the eyes.

Smear the eyes with the gall of a dog. Or smear the eyes with the eggs of a . . . . . raven. Or pound seashells (?), and the ashes of [burnt] date stones, and eggshells, and apply to the eyes. Or pound the seed of wild mint, pour it on a piece of clean wool, and soak it in the milk of a she-ass, and squeeze it in the eyes. Or steep saffron in the milk of a she-ass, and smear the eyes with the mixture. Or smear the eyes with the gall of a rook. Or take crushed sugar, and gourd sugar, and bitter herbs (?), and the gall of a cat, and the tongue of a *warânâ*, all in equal quantities, make into a powder as fine as *kohl*, and smear the eyes therewith.

For bleeding at the nose.

Dip figs in honey, and cover the region between the eyes of the patient therewith. Or crush to a powder frankincense, sulphur, and glass, work it into a paste with vinegar, and spread the mixture over the face and on the temples. Or take red dust (earth), work it into a paste with vinegar, and spread Fol. 264 *b*. | over the face of the patient as he lieth on his side (or, back). And sprinkle over him a very little exceedingly cold water. This is a certain remedy.

For foul nostrils.

Pound almonds with oil of vetches, clarify and drop some of the mixture into the nostrils.

For the splinter which entereth the ears.

Roast the root of a reed, pound it, and lay it upon it, and it will come out.

For ears which are dull of hearing and deaf.

Take pig's fat, unsalted, the fat of a turtle dove, and oil of bitter almonds, and pound bitumen (*mûmyâ*), and make it hot, and add to them, [and apply to the ears]. Or the fat of goats, and the urine of young boys, and the gall of a partridge, pound these and mix them together and apply. Or mix the gall of goats with the urine of goats, and pour into the ears. Or mix the gall of an ox with the urine of goats, and pour into the ears.

For ears which throb (*i. e.*, ache).

Heat olive oil and the gall of goats, and whilst it is hot pour it into the ears. Or heat oil of bitter almonds and the Page 560 fat of a black cock | together, and drop the mixture into the ears. Or boil raisins in vinegar and old wine, and pour into the ears. Or boil onions in olive oil, and then pour the oil into the ears. Or drop the gall of a pig into the ears. Or let him eat very large quantities of garlic, and it will benefit the patient.

For relieving ears which have ringing in them.

Squeeze out the juice from sweet pomegranates and chick-peas, and mix therewith old wine, and sesame oil, and the milk of a woman, and apply. Or mix together alkali, and myrrh, and oil of myrrh, and apply to the ears.

For the ears of children whence come blood and pus.

Pound a little spice and salt, and mix them with the milk of women, and drop the mixture into the ears. Or heat vinegar and honey together, and apply. Or boil raisins and lettuces together in vinegar, and squeeze out the juice into the ears.

The worms in the ears.

Pound . . . . and the flesh of an ox, and apply to the ears. Or boil strong onions in the urine of children, and drop the liquor into the ears. Or pound and mix together sumach, goat's milk, rinds of pomegranates, and raisins with honey, heat the mixture and drop it into the ears. Or press out the juices from the kidneys of an ox | which have been Fol. 265 a. half broiled, [mix] with salt, and apply. Or mix oil of bitter almonds with vinegar, and pour into the ears. Or mix oil and vinegar together, and apply to the ears. Or pour juice of absinthe and old oil which is cloudy into the ears, and the worms will come out.

To prevent blood coming from the ears.

Dissolve aloes and crocus compounded with vinegar, and apply. Or boil alum of the furnace in vinegar, and apply. Or apply the juice of leeks and vinegar. Or make an infusion of juice of pomegranates in vinegar, and apply. Or boil senna juice and raisins together, squeeze out the liquor, and pour into the ears.

For ears which discharge matter.

The medicine which is called "Egyptian". Take one *lîtrâ* of honey, three [*lîtrê*] of cinnabar, and three measures of vinegar, boil the honey, then pour in the vinegar, finally take the pot

off the fire and pour in the cinnabar, and treat each sore
Page 561 | according to its requirements. Smear the mixture on a piece
of rag, and apply to the ears in cases of diseases of long
standing. Or heat up the milk of an ass, and the milk of a
woman, and honey, and apply. And wash the ears with honey
water and beetroot water, and make an infusion of lentils in
water, and let the patient hold it in his mouth when he is
going to bed. Or let the patient wash the ears with the milk
of an ass and the milk of goats. Or let him use as a gargle
in his mouth vinegar and oil of roses. Or pound strong onions,
put them in wine, and macerate them well; then let him put
bread in his throat, and the onions on his neck. Or pound
the root of aristolochia into a powder, and blow it on the
neck with a reed. Or pound up the excrement of a dog to a
powder, and blow it [on the neck]. Or let him work it up
into a paste and make it into a gargle. Or take a dry crab,
pound it, mix it with cold water, and use as a gargle. Or
make an infusion in water of flax seed, peppercorns, dates,
figs, decayed (?) roses, cummin, and lentils, and let it remain in
thy mouth as thou liest on thy back. Or let the patient hold
Fol. 265 b. in his mouth the juice | of sweet pomegranates as he lieth on
his back.

For the organ of speech, that is to say, the tongue.

Take mulberries, squeeze out the juice, mix it with sugar,
boil it on the fire, and let the patient rinse out his mouth and
throat therewith. Or let him use goat's milk as a gargle.

For teeth which throb (*i. e.,* ache), and have worms in them.

Boil roses, myrtle, raisins, root of tamaris, sumach, thorns,
and olive leaves in vinegar, and let the patient hold it in his
mouth. Or work up radishes with goat's fat, and heat it, and
anoint him therewith, and let him open his mouth that the
worm may come up out of it. Or pound raisins, boil them in
olive oil, and rub the teeth therewith, and let some of it remain
in thy mouth. Or pound thorns and olive leaves with honey
and strong vinegar, and let some of the mixture remain in thy

mouth. Or burn two drachms of hyoscyamus seed, | and take Page 562
two and a half drachms of leek seed and the seed of strong
onions, and work them up with the fat of *kômê*, and make
into pills, each containing one drachm. Put them on the fire,
and then let the patient open his mouth [and inhale the smoke
thereof]. Or take a thick reed, and cut in one end of it a
·small slit, and put on the fire a rose of the rhododaphne tree,
so that the smoke thereof may enter [the reed and pass by
it] to the teeth, and the worms will die. Or place aromatic
resin on the fire, and let the patient open his mouth [and
inhale the smoke], and the worms will be drawn out.

For the teeth which have holes in them.

Place asafoetida in the hole, and the patient will have relief.
Or put in garlic. Or put in alum.

For the teeth which throb (*i. e.*, ache).

Pull out the placenta (?), which containeth a medicine for the
eyes, and take the great vein that is in it, and that hath
moisture in it, and lay it on the [aching] tooth, and the sufferer
will certainly have relief. Or rub the fat of figs on the teeth,
and crush small red roses, and lay them over it.

For teeth which are loose.

Pound raisins, boil them in olive oil, and rub the teeth with
the mixture when hot. If thou holdest some of this in thy
mouth it will strengthen every member therein. Or boil the
root of dill in water, and hold some of it in thy mouth.

To make teeth white.

Burn the horn of a stag and the horn of a goat, rub thy
teeth with the powder, and they will become white. | Fol. 266 *a*.

For pain in the mouth.

Warm wine and honey, and hold the mixture in thy mouth.
Or place in thy mouth sumach, sour grapes, *ornîthogalê* (*i. e.*,
the star of Bethlehem plant), sour pomegranates, and salt.

For putridity of the mouth (bad breath ?).

Heat dried roses, barley, and cummin over the fire, pound them together, work them into a paste with honey, and hold in thy mouth.

For the gangrene of the mouth which is called the " borer ".

Pound together one worm of a tree, one measure of raisins, one palm (?) [leaf], one olive leaf, and a little salt, and hold in thy mouth. Or take a piece of myrtle wood, wrap a piece of linen round one end; take honey and spice and put them in a vessel, and heat them over the fire; then take a very little Page 563 of the medicine on the top of the myrtle stick | and smear it over the wound.

For gangrene of the mouth.

Roast *ṣôrê*, and pomegranates, and leeks, pound them, and mix a little salt with them, and mash them up, and pound the root of aristolochia, and sprinkle it [over the wound].

For the mouths of children.

Rub them with date stones (?) and olive water.

For excess of spittle in the mouth.

Let the patient chew *bûngh* (*cannabis Indica?*) and lettuce seed, and his spittle will lessen in quantity. Or let him chew the seed of beetroot.

For cracked lips.

Pound yellow raisins, and reduce to a powder, and work up into a paste with terebinth gum and a little honey and oil, and smear the lips therewith. Or rub down yellow raisins and arsenic in oil, and smear the lips therewith. Or pound hyssop and work up into a paste with honey and smear the lips. Or treat the lips with the inner rind of the date, or with the inner skin of an egg. Or with the rind of a sweet onion. Or take one *dirham* of caryophyllus aromaticus and one *dirham* of alum, and pour on them one cup of vinegar, and boil them until they are dry and nothing but a powder remaineth. Take some of the powder, and rub thy lips therewith.

For disease of the neck.

Hold in thy mouth water, and honey, and olive oil. Or pound the plant "live for ever", and blow it on the neck of the patient. Or mix milk, and water, and salt | together, and Fol. 266b. hold them in thy mouth. Or pound garlic and peppercorns, and mix them with strong vinegar, and hold them in thy mouth. Or blow alum up the nose of the patient.

For chapping of the hands.

Work up into a paste the brain of a hare, and the blood of a *ḳĕrâyâ*, and the ashes of a crab, and a little *ḥennâ*, and smear them therewith.

For chapping of the hands and feet.

Burn olive leaves, and work up into a paste with sheep oil, and smear the hands and feet.

For the wind which dwelleth in the hands and feet.

Boil flour of barley and guimauve in the milk of a she-ass, and apply to them.

For throbbing of the fingers.

Mix the gall | of a bull with the white of an egg, and smear Page 564 on them.

For pain in the chest.

Steep well in water dates, figs, licorice root, and "ox-tongue", and clarify. Then boil the liquor until it becometh thick, and work it into a paste with barley flour, and use as a plaster.

For hardness (*i. e.,* tightness) of the chest.

Pound olive leaves, and mix with aristolochia and apply.

For pain in the breasts.

[Mix] sulphur and vinegar, and smear them therewith. Or pound well together black dried herbs, and peeled beans, and mix with oil of roses, and smear them therewith. Or boil flour of the êringion (thistle?), and mustard, and the flour of barley in vinegar and apply.

For breasts which have milk in them.

Thou shalt drink the milk of a red cow. Or make an infusion of licorice root, and dates, and figs, and sweet milk, and drink it. Or take a crab, burn it, and mix with it sesame oil, and smear on the breast. Or take the testicles of a fox and ...... fasting, and they will benefit [thee] greatly.

For a deep cough.

Boil the juice of sweet pomegranates and dates together, and let the patient drink the mixture for three days. Or take three eggs, and goat oil and honey, and a few husked peppercorns, mix them together, make hot, and use as a liniment. Or take some wine, pour into it five drachms of dried herbs, and one head of garlic, boil and clarify. Let the patient drink Fol. 267 a. it evening and morning. | Or dry thorns, pound them, mix into a paste with honey, and give the mixture to the patient to drink in hot water, and he will be relieved. Or take one part of honey, and one part of goat's butter (or, cream), and boil in a small pot. Pound sulphur, and rub down nettles and galbanum, two drachms of each, mix together with the honey and butter, and use as a liniment. Or peel *etrûgê* (quinces?) in honey, and let the patient eat them. Or boil garlic and dates in water, and let the patient drink the liquor evening Page 565 and morning. Or take four teeth of garlic, | pound them, and mix them with vinegar and smear the stomach with the mixture. It is a sure remedy.

For coughs and spitting of blood.

Put one and a half drachms of nut-galls into three eggs, make hot, and suck in the morning. Or mix dates with a little arsenic, and let the patient eat them.

For childrens' cough.

Mix fifty coriander seeds with fifty seeds of pomegranates, and put them in three figs and let the patient chew them. Or pound two drachms of licorice root and myrtle, and let the patient swallow it in dates.

For hoarseness of voice.

Boil the seeds of the carob tree in the evening, and give them to the patient to eat. Or take an egg, empty out one half of its contents, fill it up with sheep oil, make hot, and let the patient suck it. Or grind into flour licorice root and flour of vetches, and let the patient eat it in the evening. Or roast onions in the fire, and let the patient gargle with oil and honey, and let him not drink wine. Or boil well five figs, and five dates, and licorice root, and let the patient eat them.

For the heart, and liver, and internal organs wherein is some defect.

Pound chicory and anethum foeniculum, and clarify the liquor, and let the patient drink it. Or mix together cypress gum, and sumach water, and barley water, and let the patient drink a little in the morning. Or let him drink a little of the juice of the plant "live for ever" and sugar mixed with cold water.

For heat in the liver.

Mix chicory seed with water of pomegranate seeds, and let the patient drink it. Or pound sugar and mix it with a little water of berberis, and let the patient drink it. Or pound cows' gum (?), and mix it with snow, and let the patient drink | a Fol 267 *b*. little. Or mix thorn water, and water of pomegranate seed, and a few small roses, and three drachms of sumach flowers together, and let the patient drink a little.

For jaundice.

Mix seed of radishes with mulberries, and let the patient drink [an infusion of them]. Or hang the dog tooth of a bitch over the man, and the jaundice will depart.

For a hot heart.

Work barley flour up into a paste with water of *espadhrê*, and smear [the side therewith].

For thirst.

Pound white of egg with the seeds of aromatic plants, boil them in water, and let the patient drink the liquor in the

Page 566 evening. | Or pound purslane, and let the patient drink it in hot water.

For various pains in the heart (palpitation, &c.).

Boil the juices of sweet and sour pomegranates together with chamomile in a pan, and let the patient drink it. Or pound one drachm of myrtle berries, and mix it with the juice of asparagus, and a little mustard juice, and let the patient drink it three days. Or boil licorice root, mix wine therewith, pound cypress gum, mix [all] together, and let the patient drink it. Or mix a little honey with yeast water, make it hot, and let the patient drink it.

For heart disease.

Let the patient drink one drachm of the juices of asparagus, and chicory, and sour grapes, in cold water. Or pound water herbs and pomegranate seeds, and throw them into sour milk and vinegar, and add to them a little pounded mustard. Let the mixture remain all night under the stars, and let the patient drink it in the morning. Or crush to flour the inner portions of lentils, and let the patient drink it in wine. Or mix juice of pomegranates with a little pounded mustard, and let the patient drink it.

For the wind which standeth over the heart.

Boil seven *lîṭrê* of vinegar and one spoonful of honey together, and let the patient drink it cold. Or make an infusion of dill in water and clarify, and mix with it hot honey water, [and let the patient drink it]. Or make an egg hot, and put into it a little aristolochia (birthwort ?), and let the patient suck it.

Or pound cypress (or, pine) gum, and work it into a paste with wine, and let the patient drink it. Or let the patient drink *raķûthâ* in wine.

For wind of the heart.

Rub down some cummin to powder with terebinth gum, pour them into goat's | milk, and boil the mixture, and let the Fol. 268a. patient drink it in the morning when fasting. Or make an infusion in water of dill, and anethum foeniculum, and rock parsley, and clarify the liquor; mix the water with hot goat's milk, and let the patient drink the mixture.

For him who eateth dust (or, dirt).

Let him eat (?) black cummin and lettuce seed three times daily, and he will hate it.

For a child who eateth dust (or, dirt).

Make him eat crickets (or, grasshoppers) with bread without his knowing it, and he will hate dirt. Or make an infusion of licorice root, and dates, and figs in sweet milk, and give the mixture to him to eat. Or roast a yellow fox from the desert, | and feed him upon it.   Page 567

For the sick man who eateth nothing.

Pound a river crab, and make the patient drink it mixed with cabbage juice.

For the man who is sick from over-indulgence in wine, and who hath lost his appetite.

Boil rose leaves, and oil of walnuts, and wine, and let him drink it. Or let him drink licorice root in wine.

For a weak heart (or, appetite).

Rub down together honey and anethum foeniculum with vinegar and water, and let the patient drink a draught three [consecutive] mornings.

For the heart which palpitateth, faileth, and throbbeth.

Boil together in a pan the water of sweet pomegranates, and water of sumach, and honey, and let the patient drink the mixture.

For the person who is very ill.

Give him to drink the sweat of his feet [mixed with] the matter of his voluntary motions.

For him that cannot sleep.

Pound mandragora in strong vinegar, and smear the mixture over his face. Or mix together sour milk, lettuce seed, and oil of roses, smear the mixture over his forehead, and he will sleep.

For him that sleepeth overmuch without knowing it.

Pound and powder gingers (?) and cabbage seed, [mix them with water], and make the patient drink it, and he will not fall asleep.

For him that drinketh poison.

Let him drink the urine of children [mixed] with wine. Or let him drink the gall of a gazelle in goat's milk, and it wlll dissolve the poison. Or let him suck the blood from oxen.

For pain in the spleen.

Make a good infusion of willow leaves in water, and mix with it a macerated hare, and give it to the patient to drink. Or pound the leaves of the bean plant, and strain off the Fol. 268b. liquor, and let the patient drink it. | Or let him eat the spleen of a hare. Or make an infusion in water of a willow twig and its outer rind, and give it to the patient to drink, and it will destroy [the pain]. This is a well tried and sure remedy. Or let him drink a little calf's gall in cabbage water. And capture a mountain ram before sunrise in the name of him that hath the diseased spleen, and in that of his mother, and burn it in the fire, and when it is burnt let him drink the ashes thereof in wine. If he drinketh one half of them one half of his spleen will be better, and if he drinketh all of them, the ashes will destroy [the pain in] all his spleen. Or take a young swallow | from its nest, and cut off its head in the name of the man Page 568 and in that of his mother before the sun riseth, and burn it in the fire. And make the patient drink one-third of the bird's blood [pure], and one half of it with wine. And if he drinketh all of it there will remain no [pain] in his spleen. This is a sure remedy. Or if thou hangest the spleen of a fox above the patient it will destroy [the pain]. Or boil unleavened barley bread in vinegar, and apply it as a poultice. Or let a dried spleen be hung up over his left side for three days, and on the fourth day untie it, and hang it up over the fireplace, and as it drieth up the spleen of the patient will dry up. Or dry the liver of a fox, pound it, and let the patient drink it in water and honey. Or burn a stag's horn in the fire, and give it to the patient to drink in wine, and it will destroy the disease in the spleen.

For pain in the bowels and colic.

Take three eggs and six heads of garlic, pound them, and mix them with olive oil, and let the patient eat them. Or pound roasted cummin and a little salt, and mix with wine, and let the patient drink it. Or pound white acorns and put them into a ball of eggs, and let the patient eat them. Or burn the skins of garlic and work them up into a paste with olive oil, and smear on the belly.

For wind on the stomach.

Pound peppercorns, and cummin, and *sadhâb*, work up into a paste with honey, and boil a little, and let the patient eat about three drachms for two days.

For the belly which is flatulent.

Work up peppercorns with honey, and let the patient eat them for three days.

For the belly which will not receive (*i. e.*, retain) food.

Boil thorns in water, and let the patient drink the liquor.

43*

Fol. 269a. **For him who hath over-eaten, and whose food standeth | on his heart.**

Take mint, make an infusion of it in water, and let the patient drink the liquor.

For Bile.

Pound mint, and take one cup of the juice thereof, and one cup of wine, and one cup of honey, drink them all, and remain until midday without food. Or let the patient drink milk of grapes and sumach for three days.

For bile and looseness of the bowels.[1]

Mix a little of the milk of .... which doth not go dry, and a little fresh milk, and let the patient drink the mixture, and his bowels will relax and he will have a motion. Or pound **Page 569** one grain of *ḥabb* | *lĕ-mĕlûk*, and work it into a paste with red raisins or figs, and let the patient swallow it, and his bowels will be loosened. Or rub down into a powder two drachms of *salamakê*, and mix with an infusion of red grapes and a little honey, and boil the mixture, and let the patient drink it, and it will loosen his bowels.

For constipated bowels.[1]

Let the patient chew old vine twigs and his bowels will not be loose. Or pound grape stones, and let the patient drink them in wine, and his bowels will not be loose. Or burn bread and the horn of a stag, and let the patient drink the powder in wine, and the looseness will cease. Or let him drink raisins in wine. Or burn a goat's horn, and take the moisture which exudeth from it and drink it in wine. Or pound cummin, and *ḥennâ*, and crocus, and sugar well together, and mix them with honey, and let the patient eat the mixture three mornings [consecutively]. Or let him drink flowers of the red vine in wine. Or roast the seed of water-gingers and the seed of *khêbh'ĕthâ*, and let the patient drink them in cold water. Or pound birthwort, and black dried herbs, and the fat of *kômê*,

---

[1] The title is manifestly wrong.

and dates, and apply to the region of the navel, and the looseness will be stopped.

For the bowels of children.

Smear the gall of a gazelle over the region of the navel;
this is a sure remedy. Or make rue and honey into an ointment, and apply it to the fundament, and the looseness will be
stopped. · Or take a little of the insides of lentils and work it
up into a paste with honey, and let the child eat the mixture.
Or boil the tail of the fat-tailed sheep in fresh milk, spread
sumach over it, and let the child eat it, and the looseness of
his bowels will cease.

For looseness of the bowels accompanied by straining.

Boil rose | water, and juice of bitter pomegranates, and a Fol. 269*b*.
little dill, and a few lentils, and whilst they are boiling add to
them a little aniseed, and let the patient eat them. Or pound
acorns, and let the patient drink them in wine.

For the blood which cometh from the belly.

Boil purslane in water, scatter sumach over it, and let the
patient eat thereof. Or cow's milk, and honey, and broth of
hare, and broth of sheep, heat over the fire, and let the patient
drink it. Or pound cummin, and work it up into a paste with
vinegar, and smear it over the belly of the patient. Or pound
cummin, and the fat | of a goose, and the white of an egg, Page 570
and make a plug of linen [smear it with this mixture], and
apply it to the anus, and the patient will get well.

For worms in the anus and for intestinal worms.

Burn the horn of a ram, and make the patient drink the
powder [in water], and the worms will die.

For scabies of the anus.

Roast alum in the fire, mix it with spices and vinegar, and
smear the part therewith. Or pound the gall of an ox, and
garlic, and the tail of a sheep, and mix them together, and

make plugs of linen and apply the mixture to the anus therewith, and it will give relief. Or mix barley flour and wine together, and smear the anus therewith.

For a protruding anus.

Let the patient anoint it with warm sheep oil, or cow oil, and it will go back to its place. Or pound cummin and pine gum, and mix them with the gall of an ox and spice, and smear on a piece of linen and apply. Or boil garlic in bitumen, and smear the part. Or pound sea shells and pitch in equal quantities, mix them together, warm by the fire, and smear over the part three times daily. Or pound smelters' dross, and the flesh of raisins, and burnt copper (*i. e.,* oxide of copper), and apply. Or pound glass waste, and iron rust, and lime, and mix together, and apply. Or burn a spider's web and myrtle leaves in the fire, and let the patient sit over them and anoint himself with sheep oil.

For an enlarged anus.

Take water of leeks and oil of roses, mix them together, and smear the anus therewith.

For fissures in the anus.

Fol. 270a. Take the gall of a pig, | and burn it under him that hath the fissure. Or pound vegetable alkali, and hyoscyamus, and raisins, and work them up into a paste with the urine of little children, and when the anus of the patient swelleth, smear it with this mixture with the finger, and afterwards smear it with sheep oil. Or take the fat of a black serpent with a red neck and the gall of a pig, mix them together, and apply the mixture to the anus. Crush seed of wild mint, and *merdâsangh,*

Page 571 and myrtle leaves, | and mix with pig oil, make a plug of hemp [dip it in the mixture], and insert in the anus.

On a closed anus.

Mix together ashes of gourds, and ashes of fennel, and scatter on the anus. Or burn *saplûṭê* [and scatter the ashes over it]. Or burn the rinds of plantains and sprinkle the ashes over the anus. Or crush aloes and sprinkle over it. **And if there be wounds and fissures in the private parts,** boil the flesh of black raisins, and the rinds of pomegranates in wine, and smear them with the mixture.

For wasted member.

Take oil of a red cow, and *bardîlâ* leaves, pound them, and work into a paste with sheep's oil, and smear it therewith.

For ulcers (or, abscesses) of the belly.

Boil well the dung of a gazelle in water with the rinds of pomegranates, strain off the liquor, and mix with sheep oil, and apply. Or smear the ulcers with *keshnê* of the oxen of Nîsân, and wash them with vinegar. Or boil well licorice root and onions together, take the liquor, and rub the ulcers and the broken places in the skin therewith.

For the child who is swollen and yellow.

Take lentils, steep them in wine for three days, then take them out, pound and rub them down, and work into a paste with wine, and smear on a rag, and lay on his belly.

For him that hath the wind of ḳûlêngh.

Take a little dung of a white dog, and a little sugar, crush them, mix them together in water; let the patient drink the mixture, and his belly will be loosened, and he will have relief.

For pain in the interior.

Heat some wine and throw into it thorns, and boil them well, and let the patient drink a little of the liquor.

For wind in the stomach.

Pound a little | cummin, and mix it with a little salt, and Fol. 270*b*. throw it into wine, and let the patient drink it. Or pound two walnuts and mix them with a little vinegar, and let the patient drink the mixture.

For the woman who is in difficult labour, and whose child will not come down.

Tie together *tâwlê*, and hang them over the right shoulder of the woman, and when she is bringing forth, take them off, and hang them over her child. Or let her drink juice of thorns in | wine. Or boil leeks, and let her drink the water in which they are boiled. Or pound cummin, and thorns, and burnt flax stalks, and give the powder to her to drink in wine.

Page 572

For the woman whose child dieth in her belly.

Let her drink the milk of a bitch in wine. Or set on fire the hoof of a horse under her. Or burn the hoof of an ass [under her], and the child will come down. Or break up two measures of ox dung in water, and let her drink the water, and the child will come down. Or boil the root of a shepherd's staff in water, and let her drink it. Or burn thorn seed under her.

For strangury.

Rub down the seed of purslane, and let the patient drink it in wine. Or work it up into a paste with vinegar, and smear the belly therewith. Or let him drink dove's dung in wine. Or pound strong onions and the burnt stalks of flax, and let him drink them in wine. Or burn under him the seed of *hendôrê*. Or take a crab, roast it in oil until it drieth, then crush it, and make it into a powder, and let him drink about two drachms of it in wine. Or let him drink the droppings of a cock in wine. Or break the eggs of a *ḳarîtâ* bird into a vessel, and let him that hath strangury drink them, or make a plaster of them and lay them on his belly. Or boil rinds of pomegranates, and let the patient drink the water three days. Or pound the seed of dill, and let him drink it in water.

For strangury in children.

Fol. 271 a.

Make an infusion of rue, and vinegar, and carthamus tinctorius, and give it to the child and his mother to drink.

For strangury in those who micturate continually.

Roast the flesh of a crow (or, raven), pound it, put it in dough, and set it in a baking oven, and take it out, and make the child eat the bread without knowing what is in it.

For him whose urine will not come forth.

Pound narcissus leaves, press out the juice, and smear it over his member and the adjacent parts of the body. Or take mice dung, and work it into a paste with the spittle of thy mouth, and smear with it all the outside of his member.

For him that passeth blood in his urine.

Mix the gall of a horse with frankincense, that is to say, incense, and with caper berries, and cummin, and let him drink it. Or roast cummin and gingers, | and work the powders up Page 573 into a paste with honey, and let him eat it before a meal. Or boil myrtle leaves in water, and sprinkle madder on them, and let him drink the liquor. Or mix dove's dung with his urine, and heat it, and let him drink it.

For him that maketh water in the bed under him.

Burn the comb of a cock, and let him drink [the powder thereof] in wine. Or let him chew the testicles of a cock. Or roast the testicles of a cock, pound them, and let him drink them in wine. Or roast the bladder of a ram or pig, and let him eat it. Or take the brain of a hare, dry it, pound it with seed of dill and rock parsley, and let him drink it in wine. Or pound three drachms of thorns, and let him drink them in wine. Or let him chew cummin during the night. Or roast raisins and put them in an egg, and add to them the blood of sparrows, heat them and let him suck them. Or let him that maketh water in his sleep drink the dried liver of a crow, and it will do him great good. Or roast the testicles of a fox, and let him eat them.

For pains in the testicles.

Pound cleaned black raisins, and cummin, and beans that have been soaked in water and peeled, mix them together and apply. Or work up together wine lees and sheep oil and apply.

For wind of the testicles.

Let the patient drink wolf's gall in wine. Or heat one drachm each of cypress gum and "ox-tongue", and two *rìṭlê* each of wine and honey, and let him drink the mixture for three days. Or boil oil and gum, and rub the testicles with the mixture whilst the patient is asleep. Or pound dry grapes from which all the stones have been removed, and cummin, and cypress gum, and frankincense, one *mathḳâl* and two drachms each, mix all together with sesame oil, and apply to

Fol. 271*b.* the testicles in the form | of a plaster. Or mix together bean flour, wax, roses, and salt, boil them, and apply the mixture to the testicles; or the fat on the kidneys of a gazelle and unslaked lime.

For testicles which ache.

Melt down and mix together one portion each of stacte and flowers of wheat, three portions each of wax, ox fat, and pig

Page 574 fat, and | one portion of oil of roses, and apply. Or boil alkali in water, strain the liquor, pound absinthe like flour, mix the two together and apply as a plaster. Or make an infusion of chopped lettuces, and apply. Or roast pearl barley, and work up into a paste with oil, and apply as a plaster. Or pound sulphur, work up into an ointment with vinegar, and smear the testicles.

For the wind which descendeth on the testicles of boys.

Take plaster from the wall and gingers, pound them together, work them up into a paste with the urine of the sick boy, and apply in the form of a plaster.

For ulcers and wind in the body.

Melt down camphor (?) and the fat of *kômê*, and apply. Work up and mix together wine, vinegar, honey, unleavened bread, and a little pounded absinthe, and use as a lotion. Or boil well wine, vinegar, milk, the urine of children, and the flour of barley, and apply it to the region where the wind is. Or pound sweet onions, mix them with vinegar, and apply. Or pound leeks and the leaves of wild cabbages, and smear the part therewith. Or boil *kawêstâ, mêrdâsangh,* honey, licorice root, and oil of roses, and smear the part therewith.

For pain in the knees and back.

Boil bran in clean oil and water, make hot, and apply as a poultice. Or first of all anoint the arteries of the knee with oil of roses, and then take a fresh tail from the fat-tailed sheep, and lay it like a poultice on the arteries both under and over the knee. Or apply white naphtha to the knees for three days, and they will feel relief. Or pound one portion of bean flour, and three portions of "snake cucumbers", and mix them together with barley flour, sprinkle on the mixture a very little vinegar, and | work it up well in a mortar, and apply in the Fol. 272*a.* form of a poultice. Or pound ox fat and plantain leaves together, and apply.

For pain in the shoulders.

Make an infusion of alkali wood and barley, and rub them with it. Or make barley flour into a paste with the water, and apply it in the form of a poultice.

For wind of the legs.

Burn cock's feathers and terebinth gum | together, quench Page 575 them with vinegar, stir the mixture well, and use it as a liniment. Or [mix] the dust of the oven with oil of roses, and use as a liniment. Or mix arsenic and *ṣûrâ* together with the white of an egg, and smear the shoulders with the mixture.

For the red wind.

Take the small mountain leaves from a mountain which have the shape of a fig leaf, before sunrise on the fourteenth day of the month, and dry them in the shade, and keep them by thee. And when the blood cometh, rub down some leaves, and smear the milk of women who are suckling children [over the place], and spread the leaf over it. Then boil the skin of a snake, and make a plaster and lay it upon it. Or make a plaster with mountain parsley, and lay it on the place, and let it remain there until thou knowest that the place is healing. Every time thou throwest a plaster away, lay on the place the bitter little leaf (*tarpânîthâ*).

Other medicines for the red wind.

Pound red *kharpadhnakh* which groweth in gardens, and apply. Or boil fresh anethum foeniculum in milk, and apply three times, and the patient will find relief. Or mix together *'are'thâ* and *mastâ*, and apply. Or roast together the white of an egg and gourd leaves, and smear the part with the mixture. Or smear it with the juice of sweet pomegranates, and honey, and dill. Or heat together anethum foeniculum, and "ox-tongue", and mint, and apply.

For ulcers of the body.

Make an infusion of leaves of the terebinth, and leaves of the Persian apple tree, and leaves of the Egyptian willow, and leaves of the vine, and leaves of the *khûrâ* plant (?), and root of cynara, and "ox-tongue", and root of the bitter *tarpânîthâ*, and *kûzâz* root, and root of the little rose, and walnut leaves, and leaves of the "winter" plant, and leaves of *khâyâlâ*, and leaves of *kâwkhâ*, and leaves of myrtle, and plantain leaves, Fol. 272b. and the root and leaves | of beetroot and mint, and the leaves of mandragora, and almond leaves, and dill, and fennel, and hot grass; all these, I say, infuse together in water, and let the patient drink a little of the infusion, and wash in the remainder. And let him [sit] upon hot grass, and sleep upon Page 576 hot grass. Or pound cynara root, | boil it, and let him wash in the water thereof. Or make an infusion of the finest willow root, and let him drink it.

For him that hath wind in his body.

Let him eat a little *saplûṭâ* of the meadow, that is to say, the *paṭûr* plant which groweth in the mountains and in the desert and also in the plantations, with some honey.

For pleurisy.

Boil the flour of white wheat in olive oil, and apply. Or pound *sapêstâ*, and boil it in sheep oil, and apply. Or dissolve barley bread hot from the oven in wine, and apply. Or let him drink almond (?) gum mixed with cold water.

For pain in the loins.

Roast fennel from the vineyard, pound it, mix it with dill, and smear it over the loins. And administer it to the sick man in the white of an egg.

For those who are destroyed (wounded?).

Let the patient eat *kersâ* root, and *lûsh* root, and plantain skins, and root of vesicaria (?), and sesame seed, and his ailment will diminish.

For hiccoughs.

Let him that hath them drink three mouthfuls of strong vinegar, or of very cold water, or very hot water; or shake him violently.

To prevent a man snoring in his sleep.

Put the tooth of a stallion under the head of him that snoreth in his sleep, and he will not snore.

For the man who is possessed of a devil.

Tie the heart of an ass in the skin of a stag and hang it over him. Or burn the heart of a dove under him, and the devil will flee from him.

If thou wishest to know what is in the heart of thy neighbour.

Take the heart of a fig tree and say within thyself, Art thou wholly tied to me or to my neighbour? And thy neighbour himself will tell thee what is in his heart without knowing it.

For the man who wisheth to shine in the eyes | of the rulers.

Fol. 273 *a*.

Let the head of a crow be hung up over him.

For the man who would shine in the eyes of men.

Let the heart, and eye, and skin of a wolf be hung up over him, and mankind will love him.

For the face (*i. e.*, complexion) which is like wax.

Take the lung of a camel whilst warm and lay it over his face, and he shall do well.

For the image (countenance) which is changing.

Page 577   Take the leaves of bitter | herbs (?), and boil them in a pan of milk with leaves of the red *bîblâ* flower of Nîsân, and take walnut leaves and boil them by themselves. Then take the former medicine, and rub it on the patient's face, and let it dry, and then take some of the walnut leaves and rub them on the face which is changing its colour.

For the meagre man, that he may become fat and shine.

Pound galbanum, mix it with goat's milk, and let him drink it.

For a burn, before ulceration hath broken out.

Smear the burn with boiling water and the white of an egg, and this will not take place.

For a burn, et cetera.

Smear with the whites of eggs and olive oil. Or work up into a paste the ashes of burnt senna wood, and ashes of paper, and barley flour, and yellow raisins, and the rinds of pomegranates, and the whites of eggs, and use as an ointment. Or boil bean flour and crusts of bread with a cat's penis, and smear over the burn. Or roast a handful of barley at the fire, and mix it with oil of roses, and the whites of eggs, and use as an ointment.

For a burn which is becoming thick.

Roast some barley bread, pound it, and having anointed the burn scatter the powdered bread over it.

For the bite of a dog.

[Mix together] glaucium, crocus, and juice of coriander seed, and smear the place therewith. Or pound rue, and mix it with olive oil, and apply. Or pound gingers and natron, and let the patient drink them mixed in a draught. Or burn some paper, work the ashes up with oil into a paste, and apply in the form of a plaster. Or dry rose root, pound it and work it up with wine into a paste, and apply in the form of a plaster. Or pound the insides of walnuts and alum together, and apply. Or roast hair of the dog [which bit the patient], and apply. Or roast an onion whole, burn a rag from a sack, Fol. 273b. rub it down, and mix it with vinegar and spice, and apply this mixture to the wound with the onion over it.

For the bite of a mad dog.

Pound *gûftâ* and salt, and lay upon the wound. Or take the hair of a man, and macerate it in vinegar, and apply. Hang up a dog's tooth over thee, and a mad dog will not bite thee.

For the dog which goeth | after a man.   Page 578

Pound bitter almonds, put them in bread, and feed the dog on it, and he will go with thee wheresoever thou wishest.

To prevent dogs barking at thee.

Take the tail of a weasel, and sew it over thy loins, and the dogs will not bark at thee.

For the bite of a serpent and viper.

Let the man who is bitten drink the gall of a cat in wine, and let him not be divided (*i. e.,* have doubts) as to the efficacy of this medicine. Or cut the flesh round about the bite, pound onions and salt, put them in vinegar, and apply them to the wound in the form of a plaster. Or pound leeks, mix them with vinegar and wine, and lay them on the wound. Over them lay a piece of meat on a linen rag, and tie it tightly, and the poison will come out. Or bring a frog, cut it up, and lay the pieces on the wound. Or smear galbanum on the wound. Or smear it with the fat of figs. Or boil the leaves of the white Persian apple tree, and lay them on the place where the wound is. Or boil the leaves of the terebinth tree, and apply.

For the sting of a scorpion.

Pound mustard, work it into a paste with hot water and apply. Or pound garlic and the leaves of the "ox-tongue" plant, and apply. Or rub down lentils, work them into a paste with saliva, smear on a piece of rag, and apply. Or mix together unslaked lime and oil of roses, and anoint the place therewith. Or pound salt and flax stalks, and apply. Or tie a silver drachm over the wound.

To prevent a scorpion stinging thee.

Take a king's oak, that is "Shah balût", split it, and hang it over thee, and the scorpion will not sting thee.

Fol. 274a. | To [make] serpents and reptiles flee from the house.

Pound and mix together the root of galbanum, the seed of *sapestâ*, and the horn and hoof of a stag, and dregs of opium, and pour strong vinegar into the mixture, and then set it on fire inside the house, and all the reptiles will flee and make their escape from it.

To make beetles depart from a house.

Throw into it fresh roses, and they will flee. Or burn the hair and hoofs of goats in the house, and they will take to flight.

To make crickets depart from a house.

Page 579   Burn the white of an egg and the hair of a fox | in the house, and the crickets will die and fall down.

To make ants depart [from a house].

Pound cummin, and work it into a paste with water, and throw it on the nests of the ants with sulphur and hyoscyamus, which is *bûngh*. Or take sulphur, mustard, and thorns, pound them together, and scatter the mixture over the nest of the ants, and they will flee. Or place fat of *kuftâ* on the nest of the ants, and they will flee. Or if thou wilt smear this on a branch of the Egyptian willow, or of the bitter almond tree, and place it in a corner of the house, all the ants will gather together to it.

To kill fleas.

Steep cummin in water, and sprinkle on the floor of the house, and all the fleas will die. Or pound cypress leaves, and scatter them about the house, and they will die.

To keep moths from the bee-hives.

Sprinkle fresh milk and the urine of children over the legs of the hive-stands, and the bees will make honey.

To prevent the bees leaving.

Heat water and wine, and rub the mixture on the hives. Or fumigate them with burnt ass's dung.

To kill mice.

Fumigate the house with water parsley with the door shut. Or burn thoroughly the thigh of a camel, and mix it [with . . . . and throw it into their holes, and the mixture will kill them.

To drive out gnats.

Fumigate the place where thou sleepest with galbanum and sulphur, and they will fly away. Or fumigate it with colocasia, and they will perish. Or put a little cup of extract of hemp under thy head, and they will not draw nigh thee.

44

Fol. 274b.   **For sickness in sheep and cattle.** |

Take the root of the small rose, dry it, pound it, make it into a powder, mix with salt, and administer it in the food of the beasts for seven days.

To prevent chickens from flying away.

Soak wheat in the water of yellow mint, throw it to the chickens, and they will become heavy, and will not fly away.

To catch birds.

Soak wheat in urine for one day, and then boil it in sulphur until the sulphur adhereth to the wheat, separate the wheat from the sulphur, and throw it to the birds. Every bird that eateth a little of the wheat will die, and if it doth not, it cometh to the slaughter quickly.

To make hens lay eggs.

Page 580   Work up wheat flour into a paste with wine, | and give the mixture to the hens to eat.

A medicine to make the hair grow.

Pound the fat of a fox with sesame oil, and use it as an ointment on the head, and the hair will grow. Or smear on the head the marrow from the leg of a hare and the brain from his head, and the hair will grow. Or pound olive leaves thoroughly, and apply them as a plaster to the spot where there is no hair. Or let the man drink the gall of a partridge in the finest wine, in front of the moon when it is straight [overhead]. Or kill a swallow, and burn it with a shepherd's staff, whereof the leaves are withered, and mix the ashes with the urine of young men who have not known women, and smear it [on the head] and the hair will sprout and flourish beautifully. Or melt the fat of a wolf and mix it with ashes of an oak, and apply it to the place where the hair groweth not, and it will grow. Or cut off a bat's head, and boil it in olive oil, and rub it on the place where thou wishest the hair to grow, and it will grow.

For the eyebrows.

Burn spice and make it up into a paste with sesame oil, and smear on the brows. Or burn walnuts and date stones, pound them and make into a paste with oil of roses, and smear the brows, and the hair will grow.

A medicine for destroying the hair.

Take the flesh of *kômê*, and dry it in the shade, and ants' eggs, and pound well together; rub it [on the place], and the hair will not grow. | Or smear the blood of bats on the place Fol. 275 *a*. where thou wishest the hair not to grow, and it will not do so.

To prevent the hair becoming white.

Take the gall of a swallow, and pour two drops on the right side of the mouth, and one on the left side. Or take a handful of little bees, roast them in olive oil, and smear on the hair, and it will become black. Or mix the fat of a black raven with wet thorns, and rub it into the roots of the hair. Or pound ox-gall and the dung of a swallow, and mix together in water for thy head.

For the violent fever that cometh in the summer.

Take hot water and oil of roses, soak a cloth in the mixture, and lay the inside of it upon the joints of his hands and feet. Or roast mustard in the fire, pound it with oil of | roses, Page 581 and anoint all his members therewith.

For fever caused by the fatigue of travelling.

Anoint thyself with sweet oil, and sit in hot water. Or pound flax seed and "king's crown" in oil of roses and hot water, and use for him to whom come fever and headache. Or boil thorns in water, and let the patient drink it. Or pound rock parsley and *sĕdhâb*, and mix them with milk and honey, and let the patient drink the mixture. **For the second fever.**[1] Burn a pig's bone under the patient. **For the third fever.**[2]

---

[1] The fever that cometh on alternate days.
[2] The fever that cometh every third day.

44*

Hang up over him the shell of a crab, and it will cut the fever. Or hang up over him the upper lip of a mole.

For burning fever.

Fumigate the patient with wax and *tâlê*.

For a fever.

Pour gall into milk and wine, and set in the sun for a short time. Make the patient drink three drops of it, and, if the sickness cometh forth in the body like worms (?), rub him with the spleen of a sheep, and they will depart, and he will have relief.

For pain that cannot be located.

Fumigate the patient with gingers, and make them hot and apply them to him.

To make a boil burst.

Pound unleavened bread and garlic, work up into a paste with vinegar, boil, and apply in the form of a plaster. This is a certain remedy.

Fol. 275*b*.   **For an obstinate | boil that will not heal.**

Roast the nerve of a *ḳapḳâpâ* in the fire, rub it down, and work it up into a paste with cow oil; apply this to the patient, and he will get well.

For sores that run with water.

Roast the horn of an ox, pound it with the milk of young women, make the mixture into a paste, and apply. Or mix together sumach and honey, and apply.

For watery sores.

Take shells from the river, pound them fine like flour, make the powder into a paste with the white of egg, and apply.

For soft (or moist) sores.

Bring ashes of roses, and sprinkle over them.

For old sores from which run pus and water.

Pound mulberry leaves, and sprinkle [over them]. Or pound together one portion of red raisins and one portion of the fat of an antelope, and apply.

For the sore [called the] "Watcher", which cometh in the legs and back.

Pound alkali and reduce it to a powder, and work it into a paste with honey and old sheep oil, | and pour it over the Page 582 sore. And if there be in it a [piece of] broken bone, or anything else, take it out, and the wound will close. Or take a bone of a man, pound it, and scatter it over the wound.

For the "Watchers" that will not close.

Take lead, rub it down with old oil until the mixture becometh thick, and apply in the form of a plaster. Or roast an onion, mix it up with oil of roses, and apply, and it will give relief in the case of the "Watcher" even of twenty years' standing.

For the stinking sore.

Pound a fig with some rue, and apply it.

For every kind of sore.

Boil the leaf and rind of the white tree in wine, and apply.

For sores.

Pound together galbanum, and wax, and the gum of *shîdhê*, and the fat of *kômê*, and apply.

For leprosy and scabies.

Take the milk of a woman, and vinegar, and flour of trigonella, work up together, and use as an ointment. Or pound together the seed of gourds, and the seed of gingers, and the seed of cabbage, and throw the mixture into strong vinegar, and use as an ointment. And sit in the hot sun. Or dry and pound leaves of the "deaf" fig and *sĕdhâbh*, pound and work them up in vinegar. First wipe clean the spot with a napkin,

Fol. 276a. and take vinegar | and rub the place in the sun, then smear the mixture over it, and let the patient wash in hot water.

For the leprosy that is called "kharsâ" (scabies).

Pound a crab, and add it to ox dung, and use as an ointment. Or mix together the blood of black doves, and the blood of turtle doves, and spice, and bitter salt, and rub [on the body]. Or rub it with the dung of doves, and the seed of a man, and spice. Or burn a horn of an ox, and work it up into a paste with olive oil and old wine, and use the mixture as an ointment, and stand in the sun. Or boil gum of cedar, and gum of terebinth, and sulphur, with the white of an egg, and use as an ointment.

For severe gangrene.

Work up three parts of unslaked lime and one part of arsenic into a paste with strong vinegar, and make it into tablets (or, balls), and dry them in the sun. Pound [these] and sprinkle them over the gangrenous sore.

For gangrene.

Pound arsenic, and alkali, and yellow hellebore, and mix them into a paste, knead it, and make it into balls, bake them, and make them into a powder, and sprinkle [this on the sore]. Page 583 | Or alum from the furnace, and zâghâ, and white silver, and flour of lentils, pound and mix with honey, and apply.

For him that eateth his body.

Smear with the dregs of the vinegar of wine, and let the patient stand in the sun for an hour, and then let him wash himself in hot water.

For ulcers.

Physicians wash these with alkali, and soap, and mardhâ-sângh, and khabbat sôdhâ, and hennâ.

For the scabies that cometh in the back of the head and neck.

Pound the plant "warrior's hair", and mix with barley flour, and use as a plaster for seven days. Or pound the lily which groweth in cemeteries, and work up into a paste with strong vinegar, and apply. Or pound sulphur and garlic, and work up into a paste with vinegar, and apply. Or pound fresh walnut leaves, and work up into a paste with bean flour, and use as an ointment. Or pound up thorns, and kernels of walnuts, and glass, and the dung of mice, and use as a plaster. Or smear spice-unguent on paper, and dry, and pound cabbage root, and sprinkle over the diseased places.

For the man who hath | a swelling in his throat.   Fol. 276 b.

Pound together the dung of a white dog, and rinds of pomegranates, and raisins, and ashes of fennel, and let the patient hold the mixture in his mouth.

For the Edhrâ disease (boil ?).

Take raisins and the urine of oxen, and use as an ointment. Or rub the place with the urine of a dog. Or pound *merdâ-sangh* and bitter almonds, mix them with vinegar, and use as an ointment. Or pound raisins, mix them with strong vinegar, and rub the *edhrâ* (boil ?).

For ringworm and edhrâ.

Heat together alkali, and mustard, and spice-unguent, and vinegar, and use as an ointment.

For the Edhrâ of children.

Make a solution of the urine of little children, and *ḥennâ*, and *tîrêkh* (θηριακή ?), and gum of terebinth in old vinegar, pour it on the *edhrâ* and use as a liniment.

For the "mulberry" that cometh in man and beast.

Cut a branch from a mulberry tree, and burn it in the fire when it is moist, and collect the ashes. Rub the "mulberry" with spice-unguent, and then lay the ashes upon it for five days, and it will be healed. Or take a piece of the cast skin Page 584 | of a serpent four fingers [long], and tie a woollen thread to it, and then tie the skin to the "mulberry", and it will be cut. Or take the ashes of burnt mulberry wood, and the white of an egg [mix them together], and cut with a knife, and use as an ointment.

For the stinking sore in cattle.

Pound the dung of a dog, and sprinkle over the sores. Or heat [together] alkali, and spice-unguent, and sulphur, and oil, and use as an ointment. Or use the dung of a cock as an ointment. [Or] roast *se'ḳâ* root, and alkali, and lentils, and apply.

For the "moths" that come forth in the hands and feet, and are called "shîshmânyâthâ".

Rub the hands and feet with fresh garlic until they are drawn out. Or catch some of the urine of a dog in the place where he maketh water in a piece of wool or linen, and rub the "moths" with the urine, and they will dry up and disappear. Or put a piece of vine wood from the vineyard in the fire, and smear the "moths" with the juice which oozeth out when it is burning, and they will dry up. Or rub down well the flakes which come from iron, and put them in wine, Fol. 277a. and spread the mixture on a cloth, and lay it on the | *shîshmânyâthâ*. Or smear the blood of a tortoise on them, and they will be destroyed. Or rub them with the blood of a calf, and they will be destroyed. Or take a willow rod, count the *shîshmânyâthâ*, and according to their number cut notches in that willow rod, for each creature a notch. And put the rod on the fire, and as it drieth the *shîshmânyâthâ* will perish. This is a well tried and certain remedy. Or let him that hath *shîshmânyâthâ* go to the sacristan after he hath cleaned the altar, and let the sacristan take the sweepings of the altar, and collect the *shîsmânyâthâ*, [and give it to him] for three days, and they will be destroyed.

For the insects (sêsê) which come forth from the body.

Pound together cummin and black raisins (?), and smear the mixture on them. Or work up together barley flour and alkali, and smear them [therewith]. Pound roasted walnuts and terebinth nuts, sprinkle vinegar on them, and smear the mixture on them.

For the pustule.

Mix together vinegar and honey, and let the patient eat the mixture three days. Let him wait until | midday, and then Page 585 let him eat chick-peas and onions. Or mix roast alkali with honey, and smear it over his forehead. And boil sumach in old wine, and let him eat it.

For the boil.

Pound radish seed, mix it with strong vinegar, and anoint the boil.

Bean flour.

Now, bean flour cleareth away impurities (*i.e.*, eruptions) from the body in a thoroughly efficient manner. We see that the young men and the women who sell bean flour frequently use it when they wash, and that their colour (*i.e.*, complexion) is as white as snow, just as we see that other men who work in natron smear some bean flour on their faces, for it removeth pimples, and blotches, and rash from them.

Medicines which are used for wounds caused by arrows, and by swords, and by every kind of weapon.

Mahram. Bring first of all *shekrôk*, now, that which people say about this [plant] is well known, *viz.*, that a wolf killeth him that diggeth it up. [Next] apply *pedhdhĕwê* and gum of terebinth in large quantities, | and croton (or, cici), and *kar'â* Fol. 277b. (butter), and raisins, and *ḥennâ*, and flax, and madder, and the thorn-plant (*shû'altâ*), and "serpents' grass", and rock purslane, and red onions, and kernels of almonds, and spices, and, above all, apply the tail of the fat-tailed sheep. And boil [all the above] well, and clarify the liquor by means of a woollen cloth. On the first day treat the wound with honey, and salt, and madder, [and do so] until the fifth day. Then apply Marham. Do not apply alum to the head, but honey, and madder, and salt. And if the wound be over a bone, burn a piece of linen, crush a little salt, and mix with the burnt linen, and apply. And when thou art applying [thy remedy], apply onions to the mouth (*i. e.*, lips) of the wound. If it is the season of summer and there 'are no onions to be had, apply the bitter *ṭarpânîthâ*. And if it is not the season, do not apply Marham, Page 586 but gum, and fat, and croton, | and *kâr'â* (butter), are much better, and no bad smell will attack the wound.

Another Marham. Take sumach gum, the white of an egg, wax, and madder, and pour them into olive [oil], and boil them in a clean vessel on the coals, and strain (or, clarify) the mix- ture, and apply it to the wound.

Another Marham which is good for all kinds of wounds. Pound together galbanum, wax, extract of almonds, uncut edible (?) terebinth nuts, a little olive oil, and the fat of black [doves], and apply.

Another kind of Marham, which is cleansing in nature, and which eateth away proud flesh. One head of the fat of black doves, and one head of honey, and one full head of terebinth gum, and one head of soft soap. Crush and melt down these, adding the soap first and the honey afterwards. Melt all of them well [together], and apply to the wound, and it shall be healed.

White Marham, that is to say, Mâlagmâ (μάλαγμα). Boil up together wax, spice, *medsigh*, sesame oil, and *merdâsangh*, and apply.

Black Marham. Boil up together Greek beer, beer of spelt Fol. 278 a. grits from Damascus, *ratnîgh*, *henôkyâ*, | gum ammoniac, *'anzarût*, *bazrâ* seed, and *shamlâ*, and apply.

Red Marham. Boil up red lead (or, cinnabar), twigs of saffron, wax, eggs, butter, and *zûbdâ*, that is to say, butter, and apply.

Another Marham, which cleanseth the wounds caused by iron, stone, &c. Take salt, and the whites of eggs, mix them together, and cleanse the wound, and apply for one or two days, and apply to the mouth of the wound a paste containing honey. After two days take the paste from the wound, and throw it away, and bring soft soap, and rub it down, and work it into a paste with terebinth gum, and apply it to the wound twice daily, morning and evening, and at each application thou must lay over the mouth of the wound a paste made of eggs and flour which have been thoroughly well worked up together. Do thus until the wound | is healed. And anoint Page 587 the parts round about the wound daily with butter which hath no salt in it.

For the wound caused by the surgeon.

Bring soot from the caldron, and work it into a paste with oil, and apply it to the wound. And if thou wishest the wound to close, work up the ashes of croton (or, cici) into a paste with oil, and apply. Or stick on it plaster from the oven.

For a wound made by the sword.

Boil together yellow ginger, and caper berries, and ox gall, and black dried herbs (or, grapes) of *edhyâ*, and a little honey, and apply.

For the wounds caused by arrow-heads which remain in the body.

Pound together the blood of *ḳanyâ*, that is to say, the lungs, and cummin, and black dried herbs (or, grapes) of *edhyâ*, and apply it to the wounds, and they will come out.

For the wood (*i. e.*, splinter) that remaineth in the body.

Apply the fat of a hare as a plaster, and it will come out.

For the lead (bullet?) which remaineth in the body.

Apply the flesh of the *semôrâ* animal (marten?) as a plaster, and it will come out.

For wounds wherein worms come.

Sprinkle alkali, and aristolochia, and lime over them.

For him that wisheth not to get drunk when drinking wine.

Let him hold under his tongue the gum of the ṣalâmâ tree and cabbage seed, and he shall not | become drunk. Or let him say over the first cup "O merciful One" (Marhûm), and he shall not become drunk.

Fol. 278*b.*

Concerning the rule for planting vines, &c.

Know that the plant which thou shalt plant between the first and the fourth days of the mouth of Âdhâr shall take hold (*i. e.,* strike root) immediately, and shall bring forth much fruit. This is a well tried rule and is sure.

The medicines of the mountain.

We must collect the gall of a black cat and the gall of a wolf, and rub them down together, and make them like eye-paint (*koḥu*). Smear thine eyes therewith, and go forth into the mountain (or, desert), and every root in the mountain of which thou knowest the use root up, and put in its place wheat and barley.

Another, concerning things secret and things revealed.

Take the head of a black raven, empty out its brains, and place in the cavity thereof five grains of coriander seed, and bury | the head in dung. Visit it every day until the grains sprout, and then eat them, and thou shalt see whatsoever thy heart desireth to see.

Page 588

Concerning the white cock that hath a divided comb (?).

The wise men say that there are in the crop of this bird stones of all colours, and that these are good for the men who fall through Satan. Let men hang these stones on themselves and on the children who shake (have St. Vitus's dance?), and they will drive away the evil spirits.

Concerning the mole.

The wise men say that every member of the body of a mole is good for some member of a man, provided that thou dost prepare it carefully, and dost prepare it, and place it in quicksilver unguent. Roast the head of a mole with the head of a swallow, and smear the juice on part of the patient who is suffering from the "fox disease" (*alopecia*), and he will enjoy relief. Or roast the head at the fire, and mix it with sesame oil, and anoint the head of him that hath headache. Mix the brain of the mole with sesame oil, and smear the head of the man who is bald and beardless therewith, and the hair will grow. Take the upper teeth of the mole, and tie them on the arm of the child who suffereth from "shaking", and he will not "shake". Steep the heart of the mole in goat's milk, and make a man drink it, and he will become a man of understanding and knowledge. | If a man eateth mole's flesh before Fol. 279a. sunrise, he will become enlightened and intelligent. Mix the blood of a mole with the blood of a black raven, and anoint from head to foot him that is sick, and the attack of pain will pass, and he will have relief.

Concerning the cock.

Hang the large claw of the right leg upon thyself, and thou shalt obtain victory and honour in judgement (*i. e.*, thou shalt win thy case in the law court). Hang on thyself the bone of his left wing, and fever shall not come to thee. Mix the gall of the cock with honey, and smear thine eyes therewith, and they shall become exceedingly bright.

Concerning the ass of the desert.

No harm shall remain in the body of him that eateth the flesh of the wild ass in the summertime.

Concerning the camel.

If thou makest a man drink the foam from the mouth of a camel in must, when the moon is in a vertical position, he will become a bold speaker. Or take the gall of a wolf, and make a pupil (or, disciple) drink it when the moon is in a vertical position, | and anoint his face with the gall, and thou wilt Page 589 wonder at his wisdom.

The Bat.

Put its head in the covering of thy head, and thou shalt never be vanquished all the days of thy life. Hang its eyes on thy person, and thou shalt not be afraid of the scorpion. When its heart is placed on an ant's nest the ants shall not abide there.

Of the gûmarrâ, or semûrâ (*i. e.*, marten).

Burn a piece of its fur in the fire, and take the ashes and mix them with water, and sprinkle them about the house, and trouble shall be banished.

A few additional sections.

The Eagle. Every man will love him that carrieth about its heart. A man will waste away for love of him that eateth its heart. He that hangeth its eye upon his person will not sleep. The affairs of him that hangeth its tongue upon his person will be carried out satisfactorily. From him that is fumigated with the marrow of its leg, worms will be eradicated.

The Vulture. Hang on thy person its left eye, and every man shall love thee. He that hangeth on himself its beak shall not be afraid of the Governor. He that hangeth on himself its tongue shall vanquish his enemies. He that burneth Fol. 279*b*. its whole | head, and mixeth it with vinegar, and giveth it to drink to a man who is in bondage through copulation, that man shall be freed. Fumigate the person who is sick of fever with its bones, and he shall be healed.

The Hawk. Smear with its gall the eyes of the man who hath blearedness. Its blood will make the hair grow. He who eateth its heart shall shine greatly. He who eateth its spleen shall have a healthy spleen.

The Shahîn. (White Falcon.) Bury its liver in a beehive, and the bees will thrive. A snake will not harm him that drinketh its gall.

The Falcon. Make eye-paint with its gall and honey, and smear the eyes of him that hath blearedness therewith. Bury its claws under the house, and no strife shall enter therein.

The Partridge. Bury its eyes with origanum majorana under the threshold | of him whom thou seekest, and he will love Page 590 thee much. Whosoever burneth the whole bird, and worketh it into a paste with oil of asparagus, and smeareth his hair therewith, his hair will not become white. And the eyes of him that useth the gall of this bird as eye-paint will not droop in sleep.

The Hare. The liver of him that weareth the liver of the hare shall never cause him pain. Whosoever weareth its spleen and tongue [shall be protected] against the Governor, [and shall gain] the affection of all men. Mix its gall with water, and rub it on thy member, and thou shalt be able to copulate strenuously.

The Crab. Whosoever taketh its right eye and weareth it, his crops shall be full and satisfactory. And whosoever fumigateth himself with its left eye in the name of the person for whom he wisheth, that person shall love him very much. And whosoever burneth a whole crab, and worketh it up with wine, and giveth it to drink to a person who is in thrall to his wife, that person shall be set free. Whosoever giveth roast crab mixed with sugar thrice to drink to a man who hath taken poison, that man shall be made whole.

The Frog. Whosoever taketh the blood of the yellow frog and smeareth it on his head, his hair shall not grow. Whosoever burneth a frog, and scattereth its ashes about a house, from that house serpents shall flee. Dry a frog's tongue, and give it to a child to drink who cannot talk, and he will talk.

[The Ox.] Mix the gall of an ox with moist unguent, | and Fol. 280a. wash the head which hath skin disease, and [the patient] will get better.

The Fox. The wearing of the skull of a fox is good for obtaining the affection of every man. Burn fox bones under a man who is suffering from fever, and he will get well. Hang a fox's tooth over a man who is suffering from toothache, on the side of the face where the tooth acheth, [and the pain will depart]. Take the gall of a fox, and smear the eyes of him that seeth weakly, and he will see well.

The Bear. He whose eyes are smeared thrice when the moon riseth shall not be attacked by sickness. Whosoever drinketh its gall, mixed with new milk and honey, shall be cured of wind. Whosoever drinketh its moisture, the organs of his belly shall wax fat. Whosoever shall place its eye in a hive of working bees, the bees shall prosper.

The Weasel. Whosoever shall eat the testicles of this Page 591 animal | shall be able to copulate strenuously. Whosoever shall smear his eyes with its gall shall be able to see as well during the nighttime as by day. Whosoever shall drink its brain shall sleep much. Whosoever shall tie its heart to his head shall not suffer from headache. Whosoever shall partake of its blood whilst warm shall not be barked at by dogs. Whosoever shall burn its dung in the name of him that he seeketh, there shall be strife between them. If a man throweth its gall into an oven, bread will not stick therein. Of him that drinketh its brain no sores shall break out in his skin.

The heart of the Hedgehog. Whosoever shall eat the heart whilst it is warm shall see everything that he wisheth.

Of the Pelican, which is called Abû Laḥghîgh. Let the bird be put in a house for four days without food, and then after the four days let it be killed. Take its head and salt it in edible salt, and take out its brain and throw it away. Dry its head and lay it up by thee. If a man hath taken poison, thou shalt pour water on the head, and leave it there from evening till morning; then make the man who hath taken poison drink the water, and he shall be healed. This bird Fol. 280 b. hath two eggs, one destroyeth | the hair, and the other maketh it grow. Take these eggs, break them, and take two cock's feathers, and thrust each in one of the eggs. The egg which destroyeth the hair throw away, but from that which maketh it grow take away the red (i.e., the yelk), and dry it, and work it up with olive oil, and smear it on the [place of] the beard of the beardless man, and a beard will grow there. But take care that the mixture doth not touch thy fingers, so that they may not become wholly covered with hair. [If it doth touch thy fingers hair will] certainly [grow on them]. Take the right side of the bird, dry it, rub it down, and mix it with eye-paint, and smear thine eyes therewith, and every one who seeth thee will receive thee with joy. Hang the tail end of the bird ˙above thee, and thou shalt see in a dream everything which is going to happen to thee. Tie its right pinion to a piece of rag, and hang it above thee, and serpents will flee from thee. Hang its left pinion above | the man who is afraid **Page 592** lest some one will hold him bound away from his betrothed. Work up the blood of this bird with fresh lime, dry it and pound it well, reduce it to a powder, and sprinkle it over the boils which break out in the body, and they will be healed. Boil the heart of this bird in sesame oil, and let the shadow thereof fall upon the warts which come in the body, and it will eradicate them. Boil this bird's liver a little over the fire, dry it, and lay it up by thee. Then take it and work it up with honey, and throw it into a place where there are fish, and they will all float on the water, and thou wilt be able to catch them without labour. Dry the gall of the bird in the sun, pound and mix it with the ashes from a furnace, and smear with it the eyes of those whose eyelashes have fallen out, and they will grow [again]. Dry the lung of this bird, pound and mix it with food of any kind, and sprinkle it before birds, and every bird that pecketh it thou wilt be able to catch without labour. Dry the bones of these birds, both great and small, and pound them well until they are as fine as ashes. And if there be a man who hath a wound which hath been caused by a sword, or an iron weapon, or a stone, or a piece of wood, first anoint the wound with olive oil, and then sprinkle some of that powder | over it, and in a few days, by the help **Fol. 281 a.** of our Lord, it will be healed. This is a remedy which is in use among Persian physicians, and it is certain [in its effect]. And sprinkle this powder over every kind of moist sore, and it will dry up. This is certain.

Of the raven, that is, the black crow.

Its heart is good for him that seeketh to learn; eat it. Its blood is good for malignant ulcers. Its liver is good for blearedness of the eyes; dry it, pound it, make it into a powder, and use it as eye-paint. Its fat will heal every kind

45

of sore (or, ulcer) which cometh in the body. And if a man rubbeth his eyes with its brain, he will be able to see the stars in the day-time. Its lung affordeth relief in cases of pain in the spleen, which it drieth up.

Page 593   ## Concerning the bird which is called "Solomon's bird" (? hoopoe) (طير سُليمان).

The medicines which are in it.

Kill it with a piece of money on which is the Name of God. Take a new vessel, and throw the bird into it, and with it put two drachms of red lime and one drachm of wood ashes, and boil it in water until it is parboiled. Then take out the bird. Put the head on one side, and the right pinion on one side. Remove one section of it, and place it in a vessel of wine, and whosoever drinketh thereof will become drunk. The left pinion is good for the man who is in thrall to his wife; remove one section of the bird and set it in oil, and let the man drink it, and he shall be free. Tie up the right shoulder in a piece of new rag, and hang it up over a woman who is ill in her body, and she shall have relief. Steep its tongue in rose water for five days, and tie it up in buffalo skin, and hold it in thy hand, and the dogs will not bark at thee. Dry its skull, and macerate it in oil, and rub it on any part of the body thou wishest, and it will destroy the hair. Dry a bone of the back in the shade, and pound it, and pour oil of violets over it, and [if thou rubbest it on any part of the body] the hair will grow. Having salted the heart of this bird with Indian salt, tie it in a lion's skin, and give it to a woman whose child will not

Fol. 281 b. come forth from her. | Let her grasp it in her left hand, and her child will descend from her womb straightway. If a man drink the lungs of the bird in very bitter water, he will not be able to attempt to copulate all the days of his life. If its liver be pounded and mixed with crocus, it is good for the lungs and the heart, and the spleen (?). If a man hangeth up the left wing over himself on the night of his wedding, nothing will be done. Mix its right side (or, loin) with Spahnî eye-paint, and smear thine eyes therewith, and every one who seeth thee will receive thee gladly. Put the bone of its back in wine | for ten Page 594 days, and then take it out, and place it under the head of him whom thou seekest, and thou wilt not wake up until thou receivest from him thine affairs. Pound a *denkâ* of the liver of the bird, and work it up in honey, and cast it into a place where there are fish, and they will die, and will float on the surface. It is good for two people each to eat one half of the heart of this bird, for they will then love each other. And the blood of this bird is good for him that suffereth from attacks of dimness of sight. And it is good for the scrofula which cometh in the neck and head. Heat some of the blood, and smear it over the places affected by scrofula twice daily. Roast the heart of the bird, and drink it [in water], and it will be very beneficial to thee. Roast the brain of the bird, pound and mix it with sesame oil, and anoint therewith him that is vexed by devils. Roast the wings of the bird in a house, and ants will flee therefrom. And if thou roastest some of the flesh of the bird in a house, devils and sorceries (or, sorcerers) will flee therefrom. If a man hangeth the bone of a wing or of a leg of the bird upon himself when he is going into the presence of the Governor, the Governor will love him. If thou roastest the liver of the bird, and dost give it to drink to him whom thou seekest, he will love thee.

The medicine that king Solomon was wont to use.

Take the eyes of a mole, each one by itself, and mix that which is for the right eye with the right eye, and that which is for the left eye with the left eye. Put them in a glass vessel, and pour on them a little *magnâṭîs* (iron ore). Pound and smear thine eyes therewith, the right eye with the right eye of the mole, and the left eye | with the left eye of the Fol. 282 a. mole; and then thou wilt see everything in heaven and in earth as it really is. The devils will flee from thee, and thou wilt see thine enemies, and every one who seeth thee will think that thou art fire, and this [state] will continue for ten days. And thou shalt not eat the flesh of the pig.

Concerning the kahînâ root, and the answer of Dioscurus about the wonderful things which it doeth among men.

The [name of] this *Kahînâ* (*i. e.,* splendid) root, being interpreted, is "expeller of devils", and it is called "foetid of smell". Know thou that this root was the firstborn of all the roots Page 595 which came up from the earth, | and King Solomon was wont to use it. It cometh up from the earth one cubit, and its blossoms are red like [those of] a rose, and its branches and leaves are like [those of] beans. And after the flower of this root, which resembleth that of a rose, hath died away, there remain on the top thereof two little balls which are like the testicles of a man, and inside them there are black and red seeds. When thou wishest to pull up this root, cleanse thyself from impurity, and eat not bread which hath been made by women. And wash thy head, and array thyself in white apparel, and keep fasting until thou seest the stars. And come thou to this root on the sixth day of the month Îyyâr, and say thou to it, "Peace be to thee, O Kahînâ Root". It will bring an answer concerning everything that the earth maketh to grow.

The prayer for the root. "Unto Thee, O Lord, Lord God, "the Mighty One, I cry, Who art the Eternal, and the ever- "living One, and the Begetter of life in all parts [of the earth], "Who hearest and understandest; Creator of all the worlds, "the Exalted One among all things which exist, of the Sun "and Moon, and of the stars, and of the earth, and of the "sea, and of the dry land, the Creator of all creations, the "Beautiful Name, and the Giver of understanding, O give Thou Fol. 282 b. "unto us this good root for the healing of all | the children of "men."

And every evening when thou goest to it thou shalt say unto it, "Be thou well, O splendid root, with beneficent health!" And when thou hast finished this prayer, say, "Peace be unto "thee, O root, for thou reignest over all roots." Thus shalt thou do three days. And on the fourth day come in the morning before sunrise, and dig up the ground on all four sides of it, to the depth of a cubit on each side. Then bring a black dog, and tie one end of a rope round his neck, and tie the other end of the rope to that root, and then | smite the dog Page 596 until he pulleth this root up out of the ground. And when thou hast pulled it up, bring a thin plate of gold or silver, and tie it up in a piece of new, clean linen, and bury it in the place where the root was, and cover it over. And take the root, and go to [thy] house or the church, and place it in the hollow of the head of the door, or on the seat of God, and take fine incense from the church, and cense it before Him.

And recite this prayer:—" O God, Thou Sustainer of All, " Who willest all things, Who preservest all things, Thou Name, " beautiful and ornamented in all things, Thou Creator of the " Sun and Moon, give Thou unto me, O my Lord, in Thy " mercy, the actual and peaceful possession of this root, in all " its operation, together with the power thereof, so that it may " be for the healing, and cure, and relief of the souls of all the " children of men. Let it come unto me with all its working " powers. Yea and Amen."

Now thou shalt take good heed to this root, and to all the manifold power and great healing which are therein. If thou takest a part of it in thy hand, and dost go into judgement, thou shalt conquer (*i. e.*, gain thy case). And if thou goest a journey, and dost wear it on thy person, thou shalt not be afraid of devils or of thieves. And if thou goest into a house wherein there are magicians, they shall speak from inside the earth, from that place wherein they are set. And if a man stealeth anything from thee, and denieth that he hath done so, place the root under thy pillow, and it will declare and make known unto thee | by whom the object hath been stolen, and Fol. 283 a. to whom he hath sold it, and where it is, and at what hour, and on what day he bought it, and ate (*i. e.*, spent) the price thereof. And if thou takest a wife, and she tieth some of it on thy left shoulder, she shall not become with child. If a man eateth poison, and he drinketh some of the root, the poison will not hurt him. Now it will also heal the bites of snakes. And if it be placed in a village, it will drive away from it drought and hail (or, cold). And if it be placed in a ship it will drive storms away from it. And if the root be with a man it will drive away from him envy, and defilement, and

Page 597 impurity. | And if they throw it into water, and sprinkle some of the water about a house, it will preserve it from all kinds of storms and evil acts. If they place it under the threshold of a house, it will not permit any evil thing to enter therein. And if thou wishest for anything which is hidden, place this root under thy head, and recite over it the prayer given above, and that thing for which thou seekest will be seen by thee in a dream of the night.

Now, this root hath two fruits, one of which is very red, and is like unto the seed of the pomegranate, and the other is black, and is like unto a bean; and the black fruit is good for headache and aching of the eyes, and when it is placed under thy tongue, men will do good unto thee. When thou hangest it over a woman having a flow of blood, she will be healed. And if thou wishest to know [who is] thy friend, place the red fruit under thy table, and send and bid thy friends [to thy house]. Know thou that he who cometh, and stretcheth out his hand to the table, and eateth, is thy friend; and that he who cometh, and doth not stretch out his hand, is thine enemy. Take the black seed of this root, and drink it with blood. And when the root beareth fruit, take some of the seed thereof, and put some water in a vessel. And throw one grain into thy mouth, and then cast it out from thy mouth into the water which is in the vessel. And look into the water,

Fol. 283b. and if thou seest thy face, throw away that grain, for | it is not good. And take another grain, and throw it into the water which is in the vessel, and look into the water, and if thou dost not see thy face therein, it is good. Then throw that grain into thy mouth, and go whithersoever thou wishest, and thou wilt be invisible to every man. This hath been well tried, and is certain. And take some of the root and of the seed thereof, and place in the nose of the man who is in the sight of the moon, either at the middle or end of the mouth. HERE ENDETH [THE STORY OF] THE SPLENDID ROOT.

Another root, which shineth in the night, and is called Page 598 "Begetter of the Warrior". This root is not found in the places that are trodden by the children of men, but in the desert mountains.

This root riseth up out of the earth in a point, and when it hath risen up, it produceth many branches, and it hath flowers that are like unto cups. And two flowers do not blossom at the same time, but when one withereth another blossometh, and its rays shine out during the night and sparkle like the sparklings of the waters of a river in the summer, and in the day-time they are extinguished. Hang a piece of this wood over a woman whose labour is difficult, and immediately her child will descend from her womb; and if thou tiest a piece of it to her back or shoulder, it will drive a devil away from her. And if thou wishest for the devil to go out of a man, bring a vessel, and fill it with water, and place it beyond the man. Then fumigate the man with that root, and the devil will at once go forth from him, and will break the vessel in which the water is. And those who hang it upon their persons because of magicians destroy [their enchantments], and it causeth them not to see bad dreams; and there is great power of healing in this root.

Concerning another root. The leaves thereof are like those of the flax plant, and the lower part is strong, and it groweth in mountains which are difficult of access. It is good for wind, and [to cure this] a man should eat it with oil [for some] days. And it is good for the man who hath elephantiasis, and he should eat it in the morning, either with wine or oil. And it is good for the man who hath some obstinate ailment; pound and mix with it butter, | and let him eat it fasting seven days. Fol. 284a.

Concerning another root, which is called "living for ever". Dig it up on the ninth day of Îyyâr. It groweth in desert mountains, and they sow it on walls, and its leaves are like [those of] purslane. Take branches of it, | pound them, and Page 599 squeeze out the juice, and smear therewith the man who hath the red wind and the hot wind. And give it to drink, mixed with sugar, to the boy who maketh water in his bed. And it

is good for the man who hath a devil; let him drink the
weight of one drachm of it in water. And it is good for the
man who is grievously afflicted; mix it with honey and let
him eat it; and if there is no strength in him he will not
endure until the evening. And if thou placest some of it in
thy shoe, and goest into judgement, victory will be thine. It
is good for earache and headache; let the patient eat it with
pig's fat, and smear [it over him], and he will be healed. And
it is good if a man stealeth something from thee; put some of
it under his tongue: if he hath stolen it he will not be able to
speak, and if he hath not stolen it he will be able to speak.
And it is good for the woman whose children will not live;
put some of it in the skin of a stag, and hang it up over her,
and they will live. And let barren women drink it in wine.
Let the man who is afraid by night drink it in water. And
he who drinketh a little of it will become fat and well favoured.
And when thou makest those who are in subjection to devils
drink it, the devils will wander about and take to flight. And
when thou givest it to drink to a man who hath suffered from
severe illness for a very long time, God will visit him, either
for life or death.

Concerning the root that is called " Madhîn Sûhâr " (Trifolium ?). It is .... like *marķôz* (......), and its root is in-
significantly small. It is good for obstinate ailments. It is
good for wind; a man should eat it with oil for [some] days.
It is good for [obtaining] the love of animals; pound it, and
eat it with salt.

Concerning another root, which is called "Semra" (the
thorn shrub). Its leaves are strong and broad, and its blossoms
are hollow. It is good for love. Give it to a man to drink
Fol. 284b. once a month | with pastry made of flour, butter, and honey.
And it is good for wind.

Concerning another root, the flower of which is white,
Page 600 and it hath two berries (or, testicles) | in the ground, and
there are upon it drops of water. It is good for [obtaining]
the love of animals; feed dogs on it, and they will follow thee.

Concerning another root which [is called] "the Healer"(?).
It is good for wind. Pound and mix it up with a little *ṣûrâ*

(palm plant?), and make the patient eat it with pastry made of flour, butter, and honey, for [some] days.

Concerning the root sĕphînâytâ. It is yellow, and is like hyoscyamus, and it beareth berries. It is good for the man who hath jaundice. Take of it a piece of the size of a chick-pea, and throw it into wine, and give it to the patient to drink morning and evening, for two days. It is good for [the bites of] snakes. Take the seed of this root, and the seed of hyoscyamus, and the seed of mandragora, that is to say, of *tûryagh*, pound it, and drink it in wine, and thou wilt not wake up for three days. The root of this plant and its seed are good for inducing conception in women.

Concerning the root of the Daḳartâ plant. Its flowers are blue, and its leaves are like garlic. It is good for pain in the belly and sides; pound and mix it with honey and let the patient eat it. Or throw some of its twigs into wine, and make a drink thereof. It is good for wind; pound it, and administer it in draughts for two days. And it is good for the intestines (?). And it is good [when] thou art to appear before the Governor. Take six grains, and place them in thy shoe, and go forth, and thou shalt be acceptable. And it is good for love. Take a little of it, and rub between thine eyes, and on thy lips, and on thy temples (or, cheeks).

Concerning another root which groweth in the rocks. Its leaves are strong (*i. e.,* coarse), and its flowers are sticky. It is good for disease of the belly; pound it and let the patient eat it in honey. And it is good for thee to mix with it . . . . [and it will deliver thee from the bites] of serpents.

Concerning another root possessing aphrodisiac properties. Its leaves are yellow, and its stem is red, and its flowers are sticky. It is good for a man who hath a flow of blood from his belly; dry it, pound it, and give it to the patient to eat with honey. | It is good for fleshy (?) men; pound and sprinkle Fol. 285 a. it over them. It is good for bridegrooms; pound and make both bridegroom and bride drink it in wine. And it is good for every obstinate | ailment; pound and make the patient drink Page 601 it in wine.

Concerning another root. Its stem is red, and its flower is sticky. It is good for palpitation of the heart, and for scabies and wind.

Concerning another root. It is like a large lily, and its root is good for love (*i. e.,* it is an aphrodisiac).

Concerning the root Abrĕshôm. It riseth on its own root (*sic!*) and on trees. It is good for jaundice; boil, and pound, and mix it with honey, or wine, and let the patient eat it three days.

Concerning the splendid root of Edhlâ. It is good for scabies on man and beast. Pound and work into a paste with spices or sheep oil.

HERE END THE SECTIONS.

AND TO GOD BE EVERLASTING GLORY!

Ôlâph.

Aôpyôn, ὄπιον, opium. Arab. افيون.

Agdânâ. Arab. النجدان, σίλφιον, laserpitium, ferula asa foetida.

Aîrrâsê. Arab. ايرسا, ἶρις, iris florentina, the root of the lily.

Aghrîḳôn. Arab. الغاريقون, ἀγαρικόν, agaricum, boletus igniarius.

Amâraḳôn. Arab. اقحوان ابيض, ἀμάρακον, majorana, chamae-melum album, parthenium.

Asḳîl (eskîl). Arab. العنصل, بصل, σκίλλα, scilla.

Asarôn. Arab. اسارون, ἄσαρον, asarum, nardus sylvestris, nar-dus mas.

Asṭôkhôdhôs (Esṭôkhôdhôs, *i. e.,* **Estrôkhônôn).** Arab. عنب الثعلب, στρύχνος, strychnus, solanum, myrtle berries, fox grapes.

Asṭûrkâ (Esṭûrkâ). Arab. اسطرك and لبنى, στύραξ, styrax.

Âḳaḳyâ. Arab. اقاقيا, ἀκακία, acacia, juice of mimosa nilotica.

Îlânâ ḳadîshâ, "the holy tree". Pers. فنجنگشت.

Agâlôkhôn. Arab. عود هندي, ἀγάλλοχον, agallochum.

Uparpîôn. Arab. اوفربيون, εὐφόρβιον, euphorbium.

Akhsûmehlî. Arab. سكنجبين, ὀξύμελι, oxymel.

Alpôrôn (Albarôn). Arab. خَرْبَق, ἐλλέβορος, helleborus.

Alpashrâ. Arab. هزار جشان, ἡ λευκὴ ἄμπελος, vitis alba, bryonia alba.

Amûlûn, the "heart of wheat", that is نشا. Arab. ستج, ἀμύλον, amylum.

Îlasrâ. Arab. البندق, πρῖνος, ilex, chestnut.

Îrîn. Arab. زنجار, for زنجفر cinnabar.

Apthîmôn. Arab. الافتيمين, the gum of an Egyptian tree, ἐπί-θυμον, epithymum.

Ôlôgh. | Arab. عود | البخور, lignum aloes.

Ampûmâ. Arab. اسفيداج, Pers. اسفيدات, gypsum, lime-white.

Arzâ. Arab. صنوبر, cedrus, cedar nuts, pine nuts, &c.

Eshkai-gardâ. Arab. جند بادستر, κάστορος ὄρχεις, castoreum.

Ôshaḳ. Arab. اشق, gummi ammoniacum.

Âmônîḳaitâ. Arab. نوشادر, sal ammoniacum.

Edhlâ. Arab. شيطرج, that is Kâpar, lepidium latifolium.

Espîndh, Espîdh. Arab. دهن بلسان, radix santali albi, sinapi album.

Êrîsnîḳân. Arab. ازرنيخ, ἀρσενικόν, arsenicum, auripigmentum.

Enḳalh Ailâ. Arab. كثيرا, τραγάκανθα, tragacantha.

Espûgâ. A thing that is like unto the fringes of clouds, and of white wool. It floateth on the waves of the sea. Arab. اسفنج, σπόγγος, spongia.

Aghpath. Arab. الغافت, agrimonia eupatorium.

Esṭakṭê. Arab. ميعة, στακτή, stacte.

Beth.

Bîhmân. Arab. بهمن. The white = centaurea behen, and the red, salvia haematodes.

Barôthâ, that is "dîprânâ", bratum, cupressus.

Balût arʿâ. Arab. بلوط الجبل, teucrium chamaedrys.

Balût malkâ. Arab. شاه بلوط, castanea.

Basar ʿÂbhâ. Arab. الهليون, asparagus, wood-meat.

Bĕrâkhtâ. Arab. قيصوم, chamaecyparissus, abrotonum.

Berîḳân. Arab. علي الريق, jejunus, vacuus, fasting.

Bĕlîlḳê. Arab. بليلج, ...

Bĕnâth-nûrâ. Arab. بزر خردل (بزر الخزك) varioli.

Bĕnâth-herazmê. Arab. بزر الدعناع, mentha sylvestris.

Bĕnâth-arzâ. Arab. حب الصنوبر, baccae pinus strobili.

Bar-ganĕthâ, this is to say, khardhĕlâ (mustard).

Behḳîthâ, an unguent that shineth.

Bûlbâsê. Arab. بصل الزبزني, βολβός, bulbi liliorum (an emetic).

Gamal.

Gedhdhe. Arab. العلقم الحناظل, absinthium, colocynth.

Gûmê. Arab. باقلى, κύαμος, faba, phaseolus vulgare.

Gûpsîn. Arab. جَصّ بياض, γύψος, gypsum.

Gûprâ of the sea. Arab. زبد البحر, ἀλκυόνιον, spuma maris.

Gûprâ of palms. Arab. طلع, كش النخل, spatha palmae.

Gargôrînê. Arab. الحندقوقي, trifolium.

Glûskâ. Arab. كعاك, γλυκύς (?), simila.

Gâṣâ. Arab. جَصّ.

Genṭyânâ. Arab. جنطياني, γεντιανή, gentiana.

Gâwze dhĕ besmê. Arab. جوز برا, myristica moschata.

Garnîthâ of the river. Arab. مرماخور, origanum fluviatile.

| **Gargîrâ.** Arab. جرجير, εὔζωμον, brassica eruca, rocket. Page 603

Gĕnaplôs. Arab. شحم الحنظل, pulpa colocynthidis.

Gâwzê dhĕ hĕpakhĕthâ, Arab. جوز القي, nux vomica, strychnos.

Dâlâdh.

Debalĕthâ dhĕ tînê. Arab. خلايب التين, παλάθη, placenta e Fol. 286a.
ficubus aridis in massam compressis.

Dûkhnâ. Arab. دُخْنٌ, milium, holcus dochna, panicum.

Dĕyâlâ dhĕ Karsâ. Arab. انشهال وانطلاق الجوف, διάρροια, diarrhoea.

Darikônâ. Arab. دينار ونصف دينار, δαρεικός, daricus.

Drakhmâ, *i. e.,* zûzâ. Arab. الدرهم, δραχμή, drachma.

Dûsanṭeryâ is an abscess of the bowels, δυσεντερία, tormina
intestinorum.

Dapnîdin, *i. e.,* dîprânâ, (الدفران), laurus.

Hâ.

Haṭâṭâ. Arab. الغيل ام غيلان, ῥάμνος, rhamnus, a white berry
or thorn.

Hemînâ = 1¹/₂ *lîṭrê,* ἡμίνα, dimidia pars sextarii. Arab. القسط
مكيال .

Hendîk̠ôn. Arab. الفانيد, ἰνδικόν, indicum, saccharum penidium.

Hêrazmê. Arab. النعاع, mentha sylvestris.

Hûpaṭrîôn. Arab. الاغافت, succus eupatorii.

Zayin.

Zeṭî. Arab. الجرب, الحكة, leprosy.

Zeghâghâ dhĕ edhnê. Arab. الدوي الذي في الاذان, tinnitus aurium.

Zôpâ, something with a pleasant smell and like marzangûsh, and it is eaten with vegetables, Arab. زُوفا, ζοῦφα, hyssopus.

Zar'â dhĕ 'ârâ. Arab. ثمرة الطرفاء, fructus tamarisci.

Zîôn. Arab. بنفسج, "violet"; but مه also = quicksilver.

Zangîbar. Arab. الزنجبيل, zingiber.

Zûngabar. Arab. زنجفر, cinnabaris.

Zarôbhâwzâ. Arab. بقلة يمانية, βλίτον, amaranthus blitum.

Zĕr'â keṭônâ. Arab. بزر قطونا, "flax seed".

Zĕr'â pûrta'nê, that is, "seed of psyllium" (ψύλλιον).

Zûzpê. Arab. عُبيراء, zizyphus, sorbus domestica.

Page 604

Kêth.

Khebhsê. Arab. اللوبيا, phaseolus vulgaris.

Khadhbê. Arab. الهندبة, cichorium enlivia.

Khôdhkhôdh, that is the gall of the elephant, Arab. حُضَض, rhamnus infectorius.

Khûkâ. Arab. الباذروج البابونج, ocymum basilicum.

Kheṣpê dhĕ Sapâ. Arab. خزف اخضر, green earth.

Khûlyâ, Khalyâ, that is sweet wine, Arab. الشراب حلو, mustum.

Khûrbaknâ. Arab. الخَرْبَق, helleborus.

Khewarwârê, the whiteness which cometh in the eyes.

Khûrpâsê. Arab. الجزر البرّى والبستانىّ, daucus.

Khazûrbarrâ. Arab. عرا برا, bamboo, wild came.

Khazûrê Pârsâyê, that is drôḳînê دقه صبغ, Arab. الخوخ, Persian apples.

Fol. 286b. | **Khazûrê Armâyê.** Arab. مشمش, apricots.

Khazûrê paṭikhê. Arab. الفقوس, cucumeres sativi.

Khazâzîthâ. This is an abscess that existeth in a place that manifesteth sparks (?) in its colour; it is very painful, and produceth scabies.

Khelbânîthâ bassimtâ. Arab. ميعة, galbanum suave.

Khelbânîthâ sarîthâ. Arab. القنّة, galbanum foetidum.

Khalebh yârôrâ. Arab. الجاوشير, succus ferulae opopanacis.

Khalba dhĕ âpûrsĕmâ. Arab. دهن البلسان, balsam ointment.

Khamu'yâthâ. Arab. بقلة يقال لها خميضة, lapathum.

Kharûthâ, that is the top of the shoulder.

Khalbâ dhĕ 'arâbhtâ. Arab. صمغ عربى, gummi arabicum.

Khalbâ dhĕ 'egalĕthâ; pure honey which one calleth "gâlînôs", mel purissimum.

Khalbâ dhĕ yathmê. Arab. خسمررا.

Kharbaḳâḳâ. Arab. حبّ قلقل, semen cassiae torae.

Khamṣĕlâitâ, that is السورنجان, or هليلج, colchicum autumnale.

Khumrai margâ. Arab. الققنج، الكاكنج, solanum, berries of the deadly nightshade.

Ṭêth.

Ṭalla daghbhîn, mannâ, Arab. ترنجبين, ros melleus.

Ṭlaphhûn mayâ. Arab. الطاحلب, that is lûâ, lenticula stagnina.

Ṭebakhshîr. Arab. طباشير, saccharum bambusae arundinaceae.

Ṭûdhrîgh. Arab. توذري, erysimum, medicina quae coitus appetitum excitat.

Ṭûṭyâ. Arab. توتيا, lapis ex quo collyria parantur.

Ṭabh'ê dhĕ yammâ (for ܝܕܕܐ ܠܡܪܐ). Arab. طين مختوم, terra sigillata.

Ṭehhôrâ, a disease of the fundament, as one straineth at stool.

Yôdh.

Page 605

Yabhruḳhê dhĕ barrâ. Arab. التفاح البرّ, atropa mandragora.

Yabhruḳhê dhĕ shainâ. Arab. الباذنجان, solanum melongena.

Kaph.

Kûprâ dhĕ Ṭayyâyê. Arab. حنّاء مكّي, cyperus, lawsonia inermis.

Kûprâ nehôrâyâ. Arab. المقل, pix, bitumen.

Kûndôsh, or Ḳindôsh. Arab. كُندُس، خوندس, στρούθιον, struthium.

Kûshnê, that is shûlê. Arab. كوسنة, ervum, ervilia.

Kûrkĕmâ. Arab. زعفران, κρόκος, crocus sativus.

Kiyâ. Arab. المصطكي, mastix.

Kampîṭôs. Arab. مرارة الحجر, χαμαιπίτυς, chamaepitys azuga.

Fol. 287 a. **Kûndîkôn**, a drink which is prepared | among the Greeks from three things, fine wine, pure honey, and peppercorns.

Kankarzadh. Pers. كنكرزد.

Kûptâ dhĕ rûmânâ. Arab. جلنار, flos vel calyx mali punici hortensis.

Kakhlâthâ, or **Khakhrâthâ zĕʿuryâthâ.** Arab. اقراص, placentae species.

Kasyôn. Arab. كشوت رطب, cuscuta recens, vivens.

Kaspôn. Arab. سوسن برّيّ, wild lily.

Kashûbhê. Arab. القرطم, κνίκος, cnicus.

Kakhlâthâ tînyâthâ. Arab. الرامك والسك, pillulae cum musco compositae.

Kîlôs, that is "juice".

Kamlêôn. Arab. من اسماء المازريون, that is wild olive.

Lamadh.

Lâdhânôn. Arab. عرق النساء اللاذن, λάδανον, ladanum (a gum).

Lĕbhûntâ. Arab. اللبان, thus.

Lîbhâîtôn. Arab. اللبان, thus.

Lipîrôn edhlâ. Arab. شيطرج.

Mîm.

Mêrîthâ, something sweet which cometh forth from the press. Arab. العصير السلافت والشراب.

Mâspîlôn. Arab. الزعرور, μέσπιλον, mespilus.

Page 606 **Meghabyâ.** Arab. توتيا مندي | شب يماني | alumen Yamense.

Mahlôn, that is ink.

Mûrâ. Arab. يسمّا, myrrha.

Mûrâ ekrêâ . . . .

Mûrôn. Arab. دهن البلاسان, Mûrôn is prepared from many spices. Mûrôn and Nardîn are one, root and oil, and the physicians regard them as one.

Mai ketmê (κονία, lixivium). Sergius of Rêsh ʿAinâ calleth [this] "stacte".

Marwâ. Arab. المرو, origanum maru.

Mai bônê. Arab. مري, garum.

Mûshk. Arab. مسك, *musk*.

Mânôn. Aráb. نون ماء, garum.

Môn. Arab. ما ذا اي شيء هو, what is it?

Môh. Arab. الو‌ء, meum athamanticum.

Marôrâyâthâ, wild chicory.   Arab. خسمررا, طرشقوق, هندبا.

Môrîḳî.   Arab. الطرفا, μυρίκη, tamarix, i. e., 'árâ.

Malûkhâ.   Arab. اشنان القاقلي, atriplex halimus, salsa.

Marârtâ dhĕ barrâ.   Arab. المرّ, الحنظل, cucumis colocynthis.

Malîltarôn.   The Hebrew saith nigella of the garden, and the
    Greek "Malîtakhôn".

Mûḳlâ îhûdhâyâ.   Arab. المقل يهودّي, bdellium, الخمر.

Mâmîrân. | Arab. ماميران, chelidonium.   Fol. 287b.

Maḳashaitâ.   Arab. مارقشيتا.

Nûn.

Neptâyâ.   Arab. نفطا, νάφθα, naphtha.

Nardôn.   Arab. سنجرة, νάρδος, nardus.

Nârdîn.   Arab. سنبل, νάρδος κελτική.

Nethrâ.   Arab البورق وهو النطرون الذي يبيّض به الكتّان, νίτρον,
    nitrum.

Nethrâ dhĕ zaitâ.   Arab. صمغ الزيتون, baccae quae ex olivis
    defluunt.

Nethrâ Armênâyâ.   Arab. بورق ارمني, Armenian nitre.

Nârdînôn, Nârdîn.   Arab. ناردين, nardinon unguentum.

Nînâyâ.   Arab. النانجواة, ammi copticum.

Neshrâ.   Arab. سرخس, that is دواء, للديدان, اثرس.

Semkath.

Saîdhâ âwkî.   Arab. زرنيك, arsenic.

Saglê, that is sū́dhê, berries.

Saglâ.   Arab. سعد, cyperus rotundus.

Sakhânôs.   Arab. سعد, σχοίνος, juncus, scirpus.

Samûryâ, that is myrrh. | Arab. مرّ, myrrha.   Page 607

Saḳnâḳôr, the lizard of the Nile.

Sarîḳôn.   Arab. سرنيج, minium, rubrica, rubri coloris pigmentum.

Saḳôdhrîôn, wild garlic, σκόρδιον, scordium, allium sylvestre.

Sîsmbar.   Arab النّمام, σισύμβριον, sisymbrium, mentha, or nańâ.

Spîhââ, that is *galwâi.* Arab. زبد البحر, spuma maris, alcyonium'
sepia.

Sasleôs, and *sûrbhâ,* and *shlakhlôkhâ.* Arab. زهر الانجدان الرومي,
ligusticum.

Sâsâlîôs. Arab. الكاشم الزوفرا, σέσελι.

Saplûlâ, that is *nep'â.* Arab. زراوند, aristolochia.

Sarmagh (Sarmak), that is *keṭâpâ.* Arab. قطف, atriplex hortensis.

Sarîkôn of the writer. Arab. زنجفر, φῦκος, minium.

Sandîkâ, that is *zûnghôpar.* Arab. زنجفر, minium.

Sammâ dhĕ khewyâ, that is *mardârôg,* which is root of *ârêm,*
which is gentian.

Sandarkâ, Sandarkâ. Arab. زرنيك احمر, arsenic.

Sêpââ. Arab. زبد البحر, spuma maris, alcyonium.

Sa'ar gabbârâ, that is "wild *tâwlê*". Arab. كزبرة البير, capillus
gigantis, filix.

Samterîn and **Saipâ.** Arab. دمّ الاخوين, the "blood of the two
brethren".

Sandrôs. Arab. دهن بعسا, and it is called "honey of frankincense", which is opobalsamum. And opobalsamum is
the fatty extract of *mûrân,* which being burned in the
Fol. 288 a. fire emitteth a sweet smell. Others say that | opobalsamum
is the extract (or, juice) of styrax.

Sakînôn. Arab. انخر, σχίνος, pistacia lentiscus.

Sṭokhôs, that is *kohl,* stibium, eye paint.

Sâgâpînôn, or **Sagpînôn.** Arab. سكبينج, σαγάπηνον, ferula persica.

'Ayin.

'Adhrâyâ. Arab. كندس, saponaria officinalis.

'Azûpâ. Arab. الحشن.

'Azwai. Arab. العنزروت.

'Elwai. Arab. صبر, aloe.

'Asârâ dhĕ arasne. Arab. ماء الشعير, decoctum hordei. Others
say that *âresnâ* is *ṭarkînâ.*

'Ârâ. Arab. العرائ نبت يسمّى, μυρίκη, myrica, tamarix.

'Arbath wardâ. Arab. سجر مريم. Mary's tree, that is the
'arabhtâ tree, which bringeth forth flowers like roses.

Enbai khewyâ, that is "fox grapes", Arab. عنب الثعلب, solanum nigrum.

'Arâ, that is the seed of tamarix, or môrîḳi.

'Ârâ dhĕ nahrâ, that is "lily". Arab. الفنجنكشت, agnus castus
vitex.

Pê.

Pîḳôs. Arab. الطاحلب, φακός, lens.

Peghnâ. Arab. سذاب, πήγανον, ruta agresta.

Padgârê, or **Púdgârê.** Arab. النقرش, ποδόγρα, gout.

Pâw'ê. Arab. القلاع, *i. e.*, that which goeth forth from the
mouth.

Pôlîôn. Arab. الجعدة, πόλιον, teucrium polium.

Perdâ. Arab. انجرة (?), urtica.

Pithsîs. Arab. الضاق السلّ.

Paṭrâselînôn, that is the seed of rock parsley.

Pestĕkê dha-shîôl. Arab. البان, fructus moringae Arabicae.

Ṣâdhê.

Ṣeharwâ. Arab. شوكران, cicuta vivosa.

Ṣîndaragh, that is cinnamon. Arab. الدار صينيّ.

Ṣarbûḳâ. Arab. دردة القرمز, vermis sericum producens.

Ṣâshûm, Ṣarshûm, Ṣasham. Arab. الكشميزج, fructus tamaricis
orientalis.

Ṣalbûbhâ, that is *ḳardâ*, and it is like the ricinus of oxen.
Arab. خروع انحناء ركع, that is الخروع.

Ḳôp.

Ḳisûrâ. Arab. الزيسورر, arsenicum.

Ḳûrnîthâ. Arab. الفرتنج, أونية, origanum, calamintha montana.

Ḳârdamânôn, Ḳardamânâ, that is *takhlî*, κάρδαμον, lepidium
sativum.

Ḳâlâmôs, that is "reed of incense". Arab. قصب الذريرا, κάλαμος,
calamus.

Ḳâlâmantîs, one of the kinds of *ḳôrnîthâ.* Arab. فوذنج نهريّ.

Ḳaisâ dhĕ ḳisuthâ. Arab. حشب الصندل يقال عود, sandal wood.

Ḳûshtâ. Arab. قسط, κοστος, costus.

Ḳûblâ. Arab. بابونج, that is white chamomile, اقحوان ابيض.

46*

Ḳîrûthâ. Arab. الشمع, wax.

Ḳîdhâmyâ. Arab. اقليميا, καδμεία, scorio, fuligo aeris.

Ḳalyâ, that is **Ânîḳôn.** Arab. قلب, frixio tritici.

Ḳalyâthâ. Arab. سريق, triticum frixum.

Ḳâpûr. Pers. كافور, camphora.

Ḳelapĕthâ dhĕ besmê. Arab. القرفة, القرنفل, cortex caryophylli.

Page 609 **Ḳarnâkôkh.** Arab. القرنفل, cortex caryophylli. |

Ḳĕrâpalôn. Arab. قرنفل بستانيّ, caryophyllus aromaticus.

Ḳûpal. Arab. قوفل, betel nut.

Ḳantôrîôn. Arab. عربزز, centaureum.

Ḳasyà. Arab. السليخة, κασία, casia.

Ḳâḳôlâ, Ḳâkôlâgh. Arab. القاقلّة الاعار الهال, القاقلي, salsola fruti-cosa, Cardamomum vulgare.

Ḳalpônyâ. Arab. قلفونيّة راينج, that is صمخ الصنوبر, resina.

Ḳrôkô maghmâ. Arab. دمس الزعفران, saffron unguent.

Ḳaṭuth khewyâ ⎫ Arab. الحنظل, cucumis agrestis, cucumis an-
Ḳaṭûth barrâ ⎭ guinus, colocynthus citrullus.

Ḳaṭai khamârâ. Arab. قثّاء الحما, cucumis asinus.

Ḳaṭayâ. Arab. قثاء خيار, cucumis, cucumis edibilis.

Ḳanâ'ê, that is *mîlâ.* Arab. نيل, isatis tinctoria.

Ḳômîôn. Arab. كمّون, κύμινον, cuminum.

Ḳaḳbâ, that is *bûrmâ.* Arab. قطاة قدر, وتكون, κάκκαβος, cacabus.

Ḳasṭorîôn, the testicles of the *gardâ,* that is the testicles of the water-dog.

Ḳeraṭê, that is *khâghê* trees, whence come the *ḳerâṭê* berries (or, fruit).

Ḳîrâtê, that is *karâthâ,* allium porrum.

Kinnâmôn. This is not the substance which they call *ḳinnâmâ,* or در صيني, but a kind of wood that hath a pleasant smell. And if it be brought near a boiling vessel it cooleth it, and it also cooleth the water in a bath. A preparation of it is made with oil for anointing purposes.

Ḳinnâmôn, that is *esṭûrkâ,* styrax.

Ḳûrârâ, the running from the head, catarrhus, coryzâ, Arab. الزكام.

Ḳônîôn. Arab. الشوكران, κώνειον, conium maculatum, cicuta cirosa.

Ḳúʻâlâ, that is the fig-tree. Arab. جري.

Ḳimâliâ. Arab. طين رومّي, Cimolia.

Rêsh. Fol. 289a.

Râpânôn. Arab. فجل, that is *paughlè*, radishes.

Rummânâ dasheʻâlâ. Arab. الخشخاش افيون, succus papaveris Aegyptiaci.

Rummân Meṣrên. Arab. جبريل الجلّنار, flos meli punicae sylvestris.

Rúʻethâ dhĕ yammâ. Arab. زبد البحر, spuma maris, alcyonium.

Rîṭinî. Arab. صمع الصنوبر, ῥητίνη, resina.

Rêôn. Arab. راوند, rheum rhaponticum.

Raḳúthâ. Arab. خضرة, and it is called نعنع يابس, mentha siccata·

Reheṭnê, Rîṭinî, that is cedar gum. Arab. راتينج, resina, and this is صمع الصنوبر.

Rakhshúthâ perdânâitâ, called further on Rakhshúshtâ, that is *apûrsĕmâ*, balsam. Arab. دهن الباسان. Page 610

Rúʻnê, that is خُبّاز, herbae agrestes.

Rôdô mîlôn. Arab. شراب الورد, a drink made of roses.

Rôdâ meli. Arab. جلاب, ῥοδόμελι, julapium, this is "rose water".

Shîn.

Shebelbĕlâ, that is Khebelbĕlâ. Arab. لبلاب, eryngium.

Shebbeltâ. Arab. سنبلة, spica, arista.

Shebâḳ dĕmâ. Arab. اخرج الدمّ, blood-letting.

Shûwârṣĕrâ, and Shadhḳâ, and Yârúʻâ. Arab. البلنجاسف والقيسوم, artemisia. Al·shawaṣrâ are plants which grow by the sides of fields, and they have a green colour, and a sweet smell, and greenish-yellow flowers.

Shahpĕbag. Arab. نبات يشبه القيسوم شابابج, conyza odora, and this is *shûwâṣrâ*.

Sharṭôs and Shamsapram, that is Esparwê. Arab. ريحان سليمان, that is *marwâ*, المرو, ocimum basiliscum.

Shenghâr. Arab. النخوسا, anchusa tinctoria. يصبغ به الشمع احمر. And they call it السمقان, that is the "red [gum]".

Shushmîr. Arab. جوز بوا ,جوشمير بوّا ,هيل, amomum, granum paradisi.

Shîdrûgh. Arab. شادنج, lapis haematites.

Shîâķê. Arab. عذّاب.

Shishķâķûl. Arab. حب قلقل, radix arboris Indicae.

Shĕyâpê. Arab. حبوبزد, collyrium.

Shamrâ, that is Bar khalyâ, anethum foeniculum.

Shaplûlê. Arab. اشياف يتحمل بها بلاليط, tortile, ellychnium.

Shûshlê. Arab. الخراطين, "beautiful leaves", that is *spenâgh*.

Shôrṣinâ. Arab. الزعرور, mespilus germanica.

Shônyâ. Arab. ثونيز, the *'ârâ* of the river, *i. e.*, indigo.

Fol. 289*b*. Shegdhê, Sheghârê, that is bitter almonds. |
Shedhnâ, that is blood-stone.

Shekhâr dhĕ nârgâ. Arab. شكّر طبرزد, saccharum album.

Shekhâr dhĕ parṣĕnâthâ. Arab. شكّر العشر, saccharum amarum.

Shakhtîthâ. Arab. السوين.

Tâw.

Tûryâķê. Arab. ترياق, antidotus.

Tĕrîrķî. Arab. تريّاق, theriaca, medicina.

Tûrmâsâ, Egyptian beans. Arab. الترمس, lupinus, cicer.

Tâpsâyâ. Arab. البتيون, صمغ السداب, θαψία, Thapsia Asclepium.
Page 611 They say that it is | drunk in wine, and that it maketh
merry the hearts of those who drink it.

Tûdhrâ. Arab. لسان التور, buglossus, horminum.

HERE ENDETH THE BOOK OF MEDICINES.

Glory be to God Who hath given strength and hath helped
with His Grace, according to the abundance of
His Mercy, now and always, and
for ever and ever!
Amen.

The [copying] of this "Book of Medicines" was completed
on the third day of the week, which was the tenth day of
Îyyâr, in the year one thousand, eight hundred, and ninety-
four, of the birth of Christ, our Lord, our King, our Redeemer
and Lifegiver. Glory be unto Him Who maketh times and

seasons to pass away, and Who doth not Himself pass away, and Who neither cometh to an end, nor changeth, nor increaseth, nor diminisheth, for ever and ever. Yea and Amen.

It was written in the blessed village of Alḳôsh, the village of Nahum the Prophet, which is situated and built by the side of the habitation of the Persian martyr Hôrmîzd. May our Lord make him to dwell on His mighty right hand, and may He protect all the inhabitants thereof from every injury of spirit and body. Yea and Amen.

It was written in the days of the gentle Mâr Êliyâ the | Fol. 290a. Thirteenth, the Catholicus and Patriarch of Babel [and] of the East. Now, during these days (*i. e.,* whilst the book was being copied) Mâr Êliyâ died, and departed from the world of time to the world of eternity, and up to the present the throne of the | Patriarchate is empty; there is neither Patriarch nor Page 612 Governor. May the Lord in His mercy grant rest to his soul, and may He number him with the righteous and chosen ones of the Old Dispensation and of the New Dispensation. And as he ran his course and finished it gloriously, in the whole fear of God, and with strenuousness and success, and as he shepherded the rational flock of Christ, may his Lord reward him for his labour in the kingdom of the heavens. Yea and Amen.

This [copy] was written by the feeble one, the sinner, Îsâ bar-Esha‘yâ, the deacon, the son of the gentle deacon Cyriacus, who [came] from the village of Eḳrôr, which is in the country of the Sendâyê. I pray those enlightened men who read this book to remember me in their prayers—which shall be heard!—that peradventure mercy may be shewed to me before the throne of Christ our Lord.

BLESSED BE GOD FOR EVER, AND MAY HIS HOLY NAME
BE GLORIFIED TO ALL GENERATIONS.

INDEX

Bees, to prevent their leaving a place 689.

Beet, beetroot 55, 56, 59, 189, 457, 474, 508, 646, 656, 684.

Beet, juice of, 54, 148, 301.

Beet leaves 468.

Beet seed 668.

Beet water 108, 656.

Beetles, to drive out, 688.

Beetles, wild 100.

Behen 38.

Behen, the white, 716.

Behkithâ 716.

Behmân, red and white, 297, 303.

Bêl 523, 566, 604.

Beldhâ 549.

Bêldôr 345, 349.

Bêldôr, honey of, 350.

Belilkê 147, 345, 346, 348, 351, 352, 354, 355, 362, 416, 421, 437.

Belly 305, 386.

Belly and head 14.

Belly, bleeding from 225.

Belly, the cold, or dry, or hot, or wet, 320, 321.

Belly, the diseases of 475 ff.

Belly-flux 484.

Belly, the hairy 381.

Belly, looseness of 330.

Belly, the lower 19.

Belly, mouth of 14, 19, 26, 77, 78, 252, 462.

Belly, natural characteristics of, 320 ff.

Belly of Egyptian crocodile 276.

Belly-pains 18, 146, 203, 271, 273, 713.

Bendâkê 62.

Berberis 402, 671.

Berries 182, 366, 721.

Berries, incense 277, 374; see also Incense.

Berries, kûpiôn 275.

Berries, laurel 282.

Berries, roasted 362.

Berûthâ 159.

Besrânâ, kind of dropsy, 388.

Betel nut 47, 97, 192, 326, 724.

Bêth Gelîsâyâ 604.

Bêth Hamrâyê 605.

Bêth Hûzâyê 521, 653, 654.

Bêth Rhômâyê 653.

Betonica 48, 49.

Betrothals 622.

Bîblâ flower 686.

Bile 6, 7, 31, 43, 45, 321, 327, 335, 359, 372, 449, 476, 489, 496, 641, 676.

Bile, black 12, 16, 17, 45, 47, 49, 50, 53, 54, 248, 301, 304, 328, 349, 380, 381, 393, 444, 445, 450, 460, 640, 641, 646, 647.

Bile, green 380, 544.

Bile, red, 33, 36, 43, 44, 45, 49, 248, 380, 381, 448, 449, 450, 461, 646; reddish-yellow 13.

Bile, yellow 13, 445, 544.

Bile in the blood 445.

Bindweed 217.

Birds, gall of 56; to catch 690.

48*

For Product Safety Concerns and Information please contact our EU
representative GPSR@taylorandfrancis.com
Taylor & Francis Verlag GmbH, Kaufingerstraße 24, 80331 München, Germany

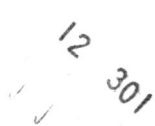